AF335571

Benchmark Papers in Genetics / 14

A BENCHMARK® Books Series

SOMATIC CELL GENETICS

Edited by

RICHARD L. DAVIDSON

**University of Illinois
College of Medicine
at Chicago**

Hutchinson Ross Publishing Company

Stroudsburg, Pennsylvania

Copyright ©1984 by **Hutchinson Ross Publishing Company**
Benchmark Papers in Genetics, Volume 14
Library of Congress Catalog Card Number: 83-10810
ISBN: 0-87933-120-8

86 85 84 1 2 3 4 5
Manufactured in the United States of America.

LIBRARY OF CONGRESS CATALOGING IN PUBLICATION DATA
Main entry under title:
Somatic cell genetics.
   (Benchmark papers in genetics; 14)
   Includes indexes.
   1. Somatic hybrids—Addresses, essays, lectures.
2. Cytogenetics—Addresses, essays, lectures.
I. Davidson, Richard L.   II. Series.   [DNLM: 1. Hybrid
cells—Collected works.   2. Hybridization—Collected
works.   3. Mammals—Collected works.   4. Mutation—
Collected works.   5. Transformation, Genetic—Collected
works.   WL BW516 v.14 / QH 425 S693]
QH451.S64   1984   599'.0873223   83-10810
ISBN 0-87933-120-8

Distributed worldwide by Van Nostrand Reinhold Company Inc.,
135 W. 50th Street, New York, NY 10020.

CONTENTS

Series Editor's Foreword   viii
Preface   ix
Contents by Author   xi

Introduction   1

PART I: SOMATIC CELL HYBRIDIZATION

Editor's Comments on Papers 1 and 2   6

1   **BARSKI, G., S. SORIEUL, and F. CORNEFERT:**   "Hybrid" Type Cells in
Combined Cultures of Two Different Mammalian Cell Strains   9
Natl. Cancer Inst. J. **26:**1269-1291 (1961)

2   **SORIEUL, S., and B. EPHRUSSI:**   Karyological Demonstration of
Hybridization of Mammalian Cells *in vitro*   27
Nature **190:**653-654 (1961)

Editor's Comments on Papers 3 Through 8   29

3   **LITTLEFIELD, J. W.:**   Selection of Hybrids from Matings of Fibroblasts
in vitro and Their Presumed Recombinants   35
Science **145:**709-710 (1964)

4   **DAVIDSON, R. L., and B. EPHRUSSI:**   A Selective System for the Isolation
of Hybrids Between *L* Cells and Normal Cells   37
Nature **205:**1170-1171 (1965)

5   **EPHRUSSI, B., and M. C. WEISS:**   Interspecific Hybridization of Somatic
Cells   39
Natl. Acad. Sci. (USA) Proc. **53:**1040-1042, (1965)

6   **OKADA, Y.:**   The Fusion of Ehrlich's Tumor Cells caused by HVJ Virus
in vitro   42
Biken's J. **1:**103-110 (1958)

7   **HARRIS, H., and J. F. WATKINS:**   Hybrid Cells Derived from Mouse and
Man: Artificial Heterokaryons of Mammalian Cells from
Different Species   53
Nature **205:**640-646 (1965)

8   **PONTECORVO, G.:**   Production of Mammalian Somatic Cell Hybrids
by Means of Polyethylene Glycol Treatment   60
Somatic Cell Genet. **1:**397-400 (1975)

Contents

Editor's Comments on Papers 9 Through 14   64

9   **DAVIDSON, R. L., B. EPHRUSSI, and K. YAMAMOTO:**   Regulation of
         Pigment Synthesis in Mammalian Cells, As Studied by
         Somatic Hybridization   70
         Natl. Acad. Sci. (USA) Proc. **56:**1437–1440 (1966)

10   **DAVIDSON, R. L.:**   Regulation of Melanin Synthesis in Mammalian
         Cells: Effect of Gene Dosage on the Expression of
         Differentiation   74
         Natl. Acad. Sci. (USA) Proc. **69:**951–955 (1972)

11   **PETERSON, J. A., and M. C. WEISS:**   Expression of Differentiated
         Functions in Hepatoma Cell Hybrids: Induction of Mouse
         Albumin Production in Rat Hepatoma-Mouse Fibroblast
         Hybrids   79
         Natl. Acad. Sci. (USA) Proc. **69:**571–575 (1972)

12   **FINCH, B. W., and B. EPHRUSSI:**   Retention of Multiple Developmental
         Potentialities by Cells of a Mouse Testicular Teratocarcinoma
         During Prolonged Culture In Vitro and Their Extinction
         Upon Hybridization with Cells of Permanent Lines   84
         Natl. Acad. Sci. (USA) Proc. **57:**615–621 (1967)

13   **KÖHLER, G., and C. MILSTEIN:**   Continuous Cultures of Fused Cells
         Secreting Antibody of Predefined Specificity   91
         Nature **256:**495–497 (1975)

14   **JOHNSON, R. T., and P. N. RAO:**   Mammalian Cell Fusion: Induction
         of Premature Chromosome Condensation in Interphase
         Nuclei   94
         Nature **226:**717–722 (1970)

Editor's Comments on Papers 15 and 16   100

15   **HARRIS, H., O. J. MILLER, G. KLEIN, P. WORST, and T. TACHIBANA:**
         Suppression of Malignancy by Cell Fusion   103
         Nature **223:**363–368 (1969)

16   **CROCE, C. M., D. ADEN, and H. KOPROWSKI:**   Somatic Cell Hybrids
         Between Mouse Peritoneal Macrophages and Simian-Virus-
         40-Transformed Human Cells: II. Presence of Human
         Chromosome 7 Carrying Simian Virus 40 Genome in Cells
         of Tumors Induced by Hybrid Cells   109
         Natl. Acad. Sci. (USA) Proc. **72:**1397–1400 (1975)

Editor's Comments on Papers 17 Through 21   113

17   **WEISS, M. C., and H. GREEN:**   Human-Mouse Hybrid Cell Lines
         Containing Partial Complements of Human Chromosomes
         and Functioning Human Genes   119
         Natl. Acad. Sci. (USA) Proc. **58:**1104–1111 (1967)

18   **NABHOLZ, M., V. MIGGIANO, and W. BODMER:**   Genetic Analysis with
         Human-Mouse Somatic Cell Hybrids   127
         Nature **223:**358–363 (1969)

19   **RUDDLE, F. H., V. M. CHAPMAN, T. R. CHEN, and R. J. KLEBE:**   Linkage
         Between Human Lactate Dehydrogenase A and B
         And Peptidase B   133
         Nature **227:**251–257 (1970)

20 MILLER, O. J., P. R. COOK, P. MEERA KHAN, S. SHIN, and M. SINISCALCO:
Mitotic Separation of Two Human X-Linked Genes in
Man-Mouse Somatic Cell Hybrids **140**
Natl. Acad. Sci. (USA) Proc. **68:**116–120 (1971)

21 DEISSEROTH, A., A. NIENHUIS, P. TURNER, R. VELEZ, W. F. ANDERSON,
F. RUDDLE, J. LAWRENCE, R. CREAGAN, and R. KUCHERLAPATI:
Localization of the Human α-Globin Structural Gene to
Chromosome 16 in Somatic Cell Hybrids by Molecular
Hybridization Assay **145**
Cell **12:**205–218 (1977)

Editor's Comments on Papers 22, 23, and 24 **160**

22 CARTER, S. B.: Effects of Cytochalasins on Mammalian Cells **164**
Nature **213:**261–264 (1967)

23 BUNN, C. L., D. C. WALLACE, and J. M. EISENSTADT: Cytoplasmic
Inheritance of Chloramphenicol Resistance in Mouse Tissue
Culture Cells **168**
Natl. Acad. Sci. (USA) Proc. **71:**1681–1685 (1974)

24 EGE, T., and N. R. RINGERTZ: Preparation of Microcells by Enucleation
of Micronucleate Cells **173**
Exp. Cell Res. **87:**378–382 (1974)

PART II: GENE TRANSFER

Editor's Comments on Papers 25 Through 29 **180**

25 SZYBALSKA, E. H., and W. SZYBALSKI: Genetics of Human Cell Lines,
IV. DNA-Mediated Heritable Transformation of a Biochemical
Trait **186**
Natl. Acad. Sci. (USA) Proc. **48:**2026–2034 (1962)

26 MUNYON, W., E. KRAISELBURD, D. DAVIS, and J. MANN: Transfer of
Thymidine Kinase to Thymidine Kinaseless L Cells by
Infection with Ultraviolet-Irradiated Herpes Simplex Virus **195**
J. Virology **7:**813–820 (1971)

27 WIGLER, M., S. SILVERSTEIN, L.-S. LEE, A. PELLICER, Y.-c. CHENG, and
R. AXEL: Transfer of Purified Herpes Virus Thymidine
Kinase Gene to Cultured Mouse Cells **203**
Cell **11:**223–232 (1977)

28 WIGLER, M., R. SWEET, G. K. SIM, B. WOLD, A. PELLICER, E. LACY, T.
MANIATIS, S. SILVERSTEIN, and R. AXEL: Transformation of
Mammalian Cells with Genes from Procaryotes and
Eucaryotes **214**
Cell **16:**777–785 (1979)

29 CAPECCHI, M. R.: High Efficiency Transformation by Direct
Microinjection of DNA into Cultured Mammalian Cells **224**
Cell **22:**479–488 (1980)

Editor's Comments on Papers 30 and 31 **235**

30 McBRIDE, O. W., and H. L. OZER: Transfer of Genetic Information by
Purified Metaphase Chromosomes **238**
Natl. Acad. Sci. (USA) Proc. **70:**1258–1262 (1973)

31 DEWEY, M. J., D. W. MARTIN, JR., G. R. MARTIN, and B. MINTZ: Mosaic Mice with Teratocarcinoma-Derived Mutant Cells Deficient in Hypoxanthine Phosphoribosyltransferase   **243**
Natl. Acad. Sci. (USA) Proc. **74:**5564–5568 (1977)

PART III: MUTANT MAMMALIAN CELLS

Editor's Comments on Papers 32, 33, and 34   **250**

32 FISCHER, G. A.:   Increased Levels of Folic Acid Reductase as a Mechanism of Resistance to Amethopterin in Leukemic Cells   **253**
Biochem. Pharmacol. **7:**75–77 (1961)

33 SCHIMKE, R. T., R. J. KAUFMAN, F. W. ALT, and R. F. KELLEMS: Gene Amplification and Drug Resistance in Cultured Murine Cells   **256**
Science **202:**1051–1055 (1978)

34 HAKALA, M. T.:   Prevention of Toxicity of Amethopterin for Sarcoma-180 Cells in Tissue Culture   **261**
Science **126:**255 (1957)

Editor's Comments on Papers 35, 36, and 37   **262**

35 SZYBALSKI, W., and M. J. SMITH: Genetics of Human Cell Lines   I. 8-Azaguanine Resistance, a Selective "Single-Step" Marker   **265**
Soc. Exp. Biol. Med. Proc. **101:**662–666 (1959)

36 LIEBERMAN, I., AND P. OVE: Enzyme Studies with Mutant Mammalian Cells   **269**
J. Biol. Chem. **235:**1765–1768 (1960)

37 CHU, E. H. Y., and H. V. MALLING: Mammalian Cell Genetics, II. Chemical Induction of Specific Locus Mutations in Chinese Hamster Cells In Vitro   **273**
Natl. Acad. Sci. (USA) Proc. **61:**1306–1312 (1968)

Editor's Comments on Papers 38 Through 42   **280**

38 PUCK, T. T., and H. W. FISHER: Genetics of Somatic Mammalian Cells I. Demonstration of the Existence of Mutants with Different Growth Requirements in a Human Cancer Cell Strain (HeLa)   **284**
J. Exp. Med. **104:**427–434, Plate 31 (1956)

39 DE MARS, R., and J. L. HOOPER: A Method of Selecting for Auxotrophic Mutants of HeLa Cells   **293**
J. Exp. Med. **111:**559–572, Plate 47 (1960)

40 PUCK, T. T., and F.-T. KAO: Genetics of Somatic Mammalian Cells, V. Treatment with 5-Bromodeoxyuridine and Visible Light for Isolation of Nutritionally Deficient Mutants   **308**
Natl. Acad. Sci. (USA) Proc. **58:**1227–1234 (1967)

41 NAHA, P. M.:   Temperature Sensitive Conditional Mutants of Monkey Kidney Cells   **316**
Nature **223:**1380–1381 (1969)

42 THOMPSON, L. H., R. MANKOVITZ, R. M. BAKER, J. E. TILL, L. SIMINOVITCH, and G. F. WHITMORE: Isolation of Temperature-Sensitive Mutants of L-Cells   **318**
Natl. Acad. Sci. (USA) Proc. **66:**377–384 (1970)

Editor's Comments on Papers 43 Through 47   326

43   **KAO, F.-T., R. T. JOHNSON, and T. T. PUCK:**   Complementation Analysis
         on Virus-Fused Chinese Hamster Cells with Nutritional
         Markers   331
              Science **164:**312–314 (1969)

44   **BAKER, R. M., D. M. BRUNETTE, R. MANKOVITZ, L. H. THOMPSON, G. F.
         WHITMORE, L. SIMINOVITCH, and J. E. TILL:**   Ouabain-
         Resistant Mutants of Mouse and Hamster Cells in Culture   334
              Cell **1:**9–21 (1974)

45   **HARRIS, M.:**   Mutation Rates in Cells at Different Ploidy Levels   347
              J. Cell. Physiol. **78:**177–184 (1971)

46   **BEAUDET, A. L., D. J. ROUFA, and C. T. CASKEY:**   Mutations Affecting
         the Structure of Hypoxanthine: Guanine
         Phosphoribosyltransferase in Cultured Chinese Hamster Cells   355
              Natl. Acad. Sci. (USA) Proc. **70:**320–324 (1973)

47   **CHASIN, L. A., and G. URLAUB:**   Chromosome-Wide Event Accompanies
         the Expression of Recessive Mutations in Tetraploid Cells   360
              Science **187:**1091–1093 (1975)

Author Citation Index   362
Subject Index   370
About the Editor   372

SERIES EDITOR'S FOREWORD

The study of any discipline assumes mastery of the literature of the subject. In many branches of science, even one as new as genetics, the expansion of knowledge has been so rapid that there is little hope of learning of the development of all phases of the subject. The student has difficulty mastering the textbook, the young scholar must tend to the literature near his own research, the young instructor barely finds time to expand his horizons to meet his class preparation requirements, the monographer copes with a wider literature but usually from a specialized viewpoint, and the textbook author is forced to cover much the same materials as previous and competing texts to respond to the user's needs and abilities.

Few publishers have the dedication to scholarship needed to serve the limited market of advanced studies. The opportunity to assist professionals at all stages of their careers has been recognized by the publishers of the Benchmark series and by a distinguished group of Benchmark volume editors knowledgeable in specific aspects of the genetics literature. Some have contributed greatly to the development of that literature, some have studied with the early scholars, and some have developed and are in the process of developing entirely new fields of genetic knowledge. In many cases the judgments of the editors become a historical document that records their opinion of the important steps in the development of the subject. The editor of this volume has selected papers and portions of papers that demonstrate both the development of knowledge and the atmosphere in which that knowledge was developed. There is no substitute for reading great papers. Here you can learn how questions are asked, how they are approached, and how difficult and essential it is to obtain definitive answers and clear writing.

Dr. Davidson traces the development of somatic cell genetics of mammalian cell systems considering papers in somatic cell hybridization, gene transfer, and mutant cells. These papers emphasize the importance of the combined approach of cell genetics and recombinant DNA to approach questions of development, regulation, and growth of normal and tumor cells. Gene mapping and gene transfer are also considered. *Somatic Cell Genetics* represents genetics as modern molecular biology, and as such is a welcome addition to the series.

DAVID L. JAMESON

PREFACE

Somatic cell genetics is a relatively young field of study; its origins date back little more than two decades. During its development somatic cell genetics drew heavily on the techniques and concepts of many other fields such as microbial genetics, biochemistry, and molecular biology. As it matured, however, the field of somatic cell genetics introduced a variety of important techniques and concepts of its own. Somatic cell genetics has provided major new approaches to the study of gene regulation and genetic organization in mammalian cells, and has made possible the manipulation of the mammalian genome in vitro, with the movement of genes even across species barriers. The techniques of somatic cell genetics have now become a standard feature of research in many diverse areas, ranging from the isolation of genes to the production of immunological reagents.

In this book I have attempted to trace the major developments in somatic cell genetics by presenting 47 classic papers. For the sake of clarity, these papers are grouped into three major areas—somatic cell hybridization, gene transfer, and mutant cells. Clearly, many more than 47 papers could be used to trace the development of somatic cell genetics. However, because of space limitations, many important papers could only be cited in the commentaries. Furthermore, in several cases, significant advances were reported simultaneously by different laboratories. Since it was possible to present only one of these papers in each case, arbitrary choices had to be made. However, I have attempted to acknowledge the contributions of the laboratories whose work could not be published in this volume. In considering papers on mutant cells, I found it necessary to establish a specific focus because the range of mutant cell studies has expanded in so many directions. My choice of papers on this topic was influenced by the relevance of the results for studies on somatic cell hybridization and gene transfer. This criterion provided a somewhat objective mechanism for choosing between papers when comparable principles were established with different mutant cell systems.

A number of important topics were intentionally not considered in order to provide a more focused presentation. Among the topics avoided were cytogenetics, X chromosome inactivation, and cell membranes. Finally, in choosing papers for this book, I have focused on mammalian cell systems. This focus reflects not only my own personal interests, but also the predominant role that mammalian systems have played in the development of somatic cell genetics.

Preface

The major problem, and also the most interesting aspect of making a compilation of papers, as in this volume, is the determination of what is a classic paper. To facilitate this determination I have tried to follow certain criteria. Clearly, papers that introduce important new techniques or establish new principles should be considered as classics. Other types of papers also may be considered as classics because of their impact on subsequent research. For example, some papers are included because they raise important questions and have stimulated much activity in other laboratories. I did not consider it necessary that these papers provide the eventual solution; raising the question is sufficient. I am also aware that the significance of some of the older papers reprinted in this volume is no longer widely acknowledged. However, such opinion does not alter the historical role of these papers since their influence, even if not currently recognized, is reflected in the present status of somatic cell genetics.

As already mentioned, one measure of a classic paper is its impact on subsequent research. By definition, therefore, it is much easier to recognize an old classic than a new one. Nevertheless, a number of recent papers have been included. These papers, for example, in the area of gene transfer, were chosen because I expect them to establish trends for the future. More time must pass before the validity of these selections can be assessed.

I would like to thank my wife Jalane for much help in writing this book. I also thank Dr. Raju Kucherlapati for critically reading the manuscript and Linda Cardenas for her help in preparing the manuscript.

RICHARD L. DAVIDSON

Aden, D., 109
Alt, F. W., 256
Anderson, W. F., 145
Axel, R., 203, 214
Baker, R. M., 318, 334
Barski, G., 9
Beaudet, A. L., 355
Bodmer, W., 127
Brunette, D. M., 334
Bunn, C. L., 168
Capecchi, M. R., 224
Carter, S. B., 164
Caskey, C. T., 355
Chapman, V. M., 133
Chasin, L. A., 360
Chen, T. R., 133
Cheng, Y.-c., 203
Chu, E. H. Y., 273
Cook, P. R., 140
Cornefert, F., 9
Creagan, R., 145
Croce, C. M., 109
Davidson, R. L., 37, 70, 74
Davis, D., 195
Deisseroth, A., 145
De Mars, R., 293
Dewey, M. J., 243
Ege, T., 173
Eisenstadt, J. M., 168
Ephrussi, B., 27, 37, 39, 70, 84
Finch, B. W., 84
Fischer, G. A., 253
Fisher, H. W., 284
Green, H., 119
Hakala, M. T., 261

Harris, H., 53, 103
Harris, M., 347
Hooper, J. L., 293
Johnson, R. T., 94, 331
Kao, F.-T., 308, 331
Kaufman, R. J., 256
Kellems, R. F., 256
Klebe, R. J., 133
Klein, G., 103
Köhler, G., 91
Koprowski, H., 109
Kraiselburd, E., 195
Kucherlapati, R., 145
Lacy, E., 214
Lawrence, J., 145
Lee, L.-S., 203
Lieberman, I., 269
Littlefield, J. W., 35
McBride, O. W., 238
Malling, H. V., 273
Maniatis, T., 214
Mankovitz, R., 318, 334
Mann, J., 195
Martin, D. W., Jr., 243
Martin, G. R., 243
Meera Khan, P., 140
Miggiano, V., 127
Miller, O. J., 103, 140
Milstein, C., 91
Mintz, B., 243
Munyon, W., 195
Nabholz, M., 127
Naha, P. M., 316
Nienhuis, A., 145
Okada, Y., 42

Ove., P., 269
Ozer, H. L., 238
Pellicer, A., 203, 214
Peterson, J. A., 79
Pontecorvo, G., 60
Puck, T. T., 284, 308, 331
Rao, P. N., 94
Ringertz, N. R., 173
Roufa, D. J., 355
Ruddle, F. H., 133, 145
Schimke, R. T., 256
Shin, S., 140
Silverstein, S., 203, 214
Sim, G. K., 214
Siminovitch, L., 318, 334
Siniscalco, M., 140
Smith, M. J., 265
Sorieul, S., 9, 27

Sweet, R., 214
Szybalska, E. H., 186
Syzbalski, W., 186, 265
Tachibana, T., 103
Thompson, L. H., 318, 334
Till, J. E., 318, 334
Turner, P., 145
Urlaub, G., 360
Velez, R., 145
Wallace, D. C., 168
Watkins, J. F., 53
Weiss, M. C., 39, 79, 119
Whitmore, G. F., 318, 334
Wigler, M., 203, 214
Wold, B., 214
Worst, P., 103
Yamamoto, K., 70

INTRODUCTION

Somatic cell geneticists have nothing against sex. However, sexual reproduction is not well suited for the genetic analysis of mammals and it is especially unsuited for the genetic analysis of man. All of the prerequisites for stringent genetic studies such as controlled matings, short generation times, and large numbers of progeny, are largely or completely missing in mammalian reproductive systems. In the last few decades, therefore, many scientists interested in mammalian genetics have turned their efforts to genetic systems that bypass sexual reproduction, that is, to what we now call *somatic cell genetics*.

In somatic cell genetics, somatic cells rather than germ cells are the basis for genetic analysis. With mammalian somatic cells growing in culture almost like microorganisms, the techniques of mutant selection, cell hybridization, and gene transfer can be used to introduce genetic variation into cells and to mix genomes in new combinations. Through the use of somatic cells it is possible to satisfy the prerequisites for high-resolution genetic analysis: genomes can be combined at will, without even respecting species barriers; the generation times of cultured cells are relatively short, about one day for a human cell in culture compared to more than twenty years for a human being; and large numbers of progeny from independent genetic events can be obtained.

By using the techniques of somatic cell genetics a variety of genetic questions that are difficult to approach via sexual reproduction are amenable to experimental analysis with cultured mammalian cells. These questions relate to gene regulation, genetic variation, and the formal genetics (chromosome mapping) of man.

Somatic cell hybridization was discovered in 1960 as an unexpected result of experiments on the cocultivation of mammalian cells of different types. Early studies on somatic cell hybridization focused on the types of cells that could be hybridized, on methods for isolating hybrid cells, and on techniques for increasing the frequency of cell hybridization. In 1964 the "HAT" selective system was introduced as the first system for the selection of somatic cell hybrids. With this selective system it became possible to isolate pure populations of hybrid cells that occurred at only very low frequencies. Soon thereafter, it was found that viable hybrids could even result from fusion between cells of different species or different tissues. The availability of selective techniques and the ability to produce hybrids between cells of different species or tissues led to a rapid expansion of activity in the area of somatic cell hybridization.

Following the early descriptive and methodological studies on somatic cell hybridization, work on cell hybrids focused on two main areas: gene regulation and gene mapping. Somatic cell hybridization made it possible to combine within a single nucleus the genomes of cells that differed in their patterns of gene expression, and this provided new opportunities for analyzing the mechanisms of genetic regulation in mammalian cells. The existence of inter-specific hybrid cells, which in certain cases segregated chromosomes of one of the parental species, allowed gene mapping studies to be performed with cultured cells. Somatic cell hybridization has proven to be the most productive way of mapping the human genome yet developed. More recently, the techniques of cell hybridization have been used to construct cells that synthesize *monoclonal* antibodies. These highly specific antibodies appear to have significant potential as immunological reagents for medical as well as scientific purposes.

Work on gene transfer began at about the same time as the early work on somatic cell hybridization. The first apparently successful transfer of a genetic trait from one mammalian cell to another was reported in 1962 and involved cellular DNA as the transforming vector. However, in contrast to the rapid progress of work on cell hybridization, work on gene transfer advanced slowly after the initial report, and it was not until a decade later that reproducible gene transfer systems for mammalian cells were developed. In 1971 a system was introduced in which Herpes simplex virus, rather than cellular DNA, was the transforming vector. This viral transformation system provided the basis for much of the gene transfer work accomplished to date, and it led to the eventual development of reliable and efficient systems for transformation by means of mammalian cell DNA.

With recent developments in recombinant DNA technology permitting the isolation of specific genes, work on gene transfer systems has become increasingly active. The introduction of foreign genes into cells, and more recently into embryos, should play a major role in attempts to elucidate the molecular mechanisms of gene regulation in higher organisms.

Work on mutant cells began well before the work on cell hybridization and gene transfer. Clearly, studies on the inheritance of variation go back to the very origins of the study of genetics. However, studies on mutant mammalian cells date back only three decades. The initial studies in this area were carried out not for the purpose of characterizing mutant mammalian cells per se, but rather to study the mechanisms by which tumors developed resistance to antitumor drugs. In 1950 the development of leukemic lines resistant to the folic acid antagonist amethopterin was reported, and the following year leukemic lines resistant to the purine analog 8-azaguanine were described. As will be discussed later, studies involving these two drugs have had a major impact on the field of somatic cell genetics. In fact, a history of somatic cell genetics could be largely centered around these two drugs. The "HAT" selective system, which was developed as a result of studies with these drugs, has made possible much of the work on cell hybridization and gene transfer accomplished to date. Furthermore, studies on the drug-resistant cells themselves have provided much of our current information on the alterations in mutant mammalian cells.

During the mid-1950s, techniques for cloning mammalian cells in culture were being developed and improved. These techniques, which permitted the isolation of pure lines of genetically marked cells, were essential to the development of somatic cell genetics. (For discussion of these techniques the reader should refer to general texts on cell culture.) In addition, the improved definition of the nutritional requirements of cultured mammalian cells quickly led to the isolation and characterization of nutritional variants of cultured cells. In 1960 the first method for isolating auxotrophic mutants of mammalian cells was described. Conditional lethal mutants, such as auxotrophic and temperature sensitive mutants, have provided much information on the physiology and genetics of cultured mammalian cells.

In all areas of somatic cell genetics, the techniques of cell genetics and of recombinant DNA are being jointly applied, and this combined approach is becoming increasingly more powerful and productive. Major developments can be anticipated in the various areas in which the techniques of somatic cell genetics and recombinant DNA have been joined, including gene regulation, gene mapping, embryology, and tumor biology.

Part I

SOMATIC CELL HYBRIDIZATION

Editor's Comments
on Papers 1 and 2

1   BARSKI, SORIEUL, and CORNEFERT
"Hybrid" Type Cells in Combined Cultures of Two Different Mammalian Cell Strains

2   SORIEUL and EPHRUSSI
Karyological Demonstration of Hybridization of Mammalian Cells in vitro

DISCOVERY OF SOMATIC CELL HYBRIDIZATION

Somatic cell hybridization is the fusion of somatic cells of two different types to form a *hybrid* cell that contains genetic elements of both *parental* cells. Somatic cell hybridization provides a method for combining mammalian genomes in vitro, completely bypassing the route of sexual reproduction. Indeed, somatic cell hybridization has permitted the combination of genomes that could never be brought together in nature. For example, it is possible to combine the genomes of cells of different tissues or different species into a hybrid.

The phenomenon of somatic cell hybridization was first observed by Barski and his co-workers in 1960 (Barski et al., 1960, and Paper 1). The initial experiments were carried out not to demonstrate somatic cell hybridization, but rather to examine the possibility of transferring genetic material between cells of two mouse lines growing in mixed culture. Although the two cell lines used in these experiments had been derived from the same cell, they had diverged during growth in culture so that they differed in a number of characteristics such as malignancy and chromosome composition.

It was the karyotypic differences between the cells that made possible the observation of somatic cell hybridization. When cells of the two lines were grown in mixed culture a new cell type was unexpectedly observed—a hybrid cell representing the karyotypic addition of the two parental cells. The hybrid cells had a chromosome number approximately equal to the sum of the chromosome numbers of the parental cells, and, in addition, contained the distinctive chromosomal markers of both parental cell lines. Although the hybrid cells occurred only infrequently in the mixed cultures, they appeared to have a selective advantage since the proportion of hybrid cells observed in karyological preparations increased progressively with continued cultivation.

Clearly, somatic cell hybrids were capable of long-term proliferation. In addition, the hybrids, which resulted from the fusion of parental cells with different levels of neoplasticity, were seen to be capable of forming tumors themselves. As discussed later (see Papers 15 and 16), the genetic analysis of malignancy by means of cell hybridization has been an area of active study and debate.

Soon after these initial observations, the occurrence of somatic cell hybridization was confirmed by Sorieul and Ephrussi (Paper 2). In their experiments the hybrids again resulted from the fusion of closely related cells, as in the experiments described previously. Subsequent experiments, however, showed that the isolation of viable hybrids was not limited to crosses between closely related cells and that hybrids could be formed between unrelated parental cells (reviewed in Ephrussi, 1965).

In the early cell hybridization experiments, all the parental cells were from permanent lines. The use of permanent lines was due largely to the necessity of having chromosomal markers in the parental cells in order for hybrids to be identified karyologically. Since the markers used to identify hybrids were the novel chromosomes occurring in the cells of permanent lines but not present in normal diploid cells, diploid cells were not used in the early hybridization experiments.

In later studies normal cells cultured from a strain of mouse carrying a cytological marker were hybridized with mouse cells of a permanent line (Scaletta and Ephrussi, 1965). The results clearly demonstrated that normal diploid cells could fuse with other cells to produce viable hybrids, at least when the other parental cell belonged to a permanent line. The parental cells used in this study differed not only in their karyotypes but also in their growth potential. The cells of the permanent line were capable of unlimited proliferation in culture while the normal cells had only a very limited capacity for proliferation in vitro. The hybrids that were isolated, however, had an unlimited capacity for in vitro proliferation, much like the permanent parental cell line. Thus, the capacity for unlimited proliferation appeared to be inherited as a dominant trait in the hybrid cells.

For detailed discussions of somatic cell hybridization the reader may refer to various books on the subject (Ephrussi, 1972; Davidson and de la Cruz, 1974; Ringertz and Savage, 1976).

REFERENCES

Barski, G., S. Sorieul, and F. Cornefert, 1960, Production dans des cultures in vitro de deux souches cellulaires en association, de cellules de caractere "hybrids," *Acad. Sci. C. R.* **251:**1825–1827.

Davidson, R. L., and F. de la Cruz, eds., 1974, *Somatic Cell Hybridization,* Raven Press, New York, 295p.

Ephrussi, B., 1965, Hybridization of Somatic Cells and Phenotypic Expression, in *Developmental And Metabolic Control Mechanisms And Neoplasia,* Williams and Wilkins, Baltimore, pp. 486–503.

Ephrussi, B., 1972, *Hybridization of Somatic Cells,* Princeton University Press, Princeton, N.J., 175p.

Ringertz, N. R., and R. E. Savage, 1976, *Cell Hybrids,* Academic Press, New York, 366p.

Scaletta, L. J., and B. Ephrussi, 1965, Hybridization of Normal and Neoplastic Cells In Vitro, *Nature* **205:**1169–1170.

Reprinted from *Natl. Cancer Inst. J.* **26**:1269–1291 (1961)

"Hybrid" Type Cells in Combined Cultures of Two Different Mammalian Cell Strains [1]

G. BARSKI, S. SORIEUL, *and* FR. CORNEFERT,[2]
Tissue-Culture and Virus Laboratory, Institut Gustave Roussy, Villejuif (Seine), France

SUMMARY

Cells of a high-cancer N1 line derived from NCTC 2472 clone of mouse fibroblasts and cells of N2 or N2T' lines derived from low-cancer NCTC 2555 clone of the same origin were cultivated together *in vitro*. The evolution of mixed-cell populations was compared with the behavior of component cells in separate cultures. Separately cultivated, the cell lines kept their essential characteristics fairly stable. N1 cells maintained a karyotype with a modal number of nearly 55 telocentric chromosomes—one extra long. N2 lines had modal chromosome numbers of 62 and 58 chromosomes, 13 biarmed. Higher polyploids in pure cultures were recognized as doubles or quadruples of stemline cells. In mixed cultures the proportion of N1 cells progressively decreased until they almost completely disappeared. At the 3d month, a new type of cell designated "M" appeared, which later reached the proportion of one third to one fourth of the whole cell population. The M cells had many "hybrid" chromosomal and biological characteristics. They showed a modal number of 115 to 116 chromosomes with 9 to 15 metacentrics, frequently with one extra long telocentric. M cells from mixed populations and secondarily isolated M-type cell clones produced, in isologous mice, malignant tumors in which M cells could be identified. The possible origin of the "hybrid" type cells is discussed.—J. Nat. Cancer Inst. 26: 1269–1291, 1961.

SANFORD *et al.* (*1*) obtained, from a cell clone originating from normal subcutaneous tissue of the C3H mouse, several sublines or clones (*2*) having different morphological, biochemical (*3*), chromosomal (*4*), and tumor-producing characteristics.

We have recently (*5*) analyzed and compared cell lines derived from either the high-cancer NCTC 2472 clone (referred to as N1) or the NCTC 2555 clone (N2) having a particularly low cancer incidence. These two clones differed considerably in their static and dynamic cell morphology *in vitro*. They were also easily recognizable by the histological structure, growth pattern, and rapidity of tumors produced in the C3H mice.

---

[1] Received for publication December 22, 1960.
[2] We wish to thank Dr. Ephrussi (Paris) and Dr. Levan (Lund) for their stimulating interest and discussion in connection with this work, Dr. Gérard-Marchant for his help in the histological preparations, and Mr. Charlier for drawing the graphs.

The N1 clone and its derived cell lines, all highly malignant, maintained constantly their pronounced fibroblast-like appearance, high mobility, and fan-shaped undulating membranes. The N1 karyotype, characterized by a modal number of nearly 55 telocentric chromosomes, appeared fairly stable during 20 months of culture *in vitro* in our laboratory. One of these chromosomes, constantly present, was very long (*5*). One metacentric, and rarely two, were present in no more than 26 percent of examined mitoses of the original N1 clone. The metacentrics disappeared nearly completely in the derived N1T line obtained after one passage of the N1 cells in the animal. Otherwise, cells recovered after from 1 to 7 passages were similar to the parental line. The N1 and derived lines had a marked tendency to produce higher polyploids which, however, maintained the typical, nearly pure telocentric chromosome pattern. These results compared fairly well with the previous chromosomal analysis of Chu, Sanford, and Earle (*4*).

Similarly, the N2 cells maintained their own features during 20 months of observation: rounded cells with regular undulating membranes or tentacular ramifications and a karyotype characterized by 54 to 67 chromosomes (modal number, 62), 9 to 19 of them biarmed (modal number, 13). These data (*5*) were on the whole similar to the results obtained with the same cells 2 years before by Chu, Sanford, and Earle (*4*), who observed, however, a somewhat higher modal number for this line: 69 chromosomes. Cells of two secondary lines, derived from the N2 clone after one animal passage, displayed some shift of chromosomal values, but remained unchanged for the essential feature of their metaphases: the modal number of 13 biarmed chromosomes. Morphologically these cell lines, despite some increase in their tumor-producing capacity, were very different from the N1 cell family, and the tumors they produced differed unequivocally from the N1 tumors in their histological aspect and slow, highly anaplastic growth.

Having accumulated data concerning the N1 and N2 cell families, especially regarding the extent of their normal variability *in vitro* and *in vivo*, we designed experiments to investigate the possibility of transfer of genetic material from one cell family to the other. To achieve this, we utilized prolonged culture *in vitro* of the 2 cell lines in close association. The behavior of combined cultures of N1 and N2 cells and the appearance in the mixed cultures of a new type of cells with "hybrid" chromosomal, morphological, and biological characteristics are presented. The results were reported in part in a preliminary note (*6*).

MATERIALS AND METHODS

Cultures of the high-cancer clone NCTC 2472 (N1) and the low-cancer clone NCTC 2555 (N2) were received from Dr. K. K. Sanford, Tissue Culture Section, Laboratory of Biology, National Cancer Institute, Bethesda, Maryland.

The newly designated N1T line, developed in our laboratory after one animal passage of the N1 cells, was in every respect similar to the parental line. The N2T' line, obtained after one animal passage of the N2 clone and showing some increase of tumor-producing capacity, remained quite similar in its chromosome characteristics to the N2 cells (5). The cultivation, transfer, establishment of new cell lines *in vitro*, techniques of observation, recording of cell behavior *in vitro*, and the procedure used for chromosome studies were previously described in detail (5). Modified synthetic medium 199 (7), completed with 15 percent horse serum, was used uniformly for all cultures. Cells in mixed cultures were transferred exclusively by mechanical dispersion without trypsinization. Special precautions were taken to exclude any risk of cell contamination between cultures: Simultaneous handling of different cell cultures was avoided, and separate labeled supplies of media were prepared for each cell strain.

To check the tumor-producing capacity, we prepared standard 0.1 ml volumes of counted cell suspensions in the culture medium and inoculated them subcutaneously into 18- to 20-g males of the inbred C3H/He mice. Tumor growth was measured and recorded as previously described (5).

RESULTS

In preliminary experiments, mixtures containing 2.5×10^6 N2 cells and 1×10^6 or 1×10^5 N1 cells were inoculated subcutaneously into C3H mice. Rapidly growing tumors resulted. Histologically these tumors were all typical, spindle-cell sarcomas, undistinguishable from N1 tumors. Morphological and chromosomal studies of cell cultures obtained by explantation *in vitro* of these tumors confirmed that they were of N1 cell origin.

After the preliminary trials, mixed cultures, established with equal numbers of N1 and N2 cells, were grown *in vitro* for prolonged periods. In the experiment designated as "MS," 10^6 cells of the N1T line were associated with the same number of N2 clone cells. In another experiment designated "MT," 10^6 cells of the N1T line were associated with 10^6 cells of N2T' line.

The mixed populations were maintained for several months in massive, actively multiplying cultures. We made observations directly by checking the appearance of the flask cultures and by qualitative and quantitative chromosomal analysis of secondary cultures prepared periodically on coverslips in flattened tubes designed in this laboratory (8). The data are represented in table 1.

If one assumes that the proportion of different mitotic types gives a satisfactory approximation concerning the different components of mixed culture, the results of these observations can be outlined as follows:

In both the MS and MT experiments, the N1 and N2 cells were present during the first 2 months of combined cultures. However, a selective advantage markedly favored the N2 cells that increased in proportion.

TABLE 1.—Evolution of cell components in mixed-cell population of MS and MT experiments

Experiment	Days of mixed cul-tures	Number of metaphases checked							
			N1 type			N2 type			M type
		Total	1 × s*	2 × s	4 × s	1 × s	2 × s	4 × s	
MS—mixed population	55	57	10	12	2	29	4	0	0
established with	104	230	0	1	1	180	27	0	21
N1T + N2 cells	139	54	0	4	1	38	3	2	6
	157	226	0	1	0	151	9	0	65
MT—mixed population	122	75	15	13	2	32	13	0	0
established with	126	29	4	6	1	12	4	0	2
N1T + N2T′ cells	147	97	12	22	1	15	23	1	23
	173	87	2	6	0	7	7	1	64
	204	115	0	6	0	23	2	0	84
	227	68	0	1	0	41	2	0	24

*s = "Stemline" hypotetraploid chromosome numbers: nearly 55 telocentrics for the N1 cells, nearly 60, with a modal number of 13 metacentrics for the N2 cells, and nearly 115, with 9 to 15 metacentrics for the M cells.

Likewise, the relative number of higher polyploids, especially of N1 type, increased. These could be easily recognized as double or quadruple stemline cells having multiples of normal chromosomal sets of either N1 or N2 complements. Thus, after 5 months of combined culture in the MS experiment, of the total 226 mitoses, only one N1 type could be found and it was a double stemline mitosis. After 7 to 8 months in the MT experiment, of 183 mitoses, only 7 higher polyploids of the N1 type were observed.

The evolution of the morphological aspect of the mixed cultures was consistent with these results. The characteristic elongated N1 cells progressively disappeared and were replaced by a population of rounded cells, some of them larger than the normal N2 cells and showing irregular, bizarre ramifications (figs. 1, 2, and 3). During the 3d month of the MS experiment and the 4th month of the MT experiment, a new cellular type appeared in the mixed cultures, which was designated the M cell. These cells, rare at the beginning, progressively increased in proportion. Their characteristics, according to detailed analysis of 48 M mitoses in the MT experiment, were as follows:

1) The chromosomes numbered between 103 and 125 but were grouped chiefly in the 115 or 116 area, which was very near the total of modal numbers of the N1- and N2-type cells (text-fig. 1a and figs. 4 through 10 and 12 through 14).

2) Biarmed chromosomes numbering 9 to 15 were present; their modal number, 12, coincides equally well with the values for the N2 cells, but was consistently below the corresponding values for the N2 polyploids (text-fig. 1b).

3) An extra long telocentric chromosome characteristic for the N1 cells was noted in 45 of the 48 checked M metaphases (figs. 8, 9, 10, and 14).

4) Once, at least, it was possible to identify singly the 11 biarmed chromosomes of an M cell with the 2 × 11 biarmed chromosomes of a

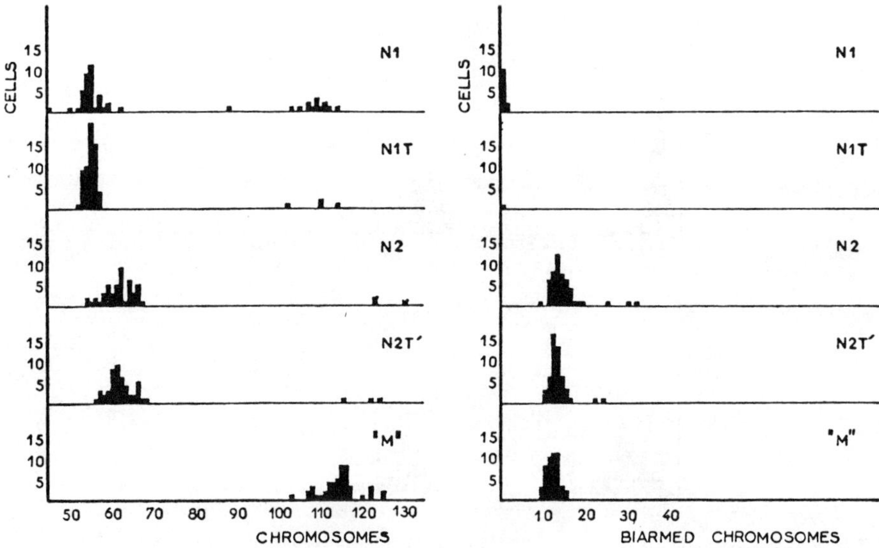

TEXT-FIGURE 1.—Frequency distribution of (a) chromosome numbers and (b) biarmed chromosome numbers of N1T, N2T', and M cells in the MT experiment; for comparison, data relative to the original N1 and N2 clones analyzed previously (5) are included.   Each text-figure records data of nearly 50 analyzed mitoses.

polyploid N2 cell found in a mixed culture (figs. 6, 8, and 15).   This particular analyzed M cell contained, as expected, the extra long chromosome, obviously absent in the polyploid N2 cell.

The M cells from the MS experiments had the same general characteristics as those from the MT series (figs. 7, and 8 through 10).

At the 8th month in the MT and the 5th month in the MS experiments, the mixed-cell populations were composed essentially of N2 and M cells. The latter reached a proportion of nearly one third of the whole population for the MT and one fourth for the MS experiment (table 1).   At this stage, samples of the cells were inoculated subcutaneously in young C3H males.   In both, rapidly growing tumors appeared at the inoculation site in high proportion of mice—6/7 for the MS and 12/12 for the MT series.

Curves representing the increase of mean diameter of these tumors approximated those obtained for the N1 tumors.   On histological sections the tumors appeared as well-vascularized, round-cell sarcomas with some lobulation around wide vascular sinuses (figs. 18 and 19).

The M tumors were easily distinguishable from the typical spindle-cell sarcomas of the N1 type as well as from the anaplastic, round, and small cell tumors of the N2 type (figs. 16 and 17).   Furthermore, the mean nuclear size of M tumor cells was, as shown in text-figure 2, significantly greater than that of either N1 or N2 tumors.

Five tumors of the MS and 2 of the MT experiments were explanted *in vitro* and the corresponding 7 established cell lines studied.   Morphologically, the cells of these sublines were, with some minor differences, close to the final MS and MT cell population (figs. 3a and 3b).   The chro-

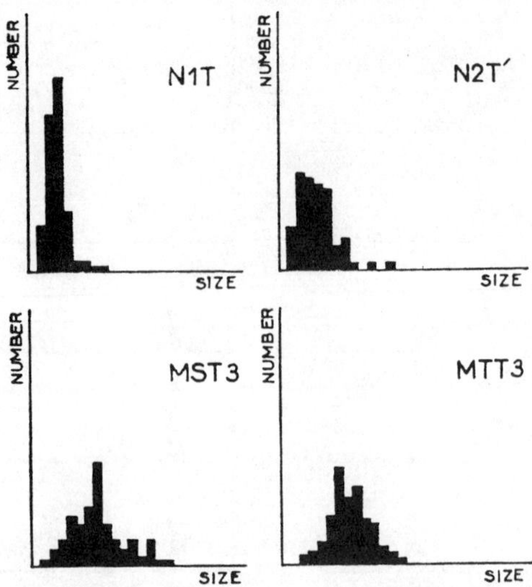

TEXT-FIGURE 2.—Distribution of nuclear size in histological sections of tumors, produced with N1 and N2T' cells and of two tumors obtained with MS and MT cell populations composed of N2 and M cells. Only M cells were found after explantation and chromosomal analysis of the MST3 and MTT3 tumors. Nuclear size is expressed in arbitrary values corresponding to planimetric measures. (*See* histological sections figs. 16 through 19.)

mosomal study of 6 of these lines always disclosed presence of cells having chromosomal numbers grouped chiefly in the 100 to 120 area (fig. 11). Of these chromosomes, 8 to 15 were biarmed. In approximately 40 percent of metaphases examined, the extra long telocentric chromosome could be identified. No correlation was found between the total number of chromosomes and the presence of the extra long telocentric chromosome. Generally, the M cells, after passage in animals, have shown a tendency toward wider dispersion and decrease of mean chromosome numbers. In 4 of the 6 analyzed lines, only the M-type cells were seen. In two others, some N1 cells were disclosed, but no persistence of N2-type cells could be found.

One of the MS tumor lines was cloned recently. Several clonal lines obtained from it are now under study. According to the results available at present, all the isolated clones have karyotypes with chromosomal numbers grouped chiefly in the 100 to 120 area, with modal numbers of 9 to 11 biarmed chromosomes. In at least one of these clones, the extra long telocentric is regularly present. All the 4 isolated M-cell-type clones so far assayed have produced tumors in 100 percent subcutaneously inoculated C3H/He mice with a standard dose of 2.5×10^6 cells per inoculum.

DISCUSSION AND CONCLUSIONS

Cultures of the NCTC 2472 (N1) and NCTC 2555 (N2) clones of mouse fibroblasts and their derived sublines were followed for many years in different laboratories, and the extent of their modifications in normal conditions *in vitro* and *in vivo*, especially in respect to their karyotypes, was outlined *(4, 5)*.

From these data one can regard the appearance of the new peculiar M cell, present in mixed but not in separate, unmixed cultures of the N1 and N2 cells, as a result of close and prolonged association of these two cell types in actively mutiplying mass cultures *in vitro*. Similar results obtained independently in different experiments bear out this conclusion.

When we consider the properties of the M cell on the whole, its morphology and cancer-producing capacity, together with the "hybrid" character of its karyotype in which joined chromosome sets of the N1- and N2-cell types can be recognized, the hypothesis of a nuclear fusion with the production of "allopolyploid" cells has to be envisaged. In support of this hypothesis one can refer to the following data and considerations:

Lewis in 1927 *(9)* described and represented with precise drawings the formation of polynucleated cells by fusion of cytoplasm of individual mononucleated cells. Since then, this phenomenon, with a variety of animal cells, has been observed by other authors *(10–13)*, and seems a frequent occurrence between cells having active ectoplasmic membranes *(14)*, as is true with N1 and N2 cells. Furthermore, it is known that the mechanisms determining biosynthesis and nuclear division in polynucleated cells are to some extent synchronized. In these conditions, simultaneous formation of many mitotic spindles is not uncommon *(15)*. As was observed and recorded by Lettré and Siebs *(16)* and Gey *(17)*, 2 spindles can then fuse and the chromosomes of 2 nuclei rearrange themselves in one common equatorial plate. The mitosis can evolve further to form 2 or more nuclei containing "mixed" chromosomal sets.

These phenomena, when occurring under normal conditions in either wild heterogenetic or uniform homogenetic cell populations, will have no easily recognizable genetic and biological consequences. In reality, heteroploid cells issued from fusion of two different but genetically similar or undetermined cells would be hardly distinguishable from polyploids produced by endomitosis or C-mitosis. In the N1 and N2 cell association, many favorable conditions were available to facilitate the demonstration of this possible phenomenon. These conditions were biological and morphologic cell and chromosome characteristics related to these two kinds of cells, recognized in advance as very distinct and relatively stable.

Another essential factor was the apparent selective advantage of the M cell, first, over the N1 component in the mixed-cell population *in vitro* and, later, over the N2 component *in vivo*.

Thus, particular circumstances aided to recognize a phenomenon that probably can occur with a certain frequency in actively multiplying cell populations and that constitutes possibly one of the mechanisms of polyploidization of somatic cells. This mechanism was envisioned many years ago by Wilson *(18)*, and recently, Schultz, discussing the malignancy of mammalian cells, again raised the possibility of nuclear fusion as one of the ways of genetic material transfer between somatic cells *(19)*.

An alternative explanation of the origin of the M cell would be a profound modification and rearrangement of the chromosomal complement of either N1 or N2 cells on an "autopolyploidization" basis. However, many considerations are against this interpretation. First, as already emphasized, the M-type cell was never observed in constantly checked, pure N1 or N2 cultures. One can hardly evisage the origin of M cells directly from the N1 cells with an abrupt formation of 10 or more metacentric chromosomes. The N2 cells, characterized as the M cells by the presence of many biarmed chromosomes, even when displaying in unmixed cultures a limited chromosomal shift and some increase of tumor-producing ability (5), never reached the high degree of malignancy proper to the M cells. Furthermore, the high polyploids in the pure N2 cultures have never shown a tendency to stabilize and to overgrow the N2 stem cells. This is consistent with the general trend manifested by high polyploids in pure, established cell lines to remain as a rather low and not increasing proportion of the whole cell population (20).

Another argument against the "autopolyploidization" hypothesis is that no transient forms between parental N1 or N2 cells and the M cell were ever seen in the primary mixed cultures. Such forms must be expected if one admits this hypothesis. According to studies by Levan and Biesele (21), Ford (22), Hsu and Klatt (23), and others, dealing with genetic evolution in cell populations *in vitro* controlled by constant karyotype analysis, chromosomal modifications in genetically homogeneous cell populations can usually be recognized by steps leading from simple polyploidization, through chromosomal rearrangements toward a stable, often aneuploid, karyotype. However, the opposite occurred with the M cells that appeared in an abrupt way and were closest to an ideal "hybrid" chromosomal complement from the beginning. Signs of rearrangement, such as loss of chromosomes and greater dispersion of their values, occurred later, especially after animal passages.

Another important point to be raised with regard to the nature of the M cell is its pronounced malignancy, expressed as a constantly high percentage of takes and rapidity of tumor growth. Therefore, the high cancer-producing capacity appears in the hybridization hypothesis as a dominant and stable character. This is especially apparent in the MS experiment in which the N2 component of the mixed culture was the original low-cancer NCTC 2555 clone.

This conclusion is supported by the recent results obtained with clones of M-type cells isolated after one animal passage of the MS mixed-cell populations. The features of these cells include high chromosome numbers grouped in the 100 to 115 area, with nearly 10 biarmed; their peculiar morphology and their invasiveness in the animal confirm the persistence of essential new properties acquired by the M cell.

REFERENCES

(1) SANFORD, K. K., LIKELY, G. D., and EARLE, W. R.: The development of variations in transplantability and morphology within a clone of mouse fibroblasts

transformed in sarcoma-producing cells *in vitro*. J. Nat. Cancer Inst. 15: 215–237, 1954.

(2) SANFORD, K. K., MERWIN, R. M., HOBBS, G. L., YOUNG, J. M., and EARLE, W. R.: Clonal analysis of variant cell lines transformed to malignant cells in tissue culture. J. Nat. Cancer Inst. 23: 1035–1059, 1959.

(3) WOODS, M. W., SANFORD, K. K., BURK, D., and EARLE, W. R.: Glycolytic properties of high and low sarcoma-producing lines and clones of mouse tissue-culture cells. J. Nat. Cancer Inst. 23: 1079–1088, 1959.

(4) CHU, E. H. Y., SANFORD, K. K., and EARLE, W. R.: Comparative chromosomal studies on mammalian cells in culture. II. Mouse sarcoma-producing cell strains and their derivatives. J. Nat. Cancer Inst. 21: 729–751, 1958.

(5) BARSKI, G., BIEDLER, J. C., and CORNEFERT, FR.: Modification of characteristics of an *in vitro* mouse cell line after an increase of its tumor-producing capacity. J. Nat. Cancer Inst. 26: 865–889, 1961.

(6) BARSKI, G., SORIEUL, S., and CORNEFERT, FR.: Production dans des cultures *in vitro* de deux souches cellulaires en association, de cellules de caractère "hybride." Compt. rend. Acad. sc. 251: 1825–1827, 1960.

(7) MORGAN, J. F., MORTON, H. J., and PARKER, R. C.: Nutrition of animal cell in tissue culture. I. Initial studies on a synthetic medium. Proc. Soc. Exper. Biol. & Med. 73: 1–8, 1950.

(8) BARSKI, G.: Tubes à lamelles applicables à différentes techniques de cultures de tissus. Ann. Inst. Pasteur 90: 512–513, 1956.

(9) LEWIS, W. H.: Binucleate cells and giant cells in tissue cultures and the similarity of the latter to giant cells in tuberculous lesions. Am. Rev. Tuberc. 15: 616–628, 1927.

(10) FAURE-FREMIET, M. E.: Etude histologique de Ficulina ficus L. (Demosspongia.) Arch. Anat. Micr. 28: 1–80, 1932.

(11) PIRES-SOARES, J. M.: Observations sur des cultures de leucocytes *in vitro*. Transformation des monocytes en macrophages, cellules géantes et cellules épithélioïdes. Compt. rend. Soc. biol. 125: 565–568, 1937.

(12) WEISS, P.: *In vitro* transformation of spindle cells of neural origin into macrophages. Anat. Rec. 88: 205–221, 1944.

(13) HSU, T. C.: Generation time of HeLa cells determined from cine records. Texas Rep. Biol. & Med. 18: 31–33, 1960.

(14) WILLMER, E. N.: Cytology and Evolution. New York, Academic Press, Inc., 1960, pp. 92–93.

(15) KUDO, R. R.: Observations on pelomyxa illinoiiensis. J. Morphol. 88: 145–184, 1951.

(16) LETTRÉ, R., and SIEBS, W.: Personal communication, 1960.

(17) GEY, G.: Personal communication, 1960.

(18) WILSON, E. B.: The Cell in Development and Heredity, 3d ed. New York, Macmillan Co., 1953, pp. 870, 884.

(19) SCHULTZ, J.: Malignancy and the genetics of the somatic cell. Ann. New York Acad. Sc. 71: 994–1008, 1958.

(20) HSU, T. C., and KELLOGG, D. S. JR.: Mammalian chromosomes *in vitro*. XII. Experimental evolution of cell populations. J. Nat. Cancer Inst. 24: 1067–1093, 1960.

(21) LEVAN, A., and BIESELE, J. J.: Role of chromosomes in cancerogenesis, as studied in serial tissue culture of mammalian cells. Ann. New York Acad. Sc. 71: 1022–1053, 1958.

(22) FORD, D. K.: Chromosomal changes occurring in Chinese hamster cells during prolonged culture *in vitro*. Proc. Third Canadian Cancer Conf. 171–188, 1959.

(23) HSU, T. C., and KLATT, O.: Mammalian chromosomes *in vitro*. X. Heteroploid transformation in neoplastic cells. J. Nat. Cancer Inst. 22: 313–319, 1959.

PLATE 189

Phase-contrast microphotographs of living cultures.   Approximately × 1,250

FIGURE 1.—Cells of high-cancer N1 line (NCTC clone 2472).

FIGURE 2.—Cells of low-cancer N2 line (NCTC clone 2555).

FIGURE 3 (*a* and *b*).—M-type cells (clone 2 and 1) derived from tumor (MST3) obtained in C3H mouse inoculated with MS mixed-cell population.   All 279 mitoses checked after explantation of MST3 tumor *in vitro* were of M type.

PLATE 190

**All figures are microphotographs of acetic-orcein-stained squash preparations.
Approximately × 3,000**

FIGURE 4.—Metaphase of typical N1T cell; 54 chromosomes, 1 extra long.

FIGURE 5.—Metaphase of N2 cell: 61 chromosomes, 12 biarmed.

FIGURE 6.—Polyploid N2 metaphase with recognizable double set of stem-cell chromosomes: 126 chromosomes, 22 biarmed.

FIGURE 7.—Metaphase of M-type cell from MS experiment: 113 chromosomes, 12 biarmed; 1 very elongated metacentric (*center*), 1 extra long telocentric (*arrow*).

PLATE 191

**All figures are microphotographs of acetic-orcein-stained squash preparations.
Approximately × 3,000**

FIGURES 8, 9, and 10.—Metaphases of M-type cells from the MT experiment, containing, respectively, 113, 122, and 124 chromosomes—11, 14, and 13 biarmed. Extra long telocentric chromosome is present in each mitosis (*arrow*).

FIGURE 11.—Metaphase of M-type cell from MTT3 strain established *in vitro* after one animal passage of mixed-cell population from the MT experiment: 113 chromosomes, 10 biarmed, 1 extra long.   All 127 mitoses checked in MTT3 strain cultures were of M type.

PLATE 192

Idiograms composed from microphotographs of acetic-orcein-stained preparations.   Approximately × 5,500

FIGURE 12.—Idiogram of N1T metaphase (same as fig. 4): 54 chromosomes, all telocentrics, 1 extra long.

FIGURE 13.—Idiogram of N2 metaphase (same as fig. 5): 61 chromosomes, 12 biarmed.

PLATE 193

Idiogram composed from microphotograph of acetic-orcein-stained preparation.   Approximately × 5,500

FIGURE 14.—Idiogram of M cell metaphase (same as fig. 8): 113 chromosomes, 11 biarmed, certain of them identifiable with metacentrics of the N2-type cell (fig. 13): 102 telocentric chromosomes, 1 extra long.

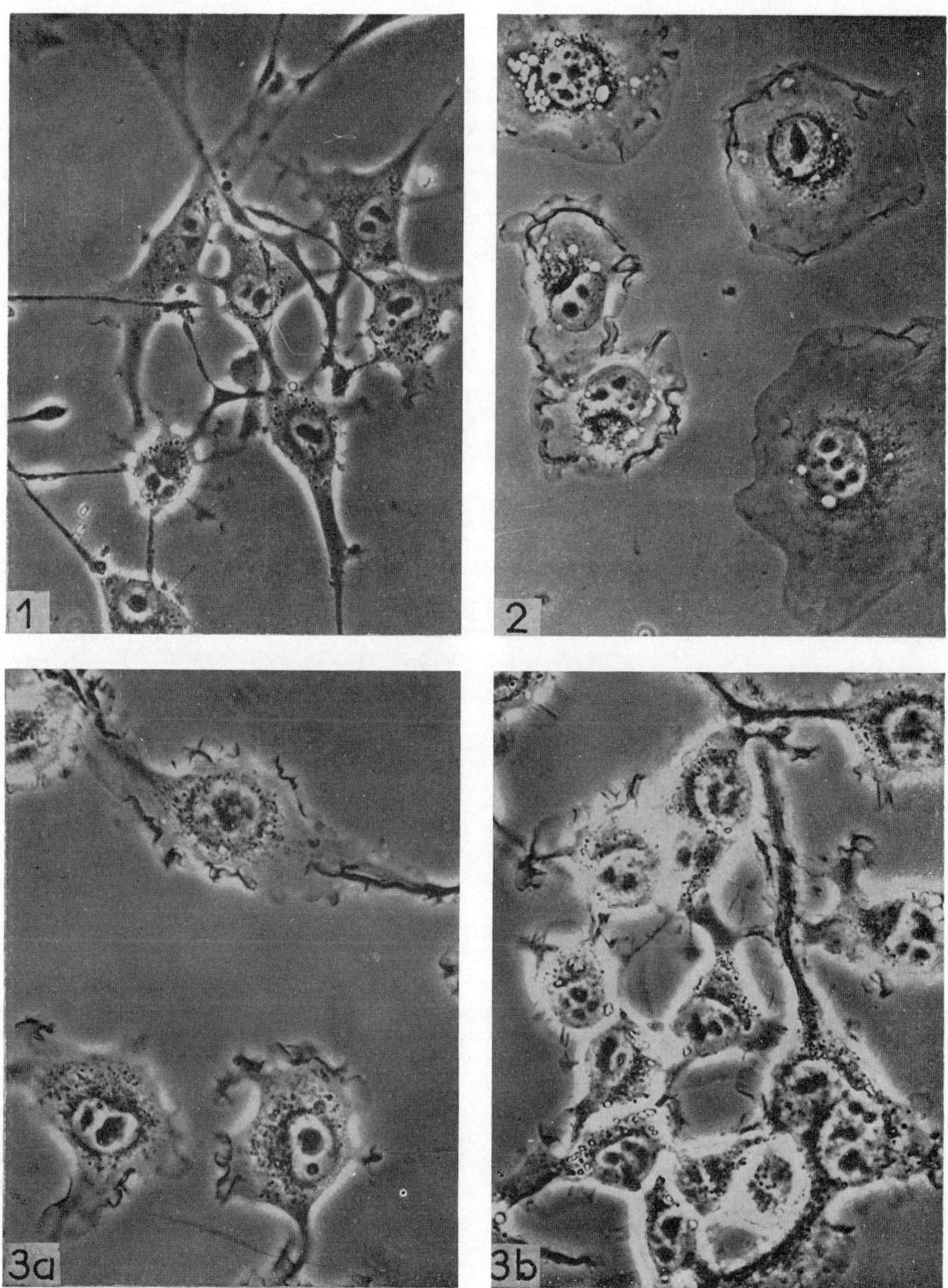

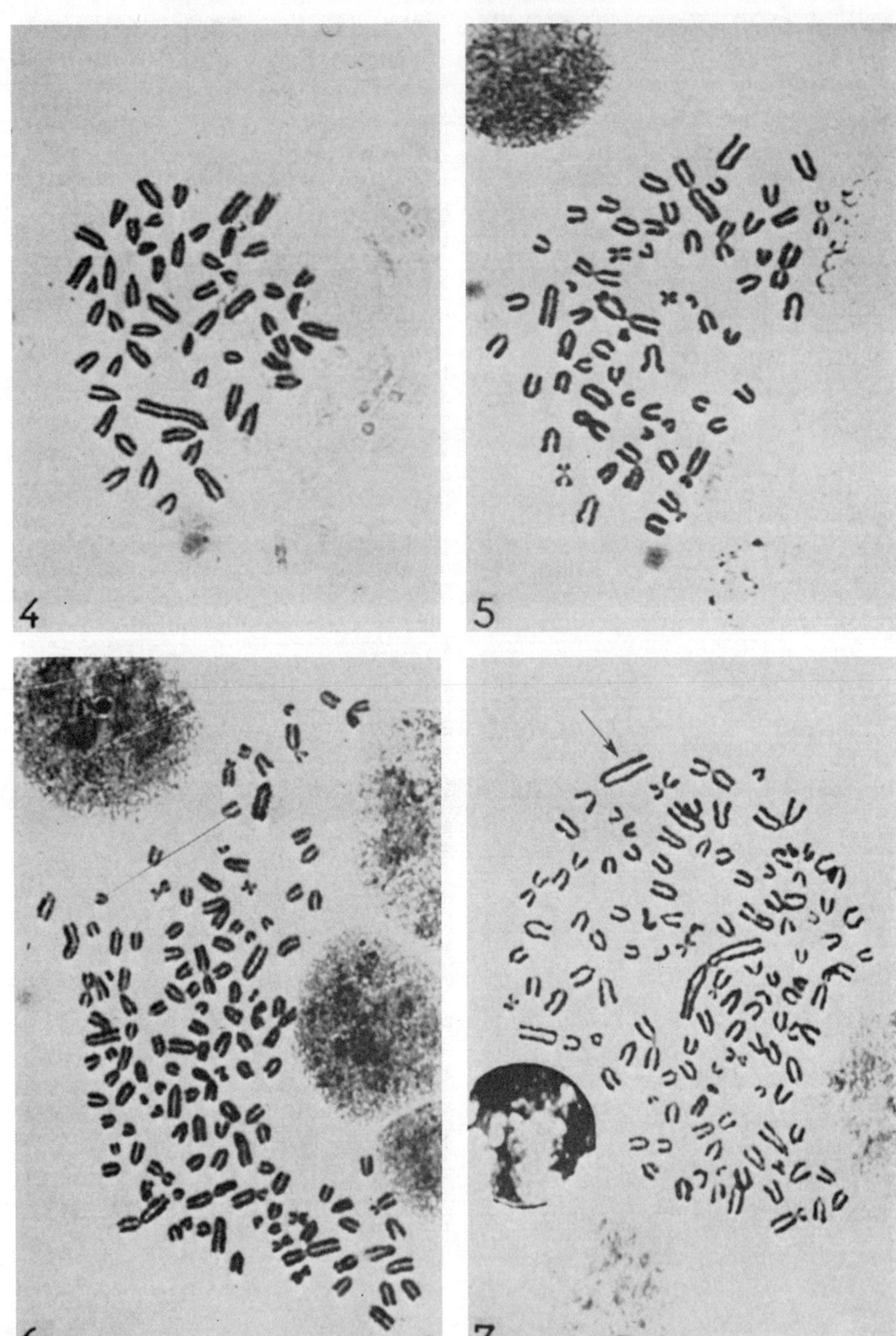

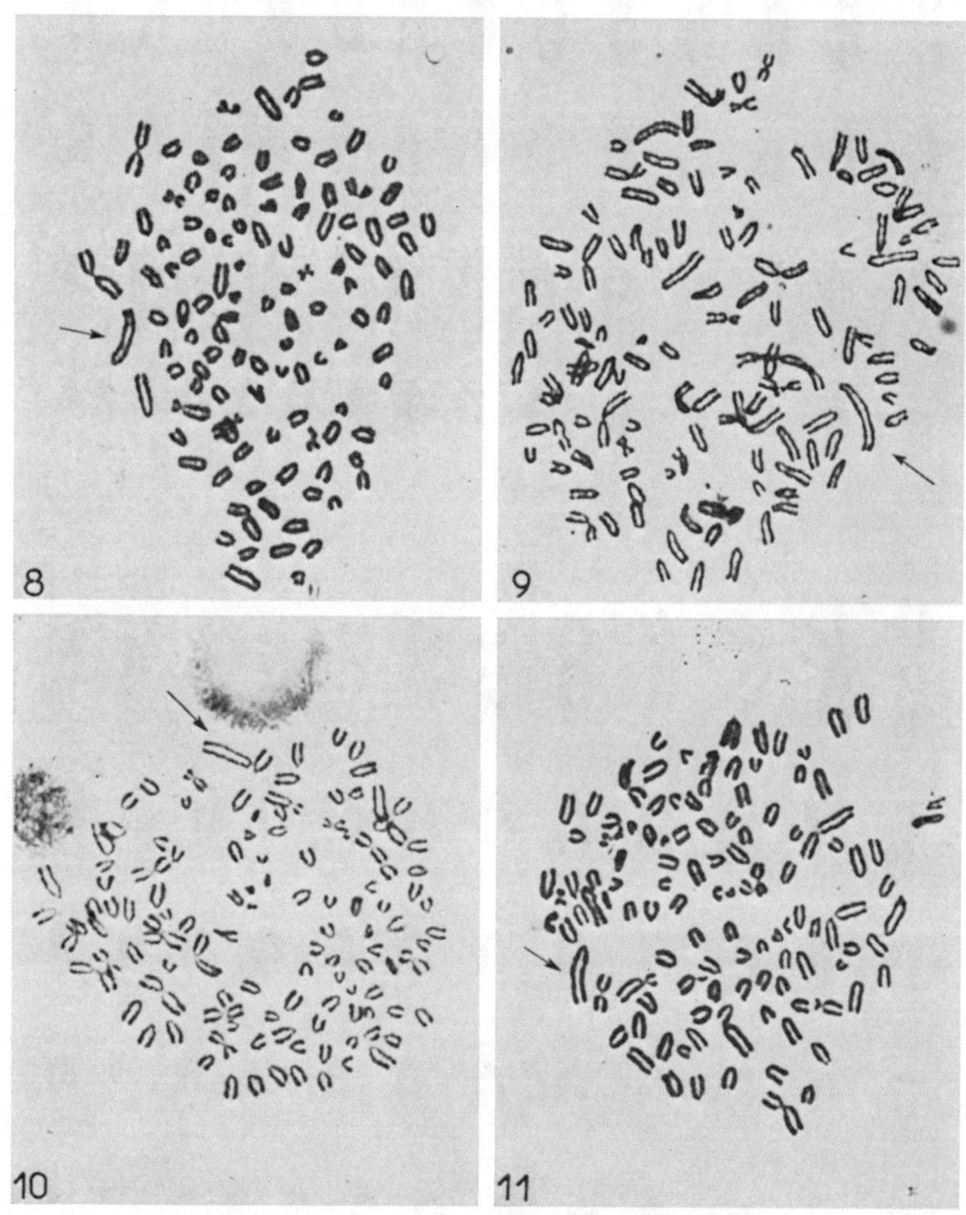

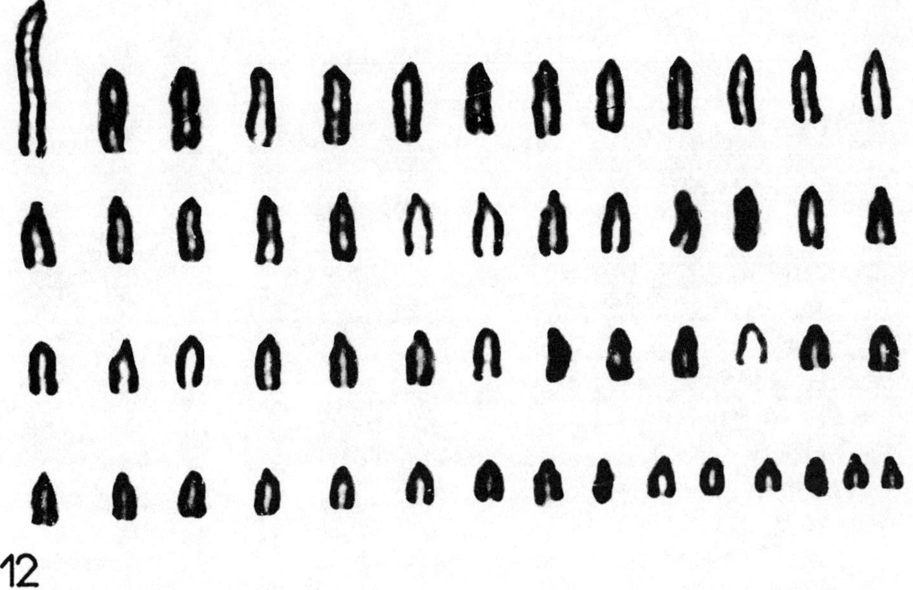

12

13

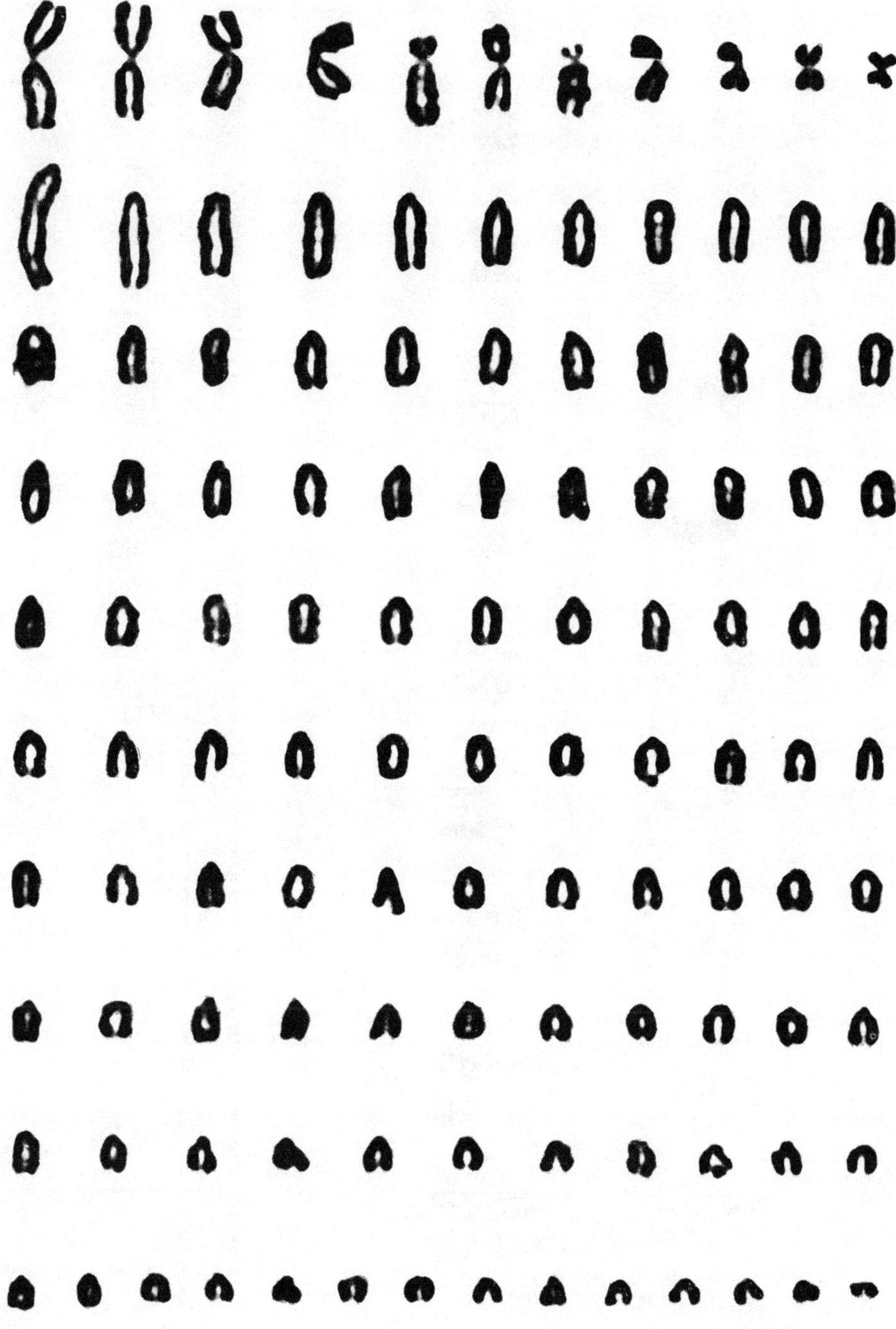

14

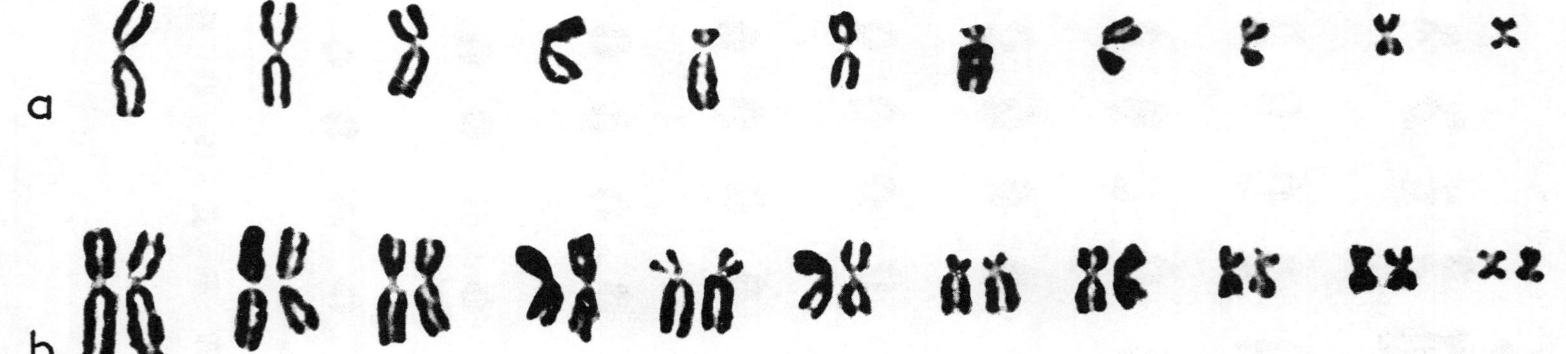

PLATE 194

Idiograms composed from microphotographs of acetic-orcein-stained preparations. Approximately $\times$ 5,500

FIGURE 15.—Comparative idiogram of biarmed chromosomes from (a) M cell (same as fig. 8) and (b) polyploid (2 $\times$ s) N2 cell (same as fig. 6). Metacentrics of these two metaphases are fairly well identifiable.

24

PLATE 195

Histological sections of tumors obtained by inoculation of C3H males with 1 million of N1T (fig. 16), N2T' (fig. 17), MS (fig. 18), and MT (fig. 19) cells. Alum hematoxylin-phloxine-saffron stain.   × 470

FIGURE 16.—Typical spindle-cell sarcoma characteristic of N1 tumors.

FIGURE 17.—Typical round-cell, anaplastic tumor of N2T' type.

FIGURE 18.—One of tumors (MST3) obtained by inoculation of MS mixed-cell population.

FIGURE 19.—One of tumors (MTT3) obtained with MT mixed-cell population.

Nuclei of MST3 and MTT3 tumor cells are strikingly bigger than in N1 and N2T tumors.   In cell lines established *in vitro* from these 2 tumors, only M-type mitoses were seen (279 and 127 mitoses checked).

25

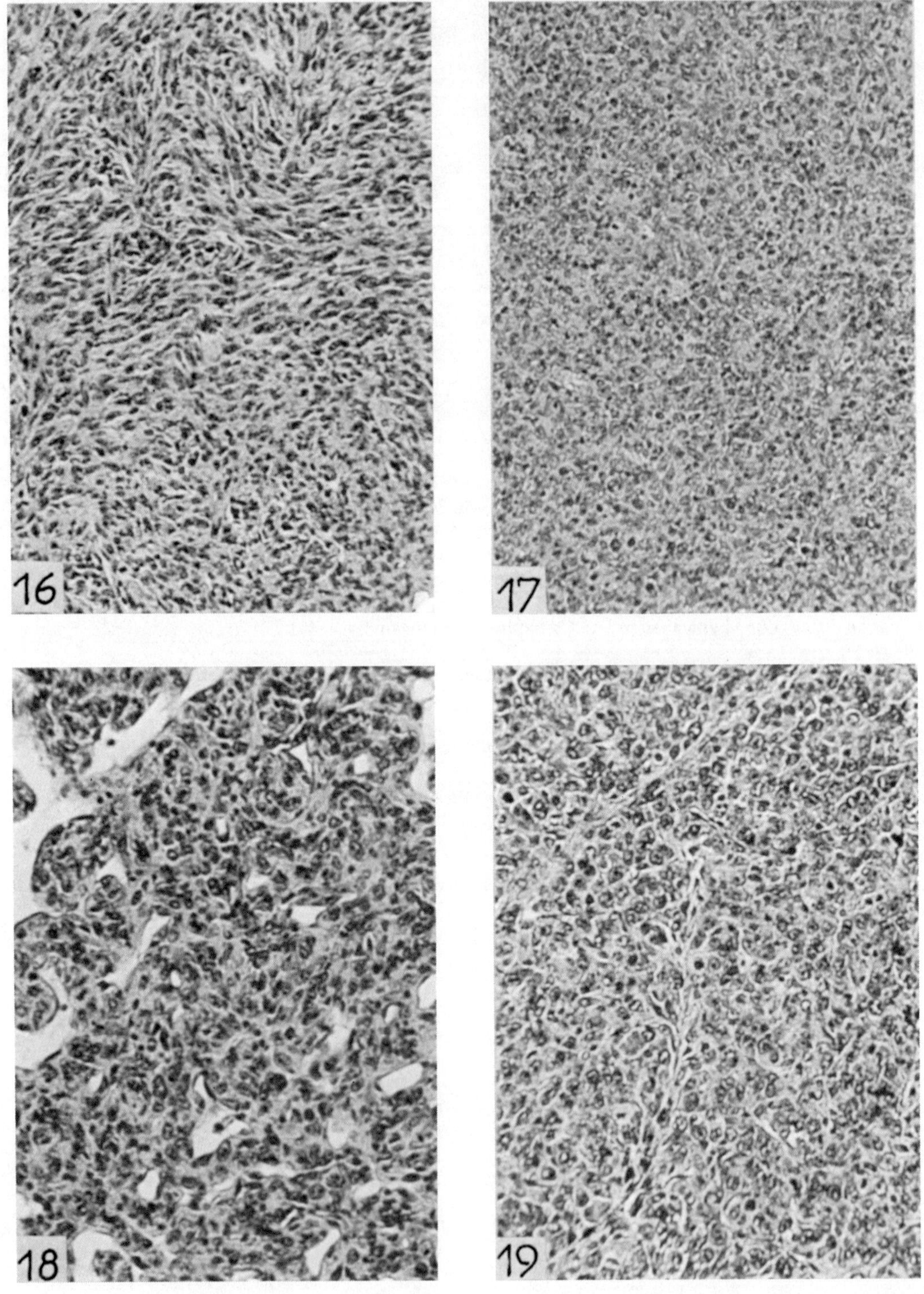

KARYOLOGICAL DEMONSTRATION OF HYBRIDIZATION OF MAMMALIAN CELLS *IN VITRO*

S. Sorieul and B. Ephrussi

In a recent publication, Barski, Sorieul and Cornefert[1] described the formation of cells with a hybrid karyotype in mixed cultures of two mouse cell strains *in vitro*. The experiments to be described show that this phenomenon is easily reproducible and that it is not limited to a pair of strains of strictly defined karyotypes.

The two cell lines used in our experiments are strains *NCTC* 2555 and *NCTC* 2472 of Sanford, Likely and Earle[2], ultimately derived from a single cell of mouse subcutaneous adipose tissue, as received from Dr. G. Barski by kind permission of Drs. Wilton R. Earle and Katherine K. Sanford. Both strains are populations composed predominantly of hypotetraploid cells but containing varying proportions of cells of higher ploidy[3]. The karyotypes of the hypotetraploid cells of the two strains, as established in the course of our experiments, are given in Table 1 and illustrated by Figs. 1.*A* and *C*.

It can be seen that the two cell types can be unmistakably identified by the presence of (*a*) a high number of bi-armed chromosomes, some of which represent good 'markers', in the cells of line *NCTC* 2555 ; (*b*) at least one extra long chromosome and a

Table 1

	NCTC 2555	*NCTC* 2472
Total number of chromosomes, modal	57	55
Total number of chromosomes, variation extremes	51–64	51–59
Number of biarmed chromosomes, modal	15	1
Number of biarmed chromosomes, variation extremes	11–16	1–3
Number of extra long telocentrics, modal	0	1
Number of extra long telocentrics, variation extremes	0	1–3
Minutes	?	0

somewhat shorter one (not shown in Table 1, but visible in Fig. 1,*C*) with a sub-median secondary construction in the cells of line *NCTC* 2472.

Mixed cultures were initiated on January 3, 1961, by inoculating equal numbers of cells of the two lines into two media (a modified 199 (ref. 4) and *NCTC* 109 (ref. 5)) containing 10 per cent of heat-inactivated horse serum. Dispersion of cells for transfer was accomplished either mechanically or by trypsinization. The cultures were incubated at 37° C. in an atmosphere of 5 per cent carbon dioxide in air. Control cultures of the two strains separately were treated

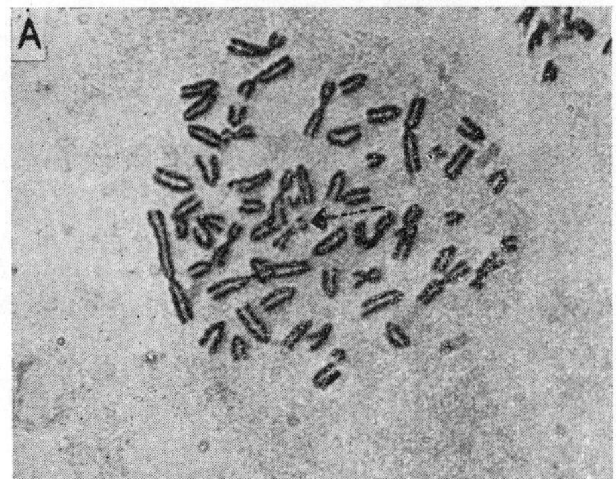

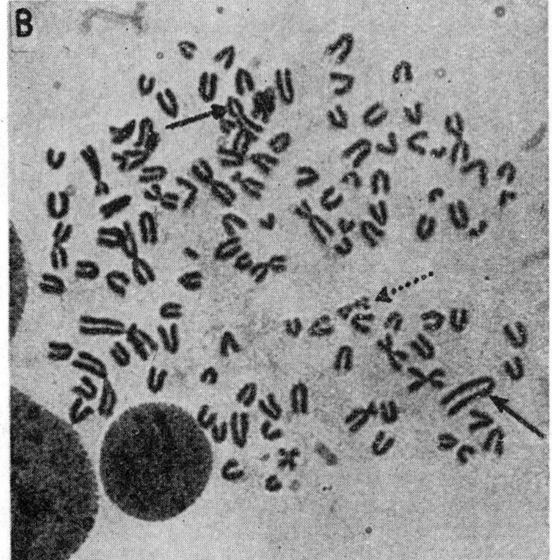

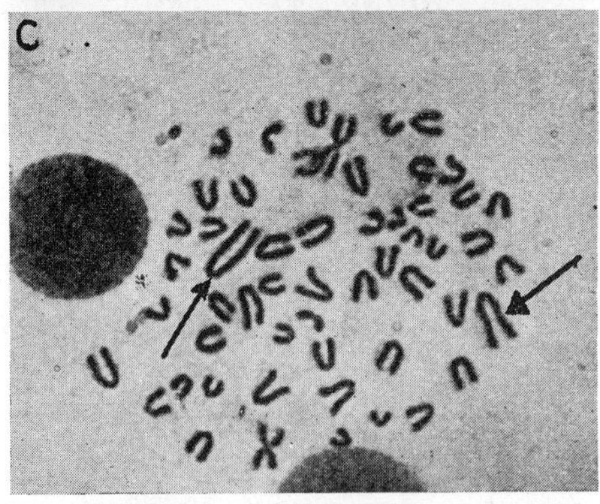

Fig. 1. *A.* Metaphase with 58 chromosomes (of which 15 are bi-armed) in a cell of strain *NCTC* 2555. Note characteristic metacentric (arrow) present in almost all cells of the strain. *B.* Representative example of hybrid cell with 112 or 113 chromosomes, of which 16 or 17 are bi-armed. Note the presence of the three 'markers' (arrows) shown in *A* and *B* *C.* Metaphase with 54 chromosomes (2 bi-armed) in a cell of strain *NCTC* 2472. Note the two 'marker' telocentrics: the extra long one and the somewhat shorter, submedially constricted one (arrows)

logically analysed mixed populations (two in medium 109 and two in medium 199). The hybrid cells are recognized by the total number of chromosomes and the number of bi-armed chromosomes expected from the fusion of hypotetraploid cells of the two strains, as well as by the presence of 'marker' chromosomes (Fig. 1,*B*). The frequencies of hybrid cells in the four mixed populations are at present approximately 1, 1, 5 and 10 per cent, the highest value having been observed in medium 109.

These results confirm the observations of Barski *et al.*, and furthermore show that the formation of hybrid cells is easily reproducible and is not restricted to two rigorously defined karyotypes : the latter is clear from the fact that strain *NCTC* 2472 used by us and strain *N* 1 *T* (derived from *NCTC* 2472) used by Barski *et al.* have somewhat different karyotypes. Thus the long constricted chromosome of line *NCTC* 2472 used by us has never been observed in Barski's line *N* 1 *T* ; the cells of the latter strain very seldom contain bi-armed chromosomes, whereas those of line *NCTC* 2472 used by us all carry at least one such chromosome.

It may be pointed out also that, of the two strains involved in the mixed cultures of Barski *et al.*, at least one was re-isolated from a tumour induced by the inoculation of the original *in vitro* line. On the other hand, the cells used in our experiments never underwent animal passage since their explantation by Sanford *et al.* It is clear, therefore, that the tendency to cell fusion is not a result of animal passage.

These observations suggest that hybridization *in vitro* of other types of somatic cells should be equally possible. If this hope is justified, hybridization may become a useful tool for the investigation of a number of problems of somatic cell genetics, of oncology and of virology[6]. Examples of problems which are common to these fields and which hybridization *in vitro* could help resolving are : the genetic mechanism of variations in resistance to drugs and other chemotherapeutic agents observed frequently in tumour cells[7]; and the possible occurrence of zygotic induction[8] in virogenic cell lines.

The practicability of the method will depend to a large extent on the establishment of selective techniques for the detection of hybrid cells. Its applicability to the study of the mechanisms of cellular differentiation will require further the establishment of techniques for overcoming barriers of tissue specificity.

The technical assistance of Mlle. M. T. Thomas is gratefully acknowledged. This work was supported by a grant from the Rockefeller Foundation.

similarly. Karyological analyses of the mixed populations and of cells of each line grown separately were carried out at intervals on aceto-orcein squashes of colchicine-treated cultures.

At the time of writing (March 28), the occurrence of hybrid cells was observed in all four of the karyo-

[1] Barski, G., Sorieul, S., and Cornefert, F., *C.R. Acad Sci., Paris*, **251**, 1825 (1960).

[2] Sanford, K. K., Likely, G. D., and Earle, W. R., *J. Nat. Cancer Inst.*, **15**, 215 (1954).

[3] Chu, E. H. Y., Sanford, K. K., and Earle, W. R., *J. Nat. Cancer Inst.*, **21**, 729 (1958).

[4] Morgan, J. F., Morton, H. J., and Parker, R. C., *Proc. Soc. Exp. Biol. Med.*, **73**, 1 (1950).

[5] Evans, V. J., Bryant, J. C., McQuilkin, W. T., Fioramonti, M. C., Sanford, K. K., Westfall, B. B., and Earle, W. R., *Cancer Res.*, **16**, 87 (1956).

[6] Lederberg, J., *J. Cell. and Comp. Physiol.*, **52**, Supp. 1, 383 (1958). Schultz, J., *Ann. N.Y. Acad. Sci.*, **71**, (6), 994 (1958).

[7] Hauschka, T. S., *J. Cell. and Comp. Physiol.*, **52**, Supp. 1, 197 (1958).

[8] Jacob, F., and Wollman, E. L., *Ann. Inst. Pasteur*, **91**, 486 (1956).

Editor's Comments
on Papers 3 Through 8

3   LITTLEFIELD
Selection of Hybrids from Matings of Fibroblasts in vitro and Their Presumed Recombinants

4   DAVIDSON and EPHRUSSI
A Selective System for the Isolation of Hybrids Between L Cells and Normal Cells

5   EPHRUSSI and WEISS
Interspecific Hybridization of Somatic Cells

6   OKADA
The Fusion of Ehrlich's Tumor Cells caused by HVJ Virus in vitro

7   HARRIS and WATKINS
Hybrid Cells Derived from Mouse and Man: Artificial Heterokaryons of Mammalian Cells from Different Species

8   PONTECORVO
Production of Mammalian Somatic Cell Hybrids by Means of Polyethylene Glycol Treatment

TECHNIQUES OF SOMATIC CELL HYBRIDIZATION

In the papers discussed in the previous section, hybrid cells were detected only by karyological analysis. Hybrids could be isolated if they had a selective advantage and overgrew the parental cells in mixed cultures or if they occurred at a sufficiently high frequency to be picked out by cloning. In a few cases hybrid cell populations could be enriched by lowering the culture temperature (Scaletta and Ephrussi, 1965). However, the inability to specifically select hybrid cells seriously limited experimentation. Therefore, the development by Littlefield in 1964 (Paper 3) of a biochemical system for the selection of hybrid cells represented a major technical advance.

This selective system involved the use of HAT medium (containing hypoxanthine, aminopterin, and thymidine), the development of which is discussed later (see Papers 25, 34, and 36). (Aminopterin inhibits denovo purine and pyrimidine synthesis.) For the hybridization experiments two biochemically marked mouse cell lines were used, one resistant to the drug 8-azaguanine and lacking the enzyme hypoxanthine-guanine phosphoribosyltransferase (HGPRT), and the other resistant to the drug 5-bromodeoxyuridine and lacking the enzyme thymidine kinase (TK). Because of their enzyme deficiencies, neither drug resistant cell line could survive in HAT medium (see also Papers 25 and 36). Thus, a selective medium was available in which neither parental cell could survive but in which hybrid cells should grow, since each parental cell expressed the enzyme activity lacking in the other. When the two biochemically marked cell lines were mixed and exposed to HAT medium, surviving colonies appeared with a very low frequency, and most of the colonies appeared to be composed of hybrid cells. Since the two parental cells were closely related and exhibited no major karyological differences, it was not possible at the time to prove that the cells growing in HAT medium were hybrids rather than polyploid revertants of one of the parental cells. However, subsequent experiments using biochemically marked mouse cell lines with distinguishable chromosome markers provided the karyological proof that the cells selected in HAT were indeed hybrids (Ephrussi and Weiss, unpublished).

Given the very low frequency at which hybrids occurred in Littlefield's experiments, it is clear that they never would have been isolated without a selective system. The HAT selective system made it possible for the first time to isolate hybrids occurring at low frequencies, and also to determine the frequency at which hybridization occurred. The isolation of hybrids in HAT medium also indicated that the mutations causing the enzyme deficiencies in the parental cells were recessive; that is, the deficiency in one parental genome did not prevent the expression of the corresponding activity of the other parental genome in the hybrid cells.

In the HAT selective system as initially described, both parental cells must carry biochemical markers. The versatility of the HAT system was expanded with the development of the *half-selective* system by Davidson and Ephrussi (Paper 4). In this system only one of the parental cells needs to carry a biochemical marker, while the other parental cell is selected against on the basis of its inherent biological properties. In the first crosses of this type, normal diploid mouse fibroblasts were used as the unmarked parental cells. Although these cells expressed both enzymes required for growth in HAT medium and could contribute such activities to hybrid cells, they were unable to grow well in culture. Thus, even in the absence of a biochemical marker in one parental cell, hybrids were selected easily.

The half-selective system also has permitted the isolation of many other types of hybrids. One important type of experiment in which the half-selective system has been used involves the isolation of hybrids between mouse cells and freshly explanted human peripheral white blood cells, which have no ability at all to grow in culture (see Paper 18). Such hybrids have played a major role in the mapping of the human genome in vitro.

The introduction of biochemical selection techniques greatly facilitated experimentation on cell hybridization and expanded the range of studies that could be carried out. Other selective systems were subsequently introduced, but the HAT system has remained the most widely used. It has been involved in the majority of cell hybridization experiments to date.

Another important development in the mid-1960s was the demonstration by Ephrussi and Weiss (Paper 5) that viable, actively proliferating hybrids could result from fusion between cells of different species. The attempt to isolate viable interspecific hybrid cells was stimulated by the report that cells of different species could be induced to fuse by treatment with a virus, resulting in the formation of multinucleate interspecific heterokaryons (see Paper 7). These multinucleate heterokaryons, however, were apparently unable to proliferate. In the experiments of Ephrussi and Weiss, cells of mouse and rat origin were mixed (without virus treatment to induce fusion) and the half-selective HAT system was used to isolate presumptive hybrids. Rat and mouse cells have markedly different karyotypes, and karyological analysis proved that the cells isolated were hybrids, containing nearly complete rat and mouse genomes. By their very existence these hybrids demonstrated that there is no incompatibility at the cellular level between the genomes of different rodent species. The interspecific hybrid cells clearly were capable of coordinating their activities to produce a regulated cell cycle, and the hybrids were capable of active and prolonged proliferation.

The isolation of viable interspecific hybrid cells represented much more than the successful production of a biological curiosity. By their nature, interspecific hybrids provided valuable tools for studies on gene regulation and also gene mapping, as will be discussed later in Part I. Because of the large karyotypic differences between cells of different species, the chromosomal contributions of the parental cells to the genome of interspecific hybrids can be accurately determined. In addition, evolution has introduced biophysical and biochemical differences into homologous enzymes of different species. This makes it possible to distinguish the activities of the parental genomes in interspecific hybrid cells.

The first studies of this type involved the enzyme lactate dehydrogenase (LDH) in rat x mouse hybrids (Weiss and Ephrussi, 1966). The rat and mouse cells each produced one form of LDH, with a clear electrophoretic difference between the rat and mouse enzymes. In the hybrid cells both the rat and mouse forms of LDH were observed, indicating that the genomes of both parental cells can be expressed in interspecific hybrids. Moreover, the hybrid cells revealed the presence of three new electrophoretic bands of LDH, migrating between the rat and mouse forms of LDH. LDH is a tetramer, and the three new bands in the hybrids represented interspecific hybrid enzyme molecules, resulting from the random association of subunits of both rat and mouse into enzymatically active tetramers.

The formation of enzymatically active interspecific hybrid molecules has since been observed for a large number of enzymes and over a broad range of species. The occurrence of species differences in homologous enzymes has been of significance in many studies by making it possible to monitor the activity of the parental genomes in interspecific hybrids. At present, with the development of species-specific molecular probes for the genes themselves (in contrast to assays for their protein products), studies with interspecific hybrids are becoming increasingly more productive.

From the time of its first observation, one of the main technical problems in studies involving cell hybridization has been the low frequency of cell fusion. In early studies a variety of approaches for increasing the recovery of hybrids were tested, but with limited success. The first widely applicable method for increasing cell fusion involved the use of a virus as a fusing agent. In 1958 Okada (Paper 6) showed that treatment of cultured Erlich's ascites cells with a myxovirus known as the hemagglutinating virus of Japan (and now commonly referred to as Sendai virus) caused extensive cell fusion and the formation of large multinucleate cells. The first step in the process of virus-induced fusion was a rapid agglutination of the cells into aggregates, followed by fusion of cytoplasmic membranes. In subsequent experiments mixtures of two different mouse tumor cell lines were treated with Sendai virus. The results, although not definitive, suggested that Sendai virus could cause fusion between cells of different types, resulting in the formation of multinucleate heterokaryons that contained nuclei from both parental cells (Okada, 1961).

In 1965 Harris and Watkins (Paper 7) used Sendai virus to induce the fusion of human and mouse cells into interspecific heterokaryons. The virus used in these studies was inactivated with ultraviolet light in order to avoid viral replication in the treated cells. By using prelabelled parental cells it was demonstrated that the multinucleate cells contained nuclei of both human and mouse origin within a common cytoplasmic membrane. These results provided the first demonstration of fusion between cells of different species, and they led to an extensive investigation of biochemical activities in heterokaryons. In the decade following these studies, inactivated Sendai virus came to be the standard agent for inducing cell fusion.

The virus-induced, interspecific heterokaryons in these studies were incapable of proliferation. However, these results led Ephrussi and Weiss (see Paper 5) to isolate viable, proliferating, interspecific hybrids without the use of virus to induce cell fusion. It was shown subsequently that inactivated Sendai virus could be used to induce the formation of viable hybrids capable of prolonged proliferation (Yerganian and Nell, 1966). Inactivated Sendai virus was used in many of the cell hybridization experiments over the next several years.

Although Sendai virus proved to be an effective agent for inducing cell fusion, it suffered from a variety of problems. Its production was cumbersome; its fusion-inducing activity was not reproducible from batch to batch; and there was always the possibility of incorporating bits of viral genetic information into treated cells, with unknown effects on the properties of the fused cells. Thus, there have been repeated attempts to find chemical agents that could fuse mammalian cells as effectively as Sendai virus but without its inherent problems. One possible agent was polyethylene glycol (PEG), which had been shown to fuse plant protoplasts (Kao and Michayluk, 1974).

The demonstration by Pontecorvo (Paper 8) that PEG can be used to induce the hybridization of mammalian cells represented an important technical advance. The use of PEG to fuse cells avoided all the problems associated with viruses, and the frequency of hybrid cells induced by PEG was comparable to the results obtained with Sendai virus. A variety of procedures have been developed for inducing cell fusion with PEG. One especially easy and efficient process has been described in which the cells are treated with PEG while attached to a substrate rather than in suspension as is usually the case (Davidson et al., 1976). Using this protocol, almost 5% of the cells treated (as the minority parent in a mixture) can be fused to form viable hybrids. This is the highest frequency of hybridization thus far observed.

In recent years PEG has become the standard agent for inducing cell fusion. Because of its immediate availability and the ease with which it can be used, PEG has made somatic cell hybridization a technique accessible to all laboratories. This accessibility has broadened the range of research problems to which the techniques of cell hybridization can be applied.

REFERENCES

Davidson, R. L., K. A. O'Malley, and T. B. Wheeler, 1976, Polyethylene Glycol-Induced Mammalian Cell Hybridization: Effect of Polyethylene Glycol Molecular Weight and Concentration, *Somatic Cell Genet.* **2:**271–280.

Kao, K. N., and M. R. Michayluk, 1974, A Method for High-Frequency Intergenetic Fusion of Plant Protoplasts, *Planta* **115:**355–367.

Okada, Y., 1961, Multinucleated Giant Cell Formation by Fusion of Two Different Strains of Cells, *Biken's J.* **4:**145–146.

Scaletta, L. J., and B. Ephrussi, 1965, Hybridization of Normal and Neoplastic Cells In Vitro, *Nature* **205:**1169–1170.

Weiss, M. C., and B. Ephrussi, 1966, Studies of Interspecific (Rat X Mouse) Somatic Hybrids. II. Lactate Dehydrogenase and Beta-Glucuronidase, *Genetics* **54:**1111–1122.

Yerganian, G., and M. Nell, 1966, Hybridization of Dwarf Hamster Cells by UV-Inactivated Sendai Virus, *Natl. Acad. Sci. (USA) Proc.* **55:**1066–1073.

SELECTION OF HYBRIDS FROM MATINGS OF FIBROBLASTS IN VITRO AND THEIR PRESUMED RECOMBINANTS

J. W. Littlefield

Barski, Sorieul, and Cornefert (*1*) first demonstrated that fusion can occur between cultured mouse cells; Ephrussi and Sorieul (*2*) have since studied this phenomenon extensively by karyotype analyses, having in mind its possible utility for the genetic analysis of mammalian cells. Gershon and Sachs (*3*) recently confirmed the occurrence of fusion, using mouse cells with different histocompatibility antigens. In the study reported here, the detection of far fewer hybrid cells than were found previously has been possible through the use of two clonal lines of mouse fibroblasts (L cells), each containing a drug-resistant marker.

Cell culture and analytical methods were the same as those used previously (*4*); the assay for thymidine kinase activity in cell extracts was a modification of that for guanylic acid–inosinic acid pyrophosphorylase (*4*). Cells of the A3-1 line lack pyrophosphorylase and are resistant to 3 μg of 8-azaguanine per milliliter (*4*), and those of the B34 line lack thymidine kinase and are resistant to 30 μg of 5-bromodeoxyuridine (BUDR) per milliliter (*5, 6*). These markers do not appear to be linked (*7*). Cultures of the B34 line contain a small number of revertant cells (*6*) and therefore are maintained in the presence of BUDR. For the mating experiments the cells were washed to remove BUDR and resuspended in medium containing 10 μg of thymidine per milliliter. Then an equal number of these cells and A3-1 cells were grown together for 4 days in a suspension culture. Four or more

replicate petri dishes containing 5×10^6 cells from this culture, and dishes containing the same number of B34 cells (similarly washed and grown with thymidine) or A3-1 cells, were incubated with medium containing $4 \times 10^{-7}M$ aminopterin, $3 \times 10^{-6}M$ glycine, $1.6 \times 10^{-5}M$ thymidine, and $1 \times 10^{-4}M$ hypoxanthine, in which only cells with both enzymes can survive to form colonies (*4, 7, 8*). The medium was replenished after 1, 3, and 6 days, and the petri dishes were examined for colonies after about 14 days.

In three experiments no colonies were present in the dishes containing A3-1 cells, while in the dishes containing B34 cells the frequency of cells (presumably revertant) which survived to form colonies averaged 8×10^{-7} (3×10^{-7}, 7×10^{-7}, and 14×10^{-7}). In the mixed cultures the frequency of viable cells averaged 5.6×10^{-6} (2.6×10^{-6}, 7.0×10^{-6}, and 7.2×10^{-6}), or 14 times the number of revertant B34 cells which would have been present in the mixed cultures.

Clonal lines were established from six such colonies from the mixed cultures. One (M3) was similar morphologically to B34, and in all other respects appeared to be a revertant of B34. Another (M11) contained two types of cells, and has not been studied extensively. The other four (M1, M4, M12, and M13) were composed of large cells similar to A3-1 morphologically, and mostly mononuclear; they contained approximately double the usual amount of DNA, RNA, and protein per cell. Chromosome counts on these lines are given in Fig. 1, and the activities of thymidine kinase and guanylic acid–inosinic acid pyrophosphorylase per cell are shown in Table 1. These data indicate that M1, M4, M12, and M13 contained about twice the usual number of chromosomes and both enzymes, presumably through the fusion of an A3-1 cell and a B34 cell.

If cell fusion is to be useful for the genetic analysis of mammalian cells, as is a comparable process in fungi (*9*), the hybrid cells must segregate to form cells of lower chromosome number. Although perhaps due to artifacts, the occasional low chromosome counts in the fused cell lines (Fig. 1) suggested segregation, as suspected also by Ephrussi and Sorieul (*2*). Furthermore, by selection with 8-azaguanine or with 8-azaguanine and BUDR it has been possible to detect in the M4 culture two cell types which might have arisen by segregation. (For technical reasons BUDR alone could not be used as a selective agent in petri dishes). In two experiments the frequencies of 8-azaguanine–resistant cells in the M4 culture were 5.9 and 7.8 $\times$ 10^{-4}, and the frequencies of doubly resistant cells, which grew quite slowly, were 2.6 and 3.8 $\times$ 10^{-6}. Some of the 8-azaguanine–resistant cells might have arisen by mutation of M4 cells (*4*), and studies of the chromosome numbers of such cells will help to answer this question. The frequency of the doubly resistant cells is considerably more than might have been expected (about 10^{-10}) from coincidental mutations in the M4 cells (*10*). Thus it seems probable that the doubly resistant cells are due to segregation with recombination. If this process can be validated, perhaps methods will be found to increase its frequency, as well as the frequency of fusion, and it may become possible to extend these techniques to diploid cultures.

Table 1. Enzymatic activities in cell extracts. Results expressed as arbitrary units per cell.

Line	Thymidine kinase	Guanylic acid–inosinic acid pyrophosphorylase
A3-1	100	1
B34	3	100
M1	98	172
M4	62	96
M12	81	109
M13	117	167

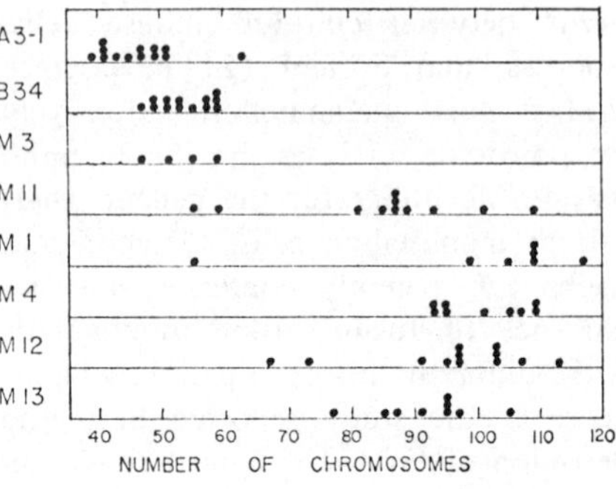

Fig. 1. Chromosome counts on the parental (A3-1 and B34) and the "mated" cell lines. No count greater than 64 was found in the parental lines.

References and Notes

1. G. Barski, S. Sorieul, F. Cornefert, *Compt. Rend.* **251**, 1825 (1960).
2. B. Ephrussi and S. Sorieul, in *Approaches to the Genetic Analysis of Mammalian Cells*, D. J. Merchant and J. V. Neel, Eds. (Univ. of Michigan Press, Ann Arbor, 1962), pp. 81–97.
3. D. Gershon and L. Sachs, *Nature* **198**, 912 (1963).
4. J. W. Littlefield, *Proc. Natl. Acad. Sci. U.S.* **50**, 568 (1963).
5. S. Kit, D. R. Dubbs, L. J. Piekarski, T. C. Hsu, *Exptl. Cell Res.* **31**, 297 (1963).
6. J. W. Littlefield and P. F. Sarkar, *Federation Proc.* **23**, 169 (1964).
7. J. W. Littlefield, unpublished material
8. W. Szybalski, E. H. Szybalska, G. Ragni, *Natl. Cancer Inst. Monograph* **7**, 75 (1962).
9. G. Pontecorvo and E. Käfer, *Advan. Genet.* **9**, 71 (1958).
10. The frequency of 10^{-10} was estimated from previous work (*4*) and unpublished studies.
11. Supported by USPHS grant CA-04670-05. This is publication No. 1173 of the Cancer Commission of Harvard University. We acknowledge the assistance of Susan J. Whitmore.

4

A SELECTIVE SYSTEM FOR THE ISOLATION OF HYBRIDS BETWEEN L CELLS AND NORMAL CELLS

By RICHARD L. DAVIDSON* and BORIS EPHRUSSI†

Department of Biology, Western Reserve University, Cleveland, Ohio

THE occurrence of mating between cultured cells of a permanent line and freshly explanted, normal cells has been recently demonstrated (previous communication). We have developed a system for the selection of such hybrids and for the determination of the rate of their formation. The purpose of this communication is to describe this system.

The permanent line used is $A9$, a sub-line of the L line, kindly supplied to us by Dr. J. W. Littlefield. $A9$ is resistant to 3 µg/ml. of 8-azaguanine. (A similar line was used by Littlefield[1] to establish a selective system involving two biochemically marked L cell lines.) Cells of line $A9$ are heteroploid and contain 51–58 chromosomes (mode = 55), 18–20 of which are bi-armed. The normal cells are

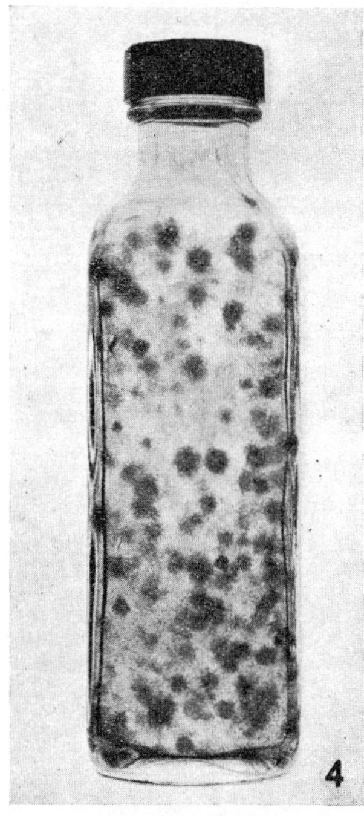

Fig. 4. Colonies of hybrid cells after two weeks in selective medium

fibroblast-like cells from secondary cultures of skin of new-born CBA mice carrying the T-6 translocation[2]. Hybrids between these two cell types are karyologically recognizable by the total number of chromosomes and the simultaneous presence of the parental markers (Figs. 1–3).

$A9$ and T-6 cells are mixed and either inoculated immediately into culture bottles containing a selective medium, or grown together for one or two days in Eagle's medium with 10 per cent calf serum, which is thereafter replaced by the selective medium. The latter is Eagle's medium with 10 per cent calf serum, supplemented with 1×10^{-4} M hypoxanthine, 4×10^{-7} M aminopterin, and $1 \cdot 6 \times 10^{-5}$ M thymidine, as in Littlefield's system[1].

This selective medium prevents the growth of $A9$ cells[1], and no revertants able to grow in this medium have been detected in our experiments. Aminopterin blocks the *de novo* synthesis of inosinic acid and thymidine. Therefore only those cells which can utilize exogenous hypoxanthine and thymidine grow in this medium. Since $A9$ lacks inosinic acid pyrophosphorylase, it cannot utilize exogenous hypoxanthine. The same medium does not interfere with the growth of T-6 cells, but since they are normal cells, they grow relatively slowly. On the basis of observations recorded in the preceding article,

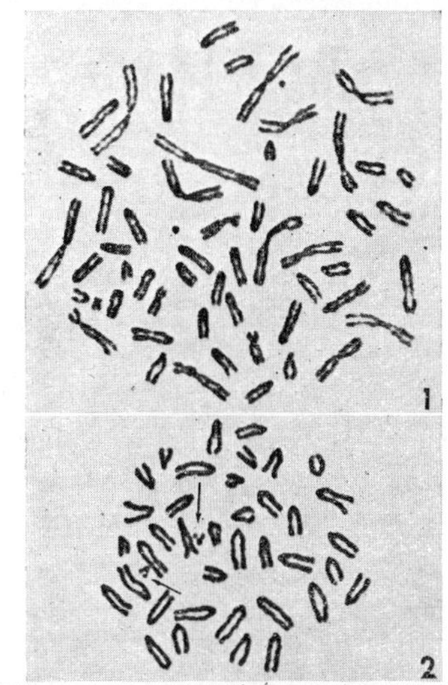

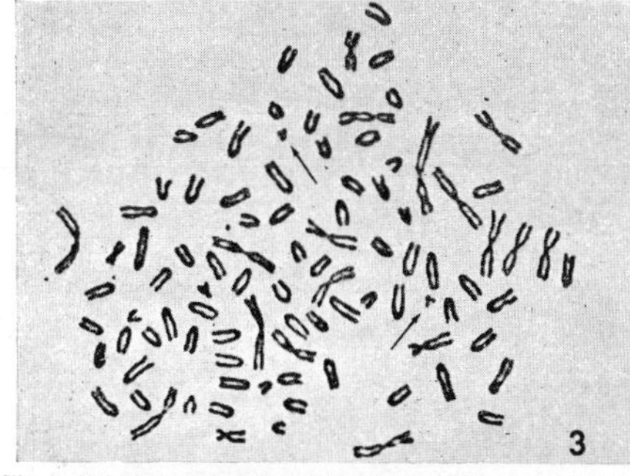

Fig. 1. Metaphase of an $A9$ cell with 55 large and 2 very small chromosomes. Note the presence of 18 bi-armed chromosomes

Fig. 2. Metaphase of a diploid cell (40 chromosomes) homozygous for the T-6 translocation (arrows)

Fig. 3. Metaphase of a hybrid cell with 94 large and 1 very small chromosomes. Note the presence of 19 bi-armed chromosomes from $A9$ and of the two T-6 markers (arrows)

* Pre-doctoral fellow of the U.S. Public Health Service.
† On leave from the University of Paris.

we expected the hybrid cells, if formed, to grow rapidly and to produce multilayered colonies on a background of a T-6 monolayer. Such colonies are indeed observed after about two weeks in selective medium (Fig. 4). The evidence that all these colonies are produced by hybrid cells is as follows. A bottle containing a large number of colonies was examined cytologically, and almost all of the mitoses were found to be of the hybrid karyotype. The remainder were mitoses of T-6 cells. In another experiment, eight colonies were isolated and grown separately. Karyological examination showed that all of them contained hybrid cells.

When the selective medium is applied 1 day after initiation of the mixed culture, the number of colonies formed indicates that as many as one in 7,000 T-6 cells had mated with A9 cells. It should be pointed out that, when the mixture is inoculated directly into the selective medium, more than half as many matings take place. Obviously, the selective medium does not immediately interfere with mating. This fact must be taken into account in calculations of mating rates.

Cultures of hybrid cells recovered from these experiments have been propagated by weekly transfer since October 1964, without decline in growth rate. Hence, the hybrids seem to have inherited from A9 this property of cells of a permanent line.

This work was supported by a grant (RG–9916) of the National Science Foundation.

[1] Littlefield, J. W., *Science*, **145**, 709 (1964).
[2] Carter, T. C., Lyon, M. F., and Phillips, R. J. R., *J. Genet.*, **53**, 154 (1955).

5

Reprinted from *Natl. Acad. Sci. (USA) Proc.* **53**:1040–1042 (1965)

*INTERSPECIFIC HYBRIDIZATION OF SOMATIC CELLS**

By Boris Ephrussi† and Mary C. Weiss‡

WESTERN RESERVE UNIVERSITY, DEVELOPMENTAL BIOLOGY CENTER, CLEVELAND, OHIO

Communicated March 24, 1965

Hybridization of somatic cells *in vitro*, first described in 1960 by Barski, Sorieul, and Cornefert,[1] has since been shown to occur in mixed cultures of many different pairs of cultured mouse cells. A recent review[2] lists 12 different "crosses" of mouse cells resulting in the formation of viable hybrids. This list includes crosses between cells of permanent lines as well as crosses between the latter and freshly explanted normal diploid cells. In this paper we describe the first interspecific cross giving rise to rapidly multiplying mononucleate somatic hybrids.[3]

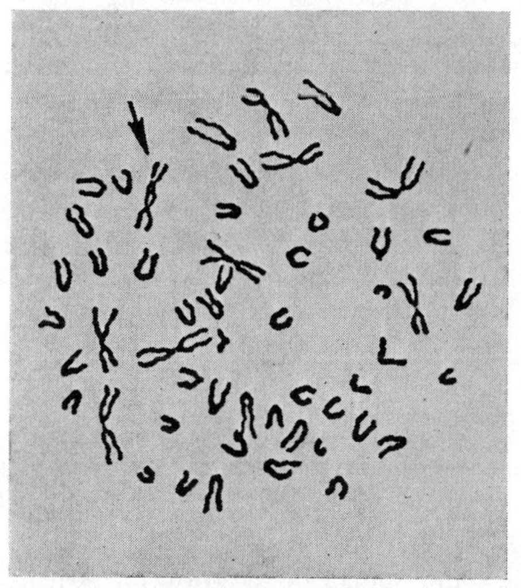

Fig. 1.—Metaphase of a cell of mouse cell line LM (TK⁻) 1D with 53 chromosomes. Note the presence of 9 long biarmed chromosomes with median or submedian centromeres; the arrow points to the "D chromosome" with a characteristic secondary constriction.

The two cell types involved are: (1) clone LM(TK⁻)1D, a 5-bromo-deoxyuridine-resistant thymidine kinase-deficient subclone of *mouse* L cells, kindly given to us by Dr. S. Kit; (2) recently explanted Wistar *rat* embryo diploid cells.

The detection and identification of the hybrids was facilitated by the use of a recently established selective technique for the isolation of hybrids between biochemically deficient L cells and normal mouse cells[5] and by the characteristically different karyotypes of the "parental" cells. The karyotype of the mouse cells (Fig. 1) is characterized by a modal number of 53 chromosomes, 7 to 11 of which are long biarmed chromosomes, which have median or submedian centromeres, and are clearly different from those of rat cells. Ninety per cent of the mouse cells carry one biarmed chromosome presenting a characteristic secondary constriction. The karyotype of the rat $(2n = 42)$ is characterized by the presence of 20 small and 4 long biarmed chromosomes, the latter with subterminal centromeres (Fig. 2).

Trypsinized suspensions of the two cell types are prepared and mixed to give a 1:1 mouse/rat cell ratio. Two million cells of this mixed population are inoculated

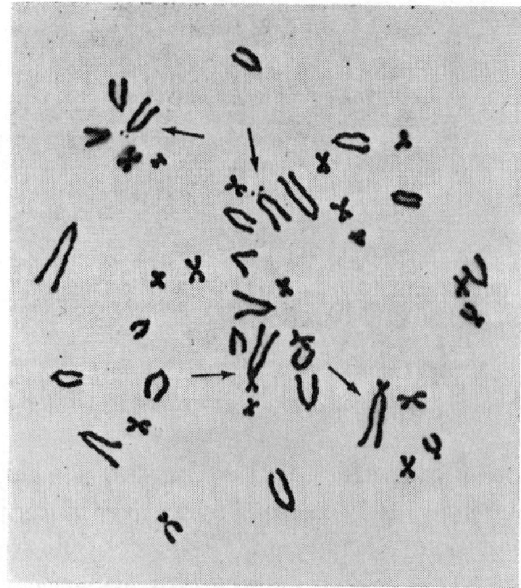

Fig. 2.—Metaphase of a diploid rat cell (42 chromosomes). Note the small biarmed chromosomes and the characteristic long ones, indicated by arrows.

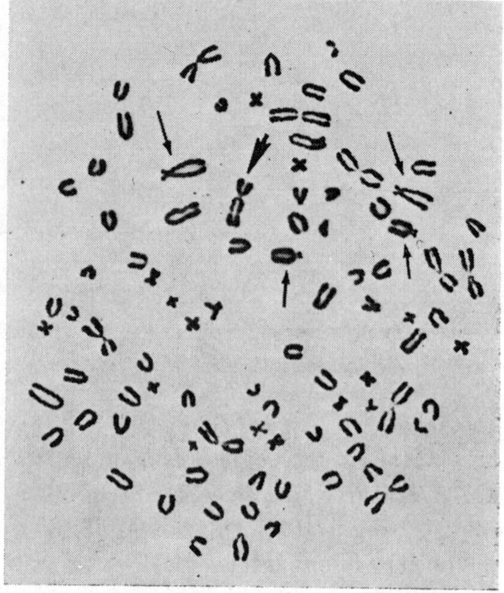

Fig. 3.—Metaphase of a hybrid cell with 89 chromosomes. Note the presence of 10 long biarmed mouse chromosomes (among them the "D chromosome," indicated by a large arrow) and of 4 long rat biarmed chromosomes (small arrows). Many small biarmed rat chromosomes can also be recognized.

into a 20-ml plastic bottle containing 5 ml of modified Eagle's minimal medium, supplemented with 5 per cent calf serum, and incubated for 24–48 hr at 37°C. During this period the cells attach to the bottom of the flask. The medium is then replaced by the selective medium, which differs from the above by the addition of $1 \times 10^{-4}\,M$ hypoxanthine, $4 \times 10^{-7}\,M$ aminopterin, and $1.6 \times 10^{-5}\,M$ thymidine, and the culture is further incubated at 37°C.[6]

In the selective medium, the LM cells degenerate (because they lack thymidine kinase and therefore cannot utilize exogenous thymidine which must be used since the *de novo* synthesis of thymidine is blocked by aminopterin) while the rat cells grow and form a monolayer. On this background, the hybrid cells, unhampered by the selective medium, form, in the course of the next few days, discrete, multilayered colonies which can easily be isolated and subcultured. Karyological examination of the subcultured hybrid cells clearly shows the presence of chromosomes characteristic of each of the parents (Fig. 3).

At the present date, the hybrid cells have been maintained under conditions of continuous multiplication for 1 month, going through at least 25 cell generations. The evolution of their karyotype and some of their enzymatic characteristics are being investigated.

We wish to thank Dr. L. J. Scaletta and Mr. M. C. Yoshida for their help in karyological analysis and for preparing the microphotographs, and Miss Patricia Menéndez for her able technical assistance.

* This work was supported by grant no. G-23129 of the National Science Foundation.
† On leave from the University of Paris.

‡ Graduate fellow of the National Science Foundation.

[1] Barski, G. S., S. Sorieul, and F. Cornefert, *Compt. Rend.*, **251**, 1825 (1960).

[2] Ephrussi, B., *19th Symposium on Fundamental Cancer Research*, Houston, Texas, in press.

[3] It has recently been shown by Harris and Watkins[4] that human and mouse cells can be caused to fuse into heterokaryons by the action of UV-inactivated Sendai virus. Whether these heterokaryons are capable of (limited or indefinite) multiplication appears to be unknown at the present time.

[4] Harris, H., and J. F. Watkins, *Nature*, **205**, 640 (1965).

[5] Davidson, R. L., and B. Ephrussi, *Nature*, **205**, 1170 (1965).

[6] Littlefield, J. W., *Science*, **145**, 709 (1964).

6

Reprinted from *Biken's J.* **1**:103–110 (1958)

The Fusion of Ehrlich's Tumor Cells caused by HVJ Virus[*1]
in vitro[*2]

YOSHIO OKADA

Department of Preventive Medicine, Research Institute for Microbial Diseases, Osaka University

(*Received for publication, November 25, 1958*)

SUMMARY

In the Ehrlich's tumor cell-HVJ system, fused cells are easily formed *in vitro* following conditions are fulfilled; 1) Concentrated viable tumor cells are used. 2) Incubation is made with mild shaking at 37°C. 3) A sufficient hemolytic activity of virus is used capable of causing a striking change or changes in the cell membrane. 4) An excess of virus over the maximal adsorption capacity of the cells is used.

The fusion phenomenon of the tumor cells seems to be a compensation for the injury of the cell membrane caused by the highly hemolytic virus.

INTRODUCTION

Recently the fusion of cells caused by viral action has been described in several reports. Fusion phenomena are classified into two categories according to their development. i) The cell fusion results from the direct action of virus on the cell membrane without any multiplication of virus. ii) The cell fusion appears in the later period of virus multiplication, i.e. a phenomenon caused by virus multiplication: e.g. giant cells in a measles virus-HeLa cell system. The fusion phenomenon treated in this paper belongs to the former case.

The author has described the appearance of many large fused cells within one hour after a large dose inoculation of HVJ into the abdominal cavity of mice implanted with Ehrlich's tumor cells. This effect of causing the fusion of cells seems to be proportional to the hemolytic activity of the virus as described in the preceding report (Okada *et al.*, 1957). These findings lead the author to a working hypothesis that the cell fusion phenomenon by HVJ is closely related to specific systematic, chemical and physical changes in the cell membrane similar to the hemolytic reaction of the red blood cell-HVJ system. However the fusion phenomenon which developes at a later period of the virus growth cycle is considered as a modified type of cytolysis. Though in such a case an autocatalytic processes may be involved in the final steps of the reaction, it is probable that there is reduction in some components of cell membrane. From the above mentioned view point the fusion phenomena are not essentially difference in nature.

This paper describes the results of experiments on the giant cell formation and the cytolytic action of HVJ virus *in vitro*.

---

*1   Synonym : Sendai virus, *Myxovirus parainfluenzae I.*

2   This paper was given at the annual meeting of the Society of Japanese Virologists in 1957.

Materials and Methods

Ehrlich's tumor cells: The ascites were withdrawn from mice 7 days after inoculation with 10^6 Ehrlich's tumor cells. Ascites and cells were separated by centrifugation. After washing three times with Hanks' solution to remove contaminating red blood cells the separated cells at suitable concentrations were suspended in the media shown in Table 1.

Table 1.　Suspending media

Mouse ascites	clarified by centrifugation at 2,500 r.p.m. for 10 minutes,	2 volumes
Hanks' solution*	without glucose	6 volumes
Heparin solution	0.1% in saline	2 volumes

Virus: The chorio-allantoic fluid infected with HVJ, Z strain, was diluted to 10^{-6} with broth and 0.2 ml. aliquots of this diluted inoculum were injected allantoically into 10 day-old embryonated eggs. After 72 hour incubation at 35.5°C the culture was harvested and centrifuged at 3.000 r.p.m. for 30 minutes and the virus in the supernatant fluid was spun down at 20,000 r.p.m. for 30 minutes in a Spinco L ultracentrifuge. The resulting pellet was resuspended in Hanks' solution. The concentrated virus suspension was 128,000 HAU./ml.. After freezing and thawing 8 times the hemolytic activity of this suspension was enhanced about 10 times without any change in hemagglutination titer. The sample was diluted with Hanks' solution to the desired concentration.

Fusion experiments in vitro and the index of cell fusion: One half ml. of virus suspension was added to 0.5 ml. of tumor cell suspension in a 20 ml. volume vaccine vial and incubated in a water bath for 1 hour at 37°C with gentle shaking. The fusion reaction was completed within 1 hour and did not proceed further under these conditions.

When an adequate concentration of tumor cells was used, a low concentration of virus produced small fused cells while a concentrated virus suspension resulted in large fused cells.

The ratio between fused cells and free cells was low in the former case and was higher in the latter. The fusion rate was represented as follows. One drop of each sample was poured into a hemocytometer. Then photomicrographs of 5 microscopic fields were taken at random with a phase contrast microscope at low magnification. The negatives were projected at a total magnification of 1000 and the largest 10 cells were traced on graph paper and their mean crose sections relative to the normal cells was expressed as the cell fusion index. This representation was used for convenience although it shows only one aspect of the fusion phenomenon quantitatively.

Determination of hemolysis: Chicken red blood cells were washed 4 times with phosphate buffered saline at pH 7.2 and suspended in the same medium. An equal amount of the red blood cell suspension and the virus sample were mixed together and incubated in a water bath. After one hour's incubation at 37°C

---

*Hanks' solution without glucose was used throughout.

the samples were centrifuged at 1,500 r.p.m. for 5 minutes and the optical density of the supernatant was determined at 540 mμ. in a Coleman Junior spectrophotometer. The turbidity of the virus suspensions were subtracted from the optical density of the supernatant and the resulting values represented the amount of hemoglobin liberated.

Manometric determination: A 1.8 ml. aliquot of the tumor cell suspension (1.5×10^7/ml.), 120,000 HAU./ml. 0.2 ml. of virus, and 0.2 ml. of 10% KOH were placed in the main chamber, side arm, and center well, respectively of a working manometer. After 15 minutes shaking at 37°C, the virus in the side arm was tipped into the main chamber and manometer readings were taken for the next 60 minutes at 10 minutes intervals.

EXPERIMENTALS

Production of fused cells by incubation with shaking: As reported in the previous paper, while fused cells appear consistently in *in vivo* experiments, difficulties are encountered in forming them in *in vitro* experiments.

When the cell-virus mixture is incubated without shakings cytolysis chiefly occurs and no fused cells are formed. When concentrated HVJ which was freeze-thawed to increase the hemolytic activity was added to the tumor cells in a small test tube and incubated without shaking at 37°C, deformation and rupture of the cytoplasm was observed in almost all cells within 15 minutes and as a result cytolysis corresponding to hemolysis in a red blood cell-HVJ system was observed. (Fig. 1A) However when the cultures were shaken many giant fused cells were formed. (Fig. 1B)

As presented in Fig. 2 the electronmicrographs of these cells showed a rough cell membrane and destruction and vacuolization of the cytoplasm in lysed cells but no abnormality in the cytoplasm of the fused cells was observed except for an increase in numbers of microvirii.

Since the fusion phenomenon occurred even under anaerobic conditions with shaking and DNP had no effect on giant cell formation at a concentration lower than the concentration causing cell degeneration, the effect of shaking is not that of simple aeration of the reaction mixture. From the author's impression it seems that the tumor cell membranes which were strongly effected by the viral action must receive some hearling effect from mechanical shaking to complete fusion of the cell membranes. If they are not shaken the injured cells are not able to keep their live state as they lack functionate-limiting membranes.

Endogeneons O_2 uptake by fused cells: Within 30 minutes of the addition of the virus more than 80% of the tumor cells were fused into giant cells. As shown in Fig. 3, O_2 uptake by the cells through this fusion reaction and that of completed fused cells was nearly equal to that of normal cells. In the cells lysed by freezing and thawing once, O_2 uptake was much reduced.

Relationship between the cell fusion index, the concentration of cells and the virus concentration: Fig. 4 illustrates the changes of fusion index at various concentration of cells and virus. At a high cell concentration (1×10^7/ml.) the mean size of the fused cells is proportional to the virus concentration and at a low cell concentra-

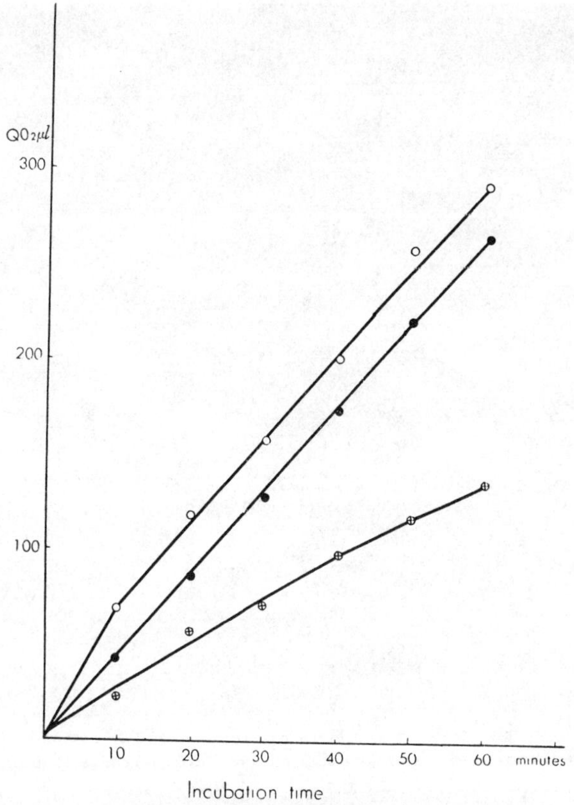

Fig. 3   O₂ uptake of Ehrlich's tumor cells fused by HVJ. The cell number at the beginning of the experiment: 3.0×10^7 ETC

○———○   Fused giant cells (in shaking culture)

●———●   Normal cells (control)

⊕·····⊕   Cells lysed by freezing-thawing once (without virus)

tion (2.5×10^6/ml. and 1.25×10^6/ml.) the size is independent of the virus concentration though it is a little larger than that of the control. The mononuclear cells remaining in the reaction mixture also increase in size when virus adsorption occurs and some show cytolysis. The higher the virus concentration, the more cytolysed cells appear.

At an intermediate concentration of cells (5×10^6/ml.) the maximum point for the mean size was found at 8192 HAU. virus titer and when a large amount of virus was used at this cell concentration the mean size decreased. Tentatively it is concluded that when a sufficient amount of virus was added to tumor cells at low concentration cytolysis was dominant and fusion occurred only at a high cell concentration.

Relation between the cell fusion index and the hemolytic activity : Figure 5 shows the relationships between the hemolytic activity of the virus and the cell fusion index when a constant amount of the cells (1×10^7/ml.) was used. The whole cell

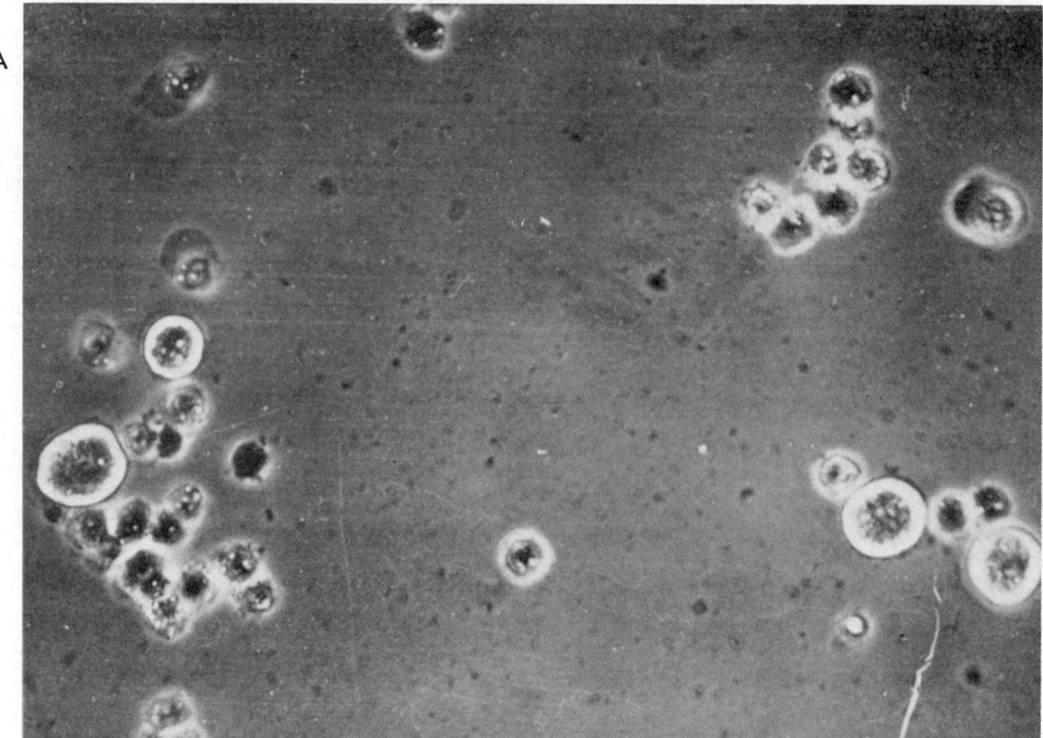

Fig. 1A. Ehrlich's tumor cells mixed with HVJ *in vitro* showing cytolysis. (Stationary culture)

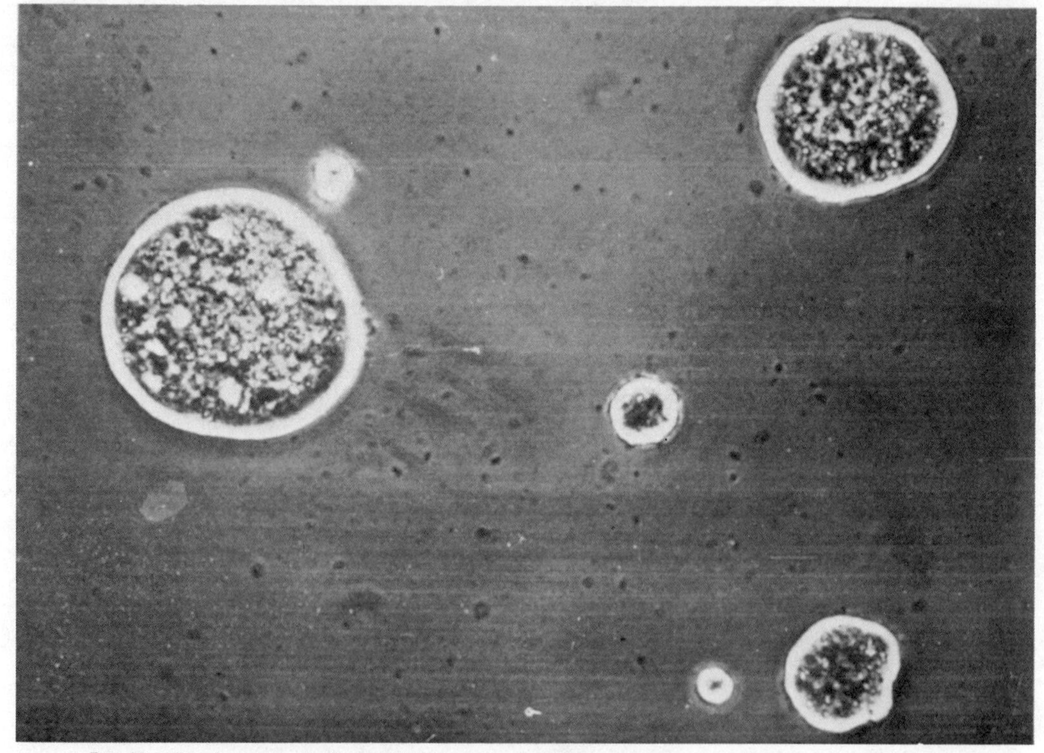

Fig. 1B. Fused giant cells (fused Ehrlich's tumor cells) produced by HVJ *in vitro*. (Shaking culture)

A

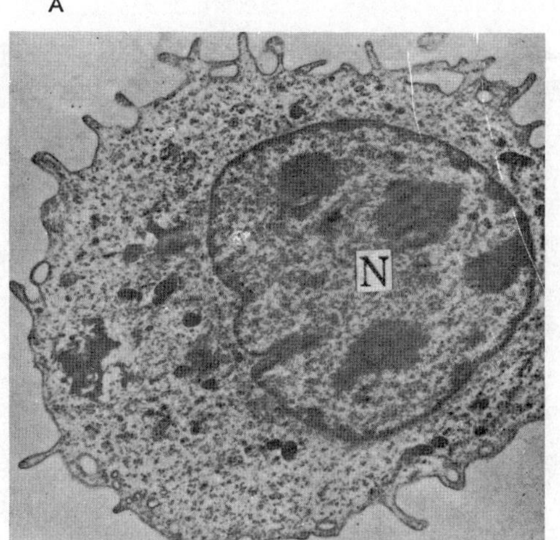

Fig. 2A.   Electronmicrographs of Ehrlich's tumor cellssh owing
the effects of HVJ *in vitro*.   Normal cell.   N : Nucleus

B

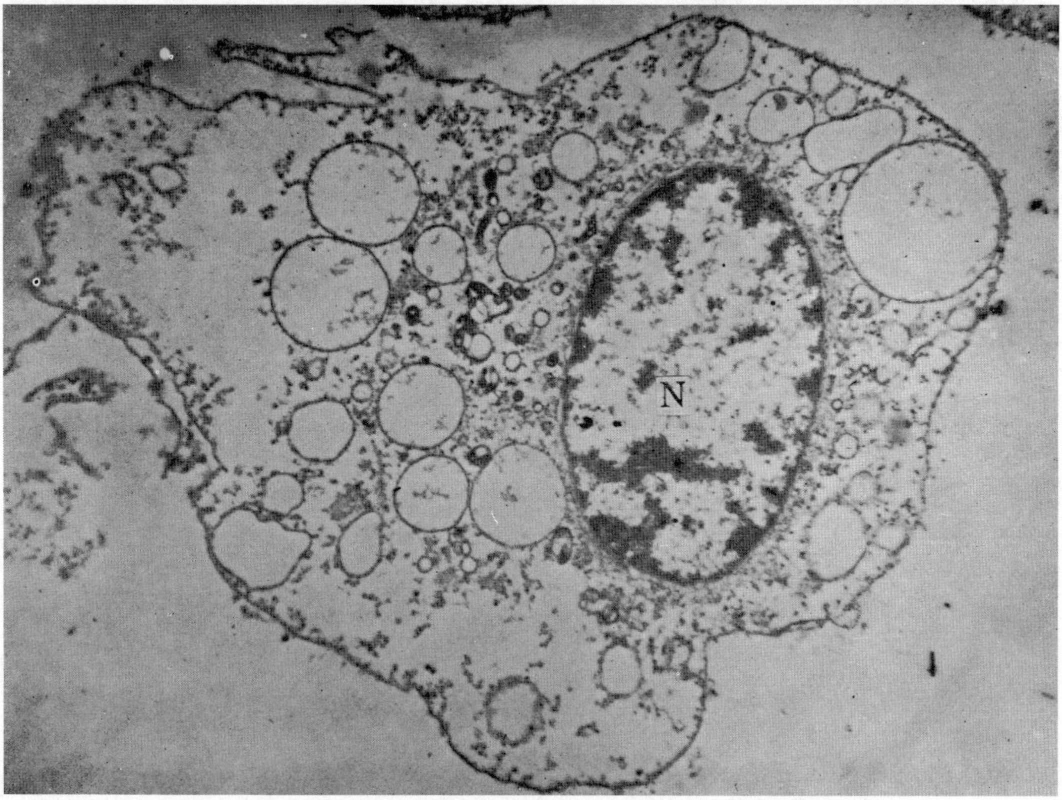

Fig. 2B.   Electronmicrographs of Ehlicher's tumor cells shownig the effects of HVJ *in vitro*.
Cell from stationary culture showing heavy damages in cytoplasm and cell membrane.

C

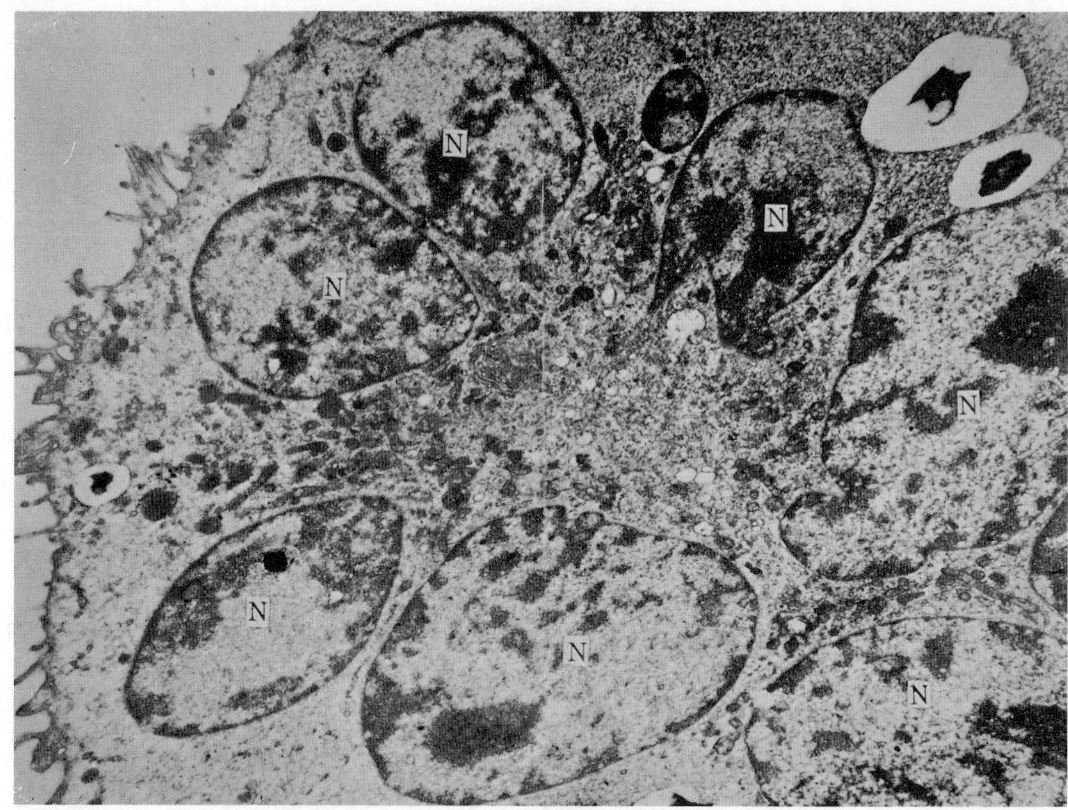

Fig. 2C. Electronmicrographs of Ehrlich's tumor cells showing the effects of HVJ *in vitro*.
A part of fused giant cell from shaking culture.

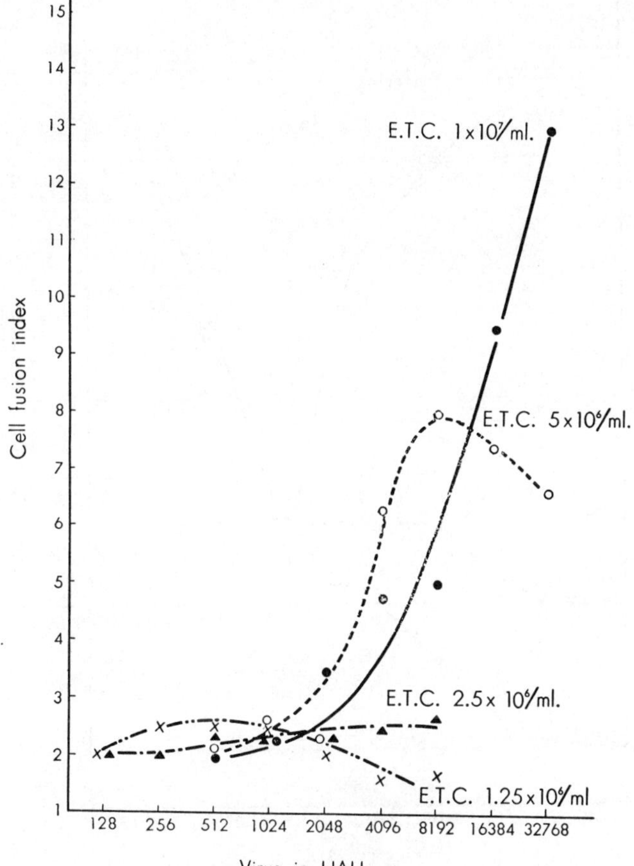

Fig. 4. Relationship between the cell fusion index, cell concentration and virus concentraticn.

number in the reaction mixture, that is the sum of the number of fused cells and non-fused cells, is inversely proportional to cell fusion index. This means that these are conversion of nonfused cells into fused cells.

As shown in Fig. 5, the cell number apparently decreases when the amount of the virus in the system is increased. The cell fusion index curve shows one inflexion point at the virus amount of 16,384 HAU.. On the left side of this point the resulted fused cells are all small, while on the right side the cells showing larger diameter are observed. This abrupt change in the cell fusion index curve means that the fused cell production by smaller amount of the virus are qualitatively different from that by larger amount (i.e. more than 16,384 HAU.) of the virus.

The direct relationships between cell fusion and the hemolytic activity of the virus was studied. The surface area of 3×10^7 red blood cells was considered equivalent to the surface area of 1×10^7 Ehrlich's tumor cells because it had been accepted that the Ehrlich's tumor cells have a surface area about three times

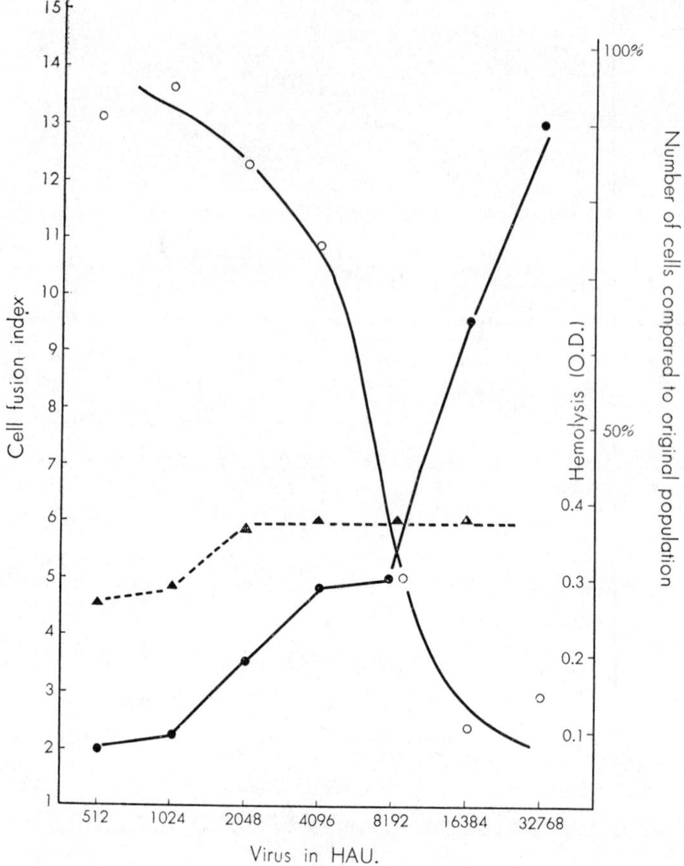

Fig. 5. Relationship between the cell fusion reaction and the hemolytic activity of HVJ.

○——○ Cell Number %   ●——● Cell fusion rate   ▲——▲ Hemolysis

that of fowl red cell and the receptors for the virus are distributed equally over both kinds of cells (Fukai, Okada and Suzuki, 1955).

When 3×10^7 red blood cells were mixed with amounts of virus from 64 to 33,000 HAU., the degree of hemolysis when less than 1024 HAU. of virus was used, was different from that with more than 2048 HAU. virus. In the former case the lysed red blood cells still held residual amounts of hemoglobin macroscopically (partial hemolysis) while in the latter case the pellet of lysed cells was white showing complete hemolysis.

In a similar experiment with Ehrlich's tumor cells it was found that for cell fusion at least 2048 HAU. of the virus was required for 1×10^7 cells.   Therefore with red blood cells, 3×10^7 cells are completely hemolysed by 2048 HAU. of the virus.

When a larger amount of the virus is given to the same number of red blood cells only 2048 HAU. are effective and the excess virus does not cause further hemolysis. However with Ehrlich's tumor cells, the threshold virus level for fusion of 1×10^7 cell is 2048 HAU. and fusion increases with an increase in virus concentration above this threshold level. It also seems as if the injuries caused in Ehrlich's tumor cells during the fusion reaction may be greater than those in red blood cells during the hemolytic reactions.

In this experiment the amount of virus adsorbed onto the Ehrlich's tumor cells could scarcely be determined from the amount of unadsorbed virus because the adsorbed virus was much less than the total virus added. To determine the exact amount of virus adsorbed some new method may be necessary.

Determination of the adsorbed virus by elution was impossible because when once absorbed onto the cells, some part of the virus formed a strong combination and disappeared as incubation proceeds at 37°C (Okada and Suzuki, 1956).

DISCUSSION

The requirements for the completion of the cell fusion reaction in an HVJ-Ehrlich's tumor cell system *in vitro* are as follows:
1) Concentrated, viable tumor cells must be used.
2) The cell virus mixture must be incubated with mild shakings at 37° C.
3) A virus having sufficient hemolytic activity to effect the cell membranes greatly must be used.
4) An amount of virus in excess of the maximal adsorption capacity of the cells must be used.

When sufficient virus and cells are mixed at 37°C, the virus particles are adsorbed onto the cell surface and agglutination occurs. Then the cell membrane is strongly attacked by the action of the virus particles and, in the case of stationary culture cytolysis results. When the system is shaken, the cell aggregates form semipermeable barriers around them, recover the activities of the individual cells and fused cells result. From this view point the fusion reaction looks like a kind of repair of the damaged cell membrane. Ultra-thin section electron-micrographs of the cells during the fusion reaction supports this view.

While there is much work on the receptor destroying activity of the MNI group viruses, studies on the hemolytic activity of mumps, Newcastle disease virus and HVJ have not yet made. As described in this paper there is a definite relationship between the hemolytic activity of HVJ and its cell fusion activity on Ehrlich's tumor cells. Therefore it may be helpful to analyse the hemolytic reaction of this virus to understand what happens in the course of cell fusion.

Recently Mobery *et al.* (1958) showed a decrease of sphingomyelin in the membrane of fowl red blood cells hemolysed by mumps virus and Hosaka (1958) showed that the degree of hemolysis by HVJ is proportional to the decrease of sphingomyelin in the membrane of fowl red blood cells. It is well known that the hemolytic reaction is generally due to changes in phospholipid components in the membrane of red blood cells. Therefore a study of the relationship between the the destruction of sphingomyelin or some other components in the cell membrane and the cell fusion reaction in the HVJ-Ehrlich's tumor cell system is of interest for the analysis of the fusion reaction.

The biggest question is why the threshold level of the virus for cell fusion is distinctly higher than the adsorption capacity of Ehrlich's tumor cells. The author is very much interested in analysing the meaning of this "excess" virus in the cell fusion reaction.

From the physical view point, great changes in the electric double layer of the cell membrane must be caused by viral action. The reason for the requirement for excess virus could be explained along these line.

The author wishes to thank Dr. Kazuo Doi for excellent electronmicrographs. Thanks are also due to Dr. Tetsuo Yoshishita for taking part in some of the experiments and Miss Kazuko Kawai for technical assistance.

REFERENCES

Fukai, K., Okada, Y. and Suzuki, T. (1955) The viral agglutination of Ehrlich's tumor cells in vitro. *Med. J. Osaka Univ.* **6**, 17–18.

Hosaka, Y. (1958) On the hemolytic action of HVJ. II. Decrease of sphingomyelin in red blood cells hemolysed by HVJ. *Biken's J.*, **1**, 90–99.

Morbery, M. L., Marinetti, G. V., Writter, R. F. and Morgan H. F. (1958) Studies of the hemolysis of red blood cells by mumps virus. III. Alteration in lipoprotein of the red blood cell wall, *J. Exp. Med.*, **107**, 87–94.

Okada, Y., and Suzuki, T. (1956) Interaction between influenza virus and Ehrlich's tumor cells. II. Fate of the virus particles adsorbed onto Ehrlich's tumor cells. *Med. J. Osaka Univ.*, **6**, 987–994.

Okada, Y., Suzuki, T. and Hosaka, Y. (1957) Interaction between influenza virus and Ehrlich's tumor cells. III. Fusion phenomenon of Ehrlich's tumor cells by the action of HVJ Z strain. *Med. J. Osaka Univ.*, **7**, 709–717.

Reprinted by permission from Nature 205:640–646 (1965)

HYBRID CELLS DERIVED FROM MOUSE AND MAN: ARTIFICIAL HETEROKARYONS OF MAMMALIAN CELLS FROM DIFFERENT SPECIES

By Prof. HENRY HARRIS and Dr. J. F. WATKINS

Sir William Dunn School of Pathology, University of Oxford

THE ability of certain animal viruses to induce the formation of multinucleate cells by fusing together single cells suggested the possibility that these viruses might be used to amalgamate different cell types and thus produce artificial animal cell heterokaryons. In this article we describe some properties of heterokaryons which we have produced by fusing together cells of human and murine origin.

The two cell types used in the present experiments were HeLa cells (originally derived from a human carcinoma) and mouse Ehrlich ascites tumour cells. These cells were chosen because they could be obtained in quantity as suspensions of single cells: the HeLa cells were grown *in vitro* in suspension culture[1]; the Ehrlich tumour was maintained in the ascitic form in the peritoneal cavity of Swiss mice. Some experiments we have made with other cell types indicate, however, that the technique can be applied to a variety of cells, including differentiated somatic cells, provided that they are susceptible to the strain of virus used to produce fusion. Of the large number of viruses which are known to induce the formation of multinucleate cells, the para-influenza I group of myxoviruses seemed the most promising, because it had been shown by Okada[2] that one strain of these viruses (*HVJ*) could induce rapid fusion in suspensions of Ehrlich ascites cells *in vitro*. The virus used in the present experiments was a strain of Sendai virus supplied by Dr. H. G. Pereira of the National Institute for Medical Research, Mill Hill.

The virus was propagated in the following way. Infected allantoic fluid with a titre of 8,000 haemagglutinating units/ml. was diluted 1 in 10^4 with phosphate-buffered saline: 0·1 ml. of this preparation was injected into the allantoic cavity of 10- or 11-day-old fertile hens' eggs, which were incubated for 3 days at 37° C. The eggs were then maintained at 4° C overnight and the allantoic fluid collected. The pooled allantoic fluid was centrifuged at $400g$ for 10 min and the haemagglutination titre of the supernatant determined. The supernatant was then centrifuged at $30,000g$ for 30 min and the deposit resuspended in one-tenth of the original volume in Hanks's solution[3] (used throughout the present experiments without glucose). The haemagglutination titre was again determined and the concentrated virus suspension stored in 1-ml. lots at $-70°$ C. This suspension, suitably diluted in Hanks's solution, was used for the experiments. Haemagglutination titrations were performed in Salkpattern haemagglutination trays. Doubling dilutions of virus were made in 0·5 ml. of phosphate-buffered saline, and approximately $2·5 \times 10^7$ guinea-pig erythrocytes suspended in 0·05 ml. of this saline were added to each cup. The smallest amount of virus which produced complete haemagglutination after 2 h at room temperature was defined as one haemagglutination unit (HAU).

Since any investigation of the physiology of the heterokaryons would be greatly complicated if these cells were engaged in virus production, the virus used to produce cell fusion was inactivated by ultra-violet light. One ml. of the concentrated suspension of virus in a watch-glass was exposed for 3 min to ultra-violet light emanating from a Philips 15-W 18-in. germicidal tube, type 'T.U.V.'. The intensity of the radiation incident on the surface of the fluid was 3,000 ergs/cm²/sec. The suspension of virus was mixed by pipetting at the end of the first and second minutes. Infectivity titrations on the irradiated virus were carried out by a modification of Fulton's method[4] in which pieces of chorio-allantoic membrane were incubated with the virus in Medium 199 (ref. 5) in a haemagglutination tray. It was found that after 3-min exposure to ultraviolet light under the present conditions, the infectivity of the virus, as measured by the production of haemagglutinin, was reduced to levels at which the measurements were obscured by the haemagglutination produced by the initial viral inoculum. This represented a reduction in infectivity of at least 3 logs. Although the ability of the virus to multiply in the chorio-allantoic membrane had thus been drastically reduced, its ability to induce cell fusion *in vitro* remained unimpaired. Evidence will be presented to show that the virus inactivated by ultraviolet light was not reactivated in the vast majority of the heterokaryons.

The technique used for inducing cell fusion was essentially similar to that described by Okada[2]. HeLa cells from a suspension culture were spun down and resuspended at a concentration of 2×10^7 cells/ml. in Hanks's solution. Ehrlich ascites cells, withdrawn from the peritoneal cavity, were washed once by centrifugation in this solution and then resuspended in it at a concentration of 2×10^7 cells/ml.: 0·5 ml. of each cell suspension was pipetted into a chilled inverted T-tube together with 1·0 ml. of the suspension of virus, diluted, if necessary, in Hanks's solution. The cells clumped immediately, and the size of the clumps was roughly proportional to the amount of virus added. The T-tube was kept at 4° C for 15 min and then shaken in a water bath at 37° C for 20 min at a rate of 100 excursions/min. During this time the cells in the clumps underwent varying degrees of fusion. Fig. 1 shows an electron micrograph of a section through a pellet of cells obtained by centrifuging a preparation at $200g$ for 10 min immediately after the period of shaking at 37° C. A trinucleate cell is shown, formed by the fusion of three discrete cells; and a fourth cell is seen in the process of fusing with this trinucleate cell.

One ml. of the cell suspension and 5 ml. of culture medium were pipetted into a 6-cm-diameter Petri dish containing 15 coverslips 1 cm in diameter. The culture medium consisted of 20 per cent calf serum and 1 per cent tryptose broth in Medium 199, to which penicillin at a concentration of 100 international units/ml. and streptomycin at a concentration of 100 µg/ml. were added. The dishes were incubated at 37° C in a gas mixture of 5 per cent carbon dioxide in air. The coverslips were transferred to fresh medium after 24 h and again after 4 days.

Four h after the cell suspension had been introduced into the Petri dishes multinucleate cells were found to have adhered to the coverslips; and within 24 h most of these had flattened out on the glass. From cell counts it could be calculated that the multinucleate cells adherent to the coverslips accounted for about 10 per cent of the single cells originally present in the suspension. Each of these multinucleate cells contained from 2 to about 20 nuclei. The nuclei were of two easily distinguishable morphological types. With May–Grünwald–Giemsa or Leishman stain one type of nucleus stained more deeply than the other. The more deeply staining nucleus contained numerous small nucleoli or coarse condensations of chromatin, while the less deeply staining nucleus con-

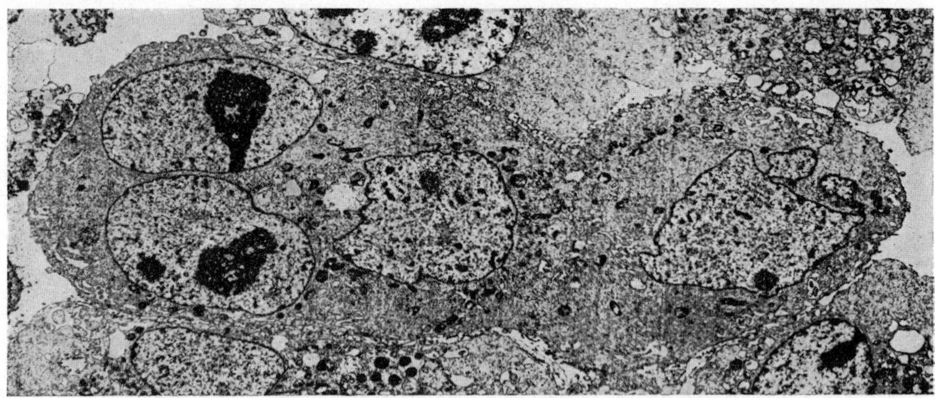

Fig. 1. An electron micrograph of a section through a preparation in which HeLa cells and Ehrlich ascites cells were fused together. A trinucleate cell is shown, formed by the fusion of three discrete cells. A fourth cell, on the right, is seen in the process of fusion with the trinucleate cell. (By courtesy of Dr. G. I. Schoefl)

tained one to three large nucleoli. The two types of nucleus are illustrated in Fig. 2, which shows a tetranucleate cell containing two nuclei of each type. Since the deeply staining pachychromatic nuclei clearly resembled the nuclei of Ehrlich ascites cells, and the less deeply staining nuclei resembled the nuclei of HeLa cells, it was difficult to avoid the conclusion that the multinucleate cells were formed by fusion of the two cell types. This was confirmed in the following way: 100 µc. of tritiated thymidine was added to a suspension culture of HeLa cells at 10 a.m., 2 p.m., 6 p.m. and 10 p.m. on one day, and again at 7 a.m. and 10 a.m. on the following day. Since the generation time of the cells under these conditions was about 20 h, it was thought that this procedure would succeed in labelling the nuclei of virtually all the cells in the culture. Autoradiographs of smears of the cells, made 2 h after the final addition of tritiated thymidine, confirmed that this was so. When a mixture of labelled HeLa cells and unlabelled Ehrlich ascites cells was treated with virus, autoradiographs of the resulting multinucleate cells revealed that the deeply staining pachychromatic nuclei were not labelled, whereas the less deeply staining nuclei with the large nucleoli were. This is illustrated in Figs. 3–5, which show a series of heterokaryons containing labelled and unlabelled nuclei in varying proportions. When the reciprocal experiment was made with labelled Ehrlich ascites cells and

unlabelled HeLa cells, it was found that the deeply staining pachychromatic nuclei were labelled and the others were not. It is thus clear that the multinucleate cells were indeed formed by the fusion of HeLa cells with Ehrlich ascites cells. At least 98 per cent of the multinucleate cells were heterokaryons when virus concentrations of 8,000 HAU/ml. or more were used to induce the cell fusion (Table 1).

Table 1. EFFECT OF VIRUS CONCENTRATION ON THE CHARACTER OF THE HETEROKARYONS PRODUCED

Counts made 24 h after formation of heterokaryons	Concentration of virus used			
	Virus inactivated by ultra-violet light			Infective virus
	800 HAU	8,000 HAU	80,000 HAU	8,000 HAU
Percentage of multinucleate cells which were hetero-karyons	77	99	98	97
	84	98	97	99
	85	97		99
Mean	82	98	98	98
No. of nuclei per heterokaryon	4·0	(18·4)	10·0	6·3
	4·3	5·9	8·5	5·1
	4·4	9·4		7·4
Mean	4·2	(11·2)	9·3	6·3
Ratio of HeLa nuclei to Ehrlich ascites nuclei	2·8	1·9	3·2	2·3
	1·9	2·1	2·7	2·6
	2·5	2·0		2·6
Mean	2·4	2·0	3·0	2·5
Ratio of multinucleate to mononucleate cells	1·0	3·5	8·4	3·1
	0·6	2·0	7·5	9·0
	0·8	3·1		8·4
Mean	0·8	2·9	8·0	6·8

The average number of nuclei per heterokaryon and the ratio of multinucleate to mononucleate cells could be varied by changing the concentration of virus used. Table 1 shows that at lower concentrations of virus the number of nuclei per heterokaryon and the ratio of multinucleate to mononucleate cells fell. Since the proportion of the original cell suspension recovered as multinucleate cells was unaffected by the concentration of virus used (Table 2), the increase in the ratio of multinucleate to mononucleate cells at higher concentrations of virus must have resulted from preferential elimination of mononucleate cells. This was apparently due to the fact that the multinucleate cells were more resistant to the cytotoxic effects of the virus than the mononucleate cells, a conclusion supported by the observation that infective virus eliminated a greater proportion of the mononucleate cells than inactivated virus at the same concentration (Table 1). With high concentrations of virus, however, and especially with infective virus, some of the multinucleate cells showed marked vacuolation. Table 1 also shows that the ratio of HeLa nuclei to Ehrlich ascites nuclei in the heterokaryons was between 2 and 3 irrespective of virus concentration. This preponderance of HeLa nuclei might be due either to the fact that HeLa cells fuse more easily than Ehrlich ascites cells or, more probably, to the fact that cells with

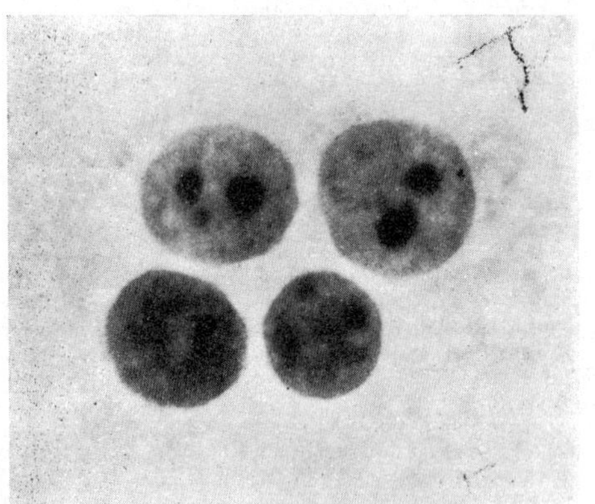

Fig. 2. A tetranucleate cell in which the two upper nuclei are derived from HeLa cells and the two lower ones from Ehrlich ascites cells (× 1,300)

Table 2. SURVIVAL TIME OF HETEROKARYONS MADE WITH DIFFERENT CONCENTRATIONS OF VIRUS

Days after production of heterokaryons	No. heterokaryons per coverslip			
	Virus inactivated by ultra-violet light			Infective virus
	80,000 HAU	8,000 HAU	800 HAU	8,000 HAU
1	$4 \cdot 2 \times 10^3$	$2 \cdot 7 \times 10^3$	$4 \cdot 9 \times 10^3$	$4 \cdot 5 \times 10^3$
	$4 \cdot 3 \times 10^3$	$3 \cdot 2 \times 10^3$	$7 \cdot 1 \times 10^3$	$5 \cdot 3 \times 10^3$
		Medium changed		
2	$3 \cdot 3 \times 10^3$	$3 \cdot 3 \times 10^3$	$5 \cdot 5 \times 10^3$	$4 \cdot 8 \times 10^3$
	$2 \cdot 9 \times 10^3$	$3 \cdot 6 \times 10^3$	$6 \cdot 2 \times 10^3$	$6 \cdot 3 \times 10^3$
3	$2 \cdot 1 \times 10^3$	$3 \cdot 2 \times 10^3$	$5 \cdot 0 \times 10^3$	$8 \cdot 5 \times 10^3$
	$2 \cdot 0 \times 10^3$	$3 \cdot 9 \times 10^3$	$4 \cdot 4 \times 10^3$	$4 \cdot 3 \times 10^3$
4	$2 \cdot 2 \times 10^3$	$5 \cdot 0 \times 10^3$	$5 \cdot 1 \times 10^3$	$2 \cdot 3 \times 10^3$
	$1 \cdot 1 \times 10^3$	$3 \cdot 7 \times 10^3$	$5 \cdot 2 \times 10^3$	$3 \cdot 5 \times 10^3$
		Medium changed		
5	$1 \cdot 8 \times 10^2$	$2 \cdot 3 \times 10^3$	$3 \cdot 2 \times 10^3$	$3 \cdot 3 \times 10^3$
	$1 \cdot 2 \times 10^3$	$2 \cdot 8 \times 10^3$	$3 \cdot 1 \times 10^3$	$3 \cdot 9 \times 10^3$
7	—	$1 \cdot 4 \times 10^3$	$1 \cdot 9 \times 10^3$	$5 \cdot 6 \times 10^2$
			$2 \cdot 2 \times 10^3$	$2 \cdot 5 \times 10^2$

surfaces containing a large HeLa contribution adhere more readily to glass. Ehrlich ascites cells adhere poorly to glass, so that heterokaryons containing a preponderance of Ehrlich ascites components might also adhere poorly. In this way, adhesion to glass might select heterokaryons with predominantly HeLa surface characteristics.

While they remained multinucleate the heterokaryons did not multiply, but, under suitable conditions, they remained alive on the coverslips for at least five days. After this time some of them began to degenerate and to lose their attachment to the glass, but others survived for

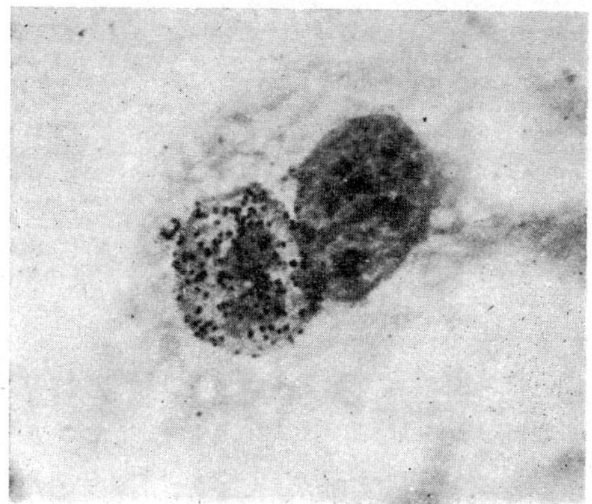

Fig. 3. Autoradiograph of a binucleate cell containing one HeLa nucleus and one Ehrlich ascites nucleus. The HeLa cells had been grown in tritiated thymidine before the heterokaryons were produced. The HeLa nucleus is labelled and the Ehrlich ascites nucleus is not. (× 1,300)

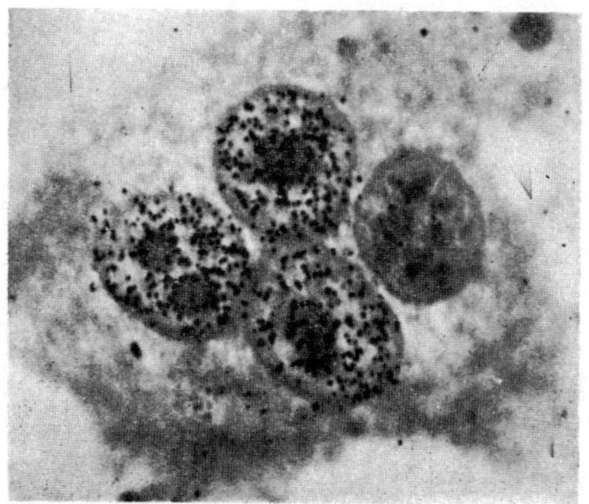

Fig. 4. A tetranucleate cell from the same population as the cell shown in Fig. 3. Three labelled HeLa nuclei and one unlabelled Ehrlich ascites nucleus are shown. (× 1,300)

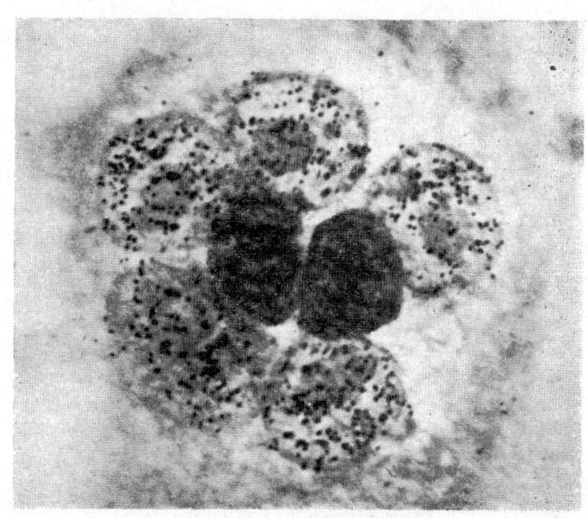

Fig. 5. An octonucleate cell from the same population as the cells shown in Figs. 3 and 4. Six labelled HeLa nuclei and two unlabelled Ehrlich ascites nuclei are shown. (× 1,300)

as long as 15 days. Time-lapse cinemicrography revealed that some of the multinucleate cells exhibited sluggish locomotion: they moved a distance approximately equal to their own length in 4 days. The survival times of the heterokaryons produced with different concentrations of inactivated virus and with infective virus are shown in Table 2. The essential reason for the failure of the heterokaryons to multiply appeared to be the fact that the cells contained more than one nucleus. Multinucleate cells containing only HeLa nuclei also failed to multiply, thus indicating that it was the multinucleate state rather than heterokaryosis which was responsible for the deficiency. In mixed cultures containing both mononucleate and multinucleate cells, the mononucleate cells did multiply. It was therefore unlikely that failure of multiplication was due to the injury produced by the virus.

Since each heterokaryon received a large number of virus particles, the possibility existed that the virus inactivated by ultra-violet light might undergo multiplicity reactivation. In order to examine this possibility the production of infective virus and haemadsorption[6] were investigated. For haemadsorption studies two coverslips were examined each day. They were washed once in Hanks's solution and placed in the cups of a haemagglutination tray: 0·5 ml. of a 3 per cent (v/v) suspension of guinea-pig erythrocytes in Medium 199 was added to each cup and the tray was left at room temperature for 2 h, during which time the erythrocytes settled evenly over the coverslips. These were then removed, washed four times in Hanks's solution, fixed in methanol and stained with May–Grünwald–Giemsa. The percentage of heterokaryons showing adsorption of one or more erythrocytes was then determined. The results are summarized in Fig. 6. This shows that when infective virus was used to make the heterokaryons, all of them exhibited the presence of surface haemagglutinin for at least 5 days. But when the heterokaryons were made with inactivated virus, surface haemagglutinin disappeared from the cells at a rate which depended on the initial virus concentration. Moreover, with infective virus, the haemadsorption was massive and involved the whole of the cell surface; with inactivated virus, haemadsorption, in the decreasing number of cells which showed it, involved a progressively smaller part of the individual cell surface. At the lowest concentration of virus (800 HAU/ml.) less than 0·05 per cent of the heterokaryons exhibited haemadsorption at 24 h. On subsequent days occasional heterokaryons (between 0·5 and 1 per cent) did show haemadsorption, but it is clear that at least 99 per cent of the cells did not produce any viral haemagglutinin during the 5-day period. The

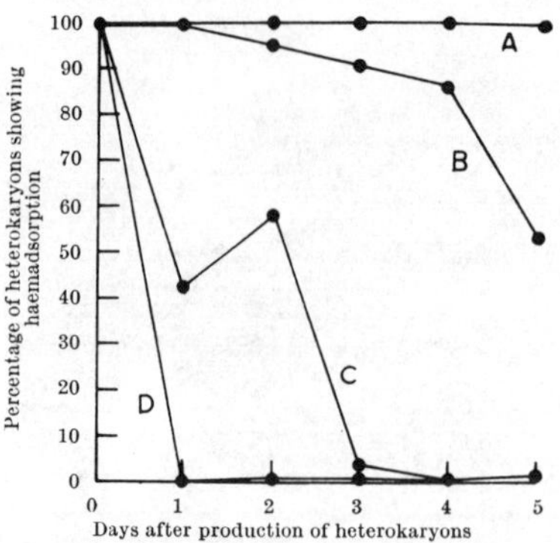

Fig. 6. Rate of loss of surface haemagglutinin in heterokaryons produced by infective virus and virus inactivated by ultra-violet light. *A*, Infective virus at a concentration of 8,000 HAU/ml. *B, C, D,* Inactivated virus at concentrations of 80,000, 8,000 and 800 HAU/ml., respectively

loss of haemagglutinin from the surface of the heterokaryons produced by inactivated virus no doubt reflects the gradual destruction or elution of the initial inoculum.

In order to examine whether heterokaryons produced by inactivated virus released any infectious virus into the medium, 10 coverslips bearing such heterokaryons were transferred to fresh medium after 24-h cultivation and were then incubated in this medium for a further 48 h. At the end of this period 10-fold dilutions of the medium were made in Hanks's solution and 0·1 ml. of each dilution inoculated into the allantoic cavity of two 10- or 11-day-old fertile hens' eggs. After incubation for 48 h at 37° C the allantoic fluids were collected and their haemagglutinin content titrated. The results of this experiment are set out in Table 3, in which the calculated amount of infectious virus produced per heterokaryon is shown. One unit of infectious virus is defined as the amount which produced 2 HAU/ml. of allantoic fluid in 48 h. It will be seen that, when the heterokaryons were made with inactivated virus at a concentration of 800 or 8,000 HAU/ml., less than one heterokaryon in a thousand produced a single unit of infectious virus. With 80,000 HAU of inactivated virus a yield of 1–10 units of infectious virus per heterokaryon was obtained. This does not, however, necessarily mean that multiplicity reactivation had occurred. Small amounts of virus probably escaped inactivation by ultra-violet light, and these might have been responsible for the yield of infectious virus. When the heterokaryons were made with 8,000 HAU/ml. of virus which had not been inactivated, the yield of infectious virus was also only 1–10 units per heterokaryon. This suggests either that auto-interference had occurred or that the virus, even when it is not inactivated, grows poorly in this system. It is, in any event, clear that with moderate or low doses of inactivated virus at least 99 per cent of the heterokaryons produce neither infectious virus nor viral haemagglutinin.

The usefulness of heterokaryons for the analysis of nucleo-cytoplasmic relationships depends to a large extent on the ability of the two sets of nuclei to synthesize RNA

and on the ability of the hybrid cell to synthesize protein. These two functions were therefore examined by studying the incorporation of tritiated uridine into RNA and tritiated leucine into protein. Heterokaryons, which had been maintained on coverslips for periods up to 5 days, were incubated for 2–6 h with tritiated uridine. The cells were then fixed, digested for 30 min at 37° C with deoxyribonuclease and extracted with 0·3 N trichloroacetic acid at 4° C. Autoradiographs of such preparations revealed that all the nuclei in the heterokaryons were labelled. The cytoplasm became labelled also, but more slowly than the nuclei, as in other nucleated cells. Fig. 7 shows an autoradiograph of a heterokaryon from a 24-h culture exposed for 2 h to tritiated uridine. The cell contains a HeLa nucleus and an Ehrlich ascites nucleus: both are labelled, and there is also slight labelling over the cytoplasm. Fig. 8 shows a similar heterokaryon exposed for 6 h to tritiated uridine: the cytoplasm of the cell is now heavily labelled. There was no marked disparity between the HeLa nuclei and the Ehrlich ascites nuclei in their ability to incorporate tritiated uridine into RNA, or, during the first 5 days, between the amount of nuclear labelling in the heterokaryons and the amount in neighbouring mononucleate cells. It thus appears that synthesis of RNA takes place in the heterokaryons in a manner comparable with that seen in mononucleate cells, and that both sets of nuclei

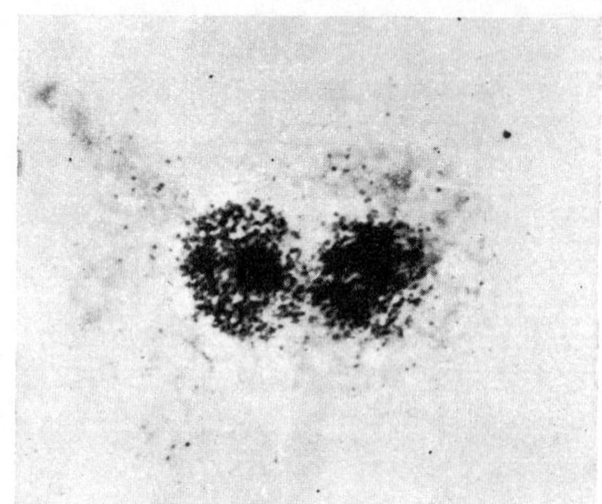

Fig. 7. Autoradiograph of a heterokaryon from a 24-h culture exposed for 2 h to tritiated uridine. The cell contains a HeLa nucleus and an Ehrlich ascites nucleus. Both are labelled, and there is also slight labelling over the cytoplasm. (× 1,100)

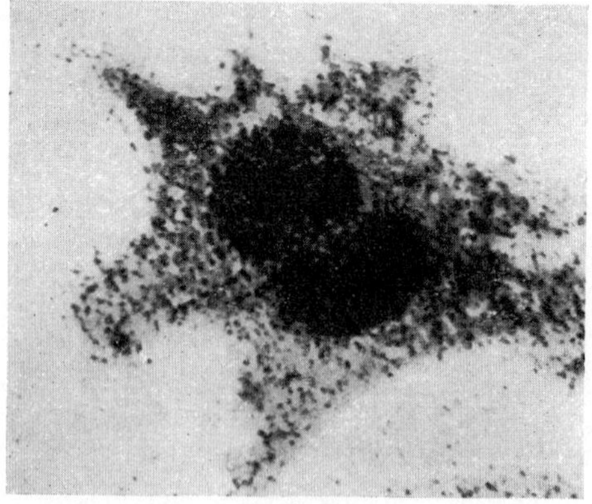

Fig. 8. A similar heterokaryon from the same population as the cell shown in Fig. 7, but exposed for 6 h to tritiated uridine. The cytoplasm of the cell is now heavily labelled. (× 1,100)

Table 3. Production of Infectious Virus by Heterokaryons

Concentration of virus used to produce heterokaryons	Haemagglutinin titre (HAU/ml. allantoic fluid) produced in 48 h by 0·1 ml. of culture medium diluted as shown						Units of infectious virus produced per heterokaryon
	10^0	10^{-1}	10^{-2}	10^{-3}	10^{-4}	10^{-5}	
Virus inactivated by ultra-violet light							
800 HAU/ml.	< 2	< 2	—	—	—	—	$< 10^{-3}$
	< 2	< 2	—	—	—	—	
8,000 HAU/ml.	< 2	< 2	< 2	—	—	—	$< 10^{-3}$
	< 2	< 2	< 2	—	—	—	
80,000 HAU/ml.	2,048	512	512	< 2	< 2	< 2	1–10
	2,048	—	512	64	< 2	< 2	
Infective virus							
8,000 HAU/ml.	512	2,048	< 2	< 2	< 2	< 2	1–10
	512	256	256	256	< 2	< 2	

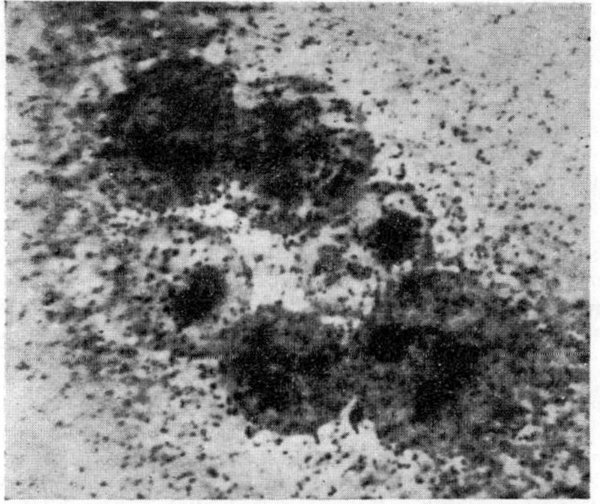

Fig. 9. Autoradiograph of a heterokaryon from a 24-h culture exposed for 6 h to tritiated leucine. There is generalized labelling over nuclei and cytoplasm. (× 1,100)

are involved in this synthesis. The genes of both mouse and man are therefore transcribed.

Autoradiographs of heterokaryons, maintained on coverslips for periods up to 5 days and then exposed for 6 h to tritiated leucine, revealed generalized labelling over nuclei and cytoplasm (Fig. 9). Failure to extract this label with 0·3 N trichloroacetic acid at 90° C confirmed that the tritiated leucine had been incorporated into protein. It is therefore clear that the heterokaryons synthesize protein as well as RNA.

Although heterokaryons in the multinucleate state did not multiply, they did synthesize DNA. When cultures which had been maintained on coverslips for 24 h were exposed for 2 h to tritiated thymidine, autoradiographs showed that about 70 per cent of the heterokaryons contained labelled nuclei. In 5 per cent of these labelled heterokaryons all the nuclei in the cell incorporated the precursor, but in the rest labelled and unlabelled nuclei were present in varying proportions. In any one cell some nuclei of each type might be labelled and others not. Fig. 10 shows a multinucleate cell in which all the nuclei are synthesizing DNA, and Fig. 11 shows a multinucleate cell in which only two of the Ehrlich ascites nuclei are synthesizing DNA. In these labelled heterokaryons about 80 per cent of the Ehrlich ascites nuclei were synthesizing DNA, but only about 30 per cent of the HeLa nuclei. The percentage of heterokaryons showing labelling after 2 h exposure to tritiated thymidine fell from 70 at 24 h to 55 at 48 h, 44 at 4 days and 31 at 7 days. But in those heterokaryons which did show some form of nuclear labelling, the proportion of Ehrlich ascites nuclei and of HeLa nuclei which were synthesizing DNA underwent little change over 7 days. An exponentially growing population of Ehrlich ascites or HeLa cells exposed to tritiated thymidine for 2 h would show labelling in about a third of the nuclei, an expression of the duration of

the phase of DNA synthesis relative to the total intermitotic time. The very high proportion of labelled Ehrlich ascites nuclei in the heterokaryons, therefore, means that these nuclei have become partially synchronized in the phase of DNA synthesis. This partial synchronization might have resulted from selection during cell fusion of those Ehrlich ascites cells in which synthesis of DNA was taking place. But the fact that 80 per cent or so of the Ehrlich ascites nuclei in labelled heterokaryons were synthesizing DNA not only at 24 h, when 70 per cent of the heterokaryons were labelled, but also at 48 h, when 55 per cent of the heterokaryons were labelled, suggests another explanation. It seems more probable that many of the Ehrlich ascites nuclei which were not in the phase of DNA synthesis at the time of formation of the heterokaryons passed into this phase during the first 24 h, and that DNA synthesis was not terminated abruptly after a few hours, as occurs in normal cells, but continued for much longer periods. Any alternative explanation would seem to require that at least some of these nuclei underwent two or more normal cycles of DNA synthesis without mitosis. Why the percentage of HeLa nuclei synthesizing DNA at 24 h should be so much lower is obscure, but, since this percentage also shows little change on subsequent days, it is likely that the phase of DNA synthesis was prolonged in at least some of the HeLa nuclei also. These findings, in any event, make one point clear. Whether or not a mammalian cell nucleus synthesizes DNA cannot be solely determined by events in the cytoplasm. Even in a common cytoplasm DNA synthesis may, at any one time, be taking place in some nuclei and not in others; and the

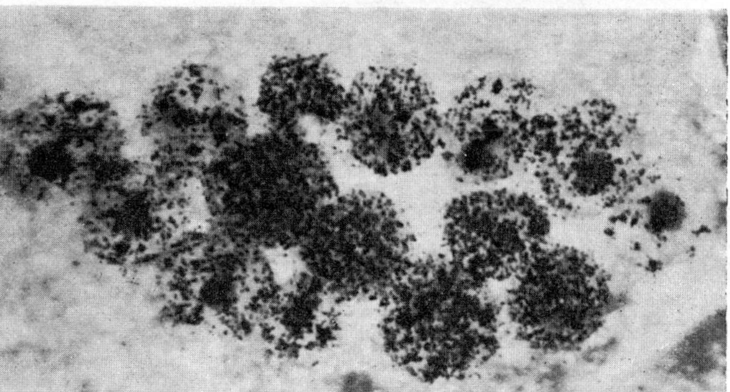

Fig. 10. Autoradiograph of a heterokaryon from a 24-h culture exposed for 2 h to tritiated thymidine. All the nuclei are synthesizing DNA. (× 1,100)

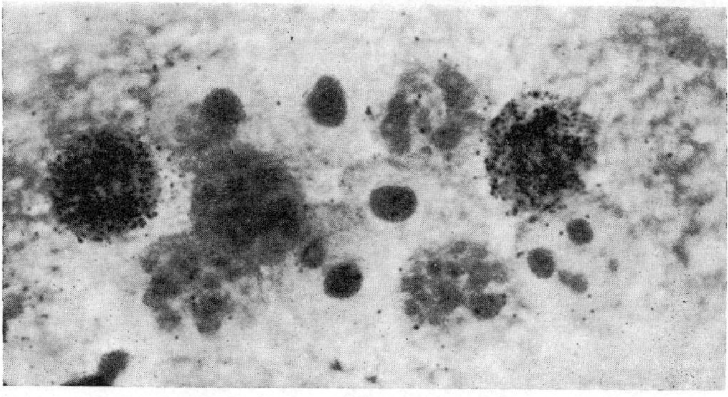

Fig. 11. A heterokaryon from the same population as the cell shown in Fig. 10. Only two of the Ehrlich ascites nuclei are synthesizing DNA. (× 1,300)

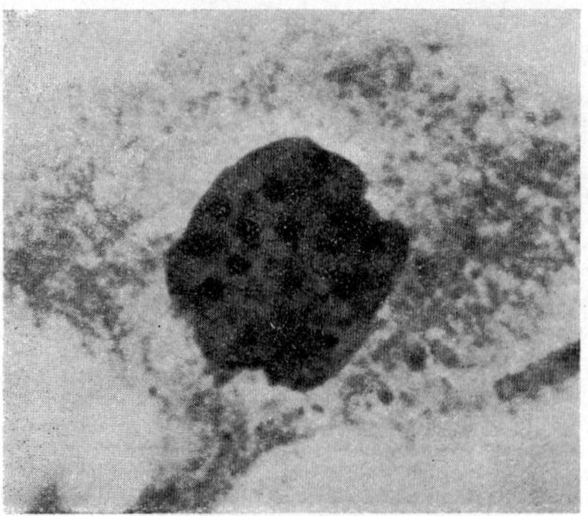

Fig. 12. A cell containing a giant nucleus produced by fusion of a number of discrete nuclei. Several nucleoli or condensations of chromatin are to be seen. (× 1,100)

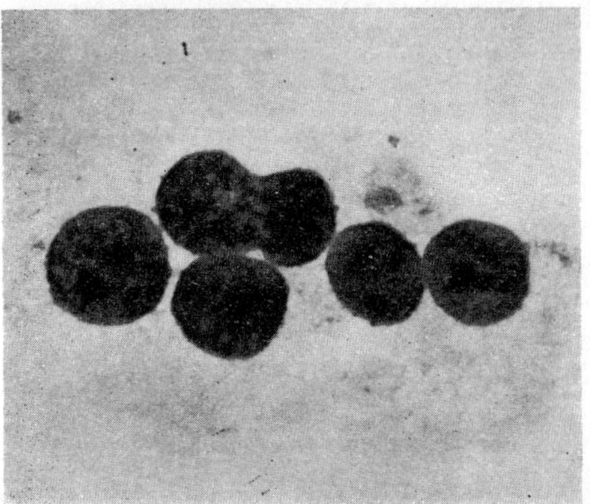

Fig. 13. A heterokaryon in which two of the Ehrlich ascites nuclei are in the process of fusion. (× 1,100)

had been grown in tritiated thymidine with unlabelled Ehrlich ascites cells, and vice versa. Autoradiographs of such preparations showed that many of the large nuclei were labelled over only one part of the nucleus. Fig. 13 shows a heterokaryon in which two of the Ehrlich ascites nuclei are in the process of fusion. Fig. 14 shows a giant nucleus in a preparation made by fusing labelled Ehrlich ascites cells with unlabelled HeLa cells: only the right-hand half of the nucleus is labelled.

It might be thought that nuclear fusion represented an extension of the activity of the virus to the nuclear membrane. But experiments on the relationship between the concentration of virus used and the amount and rate of nuclear fusion revealed that this was not the case. Fig. 15 demonstrates that the percentage of cells showing nuclear fusion on successive days is inversely related to the concentration of virus used. With inactivated virus at a concentration of 800 HAU/ml., about 25 per cent of the heterokaryons showed some degree of nuclear fusion at 24 h, and by 5 days nuclear fusion had occurred in almost all the cells. But Fig. 6 demonstrates that at least 99 per cent of the heterokaryons produced by this concentration of inactivated virus had lost all trace of virus haemagglutinin from their surface within 24 h. It therefore appeared unlikely that nuclear fusion, which progressed throughout

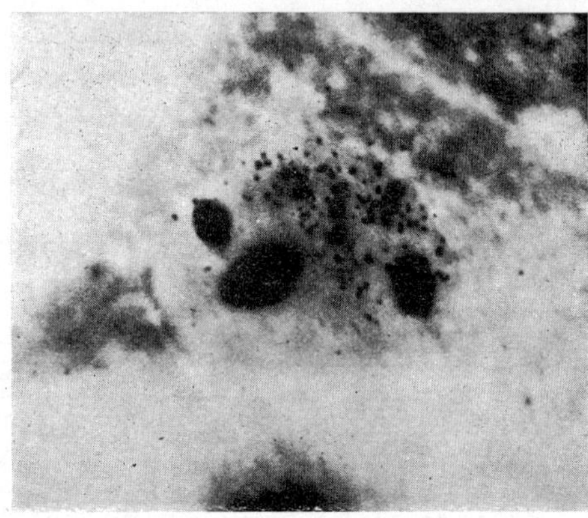

Fig. 14. Autoradiograph of a large nucleus produced by fusion of labelled Ehrlich ascites nuclei with unlabelled HeLa nuclei. The Ehrlich ascites cells were exposed to tritiated thymidine before the heterokaryons were produced. Only the right-hand half of the nucleus is labelled. (× 1,300)

nuclei from one species may be more active in the heterokaryon than the nuclei from another. The ability of the nucleus to synthesize DNA must therefore depend, at least in part, on factors operating within the nucleus.

Over a 7-day period there was a progressive fall not only in the proportion of heterokaryons showing labelling after exposure to tritiated thymidine, but also in the intensity of the labelling. At 24 h there was no great difference between the intensity of nuclear labelling in the heterokaryons and that found in neighbouring mononucleate cells; but at 7 days the intensity of labelling in the heterokaryons was greatly reduced relative to that found in mononucleate cells. It thus appears that although the phase of DNA synthesis is prolonged in the nuclei of heterokaryons, the rate of synthesis progressively falls, and eventually synthesis stops.

Under certain conditions the nuclei within the heterokaryons underwent fusion. The fused nuclei could usually be recognized by their greater size and by the fact that they contained many large nucleoli or condensations of chromatin. In many of the multinucleate cells only some of the nuclei underwent fusion, but in others all the nuclei in the cell fused together to produce one giant nucleus which sometimes had a bizarre shape (Fig. 12). Fusion took place not only between nuclei of the same kind but also between HeLa nuclei and Ehrlich ascites nuclei. This was demonstrated by fusing HeLa cells which

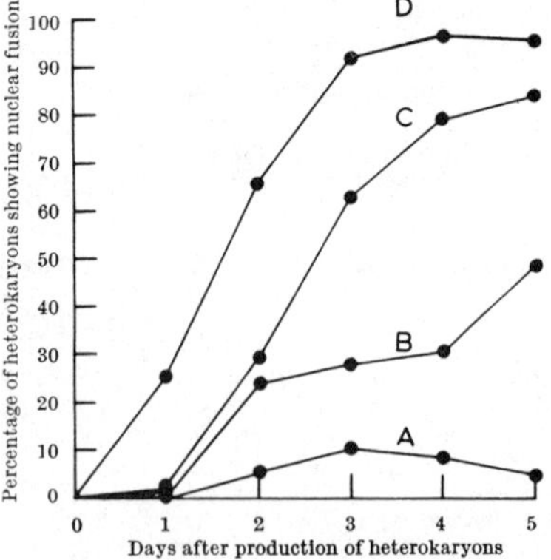

Fig. 15. Relationship between the proportion of heterokaryons showing nuclear fusion and the concentration of virus used to produce the heterokaryons. *A*, Infective virus at a concentration of 8,000 HAU/ml. *B*, *C*, *D*, Inactivated virus at concentrations of 80,000, 8,000 and 800 HAU/ml. respectively

tho 5-day period, was caused by the continued activity of the virus. Indeed, the results indicated that the presence of large amounts of virus inhibited nuclear fusion. With increasing concentrations of virus the proportion of heterokaryons showing nuclear fusion on successive days was reduced; and, at the same dose, infective virus inhibited nuclear fusion more drastically than inactivated virus. It seems as if progressive nuclear fusion is the normal course of events in the heterokaryon, and that the inhibition produced by high doses of virus is a non-specific toxic effect. Preliminary experiments indicate that other toxic substances may also inhibit nuclear fusion. Since the term heterokaryon, as it is generally used, implies a cell with more than one nucleus, multinucleate cells in which the nuclei have fused together to form a single nucleus will be called synkaryons: homosynkaryons where nuclei of the same type have fused together, and heterosynkaryons where nuclei of different types have fused together.

The synkaryons continued to incorporate tritiated uridine into both nuclear and cytoplasmic RNA for at least 5 days at a rate not greatly different from that found in single HeLa cells in the same culture. Some of them also incorporated tritiated thymidine into DNA, although, over a 7-day period, the proportion showing labelling, and the intensity of the labelling relative to single HeLa cells in the same culture, progressively fell. There was, however, reason to believe that some synkaryons, including heterosynkaryons, were able to undergo at least one mitosis. In cultures containing both multinucleate and mononucleate cells it was observed that some of the cells rounded up in mitosis were very much larger than the rest. The cultures were therefore exposed for 18 h to colchicine at a concentration of 0·0025 per cent (w/v), and the karyotypes of the accumulated metaphases examined by the method of Rothfels and Siminovitch[7]. It was found that some of the metaphase figures had up to 300 or more chromosomes, whereas the modal chromosome number for the HeLa and Ehrlich ascites cells was about 80. Moreover, some of these giant metaphases showed large numbers of both metacentric and telocentric chromosomes, including many long telocentrics. HeLa cells normally contain not more than one long telocentric chromosome, and frequently none at all[8]; and Ehrlich ascites cells normally contain very few metacentric chromosomes[9]. The presence of large numbers of both

metacentric and long telocentric chromosomes in the one cell therefore suggests that the cell originally contained both HeLa and Ehrlich ascites nuclei. Whether hetero-synkaryons can produce clones of cells containing both human and murine chromosomes is the subject of current investigation.

In the experiments which we have described, an inactivated virus has been used to impose a form of artificial sexuality on mammalian tissue cells. Since the resulting hybrid cells synthesize protein, and both sets of nuclei synthesize RNA, these hybrids can be used to investigate problems of nucleo-cytoplasmic relationship which could hitherto be studied only in the heterokaryons of fungi or yeasts. Some features of the mammalian cell system perhaps warrant special comment. The remarkable fact that viable heterokaryons can be made with cells originating from different animal species means that a very wide range of genetic markers can be used to distinguish the two cell types. Indeed, the choice is limited only by the susceptibility of the cells to the virus used to produce fusion. The average number of nuclei per heterokaryon and the proportion of nuclei of each type can, within certain limits, be controlled. Preparations can be made in which 85 per cent or more of the cells are heterokaryons, and the ratio of multinucleate to mononucleate cells can be varied. Unlike fungi or yeasts, mammalian cells use thymidine as a specific precursor of DNA. This property makes mammalian cell heterokaryons eminently suitable for investigations on the regulation of DNA synthesis, as some of the present experiments show. Finally, auto-radiographic and chemical techniques for the study of RNA metabolism in mammalian cells have now reached a moderate degree of sophistication: these procedures can be applied without modification to mammalian cell heterokaryons. There is, therefore, reason to hope that these heterokaryons will lend themselves to experiments which have hitherto not been possible in animal cells.

[1] Harris, H., and Watts, J. W., *Proc. Roy. Soc.*, B, **156**, 109 (1962).
[2] Okada, Y., *Exp. Cell Res.*, **26**, 98 (1962).
[3] Hanks, J. H., *J. Cell. Comp. Physiol.*, **31**, 235 (1948).
[4] Fulton, F., and Armitage, P., *J. Hyg. (Camb.)*, **49**, 247 (1951).
[5] Morgan, J. F., Morton, H. J., and Parker, R. C., *Proc. Soc. Exp. Biol. and Med.*, **73**, 1 (1950).
[6] Vogel, J., and Shelokov, A., *Science*, **126**, 358 (1957).
[7] Rothfels, K. H., and Siminovitch, L., *Stain Tech.*, **33**, 73 (1958).
[8] Clausen, J. J., and Syverton, J. T., *J. Nat. Cancer Inst.*, **28**, 117 (1962).
[9] Hauschka, T. S., and Levan, A., *J. Nat. Cancer Inst.*, **21**, 77 (1958).

8

Production of Mammalian Somatic Cell Hybrids by Means of Polyethylene Glycol Treatment

G. Pontecorvo

Imperial Cancer Research Fund, Lincoln's Inn Fields, London

Received 14 April 1975—Final 16 June 1975

Abstract—*Polyethylene glycol (PEG) is known to promote fusion of plant protoplasts. Various adaptations of this treatment to mammalian, including human, cell cultures are reported here. PEG is very effective in producing hybrids capable of indefinite multiplication even in cases, such as early passage human skin fibroblasts and lymphocytes, known to be highly recalcitrant to other treatments.*

INTRODUCTION

Inactivated Sendai and other viruses (1, 2) increase the rates at which mammalian somatic cell hybrids arise. However, virus treatment seems to be ineffective with some kinds of cells, production of the virus is laborious, the activity of various batches is variable and liable to decay, and the possibility of introducing fragments of viral information into the treated cells is not to be ignored. For all these reasons the search for chemical "fusogens" effective on mammalian cells has gone on for some years (see, for instance, 3–6). The results to date have not warranted dispensing with Sendai virus.

In the novel field of higher plant protoplast fusion (7) polyethylene glycol (PEG) has recently become established as a powerful fusogen (8–10). So far the published reports do not claim more than a few divisions for the products of plant protoplast fusion. Nevertheless, it seemed worthwhile to try this botanical device on mammalian cells where the techniques for determining success or failure are a matter of established routine. The results, summarized here, were quite positive. In a test system well tried before (11, 12), uninucleate mouse × Chinese hamster hybrid cells capable of indefinite multiplication arose after PEG treatment at rates at least as high as those obtained with Sendai virus. Furthermore, PEG treatment has yielded

60

hybrids even in notoriously recalcitrant combinations of cells, such as early passage human skin fibroblasts and human lymphocytes.

MATERIALS AND METHODS

In preliminary tests a number of variations of the PEG treatment have been tried, such as treating both parental cells in suspension, both in mixed monolayer or one as monolayer and the other overlaid on it. The details given below are for the treatment of mixed monolayers of equal numbers of Chinese hamster hypoxanthine guanine phosphoribosyltransferase-deficient (HGPRT$^-$) cells (strain wg, clone 1) and mouse thymidine kinase-deficient (TK$^-$) cells (strain 3T3) (for details see reference 11).

Cells of each strain in equal numbers (10^5–10^6) are inoculated as a mixed culture in a 25-cm^2 flask in Dulbecco's medium with 15% fetal-calf serum (D15) and incubated overnight. 10 g of polyethyleneglycol 6000 (either "pure" Koch-Light; or 6000–7500 mol. wt., BDH), sterilized by autoclaving, are dissolved in 10 ml of Dulbecco's medium *without serum* (D). The mixed culture, washed twice with 5 ml of phosphate buffer saline (with Ca and Mg), is overlaid for 5–15 min at 37°C with 1.5–3 ml of PEG solution. At the end of this time the cells have shrunk somewhat. The film of PEG solution, after draining off as much as possible, is very gradually diluted with D, for instance in five steps at intervals of 5 min, adding 0.3, 0.6, 1.2, 2.4, and 4.8 ml of D. (more rapid dilution kills the cells.) After this, the liquid is removed and replaced with 5 ml D15. At this juncture the morphology of the cells changes dramatically. After 30 minutes D15 medium is renewed. Bi- or multinucleate cells are formed following the introduction of D15, go on increasing in proportion for many hours, and then start to decline. After overnight incubation and, if necessary, trypsinization and plating, D15 supplemented with hypoxanthine, aminopterin, and thymidine (HAT) (13) replaces D15, and is renewed every 2–3 days. By the fifth day small colonies of possible hybrid cells begin to appear among the dying parental cells.

RESULTS

In the three hybridizations carried out between wg-1 and 3T3(TK$^-$) the yields of presumptive hybrid colonies per plated cell (of one parent) were 2.2–9.5 $\times$ 10^{-4} in the PEG-treated series and 2.6–9 $\times$ 10^{-6} in the untreated controls (see Table 1).

Early Passage Human Skin Fibroblasts. Mixed monolayers of early passage cells from a case of Lesch-Nyhan syndrome (LN) and a case of xeroderma pigmentosum (XP), treated as above, gave tetraploid presumed hybrids after successive enrichments by selection with HAT and UV light

Table 1. Presumptive hybrid colonies obtained from fusion of Chinese hamster and mouse cells[a]

	Total cells in mixed monolayers	Total hybrid colonies[b]	Hybrid colonies per 10^6 cells of one parent
Controls	6×10^6	23	7.6
PEG-6000-treated	7.9×10^6	1764	446.6

[a] Mixed confluent monolayers of equal numbers of Chinese hamster wg-1 (HGPRT⁻) and mouse 3T3 (TK⁻) cells were treated as indicated.
[b] The colonies were presumed to be hybrid cell colonies on the basis of (1) chromosome constitution on a sample of colonies, (2) morphology, and (3) ability to grow in HAT.

(30–60 ergs/mm²). With these human fibroblasts the formation of bi- or multinucleate cells is easily followed under phase contrast. Starting from about 1% before treatment it reaches a maximum (up to 42% in the experiments to date) around 40 hours after treatment and then declines over a few days. My colleague, Dr. Peter Riddle, has succeeded in recording by time-lapse microcinematography the actual fusion of cells.

Mouse or Human Fibroblasts and Human Lymphocytes. Monolayers in Falcon flasks of either human fibroblasts LN or XP or of the 3T3(TK⁻) mouse fibroblast line were overlaid for 3–5 min with the PEG solution mentioned above. The solution was then drained and a suspension in 0.3 ml phosphate buffer saline (with Ca and Mg) of washed column-purified peripheral blood lymphocytes (in some cases PHA-stimulated) was overlaid on the monolayers (10:1). Very effective agglutination occurred after a few minutes of gentle rocking. Gradual dilution with D and finally replacement with D15 led to the usual massive formation of multinucleate cells. One or more lymphocyte nuclei (?) appeared to have been taken up in a proportion of fibroblasts. Gradual enrichment by HAT or UV light, according to the kind of fibroblast used, led to cultures resistant to the selective agent. Further characterization is in progress.

ACKNOWLEDGMENTS

I am indebted to Professors E.C. Cocking and J.A. Lucy for valuable suggestions and stimulating discussions and for informing me of the success in their laboratories (14) in fusing yeast protoplasts and hen erythrocytes. To Dr. O.L. Gamborg, whose laboratory has done most of the work on protoplast fusion by PEG, I am indebted for reprints and preprints.

LITERATURE CITED

1. Harris, H., and Watkins, J.F. (1965). *Nature (London)* **205**:640–646.
2. Yerganian, G., and Nell, M.B. (1966). *Proc. Natl. Acad. Sci. U.S.A.* **55**:1066–1073.
3. Lucy, J.A., Ahkong, Q.F., Cramp, F.C., Fisher, D., and Howell, J.I. (1971). *Biochem. J.* **124**:46–47.

4. Croce, C.M., Koprowski, H., and Eagle, H. (1972). *Proc. Natl. Acad. Sci. U.S.A.* **69**:1953–56.
5. Ahkong, Q.F., Cramp, F.C., Fisher, D., Howell, J.I., and Lucy, J.A. (1972). *J. Cell Sci.* **10**:769–787.
6. Cramp, F.C., and Lucy, J.A. (1974). *Exp. Cell Res.* **87**:107–110.
7. Cocking, E.C. (1973). *Colloq. Int. C.N.R.S.* **212**:327–342.
8. Kao, K.N., and Michayluk, M.R. (1974). *Planta (Berlin)* **115**:355–367.
9. Constabel, F., and Kao, K.N. (1974). *Can. J. Bot.* **52**:1603–1606.
10. Bonnett, H.T., and Eriksson, T. (1974). *Planta (Berlin)* **120**:71–79.
11. Pontecorvo, G. (1971). *Nature (London)* **230**:367–369.
12. Pontecorvo, G. (1974). In Davidson, R.L., and de la Cruz, F. (eds.), *Somatic Cell Hybridization,* New York, pp. 65–69.
13. Littlefield, J. (1964). *Science* **145**:709–710.
14. Ahkong, Q.F., Howell, J.I., Lucy, J.A., Safwat, F., Davey, M.R., and Cocking, E.C. (1975). *Nature (London)* **255**:66–67.

Editor's Comments
on Papers 9 Through 14

9   DAVIDSON, EPHRUSSI, and YAMAMOTO
*Regulation of Pigment Synthesis in Mammalian Cells, As
Studied by Somatic Hybridization*

10   DAVIDSON
*Regulation of Melanin Synthesis in Mammalian Cells: Effect
of Gene Dosage on the Expression of Differentiation*

11   PETERSON and WEISS
*Expression of Differentiated Functions in Hepatoma Cell
Hybrids: Induction of Mouse Albumin Production in Rat
Hepatoma-Mouse Fibroblast Hybrids*

12   FINCH and EPHRUSSI
*Retention of Multiple Developmental Potentialities by Cells
of a Mouse Testicular Teratocarcinoma During Prolonged
Culture In Vitro and Their Extinction Upon Hybridization
with Cells of Permanent Lines*

13   KÖHLER and MILSTEIN
*Continuous Cultures of Fused Cells Secreting Antibody of
Predefined Specificity*

14   JOHNSON and RAO
*Mammalian Cell Fusion: Induction of Premature Chromosome
Condensation in Interphase Nuclei*

REGULATION IN SOMATIC CELL HYBRIDS

The ability to recombine genomes was one of the most important
features of studies on genetic regulatory mechanisms in microorganisms. The techniques of somatic cell hybridization have provided the opportunity for a comparable approach to the study of
genetic regulation in mammalian cells. Early studies on gene expression in hybrid cells focused on the expression of functions common to
both parental cells. Mouse cells of two lines that expressed H-2
histocompatibility antigens of different types were hybridized with

each other, and the hybrids were found to express the H-2 antigens characteristic of both parental cells (Gershon and Sachs, 1963; Spencer et al., 1964). Similarly, hybrids between mouse cells that synthesized different forms of the enzyme β-glucuronidase produced both parental forms of the enzyme (Ganshow, 1966), and hybrids between rat and mouse cells produced both parental forms of LDH (Weiss and Ephrussi, 1966). These studies demonstrated that both parental genomes are active in hybrid cells and that activities common to both parental cells prior to fusion continue to be expressed in the hybrids.

In order to investigate genetic regulatory mechanisms in hybrid cells, there obviously must be phenotypic differences between the parental cells. In the studies on microorganisms, such phenotypic differences generally were the result of mutation. In somatic cell hybridization experiments a variety of mutant mammalian cells also have been analyzed. As discussed earlier in this volume, the first studies on the biochemical selection of somatic cell hybrids involved the fusion of biochemically marked cells, each lacking a different enzyme as the result of mutation (see Paper 3). Since both enzymes were produced in the hybrids, the results provided no evidence for regulatory phenomena. This has been the general pattern in studies of this type, and the hybridization of mutant cells, therefore, has not contributed to the elucidation of mammalian regulatory mechanisms. In contrast, much information on regulation has come from the hybridization of cells in which phenotypic differences were the result of differentiation, not the result of mutation.

The first studies on the regulation of tissue-specific functions in somatic cell hybrids were reported by Davidson, Ephrussi, and Yamamoto in 1966 (Paper 9). These studies demonstrated the viability of hybrids between cells of different tissues, and they established the framework for many subsequent experiments on gene regulation in hybrid cells. In the initial experiments, pigmented Syrian hamster melanoma cells were hybridized with unpigmented mouse fibro-blasts, and the expression of the differentiated function was analyzed in the hybrids. One of the basic assumptions of these experiments (and of subsequent hybridization experiments involving other tissue-specific functions) was that the genes necessary to carry out the differentiated function were present but not expressed in the parental fibroblasts as the result of regulatory activities. It was also assumed that the results, obtained through the use of clearly abnormal (i.e., malignant) cells, could provide information on regulatory mechanisms in normal cells.

A large number of hybrids between pigmented and unpigmented cells were isolated, and they all were unpigmented. After a variety of possible artifacts (such as the nonspecific repression of the entire melanoma genome in the hybrids) were ruled out (reviewed by Davidson, 1974), the results were interpreted as evidence that the genome of the fibroblast produces a diffusible regulator substance that acts to block the expression of the pigment-synthesizing genes of the melanoma cell. Although no information on the specific mechanism of regulation was obtained in these experiments, the results provided the first evidence to suggest the existence of diffusible, trans-acting regulatory factors in mammalian cells. The extinction of pigmentation in hybrids between pigmented and unpigmented cells soon was confirmed in a cross between mouse melanoma cells and mouse fibroblasts (Silagi, 1967).

Quantitative aspects of the regulation of pigment synthesis were studied in gene dosage experiments carried out by Davidson (Paper 10) and by Ephrussi and co-workers (Fougere et al., 1972). In these experiments, hybrids containing two melanoma genomes and one fibroblast genome were isolated and characterized. (Gene dosage experiments of this type were first carried out by Davidson and Benda [1970], using glial cells, and the results, although not definitive, raised the possibility of gene dosage effects on differentiation.) In contrast to the previously isolated hybrids, which contained one melanoma and one fibroblast genome and were uniformly unpigmented, at least 50% of the hybrids containing two melanoma genomes and one fibroblast genome were heavily pigmented. Although these results could be interpreted in terms of either negatively or positively acting regulatory factors, they provided evidence for a quantitative aspect to the regulation of differentiation. The results also suggested that the factors regulating differentiation (whatever their nature) are not produced in large excess, since only a twofold change in the ratio of melanoma to fibroblast genomes was required to alter the expression of the differentiated function in hybrid cells. The gene dosage effect observed in these experiments has been seen to play a role in the expression of a variety of other differentiated functions in hybrid cells.

The expression of liver-specific functions has been extensively studied by means of cell hybridization, and some liver-specific functions appeared to be regulated in hybrids in a manner similar to pigment synthesis, as discussed previously. However, Peterson and Weiss (Paper 11) showed that one liver-specific function, albumin synthesis, is expressed in hybrid cells in a very different manner. In hybrids containing one rat hepatoma genome and one mouse fibroblast genome, albumin production continued. Immunological tests showed that the albumin produced in the hybrids was one of the rat type, corresponding to the species of the differentiated parental cell. Of greater significance, however, was the observation that some hybrids containing two hepatoma genomes plus one fibroblast genome could also synthesize mouse albumin, corresponding to the species of the undifferentiated parental cell. The expression of mouse albumin in some of the hybrids demonstrated that the albumin gene was present in the parental fibroblasts, in which the gene was not expressed as the result of differentiation. Furthermore, the albumin gene in the parental fibroblasts was maintained in such a state that it could be activated in hybrids under the influence of an appropriate balance of hepatoma genomes. These results also provided additional evidence for a gene dosage effect on differentiation.

Teratocarcinoma cells have been used in hybridization experiments in attempts to examine the potential of cells to differentiate, distinct from the expression of differentiation itself. The first studies of this type were carried out by Finch and Ephrussi (Paper 12). Teratocarcinoma cells have the potential to differentiate into many different tissue types when implanted into compatible host animals, and the initial phase of this study involved the establishment of teratocarcinoma cell culture lines that retained this multipotential character. The lines developed in this study provided the first demonstration that teratocarcinoma cells could retain their multipotential capacity through passage in culture. The multipotential mouse teratocarcinoma cells were hybridized with mouse fibroblasts. When injected into compatible hosts, the hybrids formed tumors, but these tumors lacked all the differentiated characteristics of the parental teratocarcinoma. It cannot be said whether these results are due to an effect of the fibroblast genome on the expression of differentiation itself or on the potential of the teratocarcinoma genome to differentiate. In recent years, the use of teratocarcinoma cells to study problems of development has expanded greatly, and a very different approach involving teratocarcinoma cells is presented later in this book (see Paper 31).

Many other specialized cell types have been used in cell hybridization experiments. For example, the expression of neuronal properties has been analyzed in hybrids between neuroblastoma cells and fibroblasts (Minna et al., 1971). In general, these experiments have given results consistent with the principles just discussed. Studies on the hybridization of immunologically competent cells, however, have provided results of special significance. Although the initial experiments in this area followed the pattern of the crosses just discussed, a new direction was established when immunologically competent cells were hybridized with each other. In one set of experiments, immunoglobulin-producing rat and mouse myeloma cells were hybridized with each other, and the hybrids were found to produce immunoglobulins of both parental types (Cotton and Milstein, 1973). Immunoglobulin-producing mouse myeloma cells also were hybridized with human peripheral blood lymphocytes that did not produce detectable immunoglobulin. In the hybrids, immunoglobulin production was observed, not only from the originally active mouse genome but also from the originally inactive human genome (Schwaber and Cohen, 1973).

Soon after these experiments, Köhler. and Milstein (Paper 13) isolated a new type of hybrid cell, capable of producing specific antibodies. In these studies, immunoglobulin-producing mouse myeloma cells were hybridized with spleen cells from mice previously immunized against sheep red blood cells (SRBC). A large number of hybrids were isolated and tested for antibody production, and a small percentage of the hybrids were found to produce an anti-SRBC antibody. Thus, by fusing immunoglobin-producing myeloma cells with spleen cells from a preimmunized donor, it was possible to obtain hybrids that synthesized antibodies against a predetermined antigen. Such hybrids are now commonly referred to as *hybridomas*. Since these initial experiments, a large number of hybridoma lines, which produce monoclonal antibodies against a wide variety of antigens, have been isolated. The ability to establish continuous cell culture lines producing antibodies of predefined specificity appears to have great potential for both the scientific and medical application of immunological reagents.

All of the studies described in this section involved mononucleate, actively proliferating hybrid cells. As described earlier, cell fusion can also lead to the formation of multinucleate heterokaryons, whose properties are markedly different from those of hybrid cells (see Paper 7). In contrast to hybrids, multinucleate heterokaryons are unable to replicate and they survive for relatively short times. However, heterokaryons are very active metabolically, and they have proven to be useful for studies on the localization of biochemical activities to individual nuclei. Harris and co-workers have examined interactions between nuclei in Sendai virus-induced heterokaryons (reviewed by Harris, 1970). For example, in heterokaryons between Hela cells, which synthesize both DNA and RNA, and chicken erythrocytes, which synthesize neither, the nuclei of both Hela cells and erythrocytes were seen to synthesize DNA and RNA. The significance of the activation of the erythrocyte nucleus in heterokaryons remains unclear to this day. For a detailed presentation on heterokaryons, the reader may refer to Harris (1970), Ephrussi (1972), Davidson and de la Cruz (1974), and Ringertz and Savage (1976).

Information on the regulation of the division cycle in mammalian cells has been provided by experiments carried out by Johnson and

Rao (Paper 14). In these experiments, cells in different stages of the cell cycle were fused in various combinations, and DNA synthesis and other aspects of the division cycle were monitored in the resulting multinucleate cells. When S and G2 phase cells were fused with each other, the initiation of mitosis in the G2 phase nuclei was seen to cause *premature chromosome condensation* (PCC) in the S phase nuclei. The results of these studies not only provided information on cell cycle regulation—the phenomenon of PCC also provided a new approach to a variety of genetic problems, such as chromosome identification and chromosomal organization.

REFERENCES

Cotton, R. G., and C. Milstein, 1973, Fusion of Two Immunoglobulin-Producing Myeloma Cells, *Nature* **244**:42–43.

Davidson, R. L., 1974, Gene Expression in Somatic Cell Hybrids, *Ann. Rev. Genet.* **8**:195–218.

Davidson, R. L., and P. Benda, 1970, Regulation of Specific Functions of Glial Cells in Somatic Hybrids. II. Control of Inducibility of Glycerol Phosphate Dehydrogenase, *Natl. Acad. Sci. (USA) Proc.* **67**:1870–1877.

Davidson, R. L., and F. de la Cruz, eds., 1974, *Somatic Cell Hybridization,* Raven Press, New York, 295p.

Ephrussi, B., 1972, *Hybridization of Somatic Cells,* Princeton University Press, Princeton, N.J., 175p.

Fougere, C., F. Ruiz, and B. Ephrussi, 1972, Gene Dosage Dependence of Pigment Synthesis in Melanoma X Fibroblast Hybrids, *Natl. Acad. Sci. (USA) Proc.* **69**:330–334.

Ganshow, R., 1966, Glucuronidase Gene Expression in Somatic Hybrids, *Science* **153**:84–85.

Gershon, D., and L. Sachs, 1963, Properties of a Somatic Hybrid Between Mouse Cells with Different Genotypes, *Nature* **198**:912–913.

Harris, H., 1970, *Cell Fusion,* Harvard University Press, Cambridge, Mass., 108p.

Minna, J., P. Nelson, J. Peacock, D. Glazer, and M. Nirenberg, 1971, Genes for Neuronal Properties Expressed in Neuroblastoma X L Cell Hybrids, *Natl. Acad. Sci. (USA) Proc.* **68**:234–239.

Ringertz, N. R., and R. E. Savage, 1976, *Cell Hybrids,* Academic Press, New York, 366p.

Schwaber, J., and E. P. Cohen, 1973, Human X Mouse Somatic Cell Hybrid Clone Secreting Immunoglobulins of Both Parental Types, *Nature* **244**:444–447.

Silagi, S., 1967, Hybridization of a Malignant Melanoma Cell Line with L Cells In Vitro, *Cancer Res.* **27**:1953–1960.

Spencer, R. A., T. S. Hauschka, D. B. Amos, and B. Ephrussi, 1964, Codominance of Isoantigens in Somatic Hybrids of Murine Cells Grown In Vitro, *J. Natl. Cancer Inst.* **33**:893–903.

Weiss, M. C., and B. Ephrussi, 1966, Studies of Interspecific (Rat X Mouse) Somatic Hybrids. II. Lactate Dehydrogenase and Beta-Glucuronidase, *Genetics* **54**:1111–1122.

9

Reprinted from *Natl. Acad. Sci. (USA) Proc.* **56**:1437–1440 (1966)

*REGULATION OF PIGMENT SYNTHESIS IN MAMMALIAN CELLS, AS STUDIED BY SOMATIC HYBRIDIZATION**

By Richard L. Davidson,[†] Boris Ephrussi,[‡] and Kohtaro Yamamoto

One of the main problems concerning differentiation in higher organisms is that of its genetic control. Whether positive or negative control mechanisms operate in differentiation, or indeed whether the problem can be stated in these terms borrowed from bacterial genetics, is still unknown. One of the possible approaches to these questions is to determine the effects of combining, through somatic hybridization, the genomes of two cells differing in at least one specific function. An earlier series of experiments along this line,[1] to be discussed below, gave no evidence of interaction between the genomes in hybrid cells.[2] In this communication we report the results of similarly designed experiments, involving melanin synthesis as the specialized function.

The pigment-producing cells used in all the experiments are from a subline of the Syrian hamster melanoma RPMI 3460, described by Moore,[4] into which we have introduced an 8-azaguanine resistance marker. One pigmented clone (3460-3) has been used in most of the experiments. To determine the frequency of amelanotic cells in 3460-3, we have subcloned this population on a large scale. Among the *ca.* 15,000 colonies formed (with a cloning efficiency of nearly 90%), none were unpigmented.

The melanoma cells were hybridized with each of three unpigmented mouse permanent cell lines, each resistant to 5-bromodeoxyuridine. These lines include B 82 and LM(TK$^-$) clone 1D, derived from two different sublines of L cells, and two clones of N 2, derived from NCTC 2555.[5] The karyotypes of the mouse and hamster lines are very different, making the karyological identification of the hybrids very easy. The hamster cells have a modal number of 51 chromosomes, the majority (43) of which are biarmed. All the mouse cells used have modal chromosome numbers of approximately 50, but most of their chromosomes (30–46, depending on the particular cell line) are telocentric.

The method of isolation of the hybrids is based upon the system developed by Littlefield[6] for selecting hybrids between two biochemically marked L cells. Littlefield has shown that neither an 8-azaguanine resistant cell which lacks inosinic acid pyrophosphorylase, nor a 5-bromodeoxyuridine resistant cell which lacks thymidine kinase can grow in a medium containing 1×10^{-4} M hypoxanthine, 4×10^{-7} M aminopterin, and 1.6×10^{-5} M thymidine. The hybrids inherit the synthetic abilities of both parents, and hence can grow in such a selective medium. The same medium has already been used in this laboratory to obtain viable somatic rat $\times$ mouse[7] and hamster $\times$ mouse[8] hybrids. None of the parental cell lines used in the present experiments grow in the selective medium.

For the mating experiments, 10^6 cells of one of the mouse lines were mixed with 400–20,000 melanoma cells in 6-cm Falcon plastic tissue-culture dishes containing Dulbecco's modification of Eagle's medium supplemented with 10 per cent calf serum. One day later, the medium was replaced by the above-mentioned selective medium. After about 2 weeks at 36°C with periodic feeding of cultures, large colonies of cells, morphologically different from the parental cells, were seen. Karyological analysis proved that these cells were hybrids, containing *ca.* 100 chromosomes, with the proportion of metacentric and telocentric chromosomes expected for each particular combination of melanoma and mouse cells. In a manner analogous to the Luria-Delbrück method of determining mutation rates in bacteria,[9] the following "mating rates" were obtained from the per cent of "mating dishes" inoculated with a given number of melanoma cells which did not have any hybrid colonies: one out of every 1000 melanoma cells mates with N2, one out of every 4000 with LM(TK⁻) clone 1D, and one out of every 9000 with B82. Several colonies of hybrid cells were isolated and subcultured, and at present have undergone up to 100 cell generations without any decline in growth rate.

As indicated earlier, our purpose in producing these hybrids was to establish whether they would express the specialized cell function exhibited by one of the parents. Thus far, under conditions under which melanoma cells become heavily pigmented, the hybrid cells have remained unpigmented.

In an attempt to locate the block to pigment formation, we have determined the dopa oxidase activity of the parental and hybrid cells growing in the maintenance medium. The cells were collected by trypsinization and homogenized in 0.1 M phosphate buffer, pH 6.8, containing 0.2 per cent sodium deoxycholate. For the assay, L dopa to a final concentration of 2 mg/ml was added to equal amounts of protein of each of the homogenates. The optical density was determined over a period of 3 hr at 37°C with a Klett-Summerson photoelectric colorimeter, using a blue (no. 42) filter. The results of one of the experiments in which the activities of the parental cells and of one clone of each of the hybrids were determined are given in Table 1. It can be seen that the melanoma cells have high dopa oxidase activity, whereas all the other lines tested have no such activity. (In fact, all the latter cell lines give values less than that obtained for dopa autoxidation, accounting for the negative numbers in the table. This is due to the presence in the cells of a weak inhibitor, which has negligible effect when added to a melanoma extract.)

The absence of pigment and dopa oxidase in the hybrid cells suggested that the genome of the unpigmented parent represses the expression of the melanoma genes involved in pigment synthesis. However, before this interpretation could be accepted, several alternative explanations had to be considered. First, there was the possibility that *all* hamster genes are repressed in the hybrids. This interpretation was disproved by two facts: (*a*) the production by the hybrid cells

TABLE 1

Dopa Oxidase Activity of Parental and Hybrid Lines

Cell line	Activity*
3460-3	476
B 82	−18
LM(TK⁻) clone 1D	−20
N 2	−21
3460-3/B 82	−21
3460-3/LM(TK⁻) clone 1D	−18
3460-3/N 2	−22

* Activity is expressed as the change in optical density produced by 500 μg of protein during the first 3 hr of the assay. All values are corrected for dopa autoxidation (32 in this experiment).

of both hamster and mouse subunits of lactate dehydrogenase, as demonstrated by starch gel electrophoresis; (*b*) the ability of the hybrid cells to grow in the selective medium inhibiting the growth of the parental cells, which indicates that the hamster and mouse genomes complement each other in the hybrids. (This complementation necessarily involves the production of thymidine kinase by the hamster genes, since the mouse cells (B 82,[10] LM(TK$^-$) clone 1D,[11] and, presumably, N 2) lack this enzyme.) These results are in agreement with the demonstration by Weiss and Ephrussi[12] of the continued activity in somatic rat × mouse hybrids of the genes of both parents specifying lactate dehydrogenase and β-glucuronidase. Second, the possibility had to be considered that the hybrids resulted from fusion of mouse cells with amelanotic hamster cells. This was made very unlikely by the observations mentioned above that one out of every 1000–9000 melanoma cells mates with a mouse cell, while less than one out of 15,000 melanoma cells is amelanotic. Lastly, the possibility that the hybrids had lost their pigment-forming genes appeared improbable, since the decrease in the chromosome number of the hybrids did not exceed 10 per cent. With such a small (and presumably random) loss of chromosomes, the probability of all hybrids losing the same genes is very low. On the basis of these considerations, we are inclined to conclude that a step in the process of pigment formation in the cells studied is under negative control.

This conclusion, if correct, raises several questions: (*a*) Is the repression of the pigment-forming process in the hybrids dependent upon the continued presence of certain genes of the unpigmented mouse cells? (*b*) At what point in this process is the negative control exerted? (*c*) Are other processes associated with the differentiated state of pigment-producing cells affected by hybridization in a similar manner? (*d*) Can our conclusion be extrapolated to the regulation of pigment formation in normal melanocytes? (*e*) Can it be extended to other types of differentiation?

While answers to some of these questions are expected in the near future, the following remarks are appropriate at this time. It should be recalled that two differentiated functions characteristic of another cell type have already been examined in experiments designed similarly to the above.[1] Hybrids between cells of mouse fibroblast lines differing in the rates of collagen and hyaluronic acid synthesis were isolated. The rates of production of these substances by the hybrid cells were found to be intermediate between those of the parents. The difference between these results and those reported in the present paper may indicate the operation in differentiation of a variety of control mechanisms. It may, however, be due as well to differences in experimental design, such as the use of interspecific hybrids in one case and of intraspecific hybrids in the other. Finally, in assessing the relevance of the results of the two series of experiments to differentiation of normal cells, it must be kept in mind that the parents of both sets of hybrids are cells of heteroploid, "established" lines. Therefore, it is possible that neither of the results described above reflects the regulation of differentiation as it occurs in normal cells.

Summary.—Hybrids between cells of a Syrian hamster melanoma line and each of three unpigmented mouse lines have been isolated and maintained in active proliferation *in vitro* for up to 100 cell generations. These hybrids have thus far remained unpigmented under conditions under which the melanoma cells become heavily pigmented, and they exhibit no dopa oxidase activity. It is concluded that a step in the process of pigment formation in the cells studied is under negative control.

We wish to thank Drs. Leland Hartwell, Saul Kit, and John Littlefield for kindly supplying us with cultures of RPMI 3460, LM(TK$^-$) clone 1D, and B 82, respectively, and to acknowledge the able technical assistance of Miss Linda Vikis.

* This work was supported by grant G-23129 of the National Science Foundation.

† Predoctoral fellow of the U.S. Public Health Service.

‡ On leave from the University of Paris.

[1] Green, H., B. Ephrussi, M. Yoshida, and D. Hamerman, these PROCEEDINGS, **55,** 41 (1966).

[2] Using UV-inactivated Sendai virus, Harris *et al.*[3] have produced heterokaryons between cells whose nuclei differ in DNA and/or RNA synthesis. Unfortunately, the interactions observed cannot be interpreted in terms of genetic regulation, since they were recorded before dilution of factors present in the cells at the time of fusion could have taken place.

[3] Harris, H., J. F. Watkins, C. E. Ford, and G. I. Schoefl, *J. Cell Sci.,* 1, 1 (1966).

[4] Moore, G. E., *Exptl. Cell Res.,* **36,** 422 (1964).

[5] Sanford, K. K., G. D. Likely, and W. R. Earle, *J. Natl. Cancer Inst.,* **15,** 215 (1954).

[6] Littlefield, J. W., *Science,* **145,** 709 (1964).

[7] Ephrussi, B., and M. C. Weiss, these PROCEEDINGS, **53,** 1040 (1965).

[8] Scaletta, L. J., N. Rushforth, and B. Ephrussi, in preparation.

[9] Luria, S. E., and M. Delbrück, *Genetics,* **28,** 491 (1943).

[10] Littlefield, J. W., *Exptl. Cell Res.,* **41,** 190 (1966).

[11] Kit, S., D. R. Dubbs, L. J. Piekarski, and T. C. Hsu, *Exptl. Cell Res.,* **31,** 297 (1963).

[12] Weiss, M. C., and B. Ephrussi, *Genetics,* in press.

Reprinted from *Natl. Acad. Sci. (USA) Proc.* **69**:951–955 (1972)

REGULATION OF MELANIN SYNTHESIS IN MAMMALIAN CELLS: EFFECT ON GENE DOSAGE ON THE EXPRESSION OF DIFFERENTIATION*

Richard L. Davidson

ABSTRACT Near-tetraploid (2S), pigmented, Syrian hamster cells (3460²) were hybridized with unpigmented mouse fibroblasts to produce hybrids that contained two pigment cell genomes and one fibroblast genome per cell (3²D). Half of the hybrids were pigmented and showed high activity of the enzyme dihydroxyphenylalanine oxidase; the other half of the hybrids were unpigmented and devoid of the activity of this enzyme. (In contrast, all the previously isolated hybrids containing only one pigment cell genome and one fibroblast genome were unpigmented.) Analyses of the chromosome complement and of a series of enzymes unrelated to melanin synthesis did not reveal any great differences between the pigmented and unpigmented 3²D hybrids. Upon the isolation of subclones, unpigmented 3²D hybrids produced only unpigmented colonies, whereas pigmented hybrids produced both pigmented and unpigmented colonies. Although the mechanism that results in the formation of some pigmented and some unpigmented hybrids in these experiments is not known, it is clear that the number of pigment cell genomes in a hybrid affects the expression of pigmentation.

It was shown (1–3) that hybrids between pigmented Syrian hamster melanoma cells and unpigmented mouse fibroblasts are unpigmented and lack the enzyme dihydroxyphenylalanine (DOPA) oxidase (EC 1.10.3.1), which is required for melanin synthesis. These results were interpreted as indicating the existence of a system that regulates pigment synthesis by blocking the expression of the pigment-forming genes, possibly by preventing the synthesis of DOPA oxidase. Hybrids were also isolated (4, 5) that contained one glial cell genome and one fibroblast genome, comparable to the hybrids mentioned above that contained one melanoma cell and one fibroblast genome; in addition, hybrids with two glial cell genomes and one fibroblast genome were also isolated. The study of one of the functions characteristic of the glial cells, i.e., the inducibility of the enzyme glycerol phosphate dehydrogenase (EC 1.1.1.8), suggested that hybrids with a 2:1 glial cell–fibroblast genome ratio express a differentiated function that hybrids with a 1:1 ratio do not express (4). However, the inducibility of the enzyme in the two types of hybrids was not different enough to conclusively prove an effect of genome ratios. The present ex-

periments, in which hybrids containing two pigment-cell genomes and one fibroblast genome were isolated, were therefore undertaken to determine whether the number of differentiated cell genomes in a hybrid affects the expression of the differentiated functions.

MATERIALS AND METHODS

In order to produce the desired hybrids, a pigmented near-tetraploid (2S) line of Syrian hamster melanoma cells was used. New lines of pigmented cells had been established by Dr. J. Eisenstadt in the laboratory of Dr. B. Ephrussi in an attempt to isolate new drug-resistant mutants from the pigmented Syrian hamster melanoma 3460-3, which is resistant to 3 μg/ml of azaguanine (1). I demonstrated karyologically that one of these lines was composed of 2S cells, containing about twice the number of chromosomes as did the original cell line and isolated several pigmented 2S clones that still retained their azaguanine resistance. One of these clones, called 3460-3² clone 2, was used in the present experiments. (This line will be referred to hereafter as 3460².)

The 3460² cells were hybridized with cells of the unpigmented line of mouse fibroblasts LM(TK-) clone 1D, resistant to 30 μg/ml of 5 bromodeoxyuridine. (This line will be referred to hereafter as clone 1D.)

In order to increase the frequency of cell fusion, β-propiolactone-inactivated (6) Sendai virus (7), was used according to a method of cell fusion in monolayers described elsewhere (8). In one experiment, two 35-mm (Falcon) plastic tissue-culture dishes were each inoculated with 3×10^6 clone 1D and 5×10^4 3460² cells in 2 ml of growth medium (Dulbecco's modified Eagle's medium, plus 10% fetal-calf serum). A high fibroblast to melanoma cell ratio was used to increase the frequency at which the melanoma cells fuse with fibroblasts (9). After 24 hr, the cells were treated (8) with 1000 hemagglutinating units of inactivated Sendai virus. Immediately after the virus treatment, the cells were exposed to HAT selective medium (growth medium plus 0.1 mM hypoxanthine, 0.4 μM aminopterin, and 16 μM thymidine), in which neither parental cell can grow because of the enzyme deficiencies associated with their drug resistance (10). The hybrid cells, however, are able to grow in HAT medium. 24 hr after the virus treatment, the cells from the two dishes were trypsinized and mixed, and twenty 60-mm dishes were inoculated with about 3×10^5 fibroblasts and 5×10^3 melanoma cells per dish in selective medium. After 2 weeks, large colonies of hybrid

---

* This is paper No. IV in the series, "Regulation of Melanin Synthesis in Mammalian Cells as Studied by Somatic Hybridization."

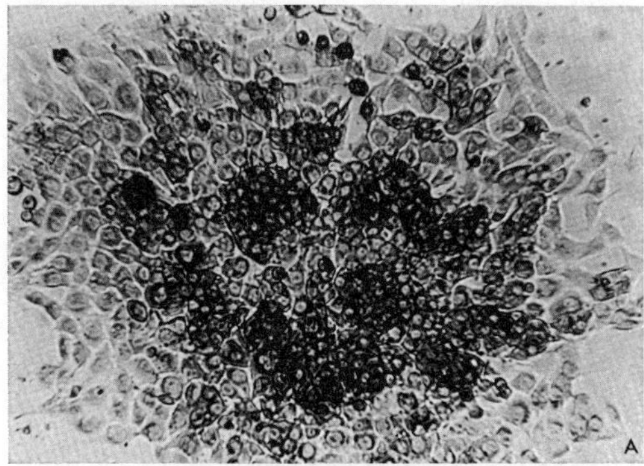

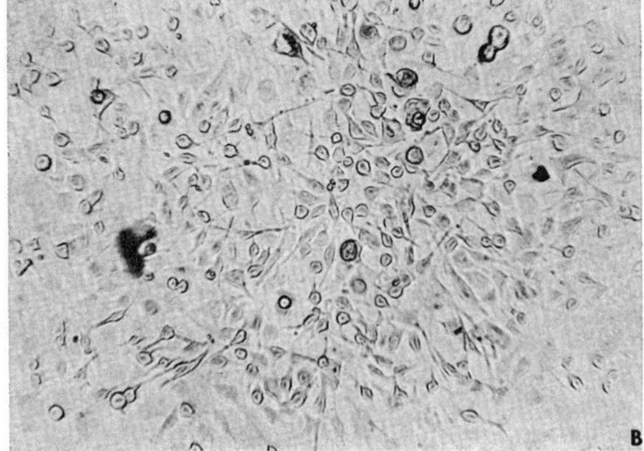

FIG. 1. Photomicrographs of living cells of 3²D hybrid colonies. Bright-field illumination, ×400. (*A*) Pigmented hybrid; (*B*) unpigmented hybrid.

cells were observed; these were subcultured in growth medium and analyzed karyologically and enzymatically. (These hybrids will be referred to below as 3²D.) All of the hybrids to be described in this paper were derived from one experiment. Similar hybrids, however, were produced in repeat experiments.

For the DOPA oxidase determinations, cell extracts were prepared and assayed as described (3). Several 100-mm tissue-culture dishes were inoculated with 10^6 cells in growth medium. The cells were harvested by trypsinization after 4 days, rinsed with 0.1 M sodium phosphate buffer (pH 6.8), and homogenized. The extracts were centrifuged at 2000 × *g* for 5 min, and the supernates were assayed. The reaction mixture contained 1 mg/ml of L-DOPA, and cell extract 500–1000 μg of protein/ml [as determined with the Folin phenol reagent (11)] in 0.1 M sodium phosphate buffer (pH 6.8). The oxidation of DOPA was followed as the increase in absorbance at 475 nm for 15 min at 37°, with a Gilford spectrophotometer. There was always a slight lag before the oxidation of DOPA began, and the activity was determined as the initial rate of increase of the absorbance after the lag. Specific activity is expressed as the increase in absorbance per mg of protein per min.

Several other enzymes not related to melanin synthesis were also analyzed in the hybrid cells. The parental cells and hybrids were harvested by trypsinization from dense cultures,

rinsed with phosphate-buffered saline (pH 7.2), and frozen at −80°. The frozen cells were sent to Dr. Frank Ruddle (Yale University, Department of Biology) who kindly performed (12) the isozyme analyses by electrophoresis.

RESULTS

About 2 weeks after the 3460² × clone 1D mixtures were treated with virus and subcultured in selective medium, large colonies of two types were observed, one containing heavily pigmented cells and the other containing unpigmented cells (Fig. 1). Pigmented colonies were present in more than half of the dishes; therefore, they occurred with a frequency of about one pigmented colony for every 5,000–10,000 melanoma cells. Unpigmented colonies occurred with about the same frequency as pigmented colonies. All of the 3²D hybrid populations described below were derived from individual colonies, except for 3²D 7 and 3²D 8, which are mixtures of two and three pigmented colonies, respectively.

Hybrid cells can be easily recognized karyologically in this cross, because the karyotypes of the mouse and hamster cells are very different. About 80% of the clone 1D chromosomes are telocentric, whereas about 80% of the 3460² chromosomes are biarmed (Table 1). The 3460-3 line from which 3460² was derived has about 50 chromosomes, 42 of which are biarmed. The 3460² cells, with about 90 chromosomes, of which 73 are biarmed, contain, therefore, almost two complete genomes, with about ten fewer biarmed chromosomes than expected. Karyological analyses revealed that all the pigmented and unpigmented colonies that had developed in the selective medium were composed of hybrid cells (Table 1). At the time of the analyses, all the hybrids were about 25 generations old. It can be seen that the hybrids have only 80–85% of the chromosome complement expected on the basis of the addition of the 3460² and clone 1D karyotypes, with 15–20% fewer biarmed chromosomes and 20–25% fewer telocentric chromosomes than expected. (In contrast, the previously described hybrids with only one melanoma genome and one fibroblast genome contained more than 95% of the expected chromosomes.) The unpigmented hybrids have, on the average, 2–3 fewer chromosomes in each class than have the pigmented hybrids, but this difference may not be significant. The hamster and mouse chromosomes within a given class are not readily distinguishable, so it cannot be said whether the loss of chromosomes within a given class is random for hamster and mouse, or whether there is some preferential loss in certain of the hybrids.

TABLE 1. *Karyological analyses of parental and hybrid cells*

Cell line	Number of chromosomes		
	Biarmed	Telocentric	Total
3460²	73* (71–76)†	18 (18–20)	91 (89–93)
Clone 1D	9 (8–11)	43 (38–46)	52 (47–54)
Expected hybrid	82 (79–87)	61 (56–66)	143 (136–147)
3²D 3 (pigmented)	70 (65–73)	52 (43–65)	124 (116–128)
3²D 8 (pigmented)	70 (59–81)	44 (31–55)	116 (93–128)
3²D 10 (unpigmented)	70 (59–77)	48 (41–57)	119 (112–128)
3²D 11 (unpigmented)	65 (60–70)	42 (39–46)	108 (105–111)

* Average chromosome number. Ten metaphases were analyzed for each culture.

† Range.

It was shown (2) that the specific activity of DOPA oxidase in hamster melanoma cells in culture varies cyclically during the growth of a culture; the activity is low just after inoculation, increases rapidly during the early part of the logarithmic phase of growth, and then declines sharply as the culture becomes dense. Therefore, the enzyme activity of 3460² cells was determined during growth in order to test whether these cells show a similar fluctuation. The 3460² cells were harvested at 2, 4, and 6 days after 100-mm dishes were inoculated with 10^6 cells. With this inoculum, the cultures become very dense and are ready for subculture after 7 days. The results of two experiments are shown in Table 2. It is clear that the specific activity of 3460² cells varies like that of 3460-3. The activity decreases by 80% as the cultures become dense, then increases again after trypsinization and subculturing. Similar experiments with one of the pigmented hybrid populations, 3²D 8, revealed that there is also a marked drop in DOPA oxidase activity in hybrids in dense cultures.

Because of this decrease in enzyme activity in dense cultures of both pigmented melanoma cells and hybrids, all determinations of DOPA oxidase activity in hybrids were performed with cells harvested 4 days after inoculation, at which time the melanoma cells have their maximum activity. Both pigmented and unpigmented hybrids were tested for DOPA oxidase activity (Table 3). The two pigmented hybrids, like the melanoma cells, have high specific activities. The pigmented hybrids have at least as much enzyme activity per cell as does the 3460² parent. The 3²D 8 hybrid has consistently been found to have 3- to 4-times more activity per cell than does 3460². Enzyme activity cannot be detected in the unpigmented hybrids and fibroblasts.

Two pigmented hybrids (3²D 3 and 3²D 8), and two unpigmented hybrids (3²D 10 and 3³D 11), as well as 3460² and clone 1D, were also analyzed for a large series of other enzymes unrelated to melanin synthesis, in order to see whether there was a correlation between pigment synthesis and the presence or absence of other mouse or hamster enzymes. The enzymes examined included lactate dehydrogenase A (EC 1.1.1.27), peptidase B, peptidase C, phosphogluconate dehydrogenase (EC 1.1.1.43), aspartate aminotransferase(EC 2.6.1.1), glucose-phosphate isomerase (EC 5.3.1.9), NAD-dependent malate dehydrogenase (EC 1.1.1.37), and NADP-dependent malate dehydrogenase (EC 1.1.1.40). For all of these enzymes, the forms synthesized by mouse and hamster can be resolved. It was found that all of the hybrids synthesize both the hamster and mouse form of all the enzymes.

One of the main questions posed by the results of these experiments involves the fact that half of the hybrids are pigmented and half are unpigmented. In an attempt to determine whether either type of hybrid produces cells of the other type, the hybrids were cloned and recloned. More than 200

TABLE 3. *DOPA oxidase of parental and hybrid cells*

Cell line	Specific activity	Activity per cell*
3460²	1.3	1.0
3²D 3 (pigmented)	0.7	1.0
3²D 8 (pigmented)	1.7	3.0
3²D 10 (unpigmented)	0	—
3²D 11 (unpigmented)	0	—
Clone 1D	0	—

* The activity per cell, as determined from the specific activity, and the amount of protein per cell. There are 0.20 ng of protein per cell for 3460², 0.36 ng per cell for 3²D 3, and 0.46 ng per cell for 3²D 8. The activities are expressed relative to that of 3460², which is assigned a value of 1.0.

subclones of the unpigmented hybrids 3²D 10 and 3²D 11 were examined; all were unpigmented. In contrast, the pigmented hybrids gave rise to unpigmented, as well as pigmented, clones. The pigmented hybrids 3²D 3 and 3²D 8 produce mainly pigmented lines, whereas the pigmented 3²D 7 and 3²D 18 produce mainly unpigmented lines. 3²D 7 and 3²D 18 hybrids also lose pigment when grown in mass cultures for prolonged periods. In order to test more critically whether pigmented hybrids give rise to unpigmented lines, four pigmented populations were cloned, and, from the mixture of pigmented and unpigmented colonies obtained, seven pigmented subclones were isolated. These pigmented subclones were then recloned; the results are shown in Table 4. It is clear that some pigmented subclones produce unpigmented lines with very high frequencies, whereas other subclones produce exclusively pigmented lines. Furthermore, the frequencies with which these pigmented subclones yield pigmented progeny are similar to the frequencies with which their pigmented parental populations yielded pigmented progeny, e.g., 3²D 7 and its two pigmented subclones 3²D 7-1 and 3²D 7-2 all produce a large majority of unpigmented lines when cloned. It thus seems that the ability of pigmented hybrids to produce pigmented progeny at high or low frequencies is, to some extent, a stably inherited characteristic. The mechanism of this pattern of inheritance is not understood.

TABLE 4. *Cloning of pigmented hybrid lines*

Hybrid cell clone	Percent pigmented colonies
3²D 7-1	0
3²D 7-2	33
3²D 18-1	40
3²D 8-2	80
3²D 3-1	95
3²D 8-1	100
3²D 3-4	100

The hybrids were cloned in (Linbro) plastic tissue-culture multi-well plates. For each line, 20 wells were inoculated with 5 cells per well and 20 wells were inoculated with 1 cell per well. The cloning efficiencies of all the hybrids are in the range of 15–20%. 20–25 Colonies were examined for each cell line.

TABLE 2. *Specific activity of DOPA oxidase of 3460²
during growth*

	Days after inoculation		
	2	4	6
Exp. 1	0.6	1.6	0.2
Exp. 2	0.7	1.1	0.2

DISCUSSION

The results of this study on hybrids containing two pigment cell genomes and one fibroblast genome are different from those of previous studies on hybrids containing only one pigment-cell genome and one fibroblast genome (1, 2). (The latter hybrids will be referred to below as 3D hybrids, as distinguished from the 3^2D hybrids.) Comparison of the two studies is facilitated, because the 2S hamster melanoma cells used in the present study were derived from the 1S melanoma cells used in the previous study, and the same mouse fibroblast line was used in both series of experiments. The results of the two studies differ in that about half of the 3^2D hybrids are pigmented, whereas all of the more than 100 independently derived 3D hybrid lines obtained were unpigmented. The isolation of pigmented 3^2D hybrids rules out the possibility that the lack of pigment in the 3D hybrids previously isolated is an artifact of hybridization, due to the act of cell fusion or to an increase in cell size and chromosome number, or to the combination of hamster and mouse genomes.

The results of the present study raise the following questions: (i) Why are any pigmented hybrids isolated when two pigment-cell genomes and one fibroblast genome are combined in the same cell, whereas only unpigmented hybrids are isolated when one pigment-cell genome and one fibroblast genome are combined? (ii) If some of the 3^2D hybrids are pigmented, why are all not pigmented?

One possibility is that all of the 3^2D hybrids are initially unpigmented, i.e., that pigment synthesis is initially repressed as in the 3D hybrids, and that some genetic event occurs that allows some of the hybrids to become pigmented. One way in which this event could occur is that the repression of pigment synthesis involves the action of regulator genes of the fibroblast genome, and that some of the hybrids have lost the chromosomes carrying these regulator genes, thus allowing the reappearance of pigmentation. Such a case has already been described in a study of the regulation of a kidney-specific enzyme in hybrid cells (13). There is, however, no data from the present study in support of this mechanism. Extensive cloning of the unpigmented 3^2D hybrids has not produced pigmented subclones. Furthermore, unless the mouse (fibroblast) chromosomes carrying the suggested regulator genes are preferentially lost, it would be expected that a very large number of mouse chromosomes would have to be lost at random to account for half of the hybrids being pigmented because of the loss of those specific chromosomes. Neither the karyological analyses nor analysis of the enzymes unrelated to melanin synthesis provided evidence of a greater loss of mouse chromosomes from the pigmented hybrids than from the unpigmented hybrids. These considerations make the above suggestion unlikely, but do not completely exclude it.

A second possibility is that all of the 3^2D hybrids are initially pigmented, and that some event occurs that causes half of the hybrids to become unpigmented. The two parts of the hypothesis will be considered separately: (i) that 3^2D hybrids are all initially pigmented, and (ii) that half of the hybrids lose the ability to make pigment. The genetic constitution of the hybrids suggests a type of gene-dosage phenomenon as an explanation for pigmented 3^2D hybrids in contrast to unpigmented 3D hybrids. For example, if the fibroblast genome produces a regulator substance that prevents pigment synthesis, it is possible that it produces enough to block the activity of a second genome in addition to itself in 3D hybrids,

but not enough to block the two additional genomes in the 3^2D hybrids. It is also possible that a factor produced by the pigment-cell genome and required for pigment synthesis might not reach a sufficient concentration in the 3D hybrids to allow pigmentation, but might reach such a concentration in the 3^2D hybrids. Given pigmented hybrids, it is then necessary to explain why half of them are initially isolated as unpigmented. It was demonstrated that some of the pigmented 3^2D hybrids produced unpigmented lines in high frequencies when cloned. There is no evidence to indicate that the initially selected unpigmented hybrids and the unpigmented lines obtained as subclones of pigmented hybrids both have lost pigmentation in the same way. However, it is possible that the type of event responsible for the loss of pigmentation when pigmented hybrids are cloned could have occurred very early in the development of some 3^2D hybrids, leading to the isolation of unpigmented colonies from the initial selection. It would be possible to account for the loss of pigmentation by the loss of some chromosomes of the melanoma genome that carry genes required for pigment synthesis. As mentioned above, however, unless there was a preferential loss of these specific chromosomes, a very large number of hamster (melanoma) chromosomes would have to be lost at random to account for half of the hybrids being unpigmented. Neither the karyological nor enzymatic analyses reveal a much greater loss of hamster genes from the unpigmented hybrids than from the pigmented hybrids. This result would seem to make chromosome loss unlikely as a mechanism for the segregation of unpigmented hybrids from pigmented ones. This last argument assumes that all of the melanoma genes of a given type would have to be lost in order to lose the function. If, however, a balance between melanoma and fibroblast genes is the critical factor, as suggested above, the loss of only a single relevant hamster chromosome could be sufficient to shift the balance in the favor of the fibroblast. Nonchromosomal events could also be involved in the production of unpigmented hybrids.

At present, neither of the above proposals, i.e., that the 3^2D hybrids are either initially all pigmented or all unpigmented, can be proven or ruled out. It is also possible that other mechanisms are operating in such a way that half of the hybrids are initially pigmented and half are initially unpigmented. The present results demonstrate conclusively, however, that the number of differentiated cell genomes in a hybrid can affect the expression of differentiated functions. These results, thus, suggest some new quantitative aspects in the regulation of gene expression in mammalian cells.

The isolation of pigmented 3^2D hybrids raises an additional question: Is the DOPA oxidase present in pigmented hybrids determined only by the genome of the originally pigmented melanoma cell, or have the genes of the originally unpigmented fibroblast been induced to begin synthesizing DOPA oxidase? It is possible that the mechanism that allows pigmentation in some 3^2D hybrids has also resulted in the activation of the previously inactive fibroblast genes involved in pigment synthesis.

I thank Dr. Frank Ruddle, who kindly performed the isozyme analyses. This research was supported in part by grants from the National Institute of Child Health and Human Development HD04807 and HD06276. After this work was begun, I learned that similar experiments are being performed in the laboratory of Dr. Boris Ephrussi. Both pigmented and unpigmented hybrids were also isolated in these experiments.

1. Davidson, R. L., Ephrussi, B. & Yamamoto, K. (1966) *Proc. Nat. Acad. Sci. USA* **56**, 1437–1440.
2. Davidson, R. L., Ephrussi, B. & Yamamoto, K. (1968) *J. Cell. Physiol.* **72**, 115–128.
3. Davidson, R. L. & Yamamoto, K. (1968) *Proc. Nat. Acad. Sci. USA* **60**, 894–901.
4. Davidson, R. L. & Benda, P. (1970) *Proc. Nat. Acad. Sci. USA* **67**, 1870–1877.
5. Benda, P. & Davidson, R. L. (1971) *J. Cell. Physiol.* **78**, 209–216.
6. Neff, J. & Enders, J. (1968) *Proc. Soc. Exp. Biol. Med.* **127**, 260–267.
7. Harris, H. & Watkins, J. F. (1965) *Nature* **205**, 640–646.
8. Davidson, R. L. (1969) *Exp. Cell Res.* **55**, 424–426.
9. Davidson, R. L. & Ephrussi, B. (1970) *Exp. Cell Res.* **61**, 222–226.
10. Littlefield, J. (1964) *Science* **145**, 709–710.
11. Lowry, O., Rosenbrough, N., Farr, A. & Randall, R. (1951) *J. Biol. Chem.* **193**, 265–275.
12. Ruddle, F., Chapman, V., Ricciuti, F., Murnane, M., Klebe, R. & Meera Kahn, P. (1971) *Nature* **232**, 69–73.
13. Klebe, R., Chen, T. & Ruddle, F. (1970) *Proc. Nat. Acad. Sci. USA* **66**, 1220–1227.

11

Reprinted from *Natl. Acad. Sci. (USA) Proc.* **69**:571-575 (1972)

Expression of Differentiated Functions in Hepatoma Cell Hybrids: Induction of Mouse Albumin Production in Rat Hepatoma–Mouse Fibroblast Hybrids*

(immunodiffusion)

JERRY A. PETERSON AND MARY C. WEISS

Centre de Génétique Moléculaire, Centre National de la Recherche Scientifique, 91 Gif-sur-Yvette, France

Communicated by Boris Ephrussi, December 27, 1971

ABSTRACT   The synthesis of serum albumin has been studied in hybrids between well-differentiated rat hepatoma cells, which synthesize serum albumin, and mouse fibroblasts (3T3) that do not synthesize albumin. By immunodiffusion techniques with noncrossreacting antisera, the production of both rat and mouse albumin by the hybrids has been examined.   Karyologically identified hybrids were produced between 3T3 cells and cells of a 1s hepatoma (Fu5) clone, and of a 2s hepatoma (2s Fu5-5cl.1E) clone. Each of the 3T3 x Fu5 hybrids produces only rat albumin. Among five 3T3 x 2s Fu5-5cl.1E hybrid clones isolated, one produces both rat and mouse albumin, two produce only mouse albumin, and two do not produce rat or mouse albumin.

It has been clearly demonstrated that a plant somatic cell contains and can express the genetic information required for the formation of an entire plant (1). An amphibian somatic cell nucleus, when transplanted into an enucleated unfertilized egg, can promote the formation of a normal adult animal (2). Although such an unequivocal demonstration cannot be cited for mammals, there is presumptive evidence that genetic information is not lost during their development; for example, there is no variation in chromosome number and structure in cells cultured from different organs (3), and the DNA content of diploid cells from various tissues is similar (4). Moreover, that genetic information is not altered or irreversibly blocked during differentiation of mammalian cells is suggested by the occasional expression by tumors originating from one tissue of characteristics limited to another tissue, as in the cases of nonendocrine tumors that produce hormones (5). However, a similar activation of dormant genes for tissue-specific functions of mammalian cells has not been accomplished experimentally.

Since hybridization of somatic cells permits the study of interactions between two different genomes within a single nucleus, a number of hybrids have been produced between mammalian cells of different ontogenetic origins in the hope of elucidating some of the mechanisms that control differentiation (for review, see ref. 6). In the first such experiments, viable hybrids between pigment-producing melanoma cells and nonpigmented fibroblasts were found to be unpigmented (7). Several other cases of extinction of specific character-

istics of differentiated cells have been described, and it appears that it is not due to loss of chromosomes (6, 8). On the other hand, there are reports that in some hybrids between cells that differ in the expression of differentiated characteristics, certain of these properties are retained, for example, the production of hyaluronic acid (9), free kappa chains of immunoglobulins (10), and electrophysiological properties (11). In hybrids between differently specialized cells, the specific differentiated characteristics disappear in most cases, but there are cases where they continue to be expressed, at least in part.

This paper describes studies of the production of serum albumin by hybrids between mouse fibroblasts and rat hepatoma cells. The hepatoma cells synthesize *in vitro* a number of proteins characteristic of the liver that include serum albumin (12), a high activity of tyrosine aminotransferase (EC 2.6.1.5) that is inducible by steroid hormones (13), and type B fructose-1,6-diphosphate aldolase (EC 4.1.2.7) (14). The subtetraploid mouse fibroblasts of line 3T3 (15), derived from a mouse embryo, synthesize collagen at a level comparable to that synthesized by normal diploid fibroblasts (16). The 3T3 cells have a very low tyrosine aminotransferase activity that is not inducible, they lack the type B aldolase and do not synthesize albumin†. In earlier reports of this series, it has been shown that rat hepatoma–mouse fibroblast hybrids fail to express two of the properties of the hepatoma parent, type B aldolase (14) and high activity and inducibility of tyrosine aminotransferase (13).

In the present experiments, using specific antisera that do not crossreact, we have examined not only whether the hepatoma genome continues to direct the synthesis of rat albumin (RSA) in the hybrids, but also whether the mouse fibroblast genome can be induced to produce mouse albumin (MSA).

MATERIALS AND METHODS

All cultures were maintained in modified (20) F12 (21) medium containing 5% fetal-calf serum (13).

The origins and karyotypes of 3T3 (clone 4E, mouse fibroblasts), Fu5 (rat hepatoma cells), and eight clones of 3T3 x Fu5 hybrids have been described (13). 2s Fu5-5cl.1E

Abbreviations: RSA, rat serum albumin; MSA, mouse serum albumin.

* This is the fourth paper in a series, *Expression of Differentiated Functions in Hepatoma Cell Hybrids*. The third paper is: Weiss, M. C. & Chaplain, M. (1971) *Proc. Nat. Acad. Sci. USA* **68**, 3026–3030.

† Most evidence indicates that serum albumin is synthesized uniquely by the liver (17). The reports that it is synthesized by chick mesenchymal cells and HeLa cells (18) have recently been seriously contested (19).

79

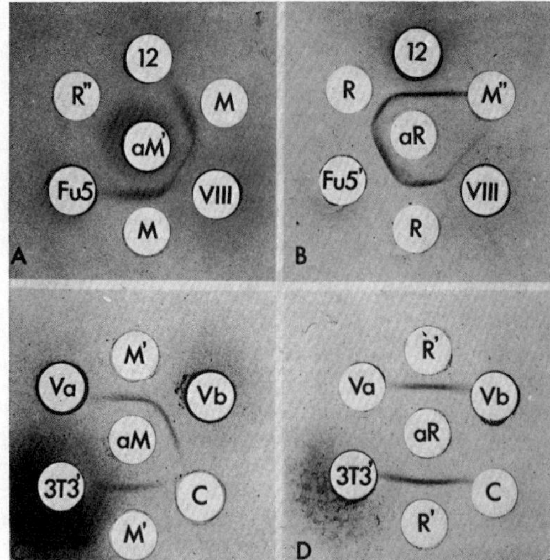

FIG. 1.   Double immunodiffusion tests of purified albumins and concentrated media (from parental and hybrid cells) with non-crossreacting rabbit anti-rat or anti-mouse albumin antisera. The central wells contain antisera: a*M* and a*M* ', anti-mouse albumin antisera obtained from two different rabbits and diluted 1:6 and 1:3, respectively; a*R*, anti-rat albumin antiserum, diluted 1:6. The outer wells contain purified albumin or concentrated growth medium from the indicated cell cultures; the abbreviations and concentrations of albumin, or factor of medium concentration, are given in parentheses. Fig. 1*A*: hybrid cells 3F12 (12; ×100) and 3-2FVIII (VIII; ×200); 1*s* hepatoma cells (Fu5; ×100); mouse albumin (M; 15 μg/ml); rat albumin (R″; 200 μg/ml). Fig. 1B: 1*s* hepatoma cells (Fu5′; ×25); mouse albumin (M″; 200 μg/ml); rat albumin (R; 15 μg/ml); others are as indicated for Fig. 1*A*. Fig. 1(*C* and *D*): hybrid cells, 3-2FVb (Vb; ×100) and 3-2FVa (Va; ×100); 3T3 cell extract (3T3′); mouse albumin (M′; 20 μg/m'); rat albumin (R′; 20 μg/ml); growth medium incubated at 37° for 72 hr without cells (C; ×100).

has the following history: cells of a subclone (Fu5-5) of Fu5 were used in a cross performed for other purposes. Among the clones isolated in this cross was 2*s* Fu5-5cl.1, which was subsequently recloned to give rise to 2*s* Fu5-5cl.1E. As indicated in its designation, cells of this clone contain twice as many chromosomes as Fu5.

Hybrids between 2*s* Fu5-5cl.1E and 3T3 were prepared by treatment of a mixture of parental cells with 500 hemagglutinating units/ml of UV-inactived Sendai virus, as described by Coon and Weiss (20); the treated suspension was inoculated at low cell density in 10-cm petri dishes in Littlefield's selective medium (22, 23), in which the thymidine kinase-deficient 3T3-4E cells are unable to proliferate. Hybrid colonies were visible after 5 days, and were isolated after 14 days. One hybrid colony was isolated from each of four different petri dishes; one of them (3-2FV) was contaminated with hepatoma cells and was therefore recloned; two subclones (3-2FVa and 3-FVb) were isolated.

For assay of albumin in the culture medium, the number of cells of different types initially seeded was adjusted so that each population would just approach confluency at 72 hr.

The hybrid cells are larger than either of the parent cells and 3T3 cells are larger than Fu5 cells. Therefore, 2 × 10⁶ Fu5, 1 × 10⁶ 2*s* Fu5-5cl.1E, 1 × 10⁶ hybrid, or 1.5 × 10⁵ 3T3 cells were seeded in 250-ml tissue culture bottles (Falcon plastics) with 16 ml of growth medium per bottle. After incubation at 37° for 72 hr, the medium was collected, and the cells were removed from the bottles with trypsin and counted. The numbers of cells per bottle at the end of 72 hr with the media shown in Figs. 1, 2, and 3 were as follows: (parental cells) Fu5, 13 x 10⁶; 2*s* Fu5-5cl.1E, 2.8 x 10⁶; 3T3, 10⁷; (Fu5 x 3T3 hybrids) 3F11, 4.1 x 10⁵; 3F12, 2 x 10⁶; 3F14 and 3F15, 2.5 x 10⁶; 3F16, 3 x 10⁶; (2*s* Fu5-5cl.1E x 3T3 hybrids) 3-2FIII; 3.7 x 10⁶; 3-2FVa, 2.4 x 10⁶; 3-2FVb, 4.5 x 10⁶; 3-2FVIII, 2.2 × 10⁶. The medium was dialyzed against 1 mM potassium phosphate buffer (pH 7.4), then lyophilized; the residue was redissolved in a small volume of distilled water. These concentrated media were examined by immunodiffusion for the presence of MSA and RSA.

The extract of 3T3 cells is a 109,000 × *g* supernatant of 41 × 10⁶ cells that were frozen and thawed three times in 0.5 ml of 25 mM barbital buffer (pH 8.2).

Commercially obtained rat and mouse serum albumins (fraction V, Pentex) were electrophoresed on a 5-mm thick slab of 0.8% agarose–6% acrylamide gel in a Tris·glycine buffer (pH 8.7), as described by Uriel (24), and the monomeric albumin band (which stained with bromophenol blue) was cut from the gel and homogenized into a paste. The proteins were freed from the acrylamide gel by three consecutive washes with distilled water; after each wash, the acrylamide gel was sedimented by centrifugation and the wash was collected. The washes were pooled, dialyzed against distilled water, and lyophilized to obtain the purified albumin. Such purified rat (RSA) and mouse (MSA) albumins showed no heterogeneity by acrylamide gel electrophoresis or by double immunodiffusion against goat anti-rat serum or goat anti-mouse serum (Hyland), respectively. All immunizations were performed with emulsions of the homogenized gel containing purified albumin and an equal volume of Freund's complete adjuvant. Primary injections of 5 mg of RSA or MSA were followed by one or two booster injections of 5 mg of albumin at monthly intervals; antisera were collected 8 days after the last injection. Since rabbit anti-RSA crossreacts with MSA, and rabbit anti-MSA crossreacts with RSA, noncrossreacting antisera were prepared by exhaustive adsorption of anti-RSA sera with polymerized proteins from mouse serum, and adsorption of anti-MSA sera with polymerized proteins from rat serum. Serum proteins were polymerized with glutaraldehyde by the method of Avrameas and Ternynck (25). Ad-

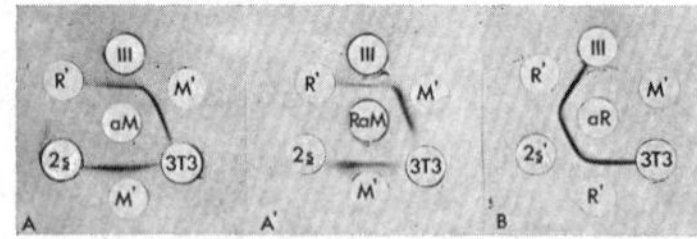

FIG. 2.   Double immunodiffusion tests of purified albumins and of concentrated media from parental and hybrid cells with non-crossreacting rabbit anti-mouse (*A*) and anti-rat (*B*) albumin, and rat anti-mouse albumin (*A* ') antisera. The contents of the wells are as described in Fig. 1, except for: rat anti-mouse albumin antiserum (*R*a*M*), diluted 1:2; medium from hybrid cells 3-2FIII (III; ×100), and parental cells 3T3 (3T3; ×100), and 2*s* Fu5-5cl.1E (2*s*; ×100; 2*s*′; ×20).

sorbed anti-RSA antisera showed no reaction by double immunodiffusion against any concentration (0.6–400 µg/ml) of the mouse albumin that was tested (see Figs. 2B and 1B for 20 and 200 µg/ml, respectively). By the same criterion, adsorbed anti-MSA antisera did not crossreact with rat albumin at concentrations between 0.6 and 400 µg/ml (see Figs. 1A and 2A). The adsorbed anti-RSA and anti-MSA antisera gave single precipitin bands with rat and mouse serum, respectively.

Rat anti-MSA antisera were prepared in Wistar rats that were immunized as follows: Day 1, 1 mg of MSA; Day 25, 0.1 mg of MSA; Day 70, 0.4 mg of MSA; Day 80, sera were collected.

Throughout this study, we used antisera that showed no detectable crossreaction.

Double immunodiffusion was performed on 1-mm thick slabs of 0.8% agarose in 25 mM barbital buffer (pH 8.2). Antigen and antibody wells were 3 mm in diameter, separated by 2 mm, and filled with 7–14 µl of each sample. After 24 hr of incubation at room temperature, the immunodiffusion slides were soaked for several days in saline, dried, and stained for protein with 0.1% amidoschwarz-B 10 (Carlo Erba) in 3.0% acetic acid.

Each sample of growth medium from parental and hybrid cells was concentrated at least 100-fold and initially tested by being allowed to diffuse from a central well of an immunodiffusion slide toward six surrounding wells arranged in a hexagonal pattern, containing undiluted and five 2-fold serial dilutions of antiserum. The minimal concentrations of purified mouse and rat albumin that gave a visible precipitin band, when 7 µl of a solution was placed in the central well, were, respectively, 2.0 µg/ml and 0.4 µg/ml (when undiluted and the first three 2-fold serial dilutions of antisera were used). Therefore, when it is stated that there is no mouse or rat albumin present in a 100-fold concentrated medium, this implies that there was less than 0.02 µg of mouse albumin and less than 0.004 µg of rat albumin per milliliter in the unconcentrated growth medium. However, since the concentration of albumin in the medium is a function of the cell number, the volume of medium, and incubation time, albumin production is expressed as µg/10⁶ cells per 72 hr (based on the number of cells present at the end of the 72-hr incubation period).

RESULTS

The two kinds of hybrids that have been studied resulted from crosses of 3T3 mouse fibroblasts with 1s (hyperdiploid) and 2s (hypertetraploid) rat hepatoma cells, respectively. The first series of hybrids (3F) was expected to contain 52 chromosomes from Fu5 (of which 25.7 are biarmed marker chromosomes and the remainder are telocentric), in addition to 75.9 chromosomes (all telocentric) from 3T3. However, these hybrids all contained fewer chromsomes than expected (by 7–16%), and in particular, fewer rat marker chromosomes than expected. It has, therefore, been concluded that this deficiency is probably the result of loss of primarily rat chromosomes (13).

In the second series of hybrids (3-2F), the 2s hepatoma parent contributed twice as many chromosomes as Fu5. Details of the parental and hybrid karyotypes are given in Table 1. It can be seen that hybrids of this series fall into three classes on the basis of their karyotypes. One hybrid

clone (3-2FVIII) has more chromosomes than expected. The fact that it has slightly fewer (8%) than the expected number of rat markers suggests that the excess chromosomes are of mouse origin. The second class, comprised of two clones (3-2FIII and 3-2FVb), has 15% fewer chromosomes than expected and has lost 20% of the rat markers. The third class, also comprised of two clones (3-2FIV and 3-2FVa), has 25% fewer chromosomes than expected and has lost 30% of the rat markers. If the rat telocentric chromosomes are lost with the same frequency as the rat markers, then the major loss in these hybrids, like in series 3F, is of the rat chromosomes. In each population of hybrid cells there are some new long metacentric chromosomes, which may have arisen by centric fusion of telocentric chromosomes (26).

The production of rat albumin by both Fu5 and 2s Fu5-5cl.1E (rat hepatoma) cells has been demonstrated by double immunodiffusion of concentrated culture medium (Fig. 1B and 2B). It can be seen that the precipitin bands formed between anti-RSA antiserum and the concentrated media

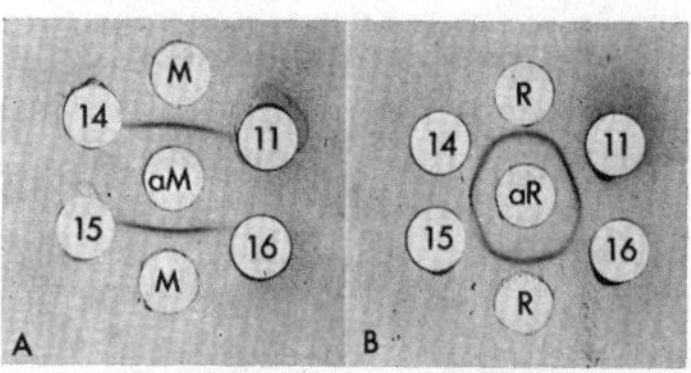

FIG. 3. Double immunodiffusion tests of concentrated media from four 3T3 × Fu5 hybrid clones with noncrossreacting rabbit anti-mouse albumin and anti-rat albumin antisera. The contents of the wells are as described in Fig. 1, except for: medium from hybrids 3F11 (11; ×100); 3F14 (14; ×75); 3F15 (15; ×100); 3F16 (16; ×100).

fuse completely with those formed between anti-RSA antiserum and purified RSA, which demonstrates that the protein produced by the hepatoma cells is serologically identical to RSA.

On the basis of quantitative assays by microcomplement fixation (which will be discussed in a more extensive publication), Fu5 produces 4–5 µg of RSA/10⁶ cells/72 hr, and 2s Fu5-5cl.1E produces twice this amount. Figs. 1A, 2A, and 2A′ show that neither of these cell lines produces material that crossreacts with the anti-MSA antisera.

Neither concentrated medium nor extracts of 3T3 cells contain material that reacts with either of the anti-albumin antisera (Figs. 1 and 2). If the sensitivity of the antisera is taken into consideration, it can be concluded that 3T3 cells produce less than 0.03 µg of MSA/10⁶ cells per 72 hr, if they produce any at all.

The 5% fetal-calf serum present in the growth medium does not crossreact with either mouse or rat anti-albumin antisera, since there is no reaction with concentrated (control) growth medium that had not been in contact with cells (Figs. 1, C and D)‡.

---

‡ The absence of cross reaction between fetal-calf serum and the antisera used in this study has been verified by testing a wide range of concentrations of fetal-calf serum (containing 0.0004–400 mg of albumin/ml). Moreover, adsorption of the antisera with polymerized proteins of fetal-calf serum (25) did not modify their specificity or reactivity.

The types of albumins produced by the different kinds of hybrid cells are given in Table 1, and the results of immunodiffusion tests of concentrated media from most of the hybrid clones are shown in Figs. 1, 2, and 3. All of the eight 3T3 x Fu5 hybrid clones that have been tested produce only rat albumin (immunodiffusion tests for five of these hybrids are shown in Figs. 1 and 3). Therefore, in these hybrids, the rat-hepatoma genome continues to direct the synthesis of rat albumin, but there is no induction of the fibroblast genome that results in the production of mouse albumin. Quite different results were obtained with the five hybrid clones that were isolated from the cross of 3T3 and 2s hepatoma cells. As shown in Table 1, the different karyotypic classes that were described above each have a unique phenotype with respect to albumin production. The first class, represented by one clone (3-2FVIII), produces both rat and mouse albumins (Fig. 1, *A* and *B*), the two clones of the second class produce only mouse albumin (Fig. 1, *C* and *D*, Fig. 2), and the two clones of the third class produce neither albumin (Fig. 1, *C* and *D*). The identification of the albumin(s) produced by the hepatoma and hybrid cells is demonstrated by the complete fusion of the precipitin bands formed between concentrated media and specific anti-albumin antisera with those between anti-albumin antisera and purified mouse or rat albumin. In no case have spurs been observed. Moreover, concentrated media that fail to show precipitin bands with anti-albumin antisera do not interfere with the formation of neighboring bands. The identity of mouse albumin in the medium from hybrid cell cultures has been verified by use of rat anti-mouse albumin antiserum (Fig. 1*A′*), as well as adsorbed anti-mouse albumin antisera from two different rabbits.

Despite the fact that the immunodiffusion test used in these experiments is only qualitative, it is possible to determine the threshold of sensitivity of the antisera (see *Methods*). Applying this criterion to the cases where only one or

neither of the albumins is detectable, we can conclude that the maximal amount of MSA that could be produced by the series 3F and 3-2F hybrids is less than 0.15 and less than 0.12 μg/10^6 cells per 72 hr, respectively. The maximal amounts of RSA that could be produced by the four 3-2F hybrid clones (where none is detected) is less than 0.025 μg of RSA/10^6 cells per 72 hr.

On the other hand, precise quantitation, obtained by microcomplement fixation, has shown that the 3F hybrid clones produce different amounts of RSA, within the range of 0.2 and 1.5 μg of RSA/10^6 cells per 72 hr. This reprerents 5–30% of the amount produced by the hepatoma parent (Fu5). Similar quantitative measurements have not yet been performed on the albumin production by 3-2F hybrids.

DISCUSSION

The induction of mouse albumin production in hybrids between rat hepatoma cells and mouse fibroblasts clearly demonstrates that in mammals a cell type that does not express a particular function as a result of the path of differentiation it has followed still has unaltered genetic information for this function, and can express it. In addition, this specific "gene activation" in interspecific hybrids indicates that the regulatory signals in the rat cells for albumin synthesis are recognized by the mouse genome. Moreover, the fact that activation of the mouse fibroblast genes for albumin synthesis occurs in hybrids in which the hepatoma gene is not expressed may be interpreted as evidence for the existence of a heritable regulatory element for the activation and maintenance of albumin synthesis that is independent of the structural gene.

When the significance of these results is considered with regard to the potential for reversal of the differentiated state in normal somatic cells, two points must be mentioned. The hybrids are obtained by fusion of cells of established heteroploid

TABLE 1. *Karyotypes* and albumin production (parental and hybrid cells)§*

Cell line	Type of albumin produced		Total no. of chromosomes	No. of telocentric chromosomes	No. of biarmed chromosomes (rat markers)	No. of new biarmed chromosomes
	Rat	*Mouse*				
Parents:						
3T3	−	−	75.9 (70–80)	75.9 (70–80)	0	0
Fu5	+	−	51.9 (51–53)	26.2 (24–27)	25.7 (24–27)	0
2s Fu5-5cl.1E	+	−	98.5 (92–104)	46.1 (41–51)	52.2 (47–56)	0
Hybrids:						
3T3 x Fu5 (3F series)	+	−	107.0–118.2†	87.0–96.8†	17.2–22.4†	0.4–2.4†
3T3 x 2s Fu5-5cl.1E (3-2F series)						
Expected:			174.4 (162–184)	122.0 (111–131)	52.2 (47–56)	
Observed:						
3-2FVIII	+	+	205.1 (193–218)	155.3 (144–165)	48.0 (41–52,70‡)	1.8 (1–3)
3-2FIII	−	+	146.0 (131–158)	102.7 (91–116)	42.1 (33–51)	1.3 (1–3)
3-2FVb	−	+	149.7 (144–154)	107.7 (103–111)	42.1 (36–51)	0.1 (0–1)
3-2FIV	−	−	131.2 (126–138)	95.2 (89–101)	36.0 (32–40)	0.5 (0–1)
3-2FVa	−	−	128.1 (120–134)	91.1 (78–104)	35.7 (27–43)	0.6 (0–3)

* Mean values (and ranges in parentheses) are given.   These numbers are based upon analysis of 10–30 metaphases of each cell line. † Ranges of mean values from eight different hybrid clones are given (13).   ‡ One metaphase.   § Hybrid clones were frozen as soon as possbile and examination of albumin production and karyotypes were always done on freshly thawed samples.

lines, and it is conceivable that neoplastic transformation may have endowed the hepatoma cells with an inductive capacity that is not possessed by normal cells. The possibility that activation of a differentiated characteristic can also be directed by normal cells is supported by the recent report that chick myosin can sometimes be detected in rat embryonic myotubes containing reactivated erythrocyte nuclei§.

The observation that mouse albumin is produced by hybrids only when the relative genetic contribution of the rat hepatoma parent is increased suggests that gene dosage plays an important role in phenotypic expression in hybrid cells. An effect of gene dosage has been observed in other types of somatic hybrids. Davidson and Benda (27) observed that while the inducibility of glycerol-3-phosphate dehydrogenase in 1s glial cells is lost when they are hybridized with fibroblasts, there is a slight inducibility of the enzyme in hybrids made between 2s glial cells and fibroblasts. Moreover, pigmentation is not systematically extinguished in hybrids between 2s melanoma cells and fibroblasts (28).

In the case of albumin, the similarities in the karyotypes of the hybrid clones that express the same phenotype, as well as the extremely high number of chromosomes in all of them, suggests that the expression of specific genes may depend on the relative numbers of regulatory and structural genes (gene or chromosome balance), rather than merely on the presence or absence of the appropriate genes. (Two well-documented cases where chromosome balance influences phenotypic expression are sex determination in *Drosophila melanogaster* and trisomy in humans). While it cannot be excluded that the absence of either type of albumin is due to loss of the appropriate structural gene, this hypothesis appears unlikely since all eight of the hybrid clones of the 3F series produce rat albumin and none produces mouse albumin, in spite of the fact that there appears to be very little loss of mouse chromosomes.

As was mentioned above, the majority of specific differentiated characteristics that have been studied are not expressed in hybrid cells. The synthesis of albumin is one of the exceptions, although it is synthesized at a level lower than that in the hepatoma cells. The observation that rat albumin is produced by all the hybrids between 1s hepatoma and 3T3 cells, while two other specific liver characteristics, production of tyrosine aminotransferase (13) and aldolase type B (14), are not expressed, indicates that the production of albumin is not coordinated with that of the two enzymes.

---

§ Carlson, S. A., Ringertz, N. R. & Savage, R. (1971) *Exp. Cell Res.* **67**, 243–244, *abstract*.

We thank Dr. B. Ephrussi for his continual support and for critically reading the manuscript, and Dr. J. Uriel, Miss. B. Nêchaud, and Dr. M. Kaminski for their generous and helpful counsel on immunological techniques. This work was conducted with the aid of grants to Dr. B. Ephrussi from the Délégation Générale à la Recherche Scientifique et Technique. J.A.P. is supported by a National Institutes of Health postdoctoral fellowship.

1. Steward, F. C. (1970) *Proc. Roy. Soc. Ser. B.* **175**, 1–30.
2. Briggs, R. & King, T. J. (1969) in *The Cell,* eds. Brachet, J. & Mirsky, A. E. (Academic Press, New York and London), Vol. 1, pp. 537–617; Gurdon, J. B. & Uehlinger, V. (1966) *Nature* **210**, 1240–1241.
3. Tjio, J. H. & Puck, T. T. (1958) *Proc. Nat. Acad. Sci. USA* **44**, 1229–1237.
4. Mirsky, A. E. & Ris, H. (1949) *Nature* **163**, 666–668; Boivin, A., Vendrely, R. & Vendrely, C. (1948) *C. R. Acad. Sci. Paris Ser. D.* **226**, 1061–1063.
5. Bower, B. F. & Gordan, G. S. (1965) *Annu. Rev. Med.* **16**, 83–118.
6. Ephrussi, B., *Hybridization of Somatic Cells* (Princeton University Press), in press.
7. Davidson, R. L., Ephrussi, B. & Yamamoto, K. (1966) *Proc. Nat. Acad. Sci. USA* **56**, 1437–1440.
8. Klebe, R. J., Chen, T. & Ruddle, F. R. (1970) *Proc. Nat. Acad. Sci. USA* **66**, 1220–1227.
9. Green, H., Ephrussi, B., Yoshida, M. & Hamerman, D. (1966), *Proc. Nat. Acad. Sci. USA* **55**, 41–44.
10. Mohit, B. & Fan, K. (1971) *Science* **171**, 75–79.
11. Minna, J., Nelson, P., Peacock, J., Glazer, D. & Nirenberg, M. (1971) *Proc. Nat. Acad. Sci. USA* **68**, 234–239.
12. Ohanian, S. H., Taubman, S. B. & Thorbecke, G. J. (1969) *J. Nat. Cancer Inst.* **43**, 397–406.
13. Schneider, J. A. & Weiss, M. C. (1971) *Proc. Nat. Acad. Sci. USA* **68**, 127–131.
14. Bertolotti, R. & Weiss, M. C., *J. Cell Physiol.*, in press.
15. Todaro, G. J. & Green, H. (1963) *J. Cell Biol.* **17**, 299–313.
16. Green, H., Goldberg, B. & Todaro, G. J. (1966) *Nature* **212**, 631–633.
17. Peters, T. Jr. (1970) *Advan. Clin. Chem.* **13**, 37–111.
18. Abdel-Samie, Y., Broda, E., Kellner, G. & Zischka, W. (1959) *Nature* **184**, 361–362.
19. Halpern, M. & Rubin, H. (1970) *Exp. Cell Res.* **60**, 86–95; Kaighn, M. E. & Prince, A. M. (1971) *Proc. Nat. Acad. Sci. USA* **68**, 2396–2400.
20. Coon, H. G. & Weiss, M. C. (1969) *Proc. Nat. Acad. Sci. USA* **62**, 852–859.
21. Ham, R. G. (1965) *Proc. Nat. Acad. Sci. USA* **53**, 288–293.
22. Littlefield, J. (1964) *Science* **145**, 709–710.
23. Davidson, R. & Ephrussi, B. (1965) *Nature* **205**, 1170–1171.
24. Uriel, J. (1966) *Bull. Soc. Chim. Biol.* **48**, 969–982.
25. Avrameas, S. & Ternynck, T. (1969) *Immunochemistry* **6**, 53–66.
26. Harris, M. (1964) *Cell Culture and Somatic Variation* (Holt, Rinehart and Winston, New York), p. 205.
27. Davidson, R. L. & Benda, P. (1970) *Proc. Nat. Acad. Sci. USA* **67**, 1870–1877.
28. Fougère, C., Ruiz, F. & Ephrussi, B. (1972) *Proc. Nat. Acad. Sci. USA* **69**, 330–334.

12

Reprinted from *Natl. Acad. Sci. (USA) Proc.* **57**:615-621 (1967)

RETENTION OF MULTIPLE DEVELOPMENTAL POTENTIALITIES BY CELLS OF A MOUSE TESTICULAR TERATOCARCINOMA DURING PROLONGED CULTURE IN VITRO AND THEIR EXTINCTION UPON HYBRIDIZATION WITH CELLS OF PERMANENT LINES*

BY BRENDA W. FINCH AND BORIS EPHRUSSI†

DEPARTMENT OF BIOLOGY, WESTERN RESERVE UNIVERSITY, CLEVELAND, OHIO

Communicated January 24, 1967

Ever since the early days of tissue culture, one of the chief debates has centered on the question of the maintenance and expression of tissue- and organ-specific functions by animal cells grown *in vitro* (cf. ref. 1–4). This debate on "differentiation versus dedifferentiation," often confused by lack of both adequate experimental data and adequate definitions, can now be considered as settled by recent demonstrations of the possibility, in at least a few cases, of obtaining cultures capable of (a) expressing, over prolonged periods of time, differentiated functions characteristic of the tissue of origin and (b) re-expressing them after growth under conditions interfering with their expression.[5–8] These experiments thus demonstrate that *in vitro* cultured cells can oscillate between a state of overt differentiation (i.e. active performance of a given function) and a state of covert maintenance of the potentiality to express the same function under appropriate conditions (called "modulation" by Paul Weiss). The mechanism underlying the two states is sought today in the type of phenomena of regulation of gene expression which has recently been uncovered in bacteria and with which they have many features in common.[3] Research along these lines is in progress in this and several other laboratories (cf. ref. 10).

The debate on "differentiation-dedifferentiation" *in vitro* has meanwhile obscured a deeper and clearly different problem, that of the mechanism of the process referred to by embryologists as "determination," i.e. the process in embryogenesis whereby in different cell lineages, different segments of the full complement of genetic information characteristic of the species are selected to be expressed under the appropriate conditions (resulting in the formation of characteristic, tissue-specific structural and metabolic profiles), while the remainder ordinarily will not be expressed. The experimental study of this fundamental process could be undertaken under well-defined conditions if cells were available which maintain a state of multipotentiality during growth *in vitro*. This consideration has led us to undertake the *in vitro* culture of cells of a mouse testicular teratoma. It must be noted that similar attempts by Pierce and Verney resulted in the establishment of four permanent lines which, however, were not multipotential;[11] and that while the present work was in progress, a brief report by Sato and Yasumura came to our attention in which the authors describe the successful establishment of teratoma cultures, whose potentialities had, however, not been tested.[12]

The primary purpose of the present paper is to describe briefly the successful long-term *in vitro* culture of clones of multipotential teratoma stem cells. Observations on the fate of the potentialities of the teratoma cells upon hybridization of these cells with cells of permanent lines will also be briefly reported.

Experimental.—Establishment of teratoma cell lines: The starting material was subline 402 AIII of a teratocarcinoma initially derived from the testis of a strain 129 mouse and serially transplanted for over 12 years.[13] It was kindly supplied to us by Dr. G. B. Pierce and has been maintained in this laboratory by the intraperitoneal injection of 0.2 ml of ascites fluid into 129/J mice. Such injections result in the production of solid peritoneal tumors containing differentiated elements as well as of ascites fluid. The latter contains numerous small "embryoid bodies," which are multipotential two-layered structures consisting of a central core of embryonal carcinoma cells surrounded by a single layer of epithelium, and thus bearing a resemblance to young mouse embryos.[14]

Clones were initiated either directly from dissociated embryoid bodies or from mass cultures derived from these. The embryoid bodies are easily removed from the ascites fluid by repeated sedimentation in Earle's saline solution. After dissociation for 15–20 minutes in 2 mM versene with 2 per cent chicken serum, the cells were gently agitated by pipetting, centrifuged, resuspended in growth medium (Dulbecco's modification of Eagle's minimal medium supplemented with 15% calf serum), and filtered twice through 15-μ Nitex monofilament nylon cloth. The cells were inoculated into 6-cm Falcon plastic Petri dishes on a feeder layer of 10^5 irradiated mouse cells in 2.5 ml growth medium. The feeder cells were derived from secondary cultures of newborn CBA mice carrying the T6 translocation.[15]

The cloning efficiency of cells from previously cultured mass populations was low (0.3%), but higher and more consistent than that of cells from freshly dissociated embryoid bodies. All cultures, whether clonal or subcultured mass populations, were homogeneous in appearance, and consisted of small, closely packed epithelial cells (Fig. 1), exhibiting a pronounced tendency to pile up and to form dense clumps. At low densities, these cells failed to grow without a feeder layer. Their attachment following transfer was generally poor (*ca.* 50% and sometimes less). The numbers of generations of growth *in vitro* calculated from inoculum sizes and final densities are therefore considerably underestimated. For the same reason, generation times (35–40 hr) must have been consistently overestimated.

Maintenance of developmental potentialities: The potentialities of the cultured cells were determined by the histological examination of tumors formed upon subcutaneous and/or intraperitoneal inoculation of 1–10 $\times$ 10^6 cells into 129/J mice. The first tests were performed on 12 clones which had undergone at least 25–50 generations *in vitro*. From 4 of these clones, a total of 23 subclones were isolated and similarly tested. In addition, a number of clonal and subclonal populations was selected to be carried for longer periods of continuous culture and tested at various intervals by repeated inoculations to check for possible alterations or differential segregation of potentialities.

At the initial inoculation of the clones, the number of types of differentiated elements, and the proportions of these, were different for the various clones. Some showed excellent differentiation (Fig. 2); others were very poorly differentiated; however, even the latter (with one possible exception) produced a minimum of five tissues, the most frequent being embryonal carcinoma, yolk sac epithelium, neuroepithelium, mesenchyme, and cartilage. One clone (no. 8), which originally produced, in addition to these tissues, bone, ciliated epithelium, keratinized squamous epithelium, and well-formed neural tubules, has been cultured for nearly a year, i.e., underwent a minimum of 140 generations. Even at the latest inoculation, per-

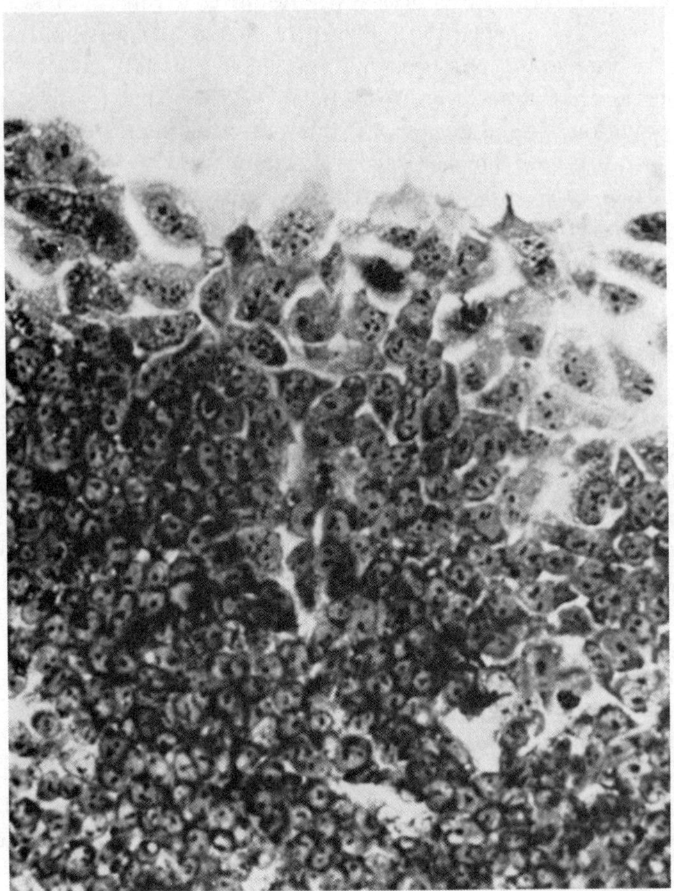

Fig. 1.—Microphotograph of a stained culture of teratoma cells *in vitro*.

formed after 100 generations *in vitro*, it gave rise to tumors showing very nearly the same differentiations, although in reduced amounts. Subclones of this clone, isolated after 38 generations and tested after 23–29 additional generations, were generally not as well differentiated as clone 8 at earlier stages.

In conclusion, it appears then that cultured teratoma cells are highly stable with regard to the retention of histogenetic potentialities even after many months of rapid proliferation in an undifferentiated state,[16] and therefore represent excellent material for the study of factors which channel the directions of differentiation.

Hybridization experiments: The cells used for hybridization experiments were (a) clone LM(TK⁻) 1D, a subclone of mouse L cells (of C3H origin), lacking thymidine kinase and resistant to 5-bromodeoxyuridine; (b) 2472-6-3, a subclone of Sanford's "high cancer" line NCTC 2472 (of C3H origin), resistant to 8-azaguanine and presumably deficient in inosinic acid pyrophosphorylase; and (c) teratoma clone 8, described in the previous section and carrying no biochemical marker.

The hybridization technique employed was that of Davidson and Ephrussi[17] for the isolation of hybrids between biochemically deficient cells and diploid cells

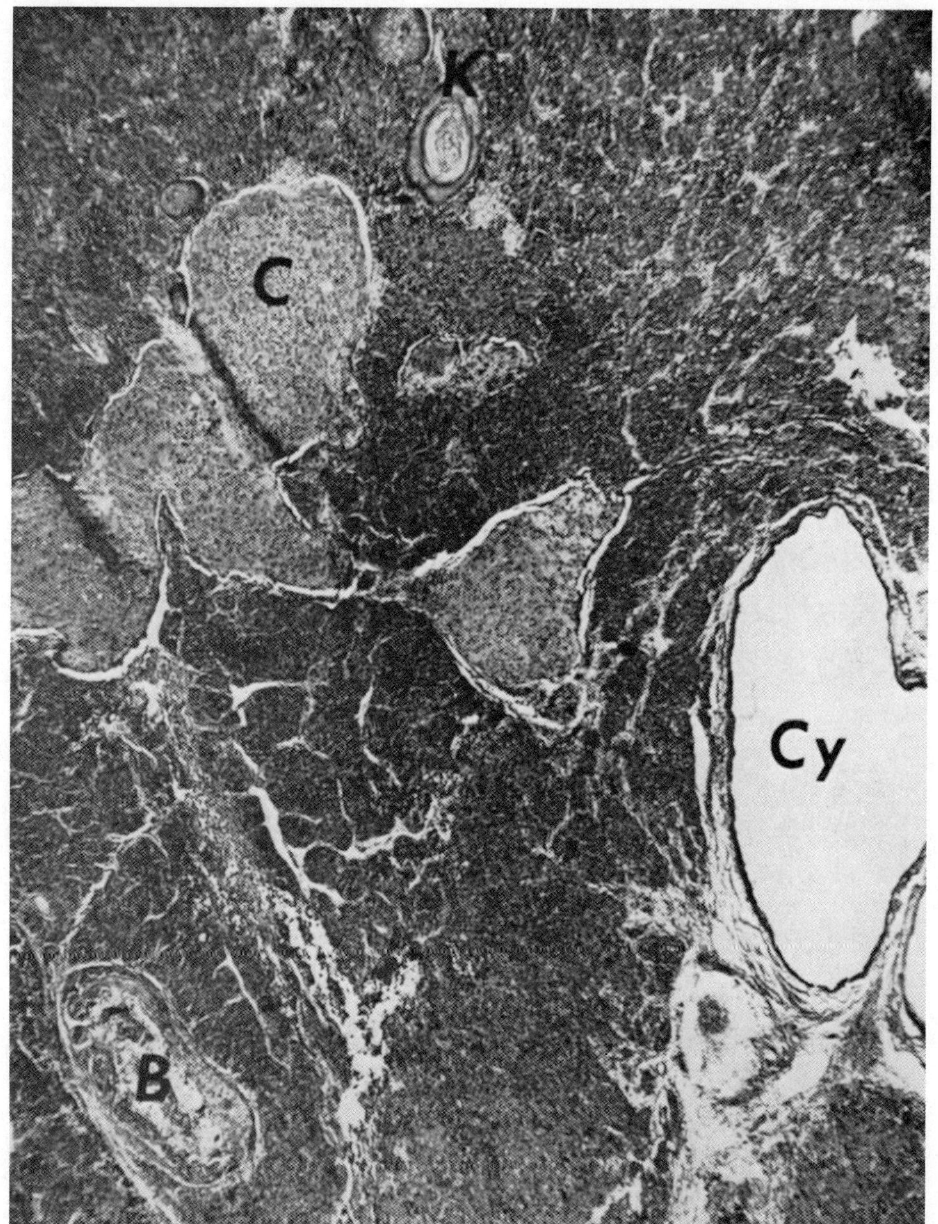

Fig. 2.—Section of a tumor produced following inoculation of subclone 5-2 of teratoma cells which had grown for 59 generations *in vitro*. *B*, bone; *C*, cartilage; *Cy*, epithelial cyst; *K*, keratin pearl. Other tissue types are present also, but cannot be distinguished on this photograph (Hematoxylin-eosin).

carrying no selective marker. The selective medium (HAT) consisted of growth medium supplemented with $1 \times 10^{-4}\ M$ hypoxanthine, $4 \times 10^{-7}\ M$ aminopterin, and $1.6 \times 10^{-5}\ M$ thymidine.

The karyotypes of LM(TK⁻) 1D and 2472-6 have been described elsewhere.[18, 19]

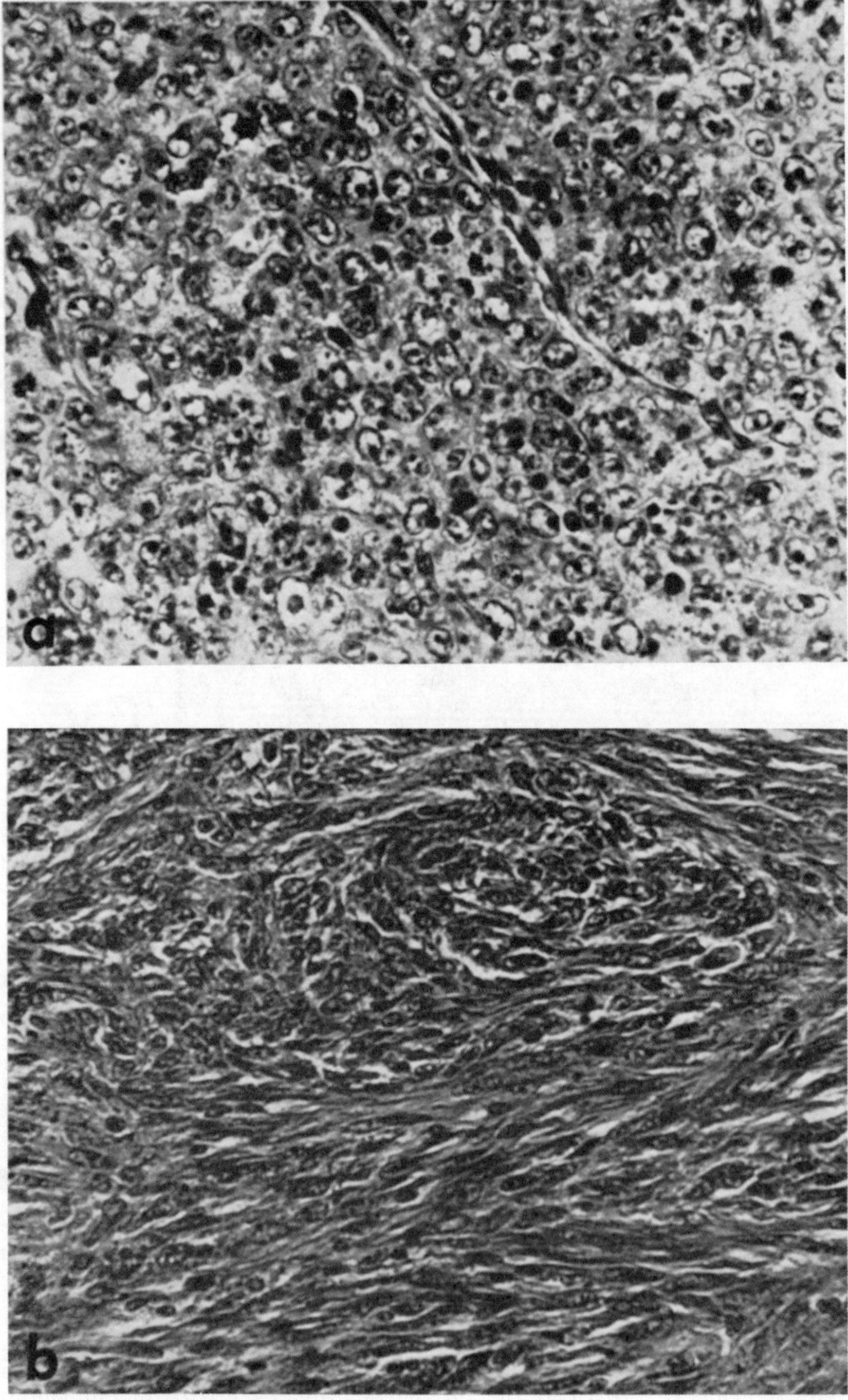

FIG. 3.—(a) Section of an undifferentiated area of a tumor produced by inoculation of teratoma cells cultured *in vitro;* (b) section of a tumor produced by inoculation of hybrid cells (teratoma $\times$ TM(LK$^-$)1D).

Both contain metacentric and other chromosomes which can be used as "markers" in "crosses" with the teratoma cells. Teratoma clone 8 has a modal number of 39 chromosomes including 38 telocentrics and 1 subtelocentric, the latter having no equivalent in the other two lines. The hybrids showed approximately the sum of the modal chromosome numbers of the two "parental" lines and the presence of chromosomes characteristic of each of the parents.

Effects of hybridization: Hybrids between cells of teratoma clone 8 and either L cells or 2472-6-3 cells were isolated from mixtures of the two cell types inoculated in growth medium and renewed one to three days later with HAT. Isolation of the hybrids is facilitated by the degeneration in HAT of the drug-resistant "parental" line and the relatively slow growth of the teratoma colonies. The hybrid cells are distinguishable from teratoma cells morphologically and by their higher growth rate. Isolated colonies of refractile, spindle-shaped cells appear after approximately 10–14 days and rapidly overgrow the teratoma cells. The hybrid nature of the cells thus isolated was confirmed by karyological analysis.

Inoculations of seven hybrid clones from the two crosses into F_1 (C3H $\times$ 129) animals resulted, one to three months later, in tumors which were totally unlike those produced by the teratoma cells. First, none of these "hybrid" tumors contained any differentiated elements, as observed in tumors produced by all 35 teratoma cultures tested thus far. Second, the hybrid tumors were not carcinomas like the undifferentiated portions of the teratoma (Fig. 3a), but typical fibrosarcomas (Fig. 3b), hardly distinguishable from those formed by inoculation of LM(TK⁻)1D, or 2472-6-3.

Thus as a result of hybridization with cells of permanent lines of connective tissue origin, the multipotentiality manifested by the teratoma cells has apparently been abolished and/or the range of expression limited to that of the former. The mechanism underlying this restriction of potentialities and its bearing on the processes of normal determination and differentiation will be further investigated.

We wish to thank Dr. G. B. Pierce for his generous help in histological analysis, and Mrs. Kornelia Veress for her able technical assistance.

* This investigation was supported by grant G-23129 of the National Science Foundation.

† Address as of July 1967: Laboratoire de Génétique physiologique, Gif-sur-Yvette, 91, Essonne, France.

[1] Grobstein, C., in *The Cell*, ed. J. Brachet and A. E. Mirsky (New York: Academic Press, 1959), vol. 1, p. 437.

[2] Konigsberg, I. R., *Science*, **140**, 1273 (1963).

[3] Ephrussi, B., in *Enzymes: Units of Biological Structure and Function*, ed. O. H. Gaebler (New York: Academic Press, 1956), p. 29.

[4] Trinkaus, J. P., *Am. Naturalist*, **90**, 273 (1956).

[5] Ephrussi, B., and H. M. Temin, *Virology*, **11**, 547 (1960).

[6] Sato, G., and V. Buonassisi, in *Retention of Functional Differentiation in Cultured Cells*, ed. V. Defendi (Philadelphia: The Wistar Institute Press, 1964), Symposium Monograph no. 1, p. 27.

[7] Coon, H. G., these PROCEEDINGS, **55**, 66 (1966).

[8] Cahn, R. D., and M. B. Cahn, these PROCEEDINGS, **55**, 106 (1966).

[9] Jacob, F., and J. Monod, in *Cytodifferentiation and Macromolecular Synthesis*, ed. M. Locke (New York: Academic Press, 1963), p. 30.

[10] Davidson, R. L., B. Ephrussi, and K. Yamamoto, these PROCEEDINGS, **56**, 1437 (1966).

[11] Pierce, G. B., and E. L. Verney, *Cancer*, **14**, 1017 (1961).

[12] Sato, G., and Y. Yasumura, *Trans. N. Y. Acad. Sci., Ser. II*, **28**, 1063 (1966).

[13] Stevens, L. C., *J. Natl. Cancer Inst.*, **20**, 1257 (1958).

[14] Stevens, L. C., *Develop. Biol.*, **2**, 285 (1960).

[15] Ford, C. E., J. L. Hamerton, D. W. H. Barnes, and J. F. Loutit, *Nature*, **177**, 452 (1956).

[16] Recently, Miss Miriam D. Rosenthal and Dr. Gordon Sato have kindly informed us in a personal communication that using a procedure of alternate culture and animal passages, followed by serial cloning, they have adapted teratoma 402 AIII, obtained from Dr. Leroy Stevens in 1964, to monolayer tissue culture. Multipotentiality of these cultures was maintained for over 6 months *in vitro*. Tumors produced by the inoculation of these cultures showed a low incidence of differentiated tissues, similar to that of the parent strain. Prolonged culture did, however, often result in progression to yolk sac carcinoma. The cell lines retained a pseudodiploid karyotype.

[17] Davidson, R. L., and B. Ephrussi, *Nature*, **205**, 1170 (1965).

[18] Ephrussi, B., and M. C. Weiss, these PROCEEDINGS, **53**, 1040 (1965).

[19] Scaletta, L. J., and B. Ephrussi, *Nature*, **205**, 1169 (1965).

13

Reprinted by permission from *Nature* **256**:495–497 (1975)

CONTINUOUS CULTURES OF FUSED CELLS SECRETING ANTIBODY OF PREDEFINED SPECIFICITY

G. Köhler and C. Milstein

THE manufacture of predefined specific antibodies by means of permanent tissue culture cell lines is of general interest. There are at present a considerable number of permanent cultures of myeloma cells[1,2] and screening procedures have been used to reveal antibody activity in some of them. This, however, is not a satisfactory source of monoclonal antibodies of predefined specificity. We describe here the derivation of a number of tissue culture cell lines which secrete anti-sheep red blood cell (SRBC) antibodies. The cell lines are made by fusion of a mouse myeloma and mouse spleen cells from an immunised donor. To understand the expression and interactions of the Ig chains from the parental lines, fusion experiments between two known mouse myeloma lines were carried out.

Each immunoglobulin chain results from the integrated expression of one of several V and C genes coding respectively for its variable and constant sections. Each cell expresses only one of the two possible alleles (allelic exclusion; reviewed in ref. 3). When two antibody-producing cells are fused, the products of both parental lines are expressed[4,5], and although the light and heavy chains of both parental lines are randomly joined, no evidence of scrambling of V and C sections is observed[4]. These results, obtained in an heterologous system involving cells of rat and mouse origin, have now been confirmed by fusing two myeloma cells of the same mouse strain, and provide the background for the derivation and understanding of antibody-secreting hybrid lines in which one of the parental cells is an antibody-producing spleen cell.

Two myeloma cell lines of BALB/c origin were used. P1Bul is resistant to 5-bromo-2'-deoxyuridine[4], does not grow in selective medium (HAT, ref. 6) and secretes a myeloma protein, Adj PC5, which is an IgG2A (κ), (ref. 1). Synthesis is not balanced and free light chains are also secreted. The second cell line, P3-X63Ag8, prepared from P3 cells[2], is resistant to 20 µg ml^{-1} 8-azaguanine and does not grow in HAT medium. The protein secreted (MOPC 21) is an IgG1 (κ) which has been fully sequenced[7,8]. Equal numbers of cells from each parental line were fused using inactivated Sendai virus[9] and samples

contining 2×10^5 cells were grown in selective medium in separate dishes. Four out of ten dishes showed growth in selective medium and these were taken as independent hybrid lines, probably derived from single fusion events. The karyotype of the hybrid cells after 5 months in culture was just under the sum of the two parental lines (Table 1). Figure 1 shows the isoelectric focusing[10] (IEF) pattern of the secreted products of different lines. The hybrid cells (samples *c–h* in Fig. 1) give a much more complex pattern than either parent (*a* and *b*) or a mixture of the parental lines (*m*). The important feature of the new pattern is the presence of extra bands (Fig. 1, arrows). These new bands, however, do not seem to be the result of differences in primary structure; this is indicated by the IEF pattern of the products after reduction to separate the heavy and light chains (Fig. 1*B*). The IEF pattern of chains of the hybrid clones (Fig. 1*B*, *g*) is equivalent to the sum of the IEF pattern (*a* and *b*) of chains of the parental clones with no evidence of extra products. We conclude that, as previously shown with interspecies hybrids[4,5], new Ig molecules are produced as a result of mixed association between heavy and light chains from the two parents. This process is intracellular as a mixed cell population does not give rise to such hybrid molecules (compare *m* and *g*, Fig. 1*A*). The individual cells must therefore be able to express both isotypes. This result shows that in hybrid cells the expression of one isotype and idiotype does not exclude the expression of another: both heavy chain isotypes (γ1 and γ2a) and both V_H and both V_L regions (idiotypes) are expressed. There are no allotypic markers for the C_K region to provide direct proof for the expression of both parental C_K regions. But this is indicated by the phenotypic link between the V and C regions.

Figure 1*A* shows that clones derived from different hybridisation experiments and from subclones of one line are indistinguishable. This has also been observed in other experiments (data not shown). Variants were, however, found in a survey of 100 subclones. The difference is often associated with changes

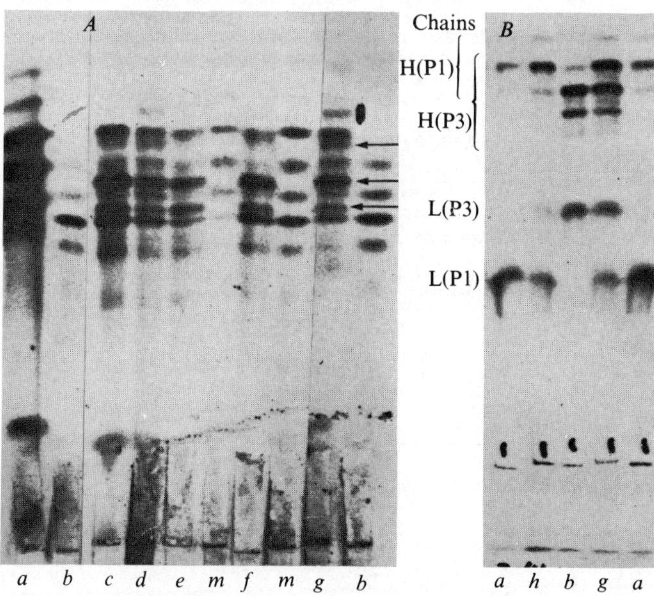

Fig. 1 Autoradiograph of labelled components secreted by the parental and hybrid cell lines analysed by IEF before (*A*) and after reduction (*B*). Cells were incubated in the presence of ^{14}C-lysine[14] and the supernatant applied on polyacrylamide slabs. *A*, pH range 6.0 (bottom) to 8.0 (top) in 4 M urea. *B*, pH range 5.0 (bottom) to 9.0 (top) in 6 M urea; the supernatant was incubated for 20 min at 37 °C in the presence of 8 M urea, 1.5 M mercaptoethanol and 0.1 M potassium phosphate pH 8.0 before being applied to the right slab. Supernatants from parental cell lines in: *a*, P1Bul; *b*, P3-X67Ag8; and *m*, mixture of equal number of P1Bul and P3-X67Ag8 cells. Supernatants from two independently derived hybrid lines are shown: *c–f*, four subclones from Hy-3; *g* and *h*, two subclones from Hy-B. Fusion was carried out[4,9] using 10^6 cells of each parental line and 4,000 haemagglutination units inactivated Sendai virus (Searle). Cells were divided into ten equal samples and grown separately in selective medium (HAT medium, ref. 6). Medium was changed every 3 d. Successful hybrid lines were obtained in four of the cultures, and all gave similar IEF patterns. Hy-B and Hy-3 were further cloned in soft agar[14]. L, Light; H, heavy.

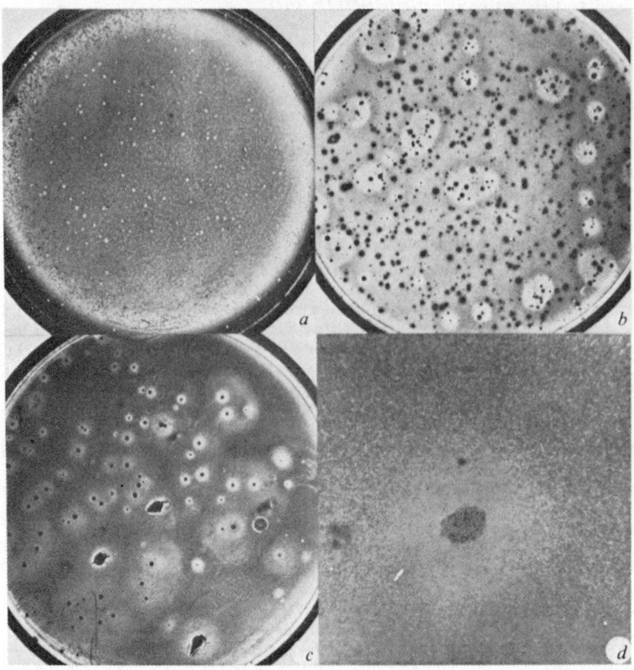

Fig. 2 Isolation of an anti-SRBC antibody-secreting cell clone. Activity was revealed by a halo of haemolysed SRBC. Direct plaques given by: *a*, 6,000 hybrid cells Sp-1; *b*, clones grown in soft agar from an inoculum of 2,000 Sp-1 cells; *c*, recloning of one of the positive clones Sp-1/7; *d*, higher magnification of a positive clone. Myeloma cells (10^7 P3-X67A g8) were fused to 10^8 spleen cells from an immunised BALB/c mouse. Mice were immunised by intraperitoneal injection of 0.2 ml packed SRBC diluted 1:10, boosted after 1 month and the spleens collected 4 d later. After fusion, cells (Sp-1) were grown for 8 d in HAT medium, changed at 1–3 d intervals. Cells were then grown in Dulbecco modified Eagle's medium, supplemented for 2 weeks with hypoxanthine and thymidine. Forty days after fusion the presence of anti-SRBC activity was revealed as shown in *a*. The ratio of plaque forming cells/total number of hybrid cells was 1/30. This hybrid cell population was cloned in soft agar (50% cloning efficiency). A modified plaque assay was used to reveal positive clones shown in *b–d* as follows. When cell clones had reached a suitable size, they were overlaid in sterile conditions with 2 ml 0.6% agarose in phosphate-buffered saline containing 25 µl packed SRBC and 0.2 ml fresh guinea pig serum (absorbed with SRBC) as source of complement. *b*, Taken after overnight incubation at 37 °C. The ratio of positive/total number of clones was 1/33. A suitable positive clone was picked out and grown in suspension. This clone was called Sp-1/7, and was recloned as shown in *c*; over 90% of the clones gave positive lysis. A second experiment in which 10^6 P3-X67Ag8 cells were fused with 10^8 spleen cells was the source of a clone giving rise to indirect plaques (clone Sp-2/3-3). Indirect plaques were produced by the addition of 1:20 sheep anti-MOPC 21 antibody to the agarose overlay.

in the ratios of the different chains and occasionally with the total disappearance of one or other of the chains. Such events are best visualised on IEF analysis of the separated chains (for example, Fig. 1*h*, in which the heavy chain of P3 is no longer observed). The important point that no new chains are detected by IEF complements a previous study[1] of a rat–mouse hybrid line in which scrambling of V and C regions from the light chains of rat and mouse was not observed. In this study, both light chains have identical C_κ regions and therefore scrambled V_L–C_L molecules would be undetected. On the other hand, the heavy chains are of different subclasses and we expect scrambled V_H–C_H to be detectable by IEF. They were not observed in the clones studied and if they occur must do so at a lower frequency. We conclude that in syngeneic cell hybrids (as well as in interspecies cell hybrids) V–C integration is not the result of cytoplasmic events. Integration as a result of DNA translocation or rearrangement during transcription is also suggested by the presence of integrated mRNA molecules[11] and by the existence of defective heavy chains in which a deletion of V and C sections seems to take place in already committed cells[12].

The cell line P3-X63Ag8 described above dies when exposed to HAT medium. Spleen cells from an immunised mouse also die in growth medium. When both cells are fused by Sendai virus and the resulting mixture is grown in HAT medium, surviving clones can be observed to grow and become established after a few weeks. We have used SRBC as immunogen, which enabled us, after culturing the fused lines, to determine the presence of specific antibody-producing cells by a plaque assay technique[13] (Fig. 2*a*). The hybrid cells were cloned in soft agar[14] and clones producing antibody were easily detected by an overlay of SRBC and complement (Fig. 2*b*). Individual clones were isolated and shown to retain their phenotype as almost all the clones of the derived purified line are capable of lysing SRBC (Fig. 2*c*). The clones were visible to the naked eye (for example, Fig. 2*d*). Both direct and indirect plaque

assays[13] have been used to detect specific clones and representative clones of both types have been characterised and studied.

The derived lines (Sp hybrids) are hybrid cell lines for the following reasons. They grow in selective medium. Their karyotype after 4 months in culture (Table 1) is a little smaller than the sum of the two parental lines but more than twice the chromosome number of normal BALB/c cells, indicating that the lines are not the result of fusion between spleen cells. In addition the lines contain a metacentric chromosome also present in the parental P3-X67Ag8. Finally, the secreted immunoglobulins contain MOPC 21 protein in addition to new, unknown components. The latter presumably represent the chains derived from the specific anti-SRBC antibody. Figure 3*A* shows the IEF pattern of the material secreted by two such Sp hybrid clones. The IEF bands derived from the parental P3 line are visible in the pattern of the hybrid cells, although obscured by the presence of a number of new bands. The pattern is very complex, but the complexity of hybrids of this type is likely to result from the random recombination of chains (see above, Fig. 1). Indeed, IEF patterns of the reduced material secreted by the spleen–P3 hybrid clones gave a simpler pattern of Ig chains. The heavy and light chains of the P3 parental line became prominent, and new bands were apparent.

The hybrid Sp-1 gave direct plaques and this suggested that it produces an IgM antibody. This is confirmed in Fig. 4 which shows the inhibition of SRBC lysis by a specific anti-IgM

Table 1 Number of chromosomes in parental and hybrid cell lines

Cell line	Number of chromosomes per cell	Mean
P3-X67Ag8	66,65,65,65,65	65
P1Bu1	Ref. 4	55
Mouse spleen cells	—	40
Hy-B (P1–P3)	112,110,104,104,102	106
Sp-1/7-2	93,90,89,89,87	90
Sp-2/3-3	97,98,96,96,94,88	95

antibody. IEF techniques usually do not reveal 19S IgM molecules. IgM is therefore unlikely to be present in the unreduced sample *a* (Fig. 3*B*) but μ chains should contribute to the pattern obtained after reduction (sample *a*, Fig. 3*A*).

The above results show that cell fusion techniques are a powerful tool to produce specific antibody directed against a predetermined antigen. It further shows that it is possible to isolate hybrid lines producing different antibodies directed against the same antigen and carrying different effector functions (direct and indirect plaque).

The uncloned population of P3–spleen hybrid cells seems quite heterogeneous. Using suitable detection procedures it should be possible to isolate tissue culture cell lines making different classes of antibody. To facilitate our studies we have used a myeloma parental line which itself produced an Ig. Variants in which one of the parental chains is no longer expressed seem fairly common in the case of P1–P3 hybrids (Fig. 1*h*). Therefore selection of lines in which only the specific antibody chains are expressed seems reasonably simple. Alternatively, non-producing variants of myeloma lines could be used for fusion.

We used SRBC as antigen. Three different fusion experiments were successful in producing a large number of antibody-producing cells. Three weeks after the initial fusion, 33/1,086

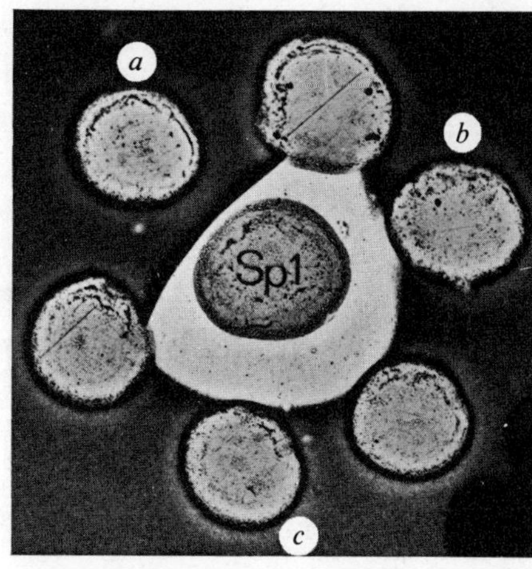

Fig. 4 Inhibition of haemolysis by antibody secreted by hybrid clone Sp-1/7-2. The reaction was in a 9-cm Petri dish with a layer of 5 ml 0.6 % agarose in phosphate-buffered saline containing 1/80 (v/v) SRBC. Centre well contains 2.5 μl 20 times concentrated culture medium of clone Sp-1/7-2 and 2.5 μl mouse serum. *a*, Sheep specific anti-mouse macroglobulin (MOPC 104E, Dr Feinstein); *b*, sheep anti-MOPC 21 (P3) IgG1 absorbed with Adj PC-5; *c*, sheep anti-Adj PC-5 (IgG2a) absorbed with MOPC 21. After overnight incubation at room temperature the plate was developed with guinea pig serum diluted 1:10 in Dulbecco's medium without serum.

clones (3 %) were positive by the direct plaque assay. The cloning efficiency in the experiment was 50 %. In another experiment, however, the proportion of positive clones was considerably lower (about 0.2 %). In a third experiment the hybrid population was studied by limiting dilution analysis. From 157 independent hybrids, as many as 15 had anti-SRBC activity. The proportion of positive over negative clones is remarkably high. It is possible that spleen cells which have been triggered during immunisation are particularly successful in giving rise to viable hybrids. It remains to be seen whether similar results can be obtained using other antigens.

The cells used in this study are all of BALB/c origin and the hybrid clones can be injected into BALB/c mice to produce solid tumours and serum having anti-SRBC activity. It is possible to hybridise antibody-producing cells from different origins[4,5]. Such cells can be grown *in vitro* in massive cultures to provide specific antibody. Such cultures could be valuable for medical and industrial use.

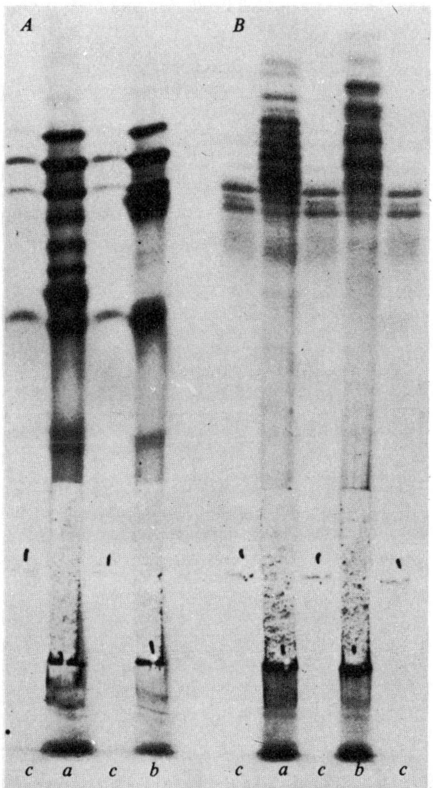

Fig. 3 Autoradiograph of labelled components secreted by anti-SRBC specific hybrid lines. Fractionation before (*B*) and after (*A*) reduction was by IEF. *p*H gradient was 5.0 (bottom) to 9.0 (top) in the presence of 6 M urea. Other conditions as in Fig. 1. Supernatants from: *a*, hybrid clone Sp-1/7-2; *b*, hybrid clone Sp-2/3-3; *c*, myeloma line P3-X67Ag8.

Received May 14; accepted June 26, 1975.

1 Potter, M., *Physiol. Rev.*, **52**, 631-719 (1972).
2 Horibata, K., and Harris, A. W., *Expl Cell Res.*, **60**, 61-70 (1970).
3 Milstein, C., and Munro, A. J., in *Defence and Recognition* (edit. by Porter, R. R.), 199-228 (MTP Int. Rev. Sci., Butterworth, London, 1973).
4 Cotton, R. G. H., and Milstein. C., *Nature*, **244**, 42-43 (1973).
5 Schwaber, J., and Cohen, E. P., *Proc. natn. Acad. Sci. U.S.A.*, **71**, 2203-2207 (1974).
6 Littlefield, J. W., *Science*, **145**, 709 (1964).
7 Svasti, J., and Milstein, C., *Biochem. J.*, **128**, 427-444 (1972).
8 Milstein, C., Adetugbo, K., Cowan, N. J., and Secher, D. S., *Progress in Immunology*, II, 1 (edit. by Brent, L., and Holborow, J.), 157-168 (North-Holland, Amsterdam, 1974).
9 Harris, H., and Watkins, J. F., *Nature*, **205**, 640-646 (1965).
10 Awdeh, A. L., Williamson, A. R., and Askonas, B. A., *Nature*, **219**, 66-67 (1968).
11 Milstein, C., Brownlee, G. G., Cartwright, E. M., Jarvis, J. M., and Proudfoot, N. J., *Nature*, **252**, 354-359 (1974).
12 Frangione, B., and Milstein, C., *Nature*, **244**, 597-599 (1969).
13 Jerne, N. K., and Nordin, A. A., *Science*, **140**, 405 (1963).
14 Cotton, R. G. H., Secher, D. S., and Milstein, C., *Eur. J. Immun.*, **3**, 135-140 (1973).

14

Reprinted by permission from *Nature* **226**:717-722 (1970)

Mammalian Cell Fusion: Induction of Premature Chromosome Condensation in Interphase Nuclei

by
R. T. JOHNSON
P. N. RAO

Eleanor Roosevelt Institute for Cancer
Research and Department of Biophysics,
University of Colorado Medical Center
Denver, Colorado 80220

Further evidence is put forward for the inducibility of chromosome condensation, particularly its consequences in interphase cells that are not ready for mitosis.

WE have shown for multinucleate HeLa cells formed by the fusion of cells from different phases of the cell cycle that DNA synthesis and mitosis are inducible[1]. Here we present further evidence for the inducibility of chromosome condensation and the consequences of such an induction in interphase ($G1$, S, and $G2$) cells which are not ready to enter mitosis. In heterophasic multinucleate cells formed by the fusion of S and $G2$ cells, the initiation of mitosis in one or more of the advanced nuclei resulted in the premature chromosome condensation (PCC) of the lagging nucleus or nuclei. Premature chromosome condensation, which is a kind of atypical mitosis without the mitotic

Table 1. SYNCHRONY IN THE FIRST MITOSIS AFTER CELL FUSION

Type of cell fusion		Binucleate		Trinucleate		Tetranucleate	
		Homophasic	Heterophasic	Homophasic	Heterophasic	Homophasic	Heterophasic
$G1/S$*	Number of asynchronous mitoses	2[a]	1[b]	0	0	0	0
	Number of synchronous mitoses	332	83	112	56	25	24
	Per cent of asynchronous mitoses	0·6	1·2	0	0	0	0
$G1/G2$*	Number of asynchronous mitoses	1[c]	0	0	1[d]	0	0
	Number of synchronous mitoses	402	70	119	60	25	25
	Per cent of asynchronous mitoses	0·25	0	0	1·6	0	0
S*$/G2$	Number of asynchronous mitoses	6[e]	6[f]	0	10[g]	0	6[h]
	Number of synchronous mitoses	313	49	127	81	51	70
	Per cent of asynchronous mitoses	1·9	10·9	0	11·0	0	7·9

Details of the various types of cells grouped under each superscript are listed below. I denotes interphase; P, prophase; M, metaphase. The digits to the left of the parentheses indicate the number of such cells.

$G1/S$* fusion: a: $2(M^*/I^*)$
          b: $1(I^*/M)$
$G1/G2$* fusion: c: $1(P/I)$
          d: $1(2M^*/P)$
S*$/G2$ fusion: e: $1(P^*/I^*)$; $2(I/M)$; $3(I/P)$
          f: $3(I^*/M)$; $3(I^*/P)$
          g: $3(2I^*/M)$; $1(I^*/P^*/P)$; $1(I^*/M^*/M)$; $2(P^*/2M)$; $1(I^*/2P)$; $2(I^*/2M)$
          h: $1(2I^*/P^*/P)$; $2(2I^*/2P)$; $1(2I^*/2M)$; $1(I^*/M^*/2M)$; $1(I^*/I/2P)$

* Indicates prelabelled nucleus.

spindle, could be readily induced in interphase cells by fusing them with mitotic cells. Induction of chromosome condensation in $G1$ and $G2$ nuclei resulted in the appearance of chromosomes with single and double chromatids respectively, whereas the S nuclei yielded unevenly condensed chromatin.

We used HeLa cells grown as suspension cultures in Eagle's minimal essential medium supplemented with sodium pyruvate, glutamine and 5 per cent foetal calf serum. The culture flasks were gassed with a mixture of carbon dioxide (5 per cent) in air. We have already described[1] the procedures for obtaining synchronous populations of $G1$, S and $G2$ and the technique of cell fusion by UV-inactivated Sendai virus. Mitotic cells prelabelled with ^{3}H-thymidine (0·05 μCi/ml.; 6·7 Ci/ mmole) were obtained by treating the cells with colcemide ($6·8 \times 10^{-7}$ M) for 19 to 24 h. The method of preparation of the multinucleate cells for autoradiography and the terminology used to describe various types of multinucleate cells are also given in ref. 1. Chromosome preparations were made by giving a hypotonic treatment before the cells were centrifuged on to the slides by cytocentrifuge (Shandon-Elliott). After fixation in 3 : 1 absolute ethanol and glacial acetic acid for 5 min and hydrolysis in 1 N HCl at 60° C for 2·5 min the cells were stained with crystal violet.

Mitotic Synchrony

The incidence of asynchronous mitoses in the first division following cell fusion in multinucleate homo- and heterophasic cells resulting from $G1/S$, $G1/G2$ and $S/G2$ fusions is shown in Table 1. It is clear that in both $G1/S$ and $G1/G2$ fusions the incidence of mitotic asynchrony in the heterophasic as well as in homophasic cells was very small. In the $S/G2$ fusion, however, the incidence of mitotic asynchrony in heterophasic cells was about 10 per cent, significantly greater than that in homophasic cells. Further, the nuclei which first entered mitosis were without exception the $G2$ and not the S phase. The small incidence of mitotic asynchrony in $G1/S$ and $G1/G2$ fusions suggests that these heterophasic components can equilibrate their resources and reach mitosis together[1]. On the other hand, for at least a portion of the population of the $G2/S$ cells, the $G2$ nucleus entered mitosis while the S nucleus remained in interphase. The asynchronous entry of the $G2$ nucleus into mitosis sometimes induced similar changes in the S nucleus producing the phenomenon of premature chromosome condensation (PCC), as shown in Fig. 1. The PCC in multinucleate cells was characterized by the presence of one or more normal sets of chromosomes derived from the $G2$ nuclei along with one or more S nuclei which had undergone irregular condensation of chromatin.

The incidence of PCC was greatest in multinucleate cells containing a large proportion of $G2$ to S nuclei. The

greater the ratio of $G2$ to S nuclei (3 : 1 and 2 : 1), the greater was the incidence of PCC of the S nuclei, both initially and throughout the period of investigation (Fig. 2). The greatest incidence of PCC in these cells

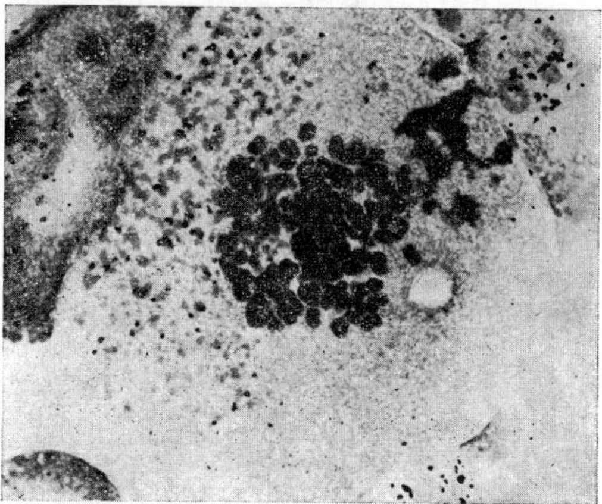

Fig. 1. Premature chromosome condensation in a heterophasic binucleate cell formed by the fusion of S and $G2$ cells. The S cells were labelled with ^{3}H-thymidine before cell fusion while the $G2$ cells were not. The initiation of mitosis in the $G2$ nucleus resulted in the premature condensation of the S-chromatin which could be identified by the silver grains in the autoradiograph.

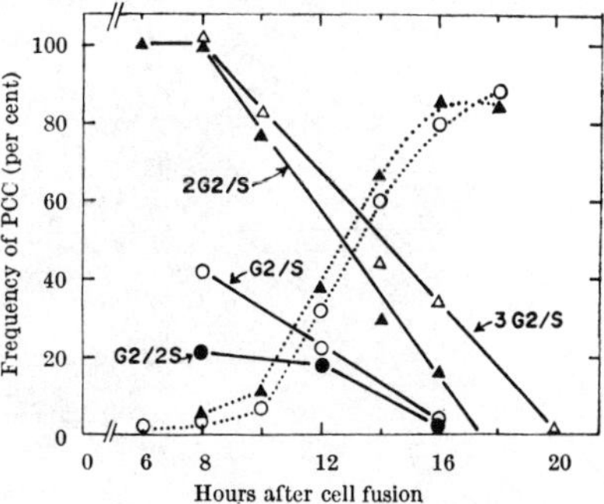

Fig. 2. The frequency of premature chromosome condensation (PCC) in various classes of heterophasic $S/G2$ cells as a function of time after the completion of cell fusion. The S nuclei were prelabelled with ^{3}H-thymidine. The mitotic accumulation functions (represented by the dotted lines) were collected from a different experiment in which colcemide was added immediately after cell fusion between S and $G2$ populations. The data on the frequency of PCC in various classes were collected from a similar experiment to which no colcemide was added following cell fusion. △—△, 3$G2$/1S; ▲—▲, 2$G2$/1S; ○—○, $G2/S$; ●—●, $G2$/2S. The mitotic accumulation functions for the classes are 2$G2/S$ (▲ - - - ▲) and $G2/S$ (○ - - - ○).

occurred when the wave of mitoses was just beginning (see the dotted lines in Fig. 2). This suggests that the most likely multinucleate cells to contain prematurely condensed chromosomes are those that entered division as the earliest of their class. These cells presumably contained an advanced $G2$ nucleus which entered mitosis asynchronously, thus inducing PCC in the lagging "S" nucleus. The decrease in the frequency of PCC as a function of time (Fig. 2) indicates that those nuclei of the heterophasic cells which remain in interphase for a longer time after cell fusion have a better chance of achieving mitotic synchrony and consequently a smaller incidence of PCC.

Fusion with Synchronized Cells

The premature condensation of chromatin of the S nuclei after the fusion between $G2$ and S phase cells strongly suggested that a mitotic inducer was present in either late $G2$ or mitotic cells and that it was capable of pulling a lagging nucleus into mitosis in a dose-dependent manner. In the following experiments synchronized mitotic cells (blocked by either colcemide or N_2O) were fused with synchronous populations of $G1$, S and $G2$ cells to see whether mitosis could be induced in these interphase nuclei. When mitotic cells were fused with $G1$ and $G2$ cells, there was a rapid condensation of the

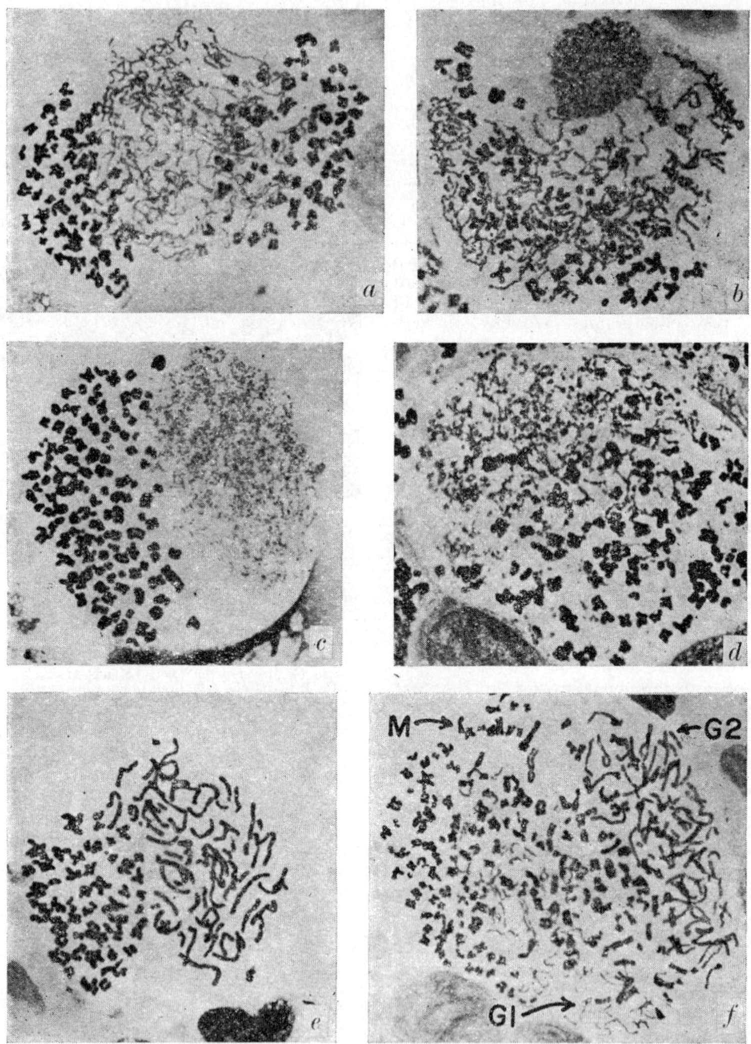

Fig. 3. Premature chromosome condensation in interphase nuclei. *a*, PCC of the $G1$ nucleus in a heterophasic $M/G1$ cell at 30 min after the addition of virus. Note the long and slender $G1$ chromosomes with single chromatids. The more condensed chromosomes are from the mitotic cell blocked by colcemide. *b*, Heterophasic cell formed by the fusion of $2G1$ cells with one mitotic cell. PCC can be observed in the $G1$ nuclei, one of which still shows its nuclear outline. *c*, Premature condensation of the S chromatin at 45 min after the fusion between mitotic and S phase cells. The chromatin of the S nucleus presents a fragmented appearance. *d*, Premature chromosome condensation of a nucleus blocked at the $G1$–S boundary by double thymidine treatment. Note that the pattern of condensation is intermediate between that of $G1$ and S nuclei. *e*, PCC of the $G2$ nucleus in an $M/G2$ heterophasic cell. The more condensed chromosomes are from the mitotic cell treated with colcemide for 19 h. *f*, A heterophasic cell formed by the fusion between $G1$, $G2$ and mitotic cells. Note the differences between $G1$, $G2$ and metaphase (M) chromosomes.

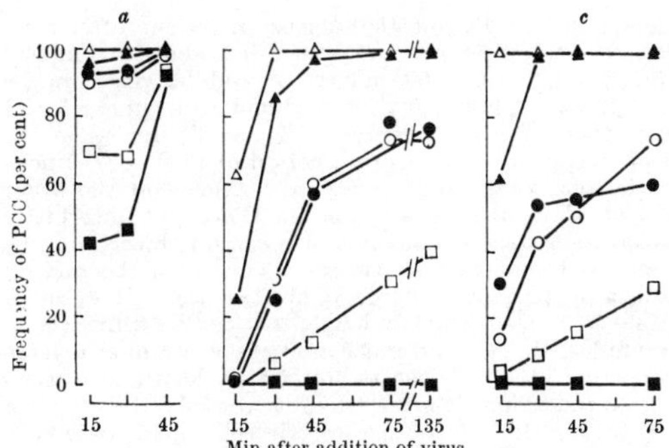

Fig. 4. The frequency of PCC in various classes of heterophasic cells obtained by fusing mitotic cells with (a) G1; (b) S; and (c) G2 cells as a function of time. △, 3M/(1G1 or 1S or 1G2); ▲, 2M/(1G1 or 1S or 1G2); ○, 1M/(1G1 or 1S or 1G2); ●, 2M/(2G1 or 2S or 2G2); □, 1M/(2G1 or 2S or 2G2); ■, 1M/(3G1 or 3S or 3G2).

interphase chromatin into chromosomal patterns. The G1 chromatin condensed into chromosomes with single chromatids, as might be expected from the unreplicated amount of DNA present (Fig. 3a, b) and the G2 chromatin condensed into chromosomes with double chromatids (Fig. 3e, f). The S chromatin, on the other hand, usually condensed less completely to form a patchwork of large and small fragments interspersed with material that is hardly condensed at all (Fig. 3c). In one of the experiments HeLa cells, blocked at the G1–S boundary by excess thymidine (2·5 mM) treatment for 18 h, were fused with mitotic cells. The pattern of chromosome condensation in these interphase nuclei was intermediate between the patterns of G1 and S phase condensation (Fig. 3d). In these nuclei the degree of chromosome condensation is much greater than in S phase nuclei, but these chromosomes have a more fragmented appearance than those from G1 nuclei.

DNA synthesis in the prematurely condensed S nuclei was studied at regular intervals by giving a 15 min pulse treatment with ^{3}H-thymidine (1·0 μCi/ml.; 6·7 Ci/mmole) starting 75 min after the addition of virus. This interval was necessary to achieve the induction of PCC in almost all the heterophasic cells. At 75 min the average number of grains per nucleus was 9·0 for the control (mononucleate S phase cells) and 4·3 for the prematurely condensed S nuclei, indicating that the rate of DNA synthesis in the PCC decreased to 48 per cent of that in the control. At 3·25 h the rate of DNA synthesis in the PCC was only 18 per cent of that in the control. The extent of DNA synthesis in the PCC, as measured by grain counts, was related to the proportion of mitotic to S phase nuclei in the heterophasic cells. The greater the ratio of mitotic to S phase nuclei, the greater was the inhibition of DNA synthesis (Table 2). These data demonstrate that the presence of metaphase chromosomes in the same cytoplasm does not immediately stop DNA synthesis in S phase nuclei, even though marked structural changes have been induced in the replicating chromosomes.

The morphological changes observed in the induction of PCC in interphase nuclei parallel exactly those seen in a normal mitosis. A prophase-like pattern was the first indication of division, and this was rapidly followed by dissolution of the nuclear membrane and the greater degree of condensation of the chromatin into whole or "fragmented" chromosomes according to the position occupied by the affected cell in the life cycle at the time

of fusion. There was no sign that the induction of chromosome condensation in these cells was accompanied by the appearance of a mitotic spindle.

Induction of PCC as a Function of Dose

The ability of a mitotic cell to induce PCC in an interphase nucleus depends largely on the ratio of mitotic to interphase nuclei in the cell at the time of fusion. The details of this dose response for three different fusions, M/G1, M/S and M/G2, are shown in Fig. 4. In each case it is clear that the greater the ratio of mitotic to interphase nuclei, the more rapid is the induction of mitosis and the greater is the proportion of cells in the class that shows induction. Compare, for example, the differences in the speed and extent of mitotic induction between the classes 3M/1G1 and 1M/3G1, 3M/1S and 1M/3S, and 3M/1G2 and 1M/3G2. From this figure it is also evident that PCC could be induced far more readily in G1 nuclei than in either S or G2 nuclei. After 45 min almost all of the cells containing 1M/3G1 nuclei were induced (Fig. 4a). In comparison, none of the 1M/3S or 1M/3G2 cells were induced even after 75 min (Fig. 4b, c). In general, the initial frequency of PCC in M/S cells was lower than that in M/G2 in almost all the combinations.

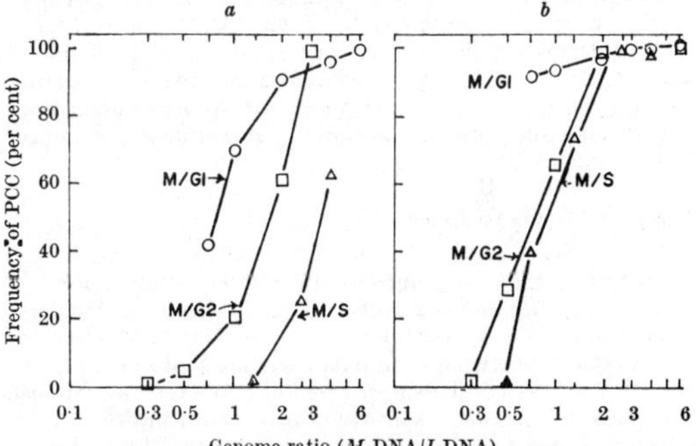

Fig. 5. The frequency of premature chromosome condensation in various heterophasic M/G1, M/S and M/G2 cells at (a) 15 and (b) 45 min after addition of virus plotted as a function of the ratio of mitotic DNA to interphase DNA in the cells (genome ratio). The genome ratio is plotted on a log scale. □, M/G2 cells; ○, M/G1 cells; △, M/S cells.

The data presented so far in Figs. 3 and 4 clearly indicate that the speed and degree of premature condensation of chromatin vary with the position of the interphase cell in the division cycle. The data presented in Fig. 4 were replotted on the basis of the ratios of mitotic to interphase DNA (genome ratios) as shown in Fig. 5. The genome ratios of mitotic DNA to interphase (inducible) DNA were obtained by assuming that a mitotic cell contained two genome units of DNA, while a cell in G1 contained one genome unit of DNA. A cell in S phase was assumed to have 1·5 units of DNA, intermediate between G1 and G2 states.

The derivation of the genome ratios for the various classes of multinucleate cells in each of the three fusions is shown in Table 3. When the frequencies of PCC are plotted against the genome ratios it becomes clear that the susceptibilities for the induction of premature condensation in the G1, S and G2 nuclei are different (Fig. 5a). The induction of PCC in G1 nuclei occurs at a smaller genome ratio than in either S or G2. At 45 min after the addition of virus the frequencies of PCC in M/S and M/G2

Table 2. DNA SYNTHESIS IN PREMATURELY CONDENSED S NUCLEI

	Time (min after addition of virus)	3M/1S	2M/1S	PCC scored 1M/1S	1M/2S	Total
Per cent incidence of DNA synthesis in PCC (number of cells scored for each class is given in parentheses)	75	20·0 (5)	45·0 (20)	70·6 (48)	100·0 (21)	69·0 (94)
	195	0 (7)	11·0 (18)	44·4 (54)	77·8 (9)	37·0 (88)
Average number of grains per PCC nucleus (number of PCC on which grains were counted is given in parentheses)	75	2·0 (1)	3·1 (9)	4·8 (34)	4·5 (42)	4·3 (88)
	195	0 (0)	1·0 (2)	3·2 (24)	3·5 (14)	3·2 (40)

Table 3. RATIO OF MITOTIC DNA/INTERPHASE DNA EXPRESSED AS GENOME UNITS*

Ratio of mitotic: interphase cells at the time of fusion	$M/G1$ (genome units)	Genome ratio	M/S (genome units)	Genome ratio	$M/G2$ (genome units)	Genome ratio
3 : 1	$3 \times 2/1$	6·0	$3 \times 2/1·5$	4·0	$3 \times 2/2$	3·0
2 : 1	$2 \times 2/1$	4·0	$2 \times 2/1·5$	2·7	$2 \times 2/2$	2·0
1 : 1	$2/1$	2·0	$2/1·5$	1·3	$2/2$	1·0
1 : 2	$2/2 \times 1$	1·0	$2/2 \times 1·5$	0·66	$2/2 \times 2$	0·5
1 : 3	$2/3 \times 1$	0·66	$2/3 \times 1·5$	0·44	$2/3 \times 2$	0·33

* One genome unit is defined as the average amount of DNA in a $G1$ cell.

cells were similar but still much smaller than in $M/G1$ cells (Fig. 5*b*). Are there any differences in the susceptibility to induction of PCC of $G1$, S, and $G2$ chromatin when the genome ratios

$$\left(\frac{M\text{–DNA}}{I\text{–DNA}} \right)$$

are identical? For example, the genome ratio of the $1M/2G1$ cells is the same

$$\left(\frac{M\text{–DNA}}{I\text{–DNA}} \right) = 1$$

as that of $1M/1G2$ cells, but their degree of susceptibility was significantly different. Differences in the susceptibilities were also observed between $2M/1G1$ and $3M/1S$ cells, both of which have a genome ratio of 4 (Table 3 and Fig. 5*a*). This suggests that the induction of premature chromosome condensation in interphase nuclei does not depend solely on the amount of interphase chromatin to be condensed.

Support for Hypothesis

The results reported here are consistent with the hypothesis that chromosomal condensation can be induced in mammalian cells. The fusion of interphase cells with those in metaphase usually leads to the condensation of the interphase chromatin within a period of 30 to 60 min. The condensation pattern of the interphase chromatin does not occur randomly but follows the sequence of events of a normal mitosis. The final pattern of condensation of the chromatin depends on exactly which stage of interphase the cell occupied at the time of fusion. The chromatin of $G1$ and $G2$ phase nuclei condensed to form chromosomes with single and double chromatids respectively, but the chromatin of S phase nuclei always condensed into a less clearly defined pattern of light and dark staining fragments. In comparison to the S phase nuclei (Fig. 3*c*), the greater degree of chromosome condensation seen in the cells blocked at the $G1$–S boundary by excess thymidine double block (Fig. 3*d*) suggests that the chromatin in these nuclei is probably primed for DNA replication but is not in an active state of DNA synthesis. Removal of the thymidine block leads to the rapid onset of DNA synthesis and incidentally to the characteristic pattern of S phase chromatin condensation of Fig. 3*c*. This difference in the patterns of condensation between S phase chromatin and either $G1$ or $G2$ chromatin suggests that the state of chromatin during the period of DNA replication is different. The physical state of replicating DNA is different from the non-replicating interphase DNA[2,3]. We show here that condensed S phase DNA is still capable of replication, although at a reduced rate. The inhibition of DNA synthesis in the PCC increased with time and with the proportion of mitotic nuclei in the cells. Presumably these two factors influence the rate and degree of condensation of the S chromatin, reducing the number of replicating sites on the DNA double helix.

The dosage effects observed in the induction of PCC may be provisionally interpreted as a titration of the inducer molecules responsible for the condensation of interphase chromatin. G1 phase nuclei are most readily induced into PCC. Even when there were three G1

nuclei and a single mitotic nucleus, at 30 min after completion of fusion 90 per cent of the cells showed complete induction in all the $G1$ nuclei. S and $G2$ phase nuclei were, however, less easily induced and, in a similar 3 to 1 ratio, there was no induction. These differences in the speed and completeness of induction of PCC in $G1$ nuclei on the one hand and the $G2$ or S nuclei on the other suggest that as the amount of DNA per interphase nucleus increases the amount of mitotic inducer must be increased to produce an effect. Even when the mitotic inducer is present in large quantities the induction of S or $G2$ nuclei requires longer periods of time to be accomplished. In multinucleate cells containing a large proportion of interphase to mitotic nuclei there was no mitotic induction. In an attempt to study the effect of PCC on the division of the mitotic cells, interphase cells were fused with mitotic cells obtained by a reversible nitrous oxide block. When the proportion of interphase to mitotic nuclei was 3 : 1 or greater in the heterophasic cells, the mitotic nucleus became pycnotic. This suggests that the presence of a large number of interphase nuclei can inhibit the completion of nuclear division of a metaphase nucleus, and that this interaction is again dose-dependent.

Mitotic : Interphase DNA Ratio

The induction of premature chromosome condensation in interphase nuclei is related, at least in part, to the proportion of the mitotic to interphase nuclei, and therefore to the relative amounts of mitotic and interphase DNA in the cell. It is probable from Fig. 5, however, that the ratio of mitotic DNA to interphase (inducible) DNA is not the only factor to determine whether the interphase chromatin condenses. It is, for example, much easier to induce a given amount of $G1$ chromatin than it is for an equivalent amount of S or $G2$ chromatin when the genome ratio of DNA is the same in either case. This implies that the mitotic inducer has a much more rapid effect on the chromatin of $G1$ than it has on either S or $G2$ chromatin, and this may be related to the physical state of DNA during the $G1$ period.

When $G2$ and S phase cells were fused together the subsequent mitosis of the heterophasic population was asynchronous in 10 per cent of the cases. In these cells the $G2$ nuclei entered mitosis while the "S" nuclei remained in interphase. In some $G2/S$ cells the entry of the $G2$ nucleus into mitosis had induced the premature condensation of the "S" phase chromatin, indicating the presence of a mitotic inducer(s). It is not clear why there was no induction of premature condensation in all the $G2/S$ multinucleate cells where asynchrony in mitosis was observed.

The premature chromosome condensation reported in the present study is similar to the phenomenon first observed by Stubblefield[5] in multinucleate Chinese hamster cells and later termed "chromosome pulverization" by other investigators[6–11]. We consider that "premature chromosome condensation" describes the phenomenon more accurately. Although PCC is most often associated with multinucleate cells it has also been reported in mononucleate permanent cell lines in culture[12,13]. The incidence of premature condensation has been found to increase with an increase in the ploidy of the cell line[13]. Premature entry of nuclei into mitosis, resulting in aberrant nuclear morphology, has also been observed in *Physarum polycephalum*[14,15]. Guttes and Guttes[14] also found that a proportion of these abnormal mitotic nuclei synthesized DNA. These findings are consistent with the observations of Takagi *et al.*[11] and with the results reported here that prematurely condensed S phase nuclei continue to synthesize DNA, although at a reduced rate.

The prematurely condensed chromatin of $G1$ and $G2$ nuclei is not fragmented or pulverized in the sense that myxoviruses normally damage chromosomes[16,17]. The morphology of the condensed *S* phase nuclei does at first seem to resemble virus-inflicted damage in its pulverized appearance. It is possible, however, that the replicative process of the *S* phase chromatin is entirely responsible for the pattern of its condensation. The inactivated virus used for the fusion seems to have no effect on the induction of premature condensation in interphase nuclei[1]. The induction of premature condensation is not accompanied by the appearance of a mitotic spindle, and it therefore remains to be seen how the prematurely condensed chromosomes are distributed during the subsequent division.

We thank Dr Theodore T. Puck for constructive criticism of our manuscript, and George Barela and David Peakman for technical assistance. R. T. J. holds a Damon Runyon postdoctoral fellowship. The investigation was aided by a US Public Health Service grant from the National Institute of Child Health and Human Development.

Received January 6, 1970.

[1] Rao, P. N., and Johnson, R. T., *Nature*, **225**, 159 (1970).

[2] Levis, A. G., Krsmanovic, V., Miller-Faures, A., and Errera, M., *European J. Biochem.*, **3**, 57 (1967)

[3] Friedman, D. L., and Mueller, G. C., *Biochim. Biophys. Acta*, **174**, 253 (1969).

[4] Rao, P. N., *Science*, **160**, 774 (1968).

[5] Stubblefield, E., in *Cytogenetics of Cells in Culture* (edit. by Harris, R. J. C.), 223 (Academic Press, New York, 1964).

[6] Sandberg, A. A., Sofuni, T., Takagi, N., and Moore, G. E., *Proc. US Nat. Acad. Sci.*, **56**, 105 (1966).

[7] Kato, H., and Sandberg, A. A., *J. Cell. Biol.*, **34**, 35 (1967).

[8] Kato, H., and Sandberg, A. A., *J. Nat. Cancer Inst.*, **40**, 165 (1968).

[9] Kato, H., and Sandberg, A. A., *J. Nat. Cancer Inst.*, **41**, 1117 (1968).

[10] Kato, H., and Sandberg, A. A., *J. Nat. Cancer Inst.*, **41**, 1125 (1968).

[11] Takagi, N., Aya, T., Kato, H., and Sandberg, A. A., *J. Nat. Cancer Inst.*, **43**, 335 (1969).

[12] Hausen, H. zur, *J. Nat. Cancer Inst.*, **38**, 683 (1967).

[13] Miles, C. P., and O'Neill, F., *J. Cell Biol.*, **40**, 553 (1969).

[14] Guttes, E., and Guttes, S., *Experientia*, **19**, 13 (1963).

[15] Rusch, H. P., Sachsenmaier, W., Behrens, K., and Gruter, V., *J. Cell Biol.*, **31**, 204 (1966).

[16] Nichols, W. W., Levan, A., Aula, P., and Norrby, E., *Hereditas*, **54**, 101 (1965).

[17] Nichols, W. W., Aula, P., Levan, A., Heneen, W., and Norrby, E., *J. Cell Biol.*, **35**, 257 (1967).

Editor's Comments
on Papers 15 and 16

15   HARRIS et al.
Suppression of Malignancy by Cell Fusion

16   CROCE, ADEN, and KOPROWSKI
*Somatic Cell Hybrids Between Mouse Peritoneal Macrophages
and Simian-Virus-40-Transformed Human Cells: II. Presence
of Human Chromosome 7 Carrying Simian Virus 40 Genome
in Cells of Tumors Induced by Hybrid Cells*

MALIGNANCY IN SOMATIC CELL HYBRIDS

Beginning with the earliest studies on somatic cell hybridization, the expression of the malignant phenotype in hybrid cells has been a subject of active investigation and controversy. In the first report on somatic cell hybridization (see Paper 1), the parental cells differed in their degree of malignancy. The hybrid cells in those experiments appeared to resemble the more malignant parent in terms of the frequency with which they formed tumors. A number of independent studies over the next several years gave similar results, suggesting that malignancy behaves as a dominant trait in hybrid cells. In contrast, experiments carried out by Harris, Klein, Miller, and co-workers (Paper 15) suggested the opposite conclusion. In these experiments, highly malignant, mouse ascites tumor cells were hybridized with relatively nonmalignant mouse L cells, and the hybrids formed tumors with as low a frequency as the less malignant parental cell. In the few hybrid cell tumors tht did develop, the hybrids had lost approximately 30% of their original chromosome complement. The authors concluded that malignancy was suppressed in the hybrids and that it was reexpressed only after the loss of chromosomes from the hybrids (presumably, the loss of chromosomes of the less malignant parental cell). The difference between these results and those of the previous studies, which suggested the dominant nature of the malignant phenotype, was attributed to the lack of detailed karyological analyses of the tumors in the earlier studies.

The expression of the malignant phenotype also has been studied in crosses involving virus-transformed cells. In studies by Croce, Aden, and Koprowski (Paper 16), hybrids between Simian Virus 40 (SV40) transformed human cells and normal mouse peritoneal macrophages were isolated and characterized. When injected into *nude* mice, these hybrids expressed the malignant phenotype of the transformed human cells. Furthermore, hybrid cells cultured from the tumors contained the full complement of chromosomes of the normal mouse parental cell. Thus, the loss of chromosomes of the normal parental cell was not required for tumor formation by the hybrids. In addition, the human chromosome number 7, which carried the SV40 genome in the transformed human parental cells, was present in 100% of the hybrid tumor cells, and it was the only human chromosome present in all the hybrid tumor cells. The expression of malignancy in hybrids containing the human chromosome number 7 plus the entire complement of normal mouse chromosomes suggested that malignancy was expressed as a dominant trait in these hybrids. The specific role of SV40 in these studies is not clear at present. (It should be noted that the viral genome is not located on chromosome 7 in all lines of SV40 transformed human cells.) The significance of *tumor formation* in *nude* mice also may be questioned, not only in these studies but in other studies using nude mice as the host animal. Nevertheless, the results of these studies stand in marked contrast to those presented in Paper 15.

Over the past two decades, many studies have focused on the expression of the malignant phenotype in somatic cell hybrids. To date, however, a consistent picture has not emerged. As just described, markedly different results have been reported in different systems, and conflicting evidence continues to accumulate. Hybrids between malignant and normal human cells were injected into nude mice, and the hybrids were nonmalignant (Stanbridge, 1976). Recently, it was suggested that the suppression of malignancy in hybrids between malignant and normal human cells may involve the *differentiation* of the hybrid cells in vivo (Peehl and Stanbridge, 1982). On the other hand, hybrids formed in vivo between malignant mouse cells and normal cells of the host were malignant, and any of the chromosomes of the normal cells could be present in the hybrid tumor cells (Avilès et al., 1977). These results suggested that none of the chromosomes of the normal cells carried information capable of suppressing malignancy.

It is possible that the apparently contradictory results obtained with different systems are due at least in part to the use of different test systems in the various studies. Conceivably, there are different modes of inheritance of the malignant phenotype in different types of tumors. Ultimately, studies utilizing other approaches may resolve this issue, for example, studies involving DNA-mediated gene transfer as discussed in Part II.

REFERENCES

Avilès, D., J. Jami, J.-P. Rousset, and E. Ritz, 1977, Tumor X Host Cell Hybrids in the Mouse: Chromosomes from the Normal Cell Parent Maintained in Malignant Hybrid Tumors, *Natl. Cancer Inst. J.* **58:**1391–1399.

Peehl, D. M., and E. J. Stanbridge, 1982, The Role of Differentiation in the Suppression of Tumorigenicity in Human Cell Hybrids, *Int. J. Cancer* **30:**113–120.

Stanbridge, E. J., 1976, Suppression of Malignancy in Human Cells, *Nature* **260:**17–20.

15

Suppression of Malignancy by Cell Fusion

by

HENRY HARRIS
O. J. MILLER
Sir William Dunn School of Pathology,
University of Oxford

G. KLEIN
P. WORST
T. TACHIBANA
Department of Tumor Biology,
Karolinska Institutet, Stockholm

Malignancy can be suppressed when malignant cells are fused with certain non-malignant ones. The hybrid cells derived from such fusions give rise to segregants in which a loss of chromosomes is associated with reversion to malignancy. The expression of histocompatibility antigens can also be suppressed when cells bearing these antigens are fused with others that express them poorly.

WE define malignancy as the ability of tumour cells to grow progressively and kill their host. The idea of using cell fusion to facilitate the genetic analysis of this phenomenon is an attractive one; but the few attempts that have been made in this direction have not been encouraging. Barski and Cornefert[1] examined hybrid cells which arose spontaneously in mixed cultures of two mouse cell lines, one highly malignant, the other less so. The ability of the hybrid cells to produce tumours resembled that of the more malignant parent. Hybrids between the more malignant of these two parent cells and normal mouse fibroblasts also appeared to be malignant[2]; and so did the hybrids which arose spontaneously in mixed cultures of non-malignant mouse cells and mouse cells rendered malignant by infection with polyoma virus[3]. These latter hybrids continued to produce antigens determined by the polyoma virus[3,4]. The conclusion drawn from these studies was that malignancy was a dominant character; and because non-malignant variants would not be detected

in vivo, there seemed little more that could be done. In none of these cases, however, were the chromosomes of the tumours analysed in detail, so that no assessment can be made of the extent to which the results might have been complicated by loss of chromosomes or by selection of atypical variants *in vivo*.

Similar conclusions have been drawn about the dominance relationships of histocompatibility antigens. Hybrids arising spontaneously in mixed cultures of three mouse cell lines which differed in the antigens determined by the H-2 locus were found to express the antigens of both parent cells; and a similar result was obtained for two surface antigens not determined by the H-2 locus[5]. It was concluded that the expression of the histocompatibility antigens was codominant; and no further analysis of the situation was made.

The introduction of the Sendai virus cell fusion technique[6] has made it possible to hybridize virtually any mammalian cells and has thus greatly increased the

"

range and power of this form of analysis. It therefore seemed worthwhile, in view of the great biological and clinical importance of both malignancy and histocompatibility, to reinvestigate these questions in a more systematic way. We are now engaged in such an investigation and shall present the findings *in extenso* at a later stage; but some of our results appear to be of sufficient general interest to warrant their publication now. We have found: (1) that malignancy can be suppressed when malignant cells are fused with certain non-malignant cells; (2) that the hybrids resulting from such fusions produce segregants in which a loss of chromosomes is associated with reversion to malignancy; (3) that, in certain circumstances, the expression of histocompatibility antigens can also be suppressed by cell fusion.

Parent Cells

Three highly malignant ascites tumours of the mouse were selected for study: the Ehrlich, the SEWA and the MSWBS tumours. The Ehrlich tumour, originally derived from a mouse mammary carcinoma, is one of the most malignant transplantable tumours of the mouse. A few cells injected into the peritoneal cavity will kill most mice within 3 weeks; and even 1 cell is lethal in about 15 per cent of recipient animals[7]. The tumour shows very weak expression of histocompatibility antigens[8] and grows in any strain of mouse. It produces lethal tumours when injected subcutaneously as well as intraperitoneally. The SEWA tumour is an ascites sarcoma originally induced in 1960 by the injection of polyoma virus subcutaneously into a newborn A.SW mouse[9]. The tumour carries the H-2^s histocompatibility antigen complex and its growth is limited to mice of the A.SW strain. It also carries the polyoma-specific transplantation antigen. Although originally an osteogenic sarcoma, the tumour now shows no osteogenic differentiation. An inoculum of 1,000 cells produces tumours in 100 per cent of genetically compatible unirradiated mice, and an inoculum of 100 cells produces tumours in 100 per cent of mice given 400 r. of total body X-irradiation[10]. The tumour was converted to the ascites form in 1968 (ref. 11) and, in this form, an inoculum of 10^6 cells kills most animals in 3–4 weeks. Like the SEWA tumour, the MSWBS tumour is an ascites sarcoma. It also carries the H-2^s histocompatibility antigen complex and is specific for the A.SW strain of mice. It was originally derived from a tumour induced by the injection of methylcholanthrene into the thigh muscle of a male A.SW × A F$_1$ hybrid mouse[12]. The variant used in the present experiments arose from cells passaged in the parental A.SW strain. This variant has lost the H-2^a antigen complex derived from the A strain parent, but has retained the H-2^s complex of the A.SW strain parent. Growth of the tumour is restricted to

A.SW mice. An inoculum of as few as fifty cells produces tumours in 100 per cent of unirradiated genetically compatible mice. The tumours are lethal within 14–20 days.

Each of these tumours was hybridized with cells of the A9 mouse fibroblast line. This line was derived by Littlefield from the "L" cell line by selection for resistance to 8-azaguanine[13]. The A9 cells lack the enzyme inosinic acid pyrophosphorylase, a defect which prevents their growth when *de novo* synthesis of purines is blocked by aminopterin[14]. Being derived from C3H mice, these cells bear the H-2^k histocompatibility antigen complex. They also carry an antigen which cross-reacts with the antigens found in the Friend, Moloney and Rauscher leukaemias (FMR antigen). This antigen is known to be present in L cells[15] and is probably connected with the virus-like particles which have been found in these cells. These particles have not, however, been shown to induce tumours. As shown in Table 2, A9 cells are not highly malignant. None of the ten unirradiated C3H mice and six unirradiated A/Sn mice inoculated with a large dose of cells developed progressive tumours. Of nine C3H mice given 400 r. of whole body irradiation before inoculation, only two developed progressive tumours; and only one of six A/Sn mice, similarly irradiated, did so. Of these three tumours, only one grew on further passage to unirradiated mice.

Hybrid Cells

The cells were fused together by the technique of H. H. and Watkins[6], and viable hybrids were selected by growing the cultures containing the fused cells in the medium devised by Littlefield[14]. This prevents the growth of A9 cells. The strain of Ehrlich ascites cells used in the present experiments does not grow *in vitro*; the SEWA cells do not adhere to glass and the MSWBS cells grow poorly *in vitro*, at least initially. The selection of the hybrid cells thus presented no difficulties. When mixed suspensions of the tumour cells and A9 cells were treated with the inactivated Sendai virus, any desired number of hybrid clones could be obtained. We tested the ability of the hybrid cells to grow progressively *in vivo* by injecting varying numbers of them subcutaneously and intraperitoneally into genetically compatible mice: C3H and A/Sn mice in the case of the A9-Ehrlich hybrids and C3H × A.SW F$_1$ and A × A.SW F$_1$ crosses in the case of the A9-SEWA and A9-MSWBS hybrids. Each group of hybrid cells was tested in adult and newborn, X-irradiated (400 r. of whole body irradiation before inoculation) and unirradiated animals. Because newborn mice given this dose of irradiation will permit the growth even of tumours bearing foreign H-2 antigens, we believe that our assays for malignancy are not complicated in any important degree by immunological rejection.

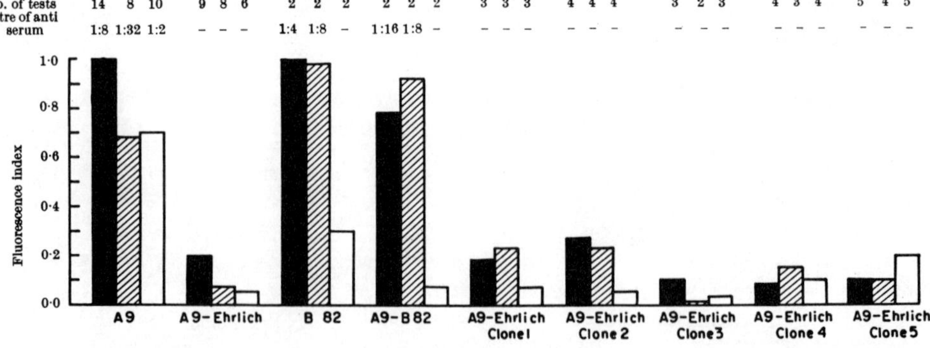

Fig. 1. Mean membrane fluorescence indices for A9 cells, B82 cells, A9-Ehrlich hybrid cells and A9-B82 hybrid cells. ■, C57Bl × DBA/2 anti C3H serum (anti H-2^k); ▨, A.CA anti A serum (contains anti H-2^k); ☐, syngeneic anti Moloney serum.

Table 1. CHROMOSOME COMPLEMENTS OF PARENTAL AND HYBRID CELL LINES

Cell type	Modal No. of chromosomes		Marker chromosomes
	Total	Bi-armed	
A9	57	24	M1, M2, M3
Ehrlich	75	1	M4, M5
SEWA	44	1	
MSWBS	29	12	M6
A9/Ehrlich	128	24	M1, M2, M3, M4, M5
A9/SEWA	94	21	M1, M2, M3
A9/MSWBS	86	36	M1, M2, M3, M6
A9/Ehrlich segregants	89, 84, 82		M1, M2, M3, M4

The expression of the H-2 and FMR antigens on the surface of the cells was studied by four techniques: membrane immunofluorescence[16,17], immune adherence[18], mixed haemadsorption[19] and cytotoxicity[20]. Some of the parent and hybrid cell lines were also compared by quantitative absorption tests against H-2 reference sera.

A9-Ehrlich Hybrids

Clones of cells derived from these two parents showed great variation in cell shape, clonal morphology and growth rate. Tight epithelial clones, clones composed of long spindle-shaped cells growing in parallel, highly dispersed clones of irregular fibroblastic cells and a whole range of intermediate categories were observed. To begin with, the most vigorous hybrids, which soon overgrew the bulk cultures, were selected for study. These hybrids grew extremely rapidly *in vitro* with a generation time of between 10 and 12 h. The hybrid nature of the cells was confirmed by karyotypic analysis. Both the Ehrlich cell and the A9 cell have specific chromosomal markers[21,22]. The markers of both parent cells were readily identified in the hybrids (Table 1). The hybrid cells contained a modal number of 128 chromosomes of which approximately 24 were bi-armed. This is what one would expect from the fusion of one Ehrlich cell and one A9 cell. There was little change in chromosome number during the first 2 months of cultivation *in vitro*.

As shown in Figs. 1 and 2 and in Table 3, membrane fluorescence, cytotoxicity, immune adherence and mixed haemadsorption tests agree in indicating that the expression of the H-2^k antigens contributed to the hybrid by the A9 cell is very largely suppressed. These antigens could be detected in the hybrids by immune adherence and by mixed haemadsorption tests, but they were present in much smaller amounts than in the A9 parent cells.

Five hybrid clones, each derived from a different primary fusion, were then examined. The results of the membrane fluorescence tests are shown in Fig. 1, and those of the mixed haemadsorption tests, carried out on 'Microtest' plates, in Fig. 3. The results of the immune adherence tests and of mixed haemadsorption tests carried out on culture plates are included in Table 3. It will be seen that small amounts of the H-2^k antigens could be detected in the various clones, especially by mixed haemadsorption, but none of the cells in these clones produced as much of these antigens as the parent A9 cells. Although there was, perhaps, some variation from clone to clone, the results with all five clones were very similar to those obtained with the A9-Ehrlich line derived from the hybrids which grew most rapidly.

It has been shown that histocompatibility antigens may be expressed less strongly in polyploid cells than in diploid cells[23]. It was therefore possible that the suppression of the H-2^k antigens in the A9-Ehrlich hybrids might have been simply a result of the increased ploidy of the hybrid cells. Hybrids were therefore made between the A9

Table 2. GROWTH OF CELLS *in vivo*

Cell type	Genotype	Host Genotype	Pretreatment	Inoculum No. cells	Route	No. takes (total No. animals with progressive tumours/total No. inoculated)	Comments
A9	C3H (H-2^k)	C3H (H-2^k)	—	2–5 × 10^5	s.c.	0/4	
,,	,, ,,	,, ,,	400 r.	2–5 × 10^5	,,	0/3	
,,	,, ,,	,, . ,,	—	1·5–5 × 10^5	i.p.	0/6	
,,	,, ,,	,, ,,	400 r.	1·5–5 × 10^5	,,	2/6	The tumours appeared after 2 and 7 weeks. Only one was transplantable
,,	,, ,,	A/Sn (H-2^a)	—	5 × 10^5	s.c.	0/2	
,,	,, ,,	,, ,,	400 r.	5 × 10^5	,,	0/2	
,,	,, ,,	,, ,,	—	1·5–5 × 10^5	i.p.	0/4	
,,	,, ,,	,, ,,	400 r.	1·5–5 × 10^5	,,	1/4	Tumour appeared after 2 weeks. Not transplantable
		A9		Total unirradiated		0/16	
				Total irradiated		3/15	
A6-Ehrlich hybrid	A9: C3H (H-2^k); Ehrlich: unknown, heterozygous	C3H (H-2^k)	—	2–5 × 10^5	s.c.	0/4	
,,	,, ,,	,, ,,	400 r.	7 × 10^4–5 × 10^5	,,	0/6	
,,	,, ,,	,, ,,	—	1·4 × 10^5–1 × 10^6	i.p.	0/8	
,,	,, ,,	,, ,,	400 r.	1·4 × 10^5–1 × 10^6	,,	4/8	Ascites tumours appeared after 3, 4 and 10 weeks. Two of them were tested and found to be transplantable. One animal developed generalized leukaemia after 18 weeks
,,	,, ,,	A/Sn (H-2^a)	—	1–5 × 10^5	s.c.	0/4	
,,	,, ,,	,, ,,	400 r.	7 × 10^4–5 × 10^5	,,	0/4	
,,	,, ,,	,, ,,	—	1·4–5 × 10^5	i.p.	0/8	
,,	,, ,,	,, ,,	400 r.	1·4–5 × 10^5	,,	1/8	Solid tumour appeared after 12 weeks. It was transplantable
		A9-Ehrlich hybrid		Total unirradiated		0/24	
				Total irradiated		5/26	The correct figure is probably 4/26 if the leukaemia is disregarded
A9-SEWA hybrid	A9: C3H (H-2^k); SEWA: A.SW (H-2^s)	A.SW × C3H F₁	—	7 × 10^4–1·9 × 10^5	s.c.	1/17	Tumour appeared after 3 weeks. Transplantability not tested
,,	,, ,,	,, ,,	400 r.	7 × 10^4	,,	4/11	Tumours appeared after 3–5 weeks. Transplantability now being tested
,,	,, ,,	A × A.SW F₁	—	3 × 10^5–1·7 × 10^6	,,	0/6	
,,	,, ,,	,, ,,	400 r.	2 × 10^5–1·7 × 10^6	,,	0/18	
,,	,, ,,	,, ,,	—	1·7 × 10^6	i.p.	0/4	
,,	,, ,,	,, ,,	400 r.	2·5 × 10^5	,,	0/4	
		A9-SEWA hybrid		Total unirradiated		1/27	
				Total irradiated		4/33	
A9-MSWBS hybrid	A9: C3H (H-2^k); MSWBS: A × A.SW F₁ (selected for compatibility with A. SW = H-2^s)	A × A.SW F₁	—	3 × 10^5	s.c.	0/12	
,,	,, ,,	,, ,,	400 r.	3 × 10^5	,,	0/8	

105

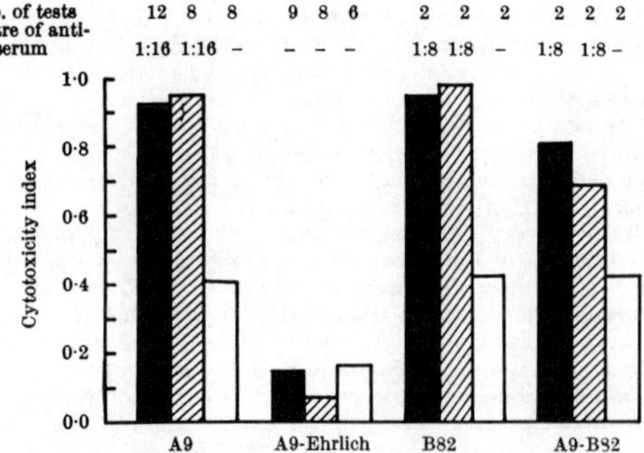

Fig. 2.   Mean cytotoxicity indices for A9 cells, B82 cells, A9-Ehrlich hybrid cells and A9-B82 hybrid cells. ■, C57Bl × DBA/2 anti C3H serum (anti H-2^k); ▨, A.CA anti A serum (contains anti H-2^k); ☐, syngeneic anti Moloney serum.

cells and B82 cells. B82 cells were also originally derived from "L" cells and bear the H-2^k antigen complex. These cells were selected for resistance to bromodeoxyuridine, and are deficient in thymidine kinase activity[24]. Hybrids between A9 and B82 cells can be selected from mixed cultures of the parent cells by the use of the selective medium devised by Littlefield[14]. The chromosomes of A9–B82 hybrids produced by inactivated Sendai virus have been extensively studied by Engel, McGee and H. H.[22]. The chromosome complement of these hybrids is approximately what one would expect from the fusion of one A9 and one B82 cell. As shown in Figs. 1 and 2, membrane fluorescence and cytotoxicity tests indicate that these hybrids express the H-2^k antigens as well as the parent cells. The suppression of these antigens in the A9-Ehrlich hybrids is thus not simply a non-specific effect of increased ploidy; this suppression is a specific effect produced by the Ehrlich component.

Although the FMR antigen is present on the surface of A9 cells in much smaller amounts than the H-2^k antigens, all four tests none the less indicate that the FMR antigen is also largely suppressed in the A9-Ehrlich hybrids (Figs. 1, 2 and 3, and Table 3). It thus appears that the Ehrlich cell, which itself expresses histocompatibility antigens very weakly, suppresses both the H-2^k antigen and the FMR antigen contributed to the hybrid cell by the A9 partner.

The results of the assays for progressive growth *in vivo* are set out in Table 2. It will be seen that the A9-Ehrlich hybrids failed to grow in the great majority of the test animals. Of fifty animals injected, forty-five produced no tumours. This low incidence of tumours is comparable with that obtained with A9 cells alone (three out of thirty-one). The tests were carried out on C3H and A mice and included adult and newborn, irradiated and unirradiated animals. Both the A9-Ehrlich line initially derived from the most rapidly growing hybrid cells and the hybrid clones derived from separate primary fusions were included in the tests. Because the hybrid cells were selected in medium containing high concentrations of aminopterin, thymidine and hypoxanthine, their ability to grow *in vivo*

Table 3.   IMMUNE ADHERENCE (IA) AND MIXED HAEMADSORPTION (MH) TESTS, DONE ON CULTURE PLATES

Serum	A9		Cell type A9-Ehrlich hybrid line		A9-Ehrlich clone 5	
	IA	MH	IA	MH	IA	MH
Tissue culture medium alone	0	0	0	0	0	0
Normal mouse serum	0	0	0	0	0	0
C57Bl × DBA/2 anti C3H (anti H-2^k)	> 90	≧ 3,000	30	500	30	2,000
ACA anti A (contains anti H-2^k)	> 90	≧ 3,000	30	200	30	600
Syngeneic anti Moloney	30	10	10	< 10	10	< 10

Figures denote the serum titres at which the tests were positive.

was also tested after the cells had been grown for some days in a medium not containing these additives. The cells still failed to produce tumours. One of the animals in this test group developed a leukaemia 18 weeks after inoculation. This leukaemia may not have been related to the injection of the hybrid cells, since the virus-like particles carried by the "L" (and presumably the A9) cells do not appear to cause leukaemia. Tumours developed in four other animals between 3 and 12 weeks after inoculation. Two of these tumours proved to be transplantable. Because the H-2^k antigens were very largely suppressed in the hybrid cell and because the Ehrlich cell is itself capable of growing in any strain of mouse, the hybrids were also injected intraperitoneally, at a concentration of 5×10^6 cells per animal, into six male and six female adult outbred Swiss mice. In eleven of these mice, no tumours developed, but in one female an ascites tumour appeared 10 weeks after inoculation. This tumour also proved to be transplantable. The karyotypes of these three transplantable tumours, developing late in the experimental animals, have been examined and are discussed in more detail later.

A9-SEWA Hybrids

The hybrid nature of these cells was again confirmed by analysis of the karyotype. The total number of chromosomes and the number of bi-armed chromosomes were again consistent with the view that the hybrids arose from the fusion of one A9 and one SEWA cell, although, in

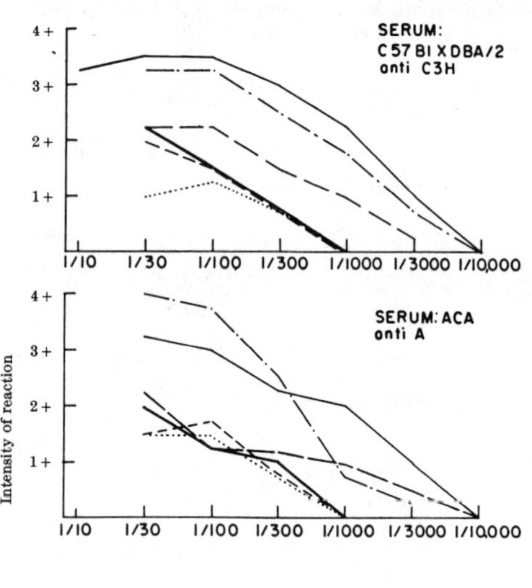

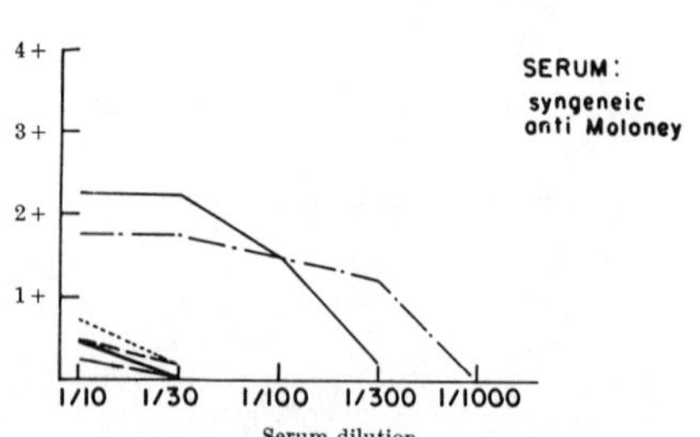

Fig. 3.   Mixed haemadsorption tests on 'Microtest' plates for A9 cells, B82 cells and different clones of A9-Ehrlich hybrid cells. ———, A9 (7 tests); — · —, B82 (3 tests); — —, A9-Ehrlich clone 2 (4 tests); · · · · ·, A9-Ehrlich clone 3 (4 tests); ▬▬, A9-Ehrlich clone 5 (2 tests); - - - -, A9-Ehrlich clone 8 (2 tests).

this case, about seven chromosomes appear to have been lost (Table 1). The A9 marker chromosomes were present.

As shown in Fig. 4, membrane fluorescence tests indicated that the H-2^k and the H-2^s antigen complexes of the two parent cells were both expressed in the hybrid. The amount of each antigen complex present in the hybrid was not greatly different from that found in the corresponding parent cell. In cytotoxicity tests, the A9 cell was found to be susceptible to the cytotoxic action of anti H-2^k, but not anti H-2^s serum, while the reverse was the case with the SEWA cells. The A9-SEWA hybrids, however, were resistant to both antisera. This effect may be related to the larger size of the hybrid cells, which reduces the density of the antigenic receptor sites on the cell surface. The total number of H-2^k receptor sites on the surface of the A9-SEWA hybrids was not very different from that on the surface of the A9 cells: with both cell types, essentially the same number of cells was required to absorb completely the reactivity of a standard A.SW anti A or a C57Bl × DBA anti C3H serum. This finding indicates that suppression of the H-2^k antigen in the A9-Ehrlich cells was not simply a result of the presence of some different histocompatibility antigens contributed by the Ehrlich component. In the A9-SEWA hybrids, the H-2^k antigen of the A9 parent is fully expressed, even though the SEWA parent contributes a different set of antigens. The suppression of the H-2^k antigen in the A9-Ehrlich hybrid is clearly a specific effect produced by the Ehrlich cell. The Ehrlich ascites tumour has been in continuous serial passage in many strains of mice since 1896 and must therefore have been subjected to prolonged selection pressure against its original histocompatibility antigens. Tumours are known to respond to such pressure by developing what has been termed "immunoresistance", a process which appears to involve a reduction in the density of antigenic receptor sites on the surface of the cell[25]. Because, in the A9-Ehrlich hybrids, not only the H-2 antigens but also the FMR antigens are suppressed, it is possible that the Ehrlich cell evades the homograft reaction by reducing the expression of all antigens on its surface. This could be achieved by the production of some component which masks the surface of the cell.

Table 2 shows that the A9-SEWA hybrids, like the A9-Ehrlich hybrids, only produced tumours in five out of sixty test animals. A.SW × C3H and A.SW × A F$_1$ hybrid mice were used and the group included adult and newborn, irradiated and unirradiated animals. Cells adapted to growth in medium which did not contain the thymidine, hypoxanthine and aminopterin used in the selective medium also failed to produce tumours. The karyotypes of the tumours which developed in animals injected with A9-SEWA hybrids have not yet been analysed.

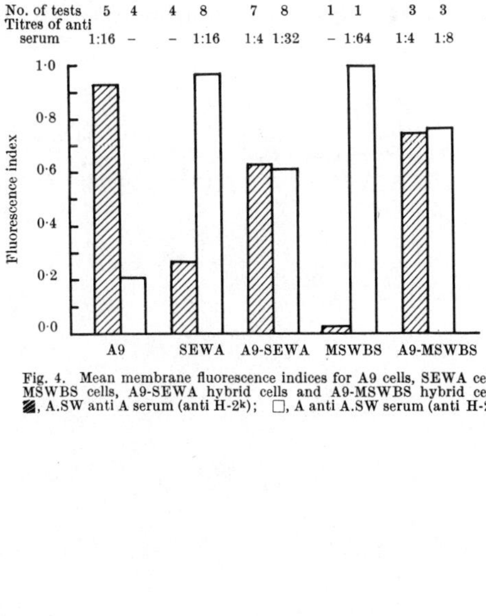

Fig. 4. Mean membrane fluorescence indices for A9 cells, SEWA cells, MSWBS cells, A9-SEWA hybrid cells and A9-MSWBS hybrid cells. ▨, A.SW anti A serum (anti H-2^k); ☐, A anti A.SW serum (anti H-2^s).

A9-MSWBS Hybrids

The MSWBS tumour is remarkable in having only twenty-nine chromosomes, of which twelve are bi-armed. Two of the bi-armed chromosomes serve as markers. A9-MSWBS hybrids were identified by their chromosome number, which approximated to the sum of the number of chromosomes in the two parent cells and by specific parental markers (Table 1). As shown in Fig. 4, A9-MSWBS hybrids, like A9-SEWA hybrids, contained both H-2^k and H-2^s antigens in normal amounts, as judged by membrane fluorescence. This confirms the specific role of the Ehrlich component in suppressing the H-2^k antigens in A9-Ehrlich hybrids. A9-MSWBS hybrids also resemble A9-SEWA hybrids in being resistant to the cytotoxic effect of anti H-2^k serum. A9-MSWBS hybrids, taken directly from the selective medium, or after a period of growth in normal tissue culture medium, failed to produce tumours in any of a group of 20 A × A.SW F$_1$ hybrid mice which included irradiated and unirradiated animals (Table 2).

Malignant Segregants

Karyotypic analysis of the three transplantable ascites tumours which developed late in mice receiving A9-Ehrlich cells intraperitoneally revealed that these tumours were indeed descended from the original hybrid cells injected. The tumours retained specific chromosomal markers from both the Ehrlich and the A9 parent (Table 1). In all three cases, however, the cells showed a marked reduction in chromosome number relative to the original A9-Ehrlich hybrids injected into the animals. The modal chromosome numbers of the three tumours were 89, 84 and 82. About forty chromosomes out of a total of nearly 130 had been lost. The loss included at least ten bi-armed chromosomes which were clearly of A9 origin. These tumours were thus segregants of the original non-malignant hybrids, in which a loss of chromosomes was associated with a reversion to malignancy. The segregant which arose in outbred Swiss mice grew very slowly and took many weeks to kill the host; the two segregants which arose in C3H mice grew much more rapidly. These two tumours also grew rapidly when transplanted to outbred Swiss mice. All three segregants initially grew very poorly *in vitro*, but eventually cells were selected from bulk populations which were able to grow well *in vitro*. A rough estimate of the frequency with which the non-malignant hybrids generate malignant segregants yields a figure of between 1 in 10^6 and 1 in 10^7. We have not yet determined whether such segregants arise *in vitro* or *in vivo* or in both situations. The work of Engel, McGee and H. H.[26] has shown that hybrids derived from very similar parent cells produce segregants in increasing numbers on continued cultivation *in vitro* ; and one might suppose that this process would be even more marked in hybrids from more distantly related parents. On the other hand, the long lag which preceded the development of these segregant tumours may indicate that, in this case, the segregants arose *in vivo*.

How is Malignancy Suppressed ?

The present experiments show that the ability of three different sorts of malignant cell to grow progressively *in vivo* can be suppressed when these cells are fused with A9 cells, and that the hybrid cells resulting from these fusions produce segregants in which a loss of chromosomes is associated with recovery of the ability to grow progressively *in vivo*. This result is most simply interpreted as indicating that the A9 cell contributes something to the hybrid which suppresses the highly malignant character of the tumour cells, and that this contribution is lost when certain chromosomes are eliminated from the hybrid. Because all malignant segregants derived from such hybrids can be selected for by growth *in vivo*, the possibilities for the genetic analysis of malignancy are obvious.

In the absence of adequate information about the chromosomal constitution of the tumours produced, we cannot at present be confident that hybrids between malignant cells and normal diploid cells are indeed malignant, as has been claimed[2,3]; but, if they are, the possibility must be considered that the suppression of malignancy observed in the present experiments is due to some special feature of the A9 cell. It may be that the metabolic consequences of prolonged selection for azaguanine resistance have conferred on the A9 cell the ability to suppress the malignant character of the mouse tumour cells. If this is so, the present experiments open a direct biochemical approach to the mechanisms by which the malignancy is suppressed; for some of the major metabolic peculiarities of the A9 cell have already been established[13,14].

The Swedish contribution to this study was supported by the Swedish Cancer Society, the Jane Coffin Childs Memorial Fund for Medical Research, the Swedish Medical Research Council and the Carl Bertel Nathhorts Fund. O. J. M. is a career scientist of the Health Research Council of the City of New York and a US National Science Foundation senior postdoctoral fellow. T. T. was supported by a travel fellowship of the International Agency for Research on Cancer, while on leave from the National Cancer Center Research Institute, Tokyo, Japan.

Received June 4, 1969.

[1] Barski, G., and Cornefert, F., *J. Nat. Cancer Inst.*, **28**, 801 (1962).

[2] Scaletta, L. J., and Ephrussi, B., *Nature*, **205**, 1169 (1965).

[3] Defendi, V., Ephrussi, B., Koprowski, H., and Yoshida, M. C., *Proc. US Nat. Acad. Sci.*, **57**, 299 (1967).

[4] Defendi, V., Ephrussi, B., and Koprowski, H., *Nature*, **203**, 495 (1964).

[5] Spencer, R. A., Hauschka, T. S., Amos, D. B., and Ephrussi, B., *J. Nat. Cancer Inst.*, **33**, 893 (1964).

[6] Harris, H., and Watkins, J. F., *Nature*, **205**, 640 (1965).

[7] Hauschka, T. S., *Trans. NY Acad. Sci.*, **16**, 64 (1953).

[8] Hauschka, T. S., and Amos, D. B., *Ann. NY Acad. Sci.*, **69**, 561 (1957).

[9] Sjögren, H. O., Hellström, I., and Klein, G., *Cancer Res.*, **21**, 329 (1961).

[10] Sjögren, H. O., *J. Nat. Cancer Inst.*, **32**, 645 (1964).

[11] Nordenskjöld, B., *Int. J. Cancer*, **3**, 628 (1968).

[12] Klein, G., and Klein, E., *J. Cell Comp. Physiol.*, **52**, suppl. I, 125 (1958).

[13] Littlefield, J. W., *Nature*, **203**, 1142 (1964).

[14] Littlefield, J. W., *Science*, **145**, 709 (1964).

[15] Leclerc, J. C., Levy, J. P., Varet, B., and Oppenheim, S., *C.R. Hebd. Séanc. Acad. Sci., Paris*, **266**, 2206 (1968).

[16] Möller, G., *J. Exp. Med.*, **114**, 415 (1961).

[17] Klein, E., and Klein, G., *J. Nat. Cancer Inst.*, **32**, 547 (1964).

[18] Nishioka, K., Irie, R. F., Kawana, T., and Takenchi, S., *Int. J. Cancer*, **4**, 139 (1969).

[19] Fagreus, A., Espmark, J. Å., and Johnsson, J., *Immunology*, **9**, 161 (1965).

[20] Gorer, P. A., and O'Gorman, P., *Transplant. Bull.*, **3**, 142 (1946).

[21] Harris, H., Watkins, J. F., Ford, C. E., and Schoefl, G. I., *J. Cell Sci.*, **1**, 1 (1966).

[22] Engel, E., McGee, B. J., and Harris, H., *J. Cell Sci.* (in the press, 1969).

[23] Hauschka, T. S., *Canad. Cancer Conf.*, **2**, 305 (1957).

[24] Littlefield, J. W., *Exp. Cell Res.*, **41**, 190 (1966).

[25] Klein, G., *Viruses Inducing Cancer, Implications for Therapy* (edit. by Burdette, W. J.), 323 (Univ. of Utah Press, Salt Lake City, 1966).

[26] Engel, E., McGee, B., and Harris, H., *Nature*, **223**, 152 (1969).

16

Reprinted from *Natl. Acad. Sci. (USA) Proc.* **72**:1397–1400 (1975)

SOMATIC CELL HYBRIDS BETWEEN MOUSE PERITONEAL MACROPHAGES AND SIMIAN-VIRUS-40-TRANSFORMED HUMAN CELLS: II. PRESENCE OF HUMAN CHROMOSOME 7 CARRYING SIMIAN VIRUS 40 GENOME IN CELLS OF TUMORS INDUCED BY HYBRID CELLS*

C. M. Croce, D. Aden, and H. Koprowski

ABSTRACT   Cells derived from tumors induced in "nude" mice after injection of cells that were hybrids between mouse peritoneal macrophages and simian virus 40 (SV40)-transformed human cells were found to retain the human chromosome 7 carrying the SV40 genome, and to express SV40-induced tumor (T) antigen. These results indicate that the presence of human chromosome 7 carrying the SV40 genome is responsible for the expression of the tumorigenic phenotype in the hybrid cells.

Through somatic cell hybridization techniques, we have previously assigned the gene for simian virus 40 (SV40) tumor (T) antigen and the SV40 genome to human chromosome 7 in SV40-transformed cells of various origins (1–3). Furthermore, we have shown that hybridization of mouse peritoneal macrophages (which are nondividing cells) with SV40-transformed human cells resulted in the formation of *transformed* somatic cell hybrids in which the human chromosome 7 carrying the SV40 genome was always present (4). Those results indicate that the human chromosome 7 that carries the SV40 genome contains gene(s) coding for transforming gene product(s) (4). In addition, we have recently shown (Koprowski, Aden, and Croce, manuscript in preparation) that somatic cell hybrids between mouse peritoneal macrophages and SV40-transformed human cells are able to induce tumors in "nude" mice, in which heterotransplantation of human tumors can be successfully achieved (5–7).

We undertook these studies to determine if the human chromosome 7 carrying the SV40 genome and coding for SV40 T antigen is present in cells derived from tumors induced in "nude" mice by the injection of hybrid cells that contain many different human chromosomes, including chromosome 7. If the tumorigenic phenotype is determined by the presence of a specific human chromosome, this chromosome should always be present in tumor cells recovered from "nude" mice. This communication reports the results of these studies.

MATERIALS AND METHODS

Hybrid Cells. The production of somatic cell hybrids between C57BL/6 mouse peritoneal macrophages and the human cells LN-SV (SV40-transformed Lesch–Nyhan fibroblasts) is described elsewhere (4). A heterogeneous population of hybrid cells containing 14 different human chromosomes

(Table 1) and a quasi-tetraploid number of mouse chromosomes was injected under the abdominal skin of 20 Balb/c *nu/nu* ("nude") mice at a concentration of 2 to 4 × 10⁷ cells per mouse.

Tumor Cells. The inoculated mice were observed daily for tumor growth, and when a palpable mass was detected, the tumor was removed aseptically and fragments were immediately frozen, fixed in acetone and stained for SV40 T antigen (8). The remaining portion was cut in small fragments and seeded either in 60 mm petri dishes or 75 cm² Falcon flasks in Eagle's minimal essential medium (MEM) containing 10% fetal calf serum. The cells derived from the tumors obtained from five "nude" mice were transferred five to eight times in culture and stained for the presence of SV40 T antigen. LN-SV SV40-transformed human cells were found to be tumorigenic in "nude" mice. On the contrary, no tumors were induced by the inoculation of C57BL/6 mouse peritoneal macrophages.

Karyotypical Analysis. A modification of the Giemsa banding technique of Seabright (9) was used in this study (10, 11). Fifty-one metaphases of the inoculated hybrid cells were analyzed (Table 1). A minimum of 26 metaphases of each "tumor" culture derived from a different "nude" mouse was also analyzed following Giemsa banding staining. This permitted identification of each human and mouse chromosome present in the cells (10, 11).

Reinoculation of Tumor Cells. Cells derived from the tumor of "nude" mouse 8 were transferred for eight passages in culture and reinjected (2 × 10⁷ cells per mouse) in 10 "nude" mice. All these animals developed tumors. Three of these tumors were transferred in tissue culture and the tumor cells were subjected to karyological analysis. Cells derived from a tumor produced in "nude" mouse 1 (Table 2), 17.8% of which contained human chromosome 5, 77.8% of which contained chromosome 6, and 100% of which contained chromosome 7, were inoculated into five "nude" mice at a concentration of 4 × 10⁷ cells per mouse. A tumor obtained from one of these mice was processed for culture as above except that a portion of the tumor was cut into small fragments (2–3 mm in diameter) and the cell fragments were implanted subcutaneously into five other "nude" mice. In addition the cultured cells derived from this tumor were injected in nine "nude" mice. All these mice developed very large masses (9–17 g in weight). Three tumors were transferred in culture and the tumor cells were subjected to karyological analysis.

Abbreviations: SV40, simian virus 40; T antigen, tumor antigen.
* Paper I of this series is ref. 4.

TABLE 1. *Presence of different human chromosomes in hybrid cells between C57BL/6 peritoneal macrophages and LN-SV human cells*

Human chromosome	Frequency in hybrids, %*	Human chromosome	Frequency in hybrids, %*
4	5.9	13	5.9
5	60.8	14	7.8
6	70.6	15	17.6
7	100†	16	2.0
9	5.9	17	60.8
11	54.9	18	3.9
12	23.5	20	5.9

* A total of 51 metaphases was analyzed.

† On the average the hybrid cells contained 2.3 chromosomes 7 per cell.

RESULTS

Karyological Analysis of the Hybrid Cells Used as Inoculum. As shown in Table 1, 14 out of the 24 different human chromosomes were present in the hybrid cells. The human chromosome 7 was the only human chromosome present in 100% of the metaphases. This frequency was followed by that of human chromosome 6 in 70.6%, human chromosomes 5 and 17 in 60.8%, and human chromosome 11 in 54.9% of metaphases. The remaining human chromosomes were encountered in less than 24% of the cells analyzed.

SV40 T Antigen in the Hybrid Cells Used as Inoculum. The hybrid cells injected in the "nude" mice were stained for the presence of T antigen and all were found to show the presence of this antigen. A minimum of 10^3 cells was analyzed. Since the gene for SV40 T antigen has been assigned to human chromosome 7 in SV40-transformed human cells (1, 2), these data confirm the results of the karyological analysis indicating the presence of human chromosome 7 in all cells of the hybrid population.

Characterization of Cells Obtained from Tumors of "Nude" Mice Injected with Hybrid Cells. The tumors that developed in five mice after the inoculation of hybrid cells were removed at the times indicated in Table 2. Examination by SV40 T antigen staining of frozen sections revealed SV40 T antigen in cells of all the tumors examined. Cells derived from each tumor grew in culture and, after five to eight transfers, they were seeded on coverslips and stained for SV40 T antigen. A mini-

mum of 10^3 cells of each culture was analyzed for the presence of SV40 T antigen. One-hundred percent of the cells of the cultures derived from each tumor showed the presence of SV40 T antigen.

Karyological Analysis of Tumor Cells Grown in Culture. As can be seen in Table 2, chromosome 7 was found in 100% of the metaphases of the cells of the cultures obtained from tumors of the "nude" mice. In addition, the cells derived from "nude" mice 7 and 8 tumors contained human chromosome 6 in less than 39% of the metaphases (Table 2). "Nude" mouse 4 tumor cells contained the human chromosome 6 in 64.1% of the cells, the human chromosome 17 in 38.4% of the cells and the human chromosome 11 in only one cell out of 39 cells examined (Fig. 1). "Nude" mouse 2 tumor cells contained human chromosome 6 in 21 out of 30 metaphases. "Nude" 1 tumor cells contained human chromosome 6 in 77.8% of metaphases and human chromosome 5 in 17.8% of the metaphases. On the average, cells derived from the tumors of the five "nude" mice contained three human chromosomes 7 per cell.

Inoculation of "Nude" Mice with Cells Derived from the "Nude" 8, 1, and 9 Tumors. Tumor cells derived from "nude" 8 were transferred for eight passages in culture and inoculated into 10 "nude" mice, all of which developed tumors. Cells of three of the 10 tumors were analyzed for the presence of SV40 T antigen and human chromosomes. The results of the analysis are reported in Table 3. The tumor cells derived from these tumors contained only one human chromosome, the 7. On the average, 2.6 chromosomes 7 per cell were retained by the hybrids. The entire complement of mouse chromosomes was present in all the cells examined. Tumor cells derived from "nude" 1 tumor were transferred for eight passages in culture and inoculated into five "nude" mice. Sixty-seven days after inoculation, a tumor mass (3.1 g) was removed from one of these "nude" mice ("nude" 9). Presence of SV40 T antigen was detected in the frozen sections of the tumors following indirect immunofluorescence staining for SV40 T antigen and in the cultured cells derived from this tumor. Karyological analysis of 41 metaphases (Table 3) of the cultured cells revealed the presence of chromosome 7 in all metaphases. On the average 3.0 chromosomes 7 per cell were retained by the hybrids. Twenty-nine out of 41 metaphases showed also the presence of one human chromosome 6. All the cells retained the entire complement of mouse parental chromosomes. The tumor cells derived from the "nude" 9 tumor were inoculated in nine "nude" mice. Very large tumor masses developed

TABLE 2. *Karyological analysis of cells recovered from tumors induced in "nude" mice by inoculation of hybrid cells*

Source of tumors	Removal, days after inoculation	Percent of cells showing SV40 T antigen	Human chromosomes*				
			5	6	7	11	17
"Nude" 1	22	100	8/45	35/45	45/45	0/45	0/45
"Nude" 2	28	100	0/30	21/30	30/30	0/30	0/30
"Nude" 4	42	100	0/39	25/39	39/39	1/39	15/39
"Nude" 7	70	100	0/26	10/26	26/26	0/26	1/26
"Nude" 8	76	100	0/33	3/33	33/33	0/33	0/33

* Numbers of metaphases showing presence of this chromosome over total analyzed.

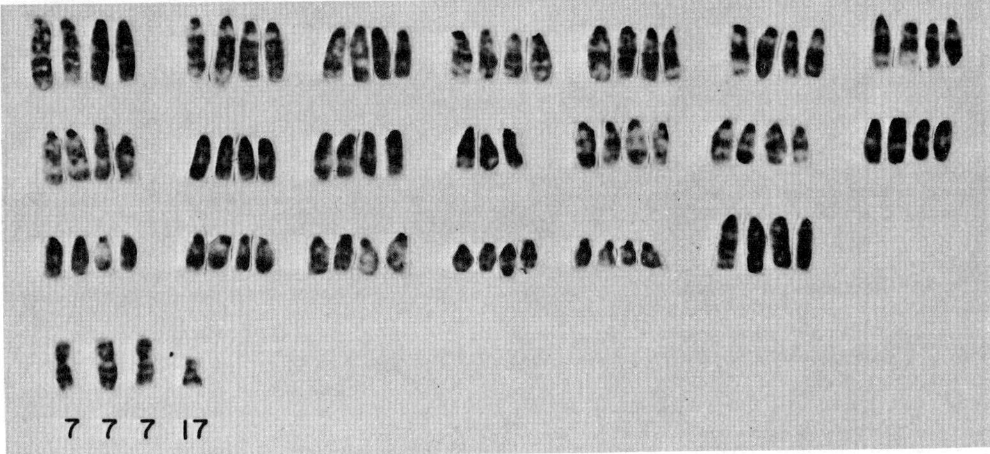

F_{IG}. 1. Karyotype of a hybrid cell derived from "nude" 4 tumor. The hybrid cell contains a quasi-tetraploid number of mouse chromosomes and four human chromosomes (three chromosomes 7 and one chromosome 17).

(9–17 g) in all of these mice. Three of the nine tumors were transferred into culture and subjected to karyological analysis (Table 3).

Implantation of Fragments of "Nude" 9 Tumor in "Nude" Mice. Five mice were implanted with fragments of the tumor derived from "nude" 9. All animals showed the presence of rapidly growing tumor masses. Tumors were removed from two mice on the 22nd day after implantation, from one mouse on the 26th day, and from the two remaining mice on the 37th day after implantation. SV40 T antigen was found in the cells of all tumors examined. Cultures of cells derived from these tumors, which have been transferred in culture for at least eight passages, again showed the presence of T antigen in 100% of the cells. Karyological analysis of the cells grown out of these five tumors in tissue culture indicated (Table 4) the presence of human chromosome 7 in all metaphases examined, human chromosome 6 in less than 70% of the meta-

phases, and in no cell could we detect human chromosome 5. Again, the entire complement of mouse chromosomes was present in all cells examined.

DISCUSSION

We have shown that somatic cell hybrids between normal diploid human cells and SV40-transformed cells behave as transformed cells *in vitro* (12). This indicates that, in hybrids between normal and SV40-transformed human cells, the transformed phenotype is dominant. We have also observed that hybrid cells between normal mouse cells (peritoneal macrophages) and SV40-transformed human cells, which retained without exception the human chromosome 7, behaved as transformed cells *in vitro* (4). This indicates that the human chromosome 7 that carries the SV40 genome contains genes coding for "transforming factors" (4).

The fact that the hybrid cells between mouse peritoneal macrophages and SV40-transformed human cells were found to produce tumors in "nude" mice made it possible to characterize the tumorigenic hybrid cells. The results of these investigations indicate that the human chromosome 7 carrying

TABLE 3. *Karyological analysis of cells of tumors derived from the inoculation of "nude" 8 and "nude" 9 tumor cells in "nude" mice*

Tumors	Removal, days after inoculation	Percent of cells showing SV40 T antigen	Human chromosomes*		
			5	6	7
"Nude" 8-1	40	100	0/32	0/32	32/32
"Nude" 8-2	40	100	0/37	0/37	37/37
"Nude" 8-3	56	100	0/37	0/37	37/37
"Nude" 9†	67	100	0/41	29/41	41/41
"Nude" 9-1	24	100	0/31	23/31	31/31
"Nude" 9-2	38	100	0/26	20/26	26/26
"Nude" 9-3	39	100	0/30	21/30	30/30

* Number of metaphases showing presence of this human chromosome over total analyzed. No other human chromosomes were present in the hybrid cells.
† Hybrid cells derived from "nude" 1 were inoculated into "nude" 9.

TABLE 4. *Karyological analysis of cells derived from tumors induced in five "nude" mice by the implantation of fragments of "nude" 9 tumor*

Tumors	Removal, days after inoculation	Percent of cells showing SV40 T antigen	Human chromosomes*		
			5	6	7
"Nude" 9-1 F†	22	100	0/31	20/31	31/31
"Nude" 9-2 F	22	100	0/33	23/33	33/33
"Nude" 9-3 F	26	100	0/33	21/33	33/33
"Nude" 9-4 F	37	100	0/26	17/26	26/26
"Nude" 9-5 F	37	100	0/29	19/29	29/29

* Number of metaphases showing presence of this chromosome over total analyzed.
† Implanted with fragments of "nude" 9 tumor.

the SV40 genome is retained by 100% of the hybrid cells recovered in culture from the "nude" tumors, whereas the other human chromosomes present in the inoculated cells are eliminated. The presence of multiple copies of human chromosome 7 in the cells recovered from the tumors may have represented a selective advantage for their growth. Thus, the presence of the human chromosome 7 carrying the SV40 genome seems to be essential not only for the expression of the transformed phenotype *in vitro* but also for the expression of the oncogenic phenotype *in vivo*.

Since in our hybrid cells the full complement of mouse chromosomes derived from the normal mouse parent was retained by the tumorigenic hybrid cells, our results contrast with the hypothesis formulated by Harris and Klein, who postulated that a hybrid between a normal and a tumorigenic cell behaves as a normal cell unless the chromosomes of the normal cell are lost from the hybrid (13–15).

Somatic cell hybridization with mouse peritoneal macrophages may be extended to cells derived from a variety of human cancers. If this results in the production of transformed somatic cell hybrids that retain specific human chromosomes, it may soon be possible to identify the human chromosomes responsible for the expression of the tumor phenotype in human malignancies.

We are very thankful to Irene Kieba, Cecilia Green, and Emma de Jesus for excellent technical assistance. This work was supported in part by Public Health Service Research Grants CA 10815 from the National Cancer Institute, RR 05540 from The Division of Research Resources, GM 20700 from the National Institute of General Medical Sciences, by a Basil O'Connor Starter Grant to CMC, and funds from the Commonwealth of Pennsylvania.

1. Croce, C. M., Girardi, A. J. & Koprowski, H. (1973) "Assignment of the T-antigen gene of simian virus 40 to human chromosome C-7," *Proc. Nat. Acad. Sci. USA* **70,** 3617–3620.
2. Croce, C. M. & Koprowski, H. (1974) "Concordant segregation of the expression of SV40 T antigen and human chromosome 7 in mouse-human hybrid subclones," *J. Exp. Med.* **139,** 1350–1353.
3. Croce, C. M., Huebner, K., Girardi, A. J. & Koprowski, H. (1974) "Rescue of defective SV40 from mouse-human hybrid cells containing human chromosome 7," *Virology* **60,** 276–281.
4. Croce, C. M. & Koprowski, H. (1974) "Somatic cell hybrids between mouse peritoneal macrophages and SV40 transformed human cells. I. Positive control of the transformed phenotype by the human chromosome 7 carrying the SV40 genome," *J. Exp. Med.* **140,** 1221–1229.
5. Flanagan, S. P. (1966) " 'Nude', a new hairless gene with pleiotropic effects in the mouse," *Genet. Res.* **8,** 295–309.
6. Rygaard, J. & Povlsen, C. O. (1969) "Heterotransplantation of a human malignant tumour of 'nude' mice," *Acta Pathol. Microbiol. Scand.* **77,** 758–760.
7. Visfeldt, J., Povlsen, C. O. & Rygaard, J. (1972) "Chromosome analyses of human tumours following heterotransplantation to the mouse mutant nude," *Acta Pathol. Microbiol. Scand.* **80,** 169–176.
8. Riggs, J. L., Takemori, N. & Lennette, E. H. (1965) "Detection of adenovirus type 12 neoantigen(s) in a continuous human amnion cell line (FL) by immunofluorescence," *Proc. Soc. Exp. Biol. Med.* **120,** 832–837.
9. Seabright, M. (1971) "A rapid banding technique for human chromosomes," *Lancet ii,* 971–972.
10. Croce, C. M., Knowles, B. B. & Koprowski, H. (1973) "Preferential retention of the human chromosome C-7 in human-(thymidine kinase deficient) mouse hybrids," *Exp. Cell Res.* **82,** 457–461.
11. Croce, C. M., Litwack, G. & Koprowski, H. (1973) "Human regulatory gene for inducible tyrosine aminotransferase in rat-human hybrids," *Proc. Nat. Acad. Sci. USA* **70,** 1268–1272.
12. Croce, C. M. & Koprowski, H. (1974) "Positive control of the transformed phenotype in hybrids between SV40 transformed and normal human cells." *Science* **184,** 1288–1289.
13. Harris, H., Miller, O. J., Klein, G., Worst, P. & Tachibana, T. (1969) "Suppression of malignancy by cell fusion," *Nature* **223,** 363–368.
14. Klein, G., Bregula, U., Wiener, F. & Harris, H. (1971) "The analysis of malignancy by cell fusion. I. Hybrids between tumour cells and L cell derivatives," *J. Cell. Sci.* **8,** 659–672.
15. Wiener, F., Klein, G. & Harris, H. (1974) "The analysis of malignancy by cell fusion. V. Further evidence of the ability of normal diploid cells to suppress malignancy," *J. Cell Sci.* **15,** 177–183.

Editor's Comments
on Papers 17 Through 21

17   WEISS and GREEN
Human-Mouse Hybrid Cell Lines Containing Partial Complements of Human Chromosomes and Functioning Human Genes

18   NABHOLZ, MIGGIANO, and BODMER
Genetic Analysis with Human-Mouse Somatic Cell Hybrids

19   RUDDLE et al.
Linkage Between Human Lactate Dehydrogenase A and B and Peptidase B

20   MILLER et al.
Mitotic Separation of Two Human X-Linked Genes in Man-Mouse Somatic Cell Hybrids

21   DEISSEROTH et al.
Localization of the Human α-Globin Structural Gene to Chromosome 16 in Somatic Cell Hybrids by Molecular Hybridization Assay

GENE MAPPING IN SOMATIC CELL HYBRIDS

It has long been recognized that sexual reproduction does not provide an adequate basis for the study of the formal genetics of mammals, and especially man. In the late 1950s, before the discovery of somatic cell hybridization, Pontecorvo (1958) suggested, by analogy with *parasexual* systems in lower organisms, that mitotic segregation in cultured somatic cells could be used to study the genetics of man, bypassing sexual reproduction entirely. However, appropriate techniques to carry out such studies were not available at the time.

Somatic cell hybridization provided the first opportunity to experimentally combine genomes of mammalian cells in vitro, and the importance of chromosome segregation for genetic analysis with hybrid cells was quickly recognized (Ephrussi et al., 1964). However, the early results were not encouraging, as the first hybrids isolated (both intra- and interspecific) were characterized by relatively stable karyotypes, with little tendency for chromosome segregation. This unpromising situation was altered in 1967 when Weiss and Green (Paper 17) isolated hybrids between human and mouse cells. Completely unexpectedly, the rapid and preferential segregation of human chromosomes was observed in these human-mouse hybrids. Within twenty generations after their formation, these hybrids had lost (on the average) approximately 85% of the chromosomes of the human parental cell. Since the hybrids resulted from the fusion of thymidine kinase (TK) deficient mouse cells with TK positive human cells and since they had been selected in HAT medium (see Papers 3 and 4), it was assumed that the hybrids growing in selective medium expressed the human TK activity. Given that very few human chromosomes were retained in the hybrids, attempts were made to correlate the presence of a specific human chromosome with the expression of TK activity. The number of human chromosomes declined when the hybrids were back-selected for TK deficiency (using the drug 5-bromodeoxyuridine), but the techniques for chromosome identification at that time were not adequate to permit the identification of the specific human chromosome carrying the TK gene.

Although these initial studies did not demonstrate the presence of human TK activity in the hybrid cells, nor could they determine the chromosomal location of the human TK gene, the results established human-mouse hybrids as a vehicle for the mapping of the human genome in vitro. Subsequent studies demonstrated that the TK activity in the hybrid cells was, indeed, of human origin (Migeon et al., 1969). With improved techniques for identifying human chromosomes, the human chromosome carrying the TK gene eventually was shown to be chromosome number 17 (Miller et al., 1971).

The analysis of human-mouse hybrids segregating human chromosomes soon became the most widely used approach for mapping human genes. One approach to mapping involved testing for synteny, or the association of two genetic loci with each other. This analysis was based on the expectation that linked loci should segregate concordantly in hybrid cells. The first application of human-mouse hybrids for testing synteny was reported by Bodmer and co-workers (Paper 18). This study involved two loci, glucose 6 phosphate dehydrogenase (G6PD), and hypoxanthine-guanine phosphoribosyltransferase (HGPRT). On the basis of family studies, both G6PD and HGPRT were known to be X linked in man. In the hybrids, there was perfect concordance between the human G6PD and HGPRT activities; in all cases the two activities were either present together or absent together. (The human HGPRT activity was not directly demonstrated in these studies but was assessed on the basis of selection with HAT medium or back-selection with 8-azaguanine.) These results confirmed the X linkage of both G6PD and HGPRT and demonstrated the value of human-mouse hybrids for testing synteny. Studies on the segregation patterns of lactate dehydrogenase (LDH) activities in the hybrids suggested, but could not prove, that the human genes for LDH-A and LDH-B are not linked.

Shortly after these studies, the use of human-mouse hybrids for determining new linkages was reported by Ruddle and co-workers (Paper 19) and by Bodmer and co-workers (Santachiara et al., 1970). In these studies, hybrids were examined for pairwise associations between unselected markers, that is, without any selection pressure for the presence or absence of the markers. The random segregation patterns of a variety of different human enzymes in a large number of independent hybrid clones were analyzed, and it was shown that human LDH-B and peptidase B consegregate and are presumably linked. This represented the first case in which a new human gene linkage relationship was established on the basis of in vitro studies with hybrid cells. Subsequent studies by Ruddle and co-workers (Boone et al., 1972) demonstrated that unselected markers could be assigned to specific human chromosomes by analyzing the random segregation patterns of both enzymes and chromosomes in human-mouse hybrid cells.

Since these early studies, many laboratories have participated in the mapping of the human genome by means of somatic cell hybridization, and a large number of loci have been assigned to specific chromosomes. It is clear that somatic cell hybridization has greatly accelerated the mapping of the human genome, and hybrid cells have proven to be the most useful tool for human gene mapping yet developed. The status of the human gene map after a decade of studies with somatic cell hybrids was reviewed by McCusick and Ruddle (1977).

In addition to the use of biochemically selectable or randomly segregating markers, a variety of other approaches have been taken for gene mapping in hybrid cells. Human-mouse hybrids that have lost specific human chromosomes can be selected through the use of many different types of agents, such as viruses, toxins, and anti-bodies. Polio virus, for example, has been used as a selective agent with human-mouse hybrids, taking advantage of the inherent difference in viral sensitivity between human and mouse cells (Kusano et al., 1970). Hybrids between human cells and Chinese hamster cells also lose human chromosomes, in some cases even more rapidly than human-mouse hybrids. Puck and co-workers have combined this extremely rapid segregation of the human chromosome with the use of auxotrophic (nutritional) markers to isolate hybrids containing single selected human chromosomes (Kao et al., 1976).

Methods also have been developed to control whether the chromosomes of one parental cell or the other will be lost in interspecific hybrids. These methods, involving either irradiation of one parental cell prior to fusion (Pontecorvo, 1971) or the use of freshly explanted cells as one of the parental cell types (Minna and Coon, 1974), are especially useful for mapping rodent genomes. All of these studies involved the segregation of chromosomes in hybrids resulting from fusion between intact cells. It is also possible to fuse cellular fragments (microcells) to transfer only a small number of chromosomes into hybrids (see Paper 24) and to use such microcell hybrids for gene mapping (Fournier and Ruddle, 1977).

In order to assign genes not only to specific chromosomes, but to specific regions of chromosomes, studies with cell hybrids have taken advantage of chromosome breakage and translocation. The use of chromosome breakage for regional gene mapping in hybrid cells was introduced by Miller, Siniscalco, and co-workers (Paper 20). In these studies, HGPRT deficient mouse cells were fused with normal human cells. Although human HGPRT activity was present in all hybrids growing in HAT medium, some of the hybrids lacked human G6PD activity. Since both of these enzymes are X linked in man, the most reasonable explanation for the presence of one enzyme, but not the other, was separation of the two loci by chromosome breakage. These results suggested that chromosome breakage in hybrid cells could be used for the subchromosomal localization of genes.

In a related approach to subchromosomal mapping, naturally occurring translocations in humans have been used to separate linked loci. The first translocation used for such a purpose was an X-autosome translocation in which much of the long arm of the X chromosome was attached to chromosome number 14 (Grzeschik et al., 1972). Human cells with this translocation were hybridized with mouse cells, and the segregation patterns of three human X linked genes were analyzed. Despite initially contradictory results from different laboratories, studies with this translocation permitted the localization of all three genes to the long arm of the X chromosome (Ricciuti and Ruddle, 1973). The results obtained with the X/14 translocation were compared with those obtained using other X-autosome translocations in which there were different breakpoints on the X chromosome, and the combination of the data from different laboratories made it possible to determine the order of the three X linked genes. In addition to the use of chromosome breakage and translocation, two other techniques that offer an opportunity for subchromosomal mapping are chromosome-mediated gene transfer and DNA-mediated gene transfer (see Part II).

All of the gene mapping studies discussed here were based on the detection of the protein products of the genes to be mapped. Thus, only genes that were expressed in hybrid cells could be mapped. However, in recent years, species-specific molecular probes for a wide variety of genes have become available. The availability of such molecular probes has made it possible to use hybrid cells to map genes that are not expressed in hybrids.

The combined use of molecular probes and hybrid cells for mapping studies was introduced by Deisseroth, Ruddle, and co-workers (Paper 21). In these studies, DNA from human-mouse hybrid cells was hybridized to human α-globin complementary DNA, and the presence of human α-globin sequences in the hybrids was correlated with the presence of specific human chromosomes. The results demonstrated that the α-globin gene is located on human chromosome number 16. More recently, several laboratories have used cloned DNA sequences as molecular probes for mapping genes in hybrid cells. The techniques of recombinant DNA are producing a rapidly increasing number of probes that can be used for gene mapping. The combination of recombinant DNA and somatic cell hybridization can be expected to further accelerate the mapping of the human genome in cultured somatic cells.

REFERENCES

Boone, C., T. Chen, and F. H. Ruddle, 1972, Assignment of Three Human Genes to Chromosomes (LDH-A to 11, TK to 17, and IDH to 20) and Evidence for Translocation Between Human and Mouse Chromosomes in Somatic Cell Hybrids, *Natl. Acad. Sci. (USA) Proc.* **69:**510–514.

Ephrussi, B., L. J. Scaletta, M. A. Stenchever, and M. C. Yoshida, 1964, Hybridization of Somatic Cells In Vitro, in *Cytogenetics of Cells in Culture,* R. J. Harris, ed., International Society of Cell Biology Symposium, vol. 3, Academic Press, New York, pp. 13–25.

Fournier, R. E., and F. H. Ruddle, 1977, Microcell-Mediated Transfer of Murine Chromosomes into Mouse, Chinese Hamster, and Human Somatic Cells, *Natl. Acad. Sci. (USA) Proc.* **74:**319–323.

Grzeschik, K. H., P. W. Allderdice, A. Grzeschik, J. M. Opitz, O. J. Miller, and M. Siniscalco, 1972, Cytological Mapping of Human X-Linked Genes by Use of Somatic Cell Hybrids Involving an X-Autosome Translocation, *Natl. Acad. Sci. (USA) Proc.* **69:**69–73.

Kao, F.-T., C. Jones, and T. T. Puck, 1976, Genetics of Somatic Mammalian Cells: Genetic, Immunologic, and Biochemical Analysis with Chinese Hamster Cell Hybrids Containing Selected Human Chromosomes, *Natl. Acad. Sci. (USA) Proc.* **73:**193–197.

Kusano, T., R. Wang, R. Pollack, and H. Green, 1970, Human-Mouse Hybrid Cell Lines and Susceptibility to Polio Virus. II. Polio Sensitivity and Chromosome Constitution of the Hybrids, *J. Virol.* **5:**682–685.

McKusick, V. A., and F. H. Ruddle, 1977, The Status of the Gene Map of the Human Chromosomes, *Science* **196:**390–405.

Migeon, B., S. Smith, and C. Leddy, 1969, The Nature of Thymidine Kinase in the Human-Mouse Hybrid Cell, *Biochem. Genet.* **3:**583–590.

Miller, O. J., P. W. Allderdice, and D. Miller, 1971, Human Thymidine Kinase Gene Locus: Assignment to Chromosome 17 in a Hybrid of Man and Mouse Cells, *Science* **173:**244–245.

Minna, J. D., and H. G. Coon, 1974, Human X Mouse Hybrid Cells Segregating Mouse Chromosomes and Isozymes, *Nature* **252:**401–404.

Pontecorvo, G., 1958, *Trends in Genetic Analysis,* Columbia University Press, New York, 145p.

Pontecorvo, G., 1971, Induction of Directional Chromosome Elimination in Somatic Cell Hybrids, *Nature* **230:**367–369.

Ricciuti, F. C., and F. H. Ruddle, 1973, Assignment of Three Gene Loci (PGK, HGPRT, G6PD) to the Long Arm of the Human X Chromosome by Somatic Cell Hybrids, *Genetics* **74:**661–678.

Santachiara, A. S., M. Nabholz, V. Miggiano, A. J. Darlington, and W. F. Bodmer, 1970, Genetic Analysis with Man-Mouse Somatic Cell Hybrids, *Nature* **227:**248–251.

17

Reprinted from *Natl. Acad. Sci. (USA) Proc.* **58:**1104–1111 (1967)

*HUMAN-MOUSE HYBRID CELL LINES CONTAINING PARTIAL COMPLEMENTS OF HUMAN CHROMOSOMES AND FUNCTIONING HUMAN GENES**

By Mary C. Weiss† and Howard Green

This paper will describe the isolation and properties of a group of new somatic hybrid cell lines obtained by crossing human diploid fibroblasts with an established mouse fibroblast line. These hybrids represent a combination between species more remote than those previously described (see, however, discussion of virus-induced heterokaryons). They are also the first reported hybrid cell lines containing human components and possess properties which may be useful for certain types of genetic investigations.

Interspecific hybridizations involving rat-mouse,[1] hamster-mouse,[2, 3] and Armenian hamster–Syrian hamster[4] combinations have been shown to yield populations of hybrid cells capable of indefinite serial propagation. Investigations of the karyotype and phenotype of such hybrids have shown that both parental genomes are present[2, 5] and functional.[6] In every case, some loss of chromosomes has been observed; this occurred primarily during the first few months of propagation, usually amounted to approximately 10–20 per cent of the complement present in newly formed hybrid cells, and involved chromosomes of both parents. Recent studies have provided evidence of preferential loss of chromosomes of one parental species in interspecific hybrids.[2, 5]

A more extreme example of this preferential loss has been encountered in the human-mouse hybrid lines to be described, in which at least 75 per cent, and in some cases more than 95 per cent, of the human complement has been lost. It has been possible to relate the human characteristics of the hybrid phenotype to the number of human chromosomes retained. This has been shown for the colonial morphology of the hybrid and for the human antigens in the hybrid cell membrane, which were detected by mixed cell agglutination. Furthermore, information has been obtained with respect to the chromosomal localization of the thymidine kinase gene.

Materials and Methods.—Cell lines: LM (TK⁻) cl 1-D (hereafter referred to as cl 1-D), a subline of mouse L cells, was isolated by Dubbs and Kit[7] and provided by Dr. B. Ephrussi. This clone, deficient in thymidine kinase, is resistant to 30 μg/ml of 5-bromodeoxyuridine (BUDR).

WI-38, a diploid strain of human embryonic lung fibroblasts,[8] was provided by the American Type Culture Collection and was propagated in monolayer culture in this laboratory for 10 to 20 cell generations before hybridization.

Culture method and selection of hybrids: All cultures were maintained in standard growth medium (Dulbecco and Vogt's modification of Eagle's minimal medium containing 10% calf serum), in some cases supplemented with 30 μg/ml of BUDR, or hypoxanthine ($1 \times 10^{-4}\,M$), aminopterin ($4 \times 10^{-7}\,M$), thymidine ($1.6 \times 10^{-5}\,M$) (HAT). The selective system used was originally described by Littlefield[9] and has been modified by Davidson and Ephrussi[10] for use with a biochemically marked cell line combined with a contact-inhibited strain of diploid fibroblasts. The mouse cell line cl 1-D, deficient in thymidine kinase, is unable to grow in medium containing HAT; no spontaneous reversion to HAT resistance has been observed in this cell line, and it seems most likely that a deletion involving this gene has occurred.[11] The human diploid strain, containing

119

thymidine kinase activity, is able to grow in this medium but it forms only a relatively thin cell layer, while hybrid cells are able to pile up against this background and form discrete colonies.

Karyotype of parental cells: Cl 1-D is characterized by the presence of 51 (50–55) chromosomes, 9 (8–10) of which are large metacentrics. Among the latter is the D chromosome[12] which, owing to the presence of a secondary constriction, appears to be dicentric and has no equivalent in the human complement. As can be seen in Figure 2a, most of the chromosomes of cl 1-D are telocentric.

WI-38 is a diploid human strain;[8] the cells contain 46 chromosomes, the pairs of which may be grouped into seven classes (Denver Classification). Of these, only two (A and G) contain chromosomes which are likely to be confused with the mouse chromosomes (Fig. 2b).

Demonstration of cell surface antigens by mixed hemagglutination: These experiments were carried out using a modification of the method described by Kelus, Gurner, and Coombs.[13] Immune sera were obtained from rabbits following injection of WI-38 cells in Freund's adjuvant. Complement was inactivated at 56° and the serum absorbed twice with a total of 6×10^6 cl 1-D cells per ml of serum. In order to test for the presence of human antigens, 2×10^5 trypsinized parental or hybrid cells were washed in buffered salt solution containing 0.1% bovine serum albumin, incubated with 0.1 ml antiserum for 1 hr at room temperature, again washed three times and mixed with 0.1 ml of a 2% suspension of washed human red blood cells. The mixture was centrifuged gently, resuspended, and a drop pipetted onto a glass slide for examination. A cover slip placed over the drop was allowed to settle for 5 min, so that erythrocytes attached to the surface of a test cell were forced to its perimeter and appeared as an encircling ring. In controls for specificity of the agglutination, mouse erythrocytes were employed in place of human. Mouse species-specific antigens were identified by the use of rabbit antimouse ascites tumor antiserum and mouse erythrocytes.

Results.—Production and identification of hybrid cells: Cultures were initiated with mixtures of 2×10^6 cl 1-D cells and 1×10^4 WI-38 cells. After four days of growth in standard medium, the cultures were placed in selective medium (HAT). The cl 1-D cells degenerated within seven days, leaving a single layer of human cells; after 14 to 21 days, hybrid colonies (Fig. 1) could be detected growing on the human cell monolayer. A number of these were isolated and grown to mass culture. In other cases the entire culture was transferred; within a few weeks the hybrid cells over-grew the remaining human fibroblasts and in the course of serial cultivation all human cells disappeared from the population.

Of three independent experiments performed, all yielded hybrid colonies, with a frequency of approximately one per 2×10^4 WI-38 cells. In all hybrids examined 20 generations after their formation, the same karyotypic pattern was found; all (or nearly all) of the expected mouse chromosomes were present, but of the human chromosomes only a minority remained, varying in number from 2 to 15 in the

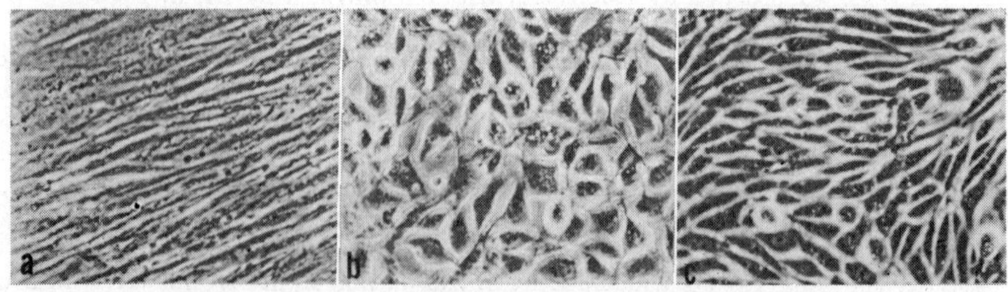

Fig. 1.—Phase contrast photomicrographs of living cells of (a) WI-38; (b) cl 1-D; (c) hybrid clone HM-2 P2. The parallel orientation of the highly elongated WI-38 cells is in contrast to the random orientation of the highly refractile and less fusiform cells of cl 1-D. The cells of the hybrid clone, while highly refractile, have a degree of orientation intermediate between cl 1-D and WI-38.

various populations.   The mean numbers of human chromosomes found in six independent hybrid clones were as follows: 6.5, 2.0, 11.6, 10.9, 3.0, and 3.1.   An occasional hybrid mitosis contained more than one copy of the same human chromosome, but most contained only one member of a given pair.   The karyotype of a hybrid cell of clone HM-2 is shown in Figure 2c.

Loss of human thymidine kinase gene(s) from hybrid cells: Since survival of cells in HAT medium requires the presence of thymidine kinase, all hybrid cells selected in these experiments presumably contain the human gene(s) for this enzyme.   This conclusion is supported by the fact that the cells can be killed by growth in the presence of BUDR.   During continued propagation in HAT, any variants which may have lost this gene are eliminated.   However, such variants occur with high frequency and were obtained selectively in a single step by transfer of the population to a medium containing BUDR, which eliminated most of the population and permitted the growth only of cells without thymidine kinase activity. As expected, these BUDR-resistant variant hybrid cells were not able to grow in HAT medium.   They did retain many, but not all, of the human chromosomes which were present before the selection (see below).   Cells resistant to BUDR also appeared in cultures propagated in standard medium without HAT and seemed to enjoy some selective advantage, for after several months of serial propagation they amounted to about 50 per cent of the population, as measured by HAT sensitivity.   As no revertants to HAT resistance were obtained, BUDR resistance probably occurred by deletion.

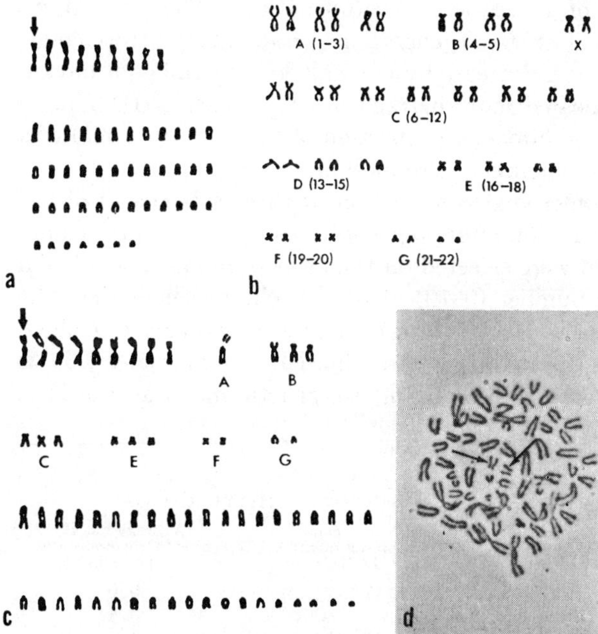

Fig. 2.—Chromosomes of Cl 1-D, WI-38, and hybrids.   (*a*) Karyotype of the mouse parental line, cl 1-D.   Nine long metacentric chromosomes are present, one of which (*arrow*, the D chromosome) is identifiable in all chromosome preparations. The karyotype also shows 43 telocentric chromosomes, none of which contains distinguishable satellites.   (*b*) Karyotype of a cell of the WI-38 strain (see also ref. 8).   Comparison with (*a*) shows that only chromosomes of groups A (long metacentrics) and G (small acrocentrics) are likely to be confused with chromosomes of cl 1-D.   (*c*) Karyotype of hybrid clone HM-2.   All chromosomes have been classified as to their origin, 14 of them being identified as human and 48 as mouse.   In this karyotype, representatives from each of the major human chromosome groups except D are present. However, in other mitoses of HM-2, group D chromosomes were present.   The A group chromosome is the only large bi-armed chromosome with a nonmedian centromere and is therefore probably a member of the no. 2 pair.   (*d*) Mitosis of a hybrid cell of HM-1 after 60 generations of growth in HAT medium.   The human chromosomes indicated by arrows are of group 6-12-X and are thought to carry the thymidine kinase gene.   Two other human chromosomes are present, one of group D and one of group G.

Since most of the hybrid lines carried few human chromosomes, almost all of which could be distinguished from those of mouse origin, we attempted to correlate thymidine kinase activity with the presence of a specific human chromosome. Table 1 shows that nearly every cell in populations grown in HAT medium contains at least one chromosome of the group 6–12 and X, whereas chromosomes of this group are relatively rare in hybrid cells propagated in BUDR. In some of the hybrid metaphases, a submetacentric chromosome of this group has a peculiar configuration (Fig. 2d), in that there is slight separation of the chromatids in the region of the centromeres.[14] As shown in Table 1, this chromosome was identified in 15 of the 40 mitoses examined from HAT-grown populations and was not detected in any of the 34 BUDR-grown cells. This evidence would suggest that loss of the thymidine kinase gene occurs by loss of the relevant chromosome, and that this chromosome belongs to the 6–12-X group.

Although very soon after their formation human-mouse hybrids already showed extensive loss of human chromosomes, hybrid clones isolated later were found to be relatively stable karyotypically. However, it was expected that their human chromosome content would depend to some extent upon the selective conditions employed; therefore after 30 generations in HAT medium, hybrid cells were propagated in (1) HAT, (2) standard medium, and (3) BUDR. Two clonal hybrid populations were chosen for this study, HM-1 and HM-2, characterized by the presence of 6 and 11 human chromosomes, respectively.

Figure 3 shows the chromosomal changes which occurred during four to six months of serial culture. In both lines there was extensive loss of human chromosomes correlated with number of generations of propagation. The populations grown in HAT medium, checked after 20–60 generations, showed very little change, but after 85 generations there was a decrease by one half in the number of human chromosomes. This change appeared more marked in the absence of HAT or in BUDR medium. Under these conditions, a proportion of the cells (10–25%) of a number of clones lost all identifiable human chromosomes.

Figure 3b also shows the chromosomal composition of three subclones isolated from the HM-2 HAT-grown population after 85 generations. Two of these clones (HM-2 P5 and HM-1 P2), which were selected on the basis of humanlike colonial morphology, contained a larger number (9–12) of human chromosomes than the average of the parent population, while the third clone, which strongly resembled cl 1-D, nevertheless contained approximately six human chromosomes. It is clear that in all hybrid populations very little mouse genetic material was lost;

TABLE 1

CHROMOSOMES OF HAT-RESISTANT AND BUDR-RESISTANT HYBRID CLONES

Hybrid line	Total no. of cells counted	No. of cells containing member of 6-12-X group	Average no. of human chromosomes per cell
HM-1 HAT	20	17* (7)	6.0
HM-1 BUDR	20	5 (0)	3.4
HM-2 HAT	20	18* (8)	4.9
HM-2 BUDR	14	2 (0)	1.8

In parentheses are the number of cells containing the chromosome of distinctive appearance described in text.

* Failure to identify 6-12-X chromosome(s) in all cells of these HAT-grown populations could be due to chromosomal rearrangements, such as translocations. A cell which has lost the thymidine kinase-bearing chromosome might still be able to complete one or two cell divisions in HAT medium.

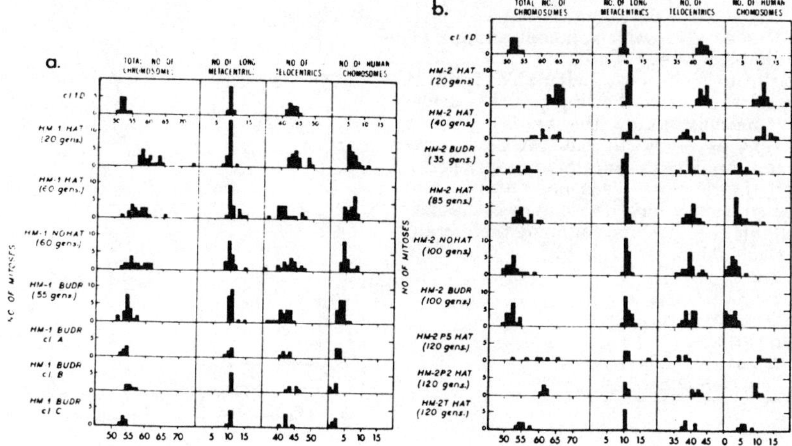

FIG. 3.—Histograms showing evolution of the karyotype of (*a*) HM-1 and (*b*) HM-2.   The chromosomes of the hybrids are grouped into the following classes: (1) total number of chromosomes;   (2) number of long metacentrics (this class is composed almost entirely of chromosomes of mouse origin but might also include one or two group A human chromosomes); (3) mouse telocentrics (in some cases, members of the group G human chromosomes would be placed in this class if the chromatin distal to the centromere were not visible as two distinct masses); and (4) number of human chromosomes.   It is likely that these values are low by perhaps one or two chromosomes, due to failure to distinguish group A, and perhaps group G, chromosomes.   Numbers on abscissa refer to bars immediately to their right.

on the average there was a decrease in number of mouse telocentrics by three or four and no decrease in number of long metacentrics.   Some chromosomal rearrangements may have occurred among the mouse telocentrics in a minority of the population of HM-1, since those cells which contain fewer than expected telocentric chromosomes contain more than the expected number of long metacentrics and it appears likely that these changes were the result of centric fusion of telocentrics. (For a more detailed discussion of this process see ref 2.)   However, in most hybrid populations no evidence of chromosomal rearrangement was found.

Cell membrane antigens and hybrid karyotype: The presence of human species-specific antigens in the cell membrane of hybrid cells was determined by mixed cell agglutination.   A cell was considered positive if two thirds of its perimeter was covered with erythrocytes.   By this criterion 60 per cent of the cells of WI-38 were positive under the standard conditions.   Values given by hybrid cells are shown in Figure 4 on an arbitrary scale denoted as relative agglutination index (R.A.I.; WI-38 = 1.0) plotted against the mean number of human chromosomes per cell.   Cl 1-D gave a value of 0.005, which can be considered as background due to nonspecific adherence of erythrocytes.   All hybrids gave agglutination values in excess of this, and the values increased with increasing number of human chromosomes up to 12, at which point the R.A.I. became equal to that of the WI-38 parent. The R.A.I. rose very slowly with chromosome number up to about five chromosomes per cell, and then more sharply until the maximum was reached at 12 chromosomes. Similar results were obtained when less stringent criteria were used for scoring agglutinations; under these conditions a more nearly linear relation was observed

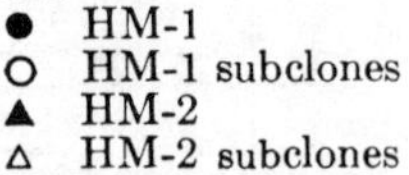

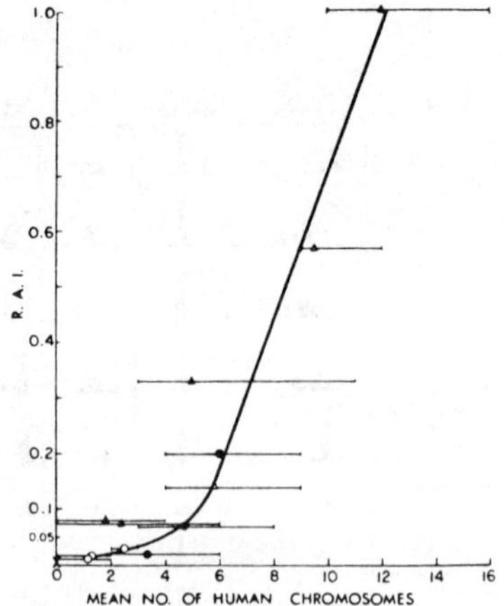

FIG. 4.—Human cell membrane antigens in hybrid lines. Ordinate gives relative agglutination index calculated from the fraction of cells showing positive hemagglutination, taking the values given by WI-38 as equal to 1.0. Abscissa gives mean human chromosome number per cell and the range for each population. Since the subclones have the narrowest range of human chromosome number, their agglutination values are probably the most significant.

- ● HM-1
- ○ HM-1 subclones
- ▲ HM-2
- △ HM-2 subclones
- ✕ cl 1-D control

between human chromosome number and R.A.I., but the background values given by cl 1-D were much higher.

The agglutination values obtained from the hybrids using antimouse cell serum and mouse erythrocytes were practically identical with those given by cl 1-D under the same conditions. This is consistent with the fact (Fig. 3) that virtually the entire cl 1-D genome is present in all the hybrids.

In control experiments no mixed agglutination was observed when hybrid cells were tested with antimouse serum and human erythrocytes, or with antihuman serum and mouse erythrocytes.

Poliovirus infection of hybrid cells: Somatic cell hybrids may be useful for the study of the process of viral infection. For example, it has been shown in heterokaryons made between cells of two different species that virus unable to multiply in one parental cell type is able to do so in the common cytoplasm.[15] However, in our hybrids containing up to 12 human chromosomes poliovirus infection at high multiplicity did not lead to any detectable cytocidal effect, under conditions in which the parental strain WI-38 was completely destroyed within 24–36 hours. While poliovirus itself does not successfully infect mouse cells, L-cells, from which cl 1-D is derived, are known to support viral multiplication very well following infection with poliovirus RNA.[16] Therefore, it seems most likely that failure of the virus to kill the hybrid cells is due to inability of the virus to penetrate the cell membrane, due either to lack of essential human elements or to interference by mouse elements.

Discussion.—The production and properties of a number of different interspecies hybrid lines, including some made between diploid strains and established cell lines, have been summarized by Ephrussi.[17] It has also been shown by Harris and Watkins[18] that virus-induced heterokaryons between HeLa (human) and Erhlich ascites (mouse) cells underwent at least one or two divisions as some hybrid mitoses could be identified. Though no established lines of hybrid cells developed from these heterokaryons, Yerganian and Nell[4] were able by this method to produce established lines of rodent cell hybrids. It might have seemed doubtful whether a permanent hybrid line could be made between a human fibroblast strain and an established mouse line since, in addition to the remoteness of the two species, human fibroblasts do not spontaneously develop into established lines as do those of rodents, but uniformly die out after 50–100 cell generations.[8, 19] Though it turned out that these factors did not prevent the formation of hybrid lines, they may have been the cause of the very extensive loss of human chromosomes from the hybrid cells.[20] This loss began before 20 generations after fusion and most likely during the first few divisions, as the colonial morphology did not change appreciably from the time colonies were first identified to the time pure cultures were obtained.

After the development of hybrid populations, continued subculture results in continuing though slow elimination of human chromosomes. Since nearly the entire cl 1-D complement is retained, the human genes are probably not essential for viability, with the exception of that providing thymidine kinase activity, which is necessary as long as the cells are cultivated in HAT. Omission of HAT from the medium permits variants lacking thymidine kinase to survive, as indicated by the appearance of HAT-sensitive cells. When a hybrid population is placed in BUDR, cells bearing thymidine kinase are eliminated and the entire population becomes HAT-sensitive.

These losses of the thymidine kinase gene appear to occur by loss of the relevant chromosome(s) since: (1) there is a small decline in mean human chromosome number associated with continued growth in the absence of HAT, and in the BUDR-resistant populations; (2) there is a rather strong correlation between the absence of a definite chromosome of the 6-12-X group, BUDR resistance, and HAT sensitivity; and (3) there are no obvious chromosomal rearrangements by which genetic material of that chromosome might have been retained.

It has been shown in a number of cases that antigens characteristic of the parent cells are expressed in the hybrids. Among the antigens which follow this rule are H-2 antigens of the mouse,[21] and polyoma virus-induced T-antigens and surface antigens.[22] It is therefore not surprising that human species-specific cell membrane antigens could be detected in all of the hybrids described here. Even the presence of a very small number of human chromosomes (<4) produced an R.A.I. of 2–15 times background. The genes for surface antigens appear to be widely distributed among the different human chromosomes, since agglutinability of the hybrid cells increased progressively with number of human chromosomes.

Clones of hybrid cells selected on the basis of humanlike morphology (HM-2P2 and HM-2P5) contained the largest number of human chromosomes (9.5 and 12) and gave the highest agglutination values. On the other hand, a clone selected on the basis of absence of humanlike morphology (HM-2T) had six human chromosomes and a corresponding R.A.I. This suggests that some human chromosomes may have little to do in determining colonial morphology, though they do determine cell membrane composition. Similar hybrids may permit the chromosomal localization of other human genes expressed at the cellular level, such as those for blood group antigens and enzymes whose physical properties are different from those of mouse determination. It should also be possible to isolate clones of hybrid cells which have lost all human chromosomes and which might be useful for the investigation of nonchromosomal genes.

Summary.—Cocultivation of a human diploid cell strain with a thymidine kinase deficient mouse cell line has led to the formation of hybrid cell lines, which were isolated in selective medium. The hybrid lines contained substantially the entire mouse genome and a greatly reduced complement of human chromosomes.   The functioning of the human genes was shown by the presence of human antigens on the surface of the hybrid cells.   The agglutinability of the cells in mixed hemagglutination tests depended on the number of human chromosomes contained (up to 12), indicating that the genes for surface membrane antigens are widely distributed among the human chromosomes.   The human gene for thymidine kinase permits the hybrid cells to grow in medium containing aminopterin, but variants lacking this function may be selected in medium containing 5-bromodeoxyuridine.   These variants appear to have lost a human chromosome of the 6-12-X group and it is suggested that this chromosome contains the thymidine kinase gene.   Continued growth of hybrid lines results in slow elimination of human chromosomes.   Study of clones containing a small number of human chromosomes should permit the localization of other human genes.

It is a pleasure to acknowledge the kind assistance of Mr. C. deSzalay and Dr. L. J. Scaletta and the valuable advice of Drs. G. J. Todaro and Z. Ovary.

* Aided by grant CA 06793, postdoctoral fellowship 1-F2-GM-34,679 (M.W.), and award 4-K6-CA-1181 (H.G.), all from the U.S. Public Health Service.

† Present address: Carnegie Institute of Washington, Department of Embryology, 115 W. University Parkway, Baltimore, Maryland.

[1] Ephrussi, B., and M. C. Weiss, these PROCEEDINGS, **53**, 1040 (1965).

[2] Scaletta, L. J., N. Rushforth, and B. Ephrussi, *Genetics*, in press.

[3] Davidson, R. L., B. Ephrussi, and K. Yamamoto, these PROCEEDINGS, **56**, 1437 (1966).

[4] Yerganian, G., and M. B. Nell, these PROCEEDINGS, **55**, 1066 (1966).

[5] Weiss, M. C., and B. Ephrussi, *Genetics*, **54**, 1095 (1966).

[6] *Ibid.*, p. 1111.

[7] Dubbs, D. R., and S. Kit, *Exptl. Cell Res.*, **33**, 19 (1964).

[8] Hayflick, L., and P. S. Moorhead, *Exptl. Cell Res.*, **25**, 585 (1961).

[9] Littlefield, J., *Science*, **145**, 709 (1964).

[10] Davidson, R., and B. Ephrussi, *Nature*, **205**, 1170 (1965).

[11] Kit, S., D. R. Dubbs, L. J. Piekarski, and T. C. Hsu, *Exptl. Cell Res.*, **31**, 297 (1963).

[12] Hsu, T. C., *J. Natl. Cancer Inst.*, **25**, 1339 (1960).

[13] Kelus, A., B. W. Gurner, and R. R. A. Coombs, *Immunology*, **2**, 262 (1959).

[14] This configuration was not seen in the human parent cell and may have occured as a consequence of the new environment of the chromosome in the hybrid cell.   It is also possible that this chromosome arises through translocation and is only part human, but this seems unlikely as it was seen in three hybrid populations of independent origin.

[15] Koprowski, H., F. C. Jensen, and Z. Steplewski, these PROCEEDINGS, **58**, 127 (1967).

[16] Holland, J. J., B. H. Hayer, L. C. McLaren, and J. T. Syverton, *J. Exptl. Med.*, **112**, 821 (1960).

[17] Ephrussi, B., "Phenotypic Expression," *In Vitro* (Baltimore: Williams and Wilkins, 1967), vol. 2, p. 40.

[18] Harris, H., and J. F. Watkins, *Nature*, **205**, 640 (1965).

[19] Todaro, G. J., and H. Green, *Proc. Soc. Exptl. Biol. Med.*, **116**, 688 (1964).

[20] A more general discussion of some of the mechanisms of loss of chromosomes in interspecific somatic hybrids has been given in: Ephrussi, B., and M. C. Weiss, in *Control Mechanisms in Developmental Processes*, ed. M. Locke (New York: Academic Press, in press).

[21] Spencer, R. A., T. S. Hauschka, D. B. Amos, and B. Ephrussi, *J. Natl. Cancer Inst.*, **33**, 893 (1964).

[22] Defendi, V., B. Ephrussi, H. Koprowski, and M. C. Yoshida, these PROCEEDINGS, **57**, 299 (1967).

Genetic Analysis with Human–Mouse Somatic Cell Hybrids

by

M. NABHOLZ
V. MIGGIANO*
W. BODMER
Department of Genetics,
School of Medicine,
Stanford University,
Stanford, California

Somatic hybridization between mouse cells and human leucocytes confirms the X-linkage of 8-azaguanine resistance in man and suggests that human lactate dehydrogenase A and B genes are not linked.

HYBRIDIZATION between somatic cells *in vitro*[1,2,16] has provided the basis for genetic analysis of somatic cells. The use of human–mouse hybrid cell lines for the assignment of human genes to their chromosomes has been pioneered by Green and his co-workers[3,4]. Using cell fusion mediated by Sendai virus[5] we have developed a system for the hybridization of human peripheral white blood cells with mouse cells[6]. The mouse cell line (1R) used in these experiments is an 8-azaguanine (8-AZG) resistant derivative of a subline of L cells, which does not grow in Littlefield's HAT selective medium[7,8] containing hypoxanthine, amethopterin, thymidine and glycine, and presumably lacks the enzyme hypoxanthine-guanine phosphoribosyl transferase (HPRT). Partially purified peripheral white blood cells and 1R cells in a ratio of about 4 to 1 are mixed with Sendai virus to mediate fusion and then plated. Normal medium is changed to HAT selective medium after 2–4 days' incubation. The human white cells do not proliferate and the mouse cells are eliminated by the selective medium. Only hybrid cells containing the human HPRT + gene are able to grow and form colonies. A typical yield from such a hybridization experiment is one colony per 2 to 3×10^4 1R cells plated. Further details of the technique are described by Miggiano *et al.*[6]. Here we summarize some of the properties of five hybrid lines we have studied in detail, describe the evidence for locating the HPRT gene on the X-chromosome and discuss some evidence for the segregation of the human lactate dehydrogenase genes in the hybrids.

* On leave of absence from Istituto di Genetica Medica, Università di Torino, Italy.

Karyotypic Evolution of Hybrid Lines

Fig. 1a shows the karyotype of a typical 1R cell. In all cells that have been examined, the smallest biarmed chromosome is larger than or at least the same size as the largest telocentric chromosome. The mean total number of chromosomes is 58·2 (standard deviation 2·8), while the corresponding means and standard deviations for the numbers of telocentric and biarmed chromosomes are $40·2 \pm 3·4$ and $18 \pm 1·8$ respectively. The hybrid lines have, 40 to 60 days after fusion, a significantly larger number of chromosomes than 1R, comprising most of the 1R chromosomes and 4 to 14 extra chromosomes, presumably of human origin. The chief evidence for some human chromosomes in the hybrids is the occurrence in all hybrids, at least at the earliest times, of biarmed chromosomes which are appreciably smaller than the largest telocentrics. Positive identification of most of the human chromosomes is, however, impossible, especially for the larger metacentrics, because of confusion with the variable number of biarmed chromosomes normally present in 1R. Figs. 1b, c and d show representative karyotypes of cells from two different hybrid lines at different times. The small metacentric chromosomes present in some of the hybrids (Figs. 1c and 1d) are most probably chromosomes of the human F group (Denver Classification). The larger hybrid biarmed chromosomes, which are comparable in size with the largest 1R telocentric, have approximately the dimensions of the largest of the human C group, namely, chromosome 6 or the X-chromosome.

Fig. 2 shows the change in the number of biarmed

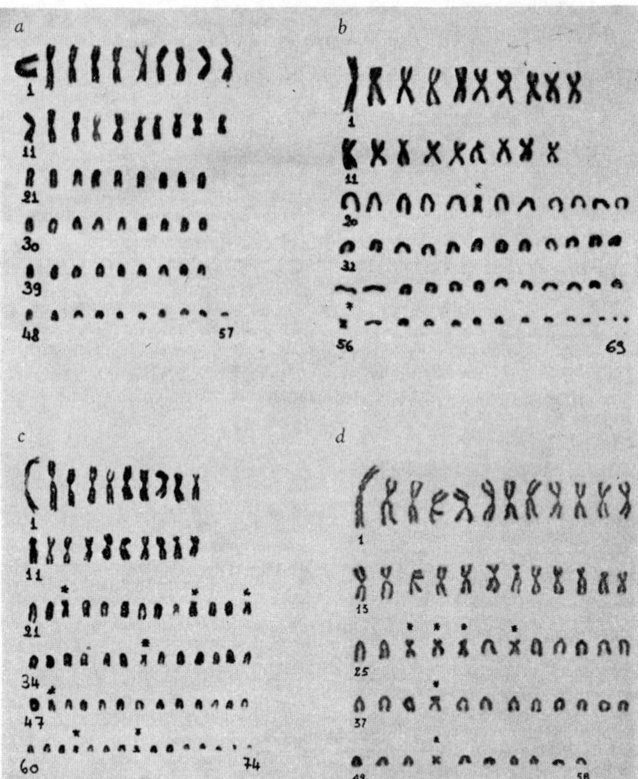

Fig. 1.   Karyotypes of the parental mouse line and hybrid derivatives.   *a*, 1R, parental line; *b*, line 1W1, 68 days after cell fusion; *c*, line 4W10, 99 days after cell fusion; *d*, line 4W10, 240 days after cell fusion.   Stars designate chromosomes which are assumed to be of human origin.

chromosomes with time after fusion for the lines 1W1 and 3W4 assuming that the complete human complement is present initially.   This assumption seems reasonable because Harris *et al.*[5] have found that the cells most likely to give rise to hybrid lines are heterokaryons containing two nuclei that undergo synchronized DNA synthesis and mitosis during which nuclear fusion occurs.   So far the earliest time point we have examined is 44 days after fusion, by which time there must have already been considerable karyotypic evolution[3,4].   The hybrids tend to lose more chromosomes with continued growth and each of the five lines we have studied presents a different picture[6].

After about 100 to 150 days, the karyotypes are already relatively stable.   All the presumptive hybrid lines except 1W1 show strong positive reactions at 110 or 163 days with a specific rabbit anti-human serum[6], which confirms the presence of at least a part of the human genome in those lines.   The mouse H2-k antigen is present on 1R and all the hybrids, as expected, because L cells were derived from the C3H inbred strain, which carries H2-k.

Glucose-6-phosphate Dehydrogenase (G6PD) Activity of Hybrids and Clones derived from Them

Human G6PD was present in all the hybrid lines (Fig. 3).   The difference in mobility of the extracts of fibroblasts from a Caucasian donor and of HeLa, which is known to have the A+ phenotype common in Africans, and the correspondence of the activities shown by fibroblasts and red cell haemolysates, confirms that the detected activity is the X-linked G6PD.   The multiple bands of G6PD activity are probably the result of differences in the number of cofactor molecules bound per G6PD molecule because mercaptoethanol treatment results in a shift of the activity toward the slower moving bands, and their relative migration rates remain constant in gels with increased acrylamide concentrations.   Human and mouse extracts show two predominant strong bands. The hybrid extracts show, in addition to the human and mouse bands, two strong bands between the two parental activities, presumably hybrid molecules.   The presumptive hybrid enzyme activity is greater than either parental activity because the enzyme is probably a dimer, and in the hybrid lines molecules are probably formed by random combination of approximately equal amounts of human and mouse subunits.   The presence of the human G6PD in all the hybrids indicates, of course, the presence of the human X-chromosome, or at least part of it.

Deficiency of hypoxanthine–guanine phosphoribosyl transferase (HPRT) activity—the basis for the sensitivity of the 8-AZG-resistant cells to the HAT selective medium[7-10]—is the primary basis for a human X-linked recessive neurological disorder discovered by Lesch and Nyhan[11,12].   The simplest explanation, therefore, for the presence of the human X-chromosome in our hybrids is that the normal human HPRT activity, controlled by an X-linked gene, is required to overcome the sensitivity of 1R cells to the HAT selective medium, and allow growth of the human–mouse hybrids.   This hypothesis was confirmed by the results of a back selection experiment in

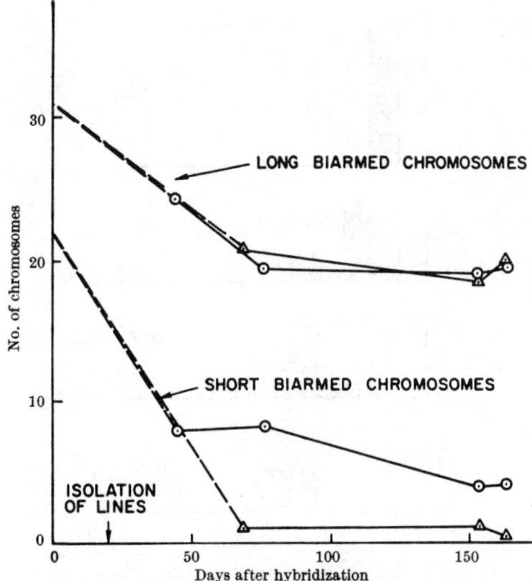

Fig. 2. Loss of human chromosomes: the initial number of the two types of biarmed chromosomes in the hybrid lines ($\bigcirc$: 3W4, $\triangle$: 1W1) are calculated as follows: Long biarmed chromosomes = 18 for 1R + human pairs 1 to 6 + human X = 31. Short biarmed chromosomes = human pairs 7 to 12, 16 to 20 = 22 chromosomes. The dashed line indicates the absence of actual data during the initial period.

which 8-AZG resistant clones were selected from the hybrid lines and assayed for their G6PD activity. 8-AZG resistant clones were selected by seeding plates with about 200 cells about 200 days after fusion and following growth of the mass cultured lines in 8-AZG for about 30 days. The isolated colonies[6] had the properties expected of resistant clones; they grow in high concentrations of 8-AZG and are sensitive to HAT selective medium. Control clones were obtained by plating in non-selective or in HAT medium. The results (Table 1) show absolute linkage between 8-AZG resistance and the loss of human G6PD, a striking confirmation of the presence of the human HPRT gene on the X-chromosome.

Clonal Analysis of Lactate-dehydrogenase (LDH)

By contrast to the relative constancy of the G6PD patterns in the hybrid lines, LDH zymograms of the hybrids vary considerably. Fig. 4 shows the LDH zymograms for 1R, HeLa cells and four patterns obtained from hybrid clones. In these zymograms, bands 5 to 1, tetramers with the configurations A_4, A_3B_1, A_2B_2, A_1B_3, B_4 respectively[14], should normally occur in sequence starting from the top of the gel. The mouse parent line 1R shows bands 5 and 4 and a trace of band 3. This was confirmed in slab gel runs using mouse diaphragm extract run alongside 1R extract. The prevalence of the A subunit in other L cell sublines has been described by Weiss and Ephrussi[15] and we have also observed the same pattern in 3T3, a line derived from Swiss mice. Zymograms from HeLa cells show bands 1 to 4. The fifth human band does not migrate in this system. The first bands from mouse and human cell extracts migrate at the same position. This means that we cannot unequivocally distinguish the presence of the human B subunit in our extracts, although the more prevalent A subunits are readily distinguishable.

Only hybrid 2W1 has a zymogram essentially identical to 1R showing no evidence for any human LDH subunits. The patterns 1 to 4 in Fig. 4 are the chief ones we have observed in the other four hybrid lines and in clones derived from them. Line 1W1 shows predominantly the first pattern, with two bands cathodal to the fifth mouse band, which presumably represent different combinations of human and mouse A subunits. The second pattern, with an additional band between the mouse fifth and

Table 1. REVERSION TO 8-AZG RESISTANCE AND JOINT DISTRIBUTION OF HUMAN G6PD AND LDH

Cloning medium*	Number of clones†	Human G6PD‡	LDH patterns 0	1 or 2	3 or 4
KKL	27	+	7	13	7
KK	15	+	3	5	7
AZO	9	−	4	2	3

Data from all lines are combined (see Fig. 5 for distribution of G6PD among lines).

* KK = standard growth medium. KKL = HAT selective medium. AZO = KK + 10⁻⁴ M 8-AZG.

† Each class of clones (KKL, KK, AZO) contains at least one representative from each of the five lines.

‡ All clones isolated in KKL or KK were human G6PD +, while all those isolated in AZO were human G6PD −.

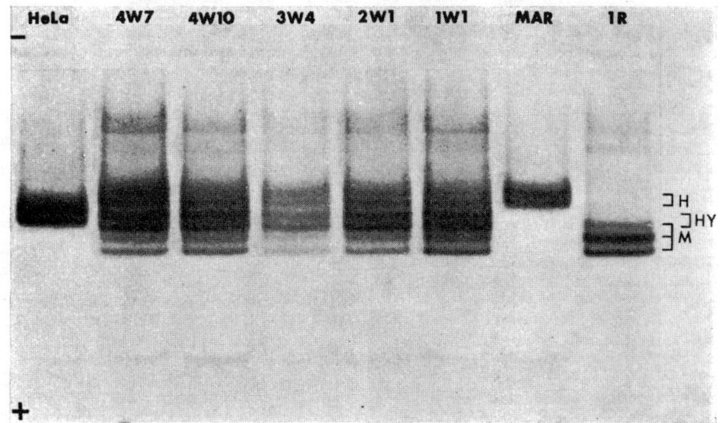

Fig. 3. G6PD zymograms of 1R, fibroblasts derived from the lymphocyte donor for the hybridization (MAR), HeLa cells and the hybrid lines. M denotes bands present in the zymograms from mouse cells, H bands present in zymograms from the donor fibroblasts and HY bands detected only in zymograms for the hybrids. Electrophoresis was carried out in an EC–470 (E–C. Apparatus Corp.) cell. A discontinuous *tris* tricine/*tris* HCl buffer system with a resolving and a stacking gel was used. The details are described by Miggiano *et al.*[4].

fourth bands, which probably represents molecules consisting of two mouse A, one mouse B and one human A subunit, is the predominant pattern for lines 4W7 and 4W10. The third pattern is characteristic of line 3W4. This has extra anodal bands indicating either increased synthesis of mouse B subunits or presence of appreciable amounts of human B subunits. The latter possibility seems likely because of the observed consistency of the 1R-like patterns in mouse lines of different origin. The fourth pattern was obtained from a single clone derived from 3W4. The patterns displayed by the hybrid lines remained quite constant over a period of at least 100 days and after growth in different media; this is remarkable because the LDH patterns of different clones derived from any given line were heterogeneous (Fig. 5).

Because line 2W1, which has human G6PD and presumably therefore the human X-chromosome, lacks human LDH, the human LDH A and probably B genes are not on the X-chromosome. This agrees with pedigree data on human LDH variants[20]. The heterogeneity within lines contrasts sharply with the relative uniformity of the G6PD patterns. The joint distribution of LDH and G6PD activities in the clones (Fig. 5 and Table 1) emphasizes the independence of the G6PD and LDH genes. The loss of the human X-chromosome apparently does not destabilize the karyotype of a hybrid cell line sufficiently to lead to a clear-cut non-specific loss of other human chromosomes.

Genetic Interpretation of LDH Patterns

The distinct LDH zymogram patterns, together with their stability between lines, suggest that each pattern may correspond to a particular genotype for the genes which determine the LDH A and B subunits. The chief difference between patterns one and two is the amount of the human A subunit, while the main additional bands in the third pattern most probably result from the presence of the human B subunit. The most plausible basis for a genetically determined quantitative difference is a dosage effect related to the number of human genes present in the hybrid (Table 2).

Table 2 has two important implications. First, the existence of a dosage effect indicates segregation of

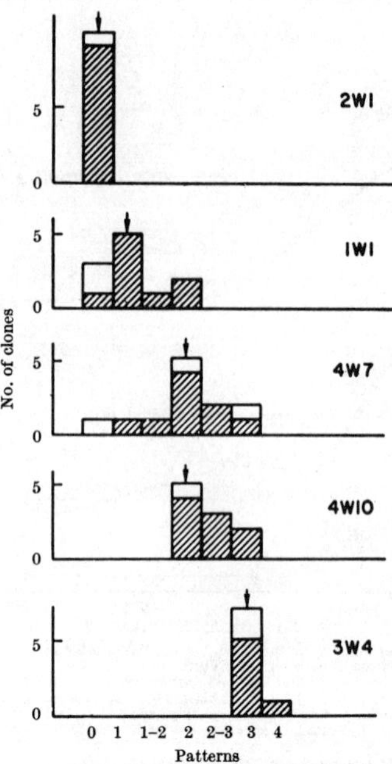

Fig. 5. Distribution of LDH patterns among the clonal offspring in the hybrid lines. The patterns correspond to those defined in Fig. 4. Where unequivocal assignment was impossible an intermediate pattern (1–2, 2–3) is indicated. The arrows indicate the pattern in the line from which the clones were derived. The clones are the same as those described in Table 1. In each column the distribution of clones containing or lacking human G6PD is indicated by shaded and clear areas respectively.

homologous chromosomes in the hybrid; indeed, the dosage effect may well provide a general and convenient assay for such segregation. The high proportion of lines with predominant patterns two and three, however, might

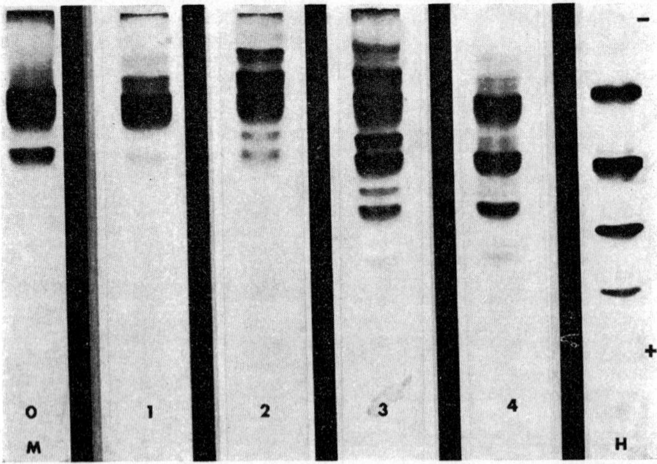

Fig. 4. LDH patterns: 0, pattern produced by extracts from 1R and other mouse lines, also by hybrid line 2W1; H, pattern obtained from HeLa cell extracts, band 5 does not migrate in this system; 1–4, patterns observed in hybrid lines and clones derived from these lines. Electrophoretic runs were carried out in acrylamide disc gels. The buffer system is a slight modification of the original Davis–Ornstein[17] system. As described by Fritz and Jacobson[18] there is a tendency of the LDH bands in the mouse to migrate as a cluster of subbands. As a band we designate here the whole group of subbands; for example, in pattern two, each of the two bands cathodal of mouse band 5 clearly consists of at least two subbands.

Table 2. GENETIC INTERPRETATION OF LDH PATTERNS

LDH pattern*	Human LDH A	Genotype† B
0	$-/-$	$-/-$
1	$+/-$	$-/-$
2	$+/+$	$-/-$
3	$+/+$	$+/-$ or $+/+$
4	$+/-$	$+/-$ or $+/+$

* For the definition of the patterns see Fig. 4.

† $+$ and $-$ indicate presence and absence of the respective human chromosomes in the hybrids.

suggest that homologues are not retained at random in the hybrids. Second, the existence of clones with human A but not B subunits suggests that the genes controlling these subunits are not on the same chromosome. It is worth noting, however, that we have not observed any clones with the presumptive human B but not A subunits. One important prediction of the genetic model is that clones of types one or two should never give rise to those of types three or four, while the reverse transition could happen. Of course, further characterization of the LDH activities in the hybrids is needed to confirm our hypothesis.

Joint Segregation of Human Antigens and LDH Subunits

Table 3 shows the joint distribution of LDH types and reaction to anti-human serum. Because cytotoxicity is used as the serological assay, minor admixtures of negative with positive or positive with negative cells are not readily detected in the mass cultured lines. Thus, for example, although 1W1 showed no gross positive reaction to the anti-human serum it gave rise to both positive and negative clones. The fact that all clones with patterns three or four react to the anti-human serum is consistent with the hypothesis that one or more of the genes determining reaction to the anti-human serum are on the same chromosome as the LDH locus (or loci) which determines these patterns. We do not, of course, have any information about the specificities contained within the anti-human serum, although it seems likely that the reactions to the hybrid lines, which contain only a few human chromosomes, may not be unduly complex. This association between LDH patterns and human antigens is analogous to the correlation between decreasing numbers of human chromosomes and decreasing reaction to an anti-human serum described by Weiss and Green[3]. There is no significant association between reaction to the anti-human serum and the presence of human G6PD.

Clonal Analysis of Karyotypes

The relatively large number of larger biarmed chromosomes in the karyotype of the parent mouse line 1R makes unequivocal identification of the human X-chromosome in the hybrids impossible. Nevertheless the fact that all the hybrid lines contain at least one or two extra biarmed chromosomes in the correct size range for the human X, at the smaller end of the large biarmed chromosomes, is at least consistent with the linkage between 8-AZG resistance and G6PD.

A karyotype analysis of nineteen clones was carried out to compare chromosomal variation within and between clones. There was some variation between clones, both within and between lines, with respect to the recognizable human F group chromosomes. Because there were clones of all LDH patterns which lacked the F-type chromosome, the LDH genes are presumably not

Table 3. JOINT CLONAL DISTRIBUTION OF LDH PATTERNS AND REACTION TO ANTI-HUMAN SERUM

Reaction to anti-human serum	LDH patterns			
	0	1	2	3 and 4
$+$	2	1	6	9
$-$	3	4	2	0

Data from all lines are combined. Of the clones which do not react to anti-human serum, all three LDH 0's are from 2W1, all four LDH 1's are from 1W1 and one LDH 2 clone comes from each of 4W7 and 4W10. The difference between zero, one and two versus three and four is significant at the 1 per cent (one-tail) level using Fisher's exact 2×2 test.

on an F group chromosome. As there were no other clearly identifiable human chromosomes in the hybrids we made a hierarchical analysis of variance of the numbers of big and small biarmed and telocentric chromosomes. The big biarmed chromosomes are those which are larger than the largest telocentric, and are probably mostly of mouse origin. The short biarmed chromosomes, shorter than the largest telocentric, are most probably of human origin. The aim of the hierarchical analysis of variance is to assess the relative contributions of differences between cells within clones, between clones within lines and between lines to overall chromosomal variation (Table 4).

Table 4. HIERARCHICAL ANALYSIS OF VARIANCE OF NUMBERS OF DIFFERENT CATEGORIES OF CHROMOSOMES BETWEEN CELLS, WITHIN CLONES, WITHIN LINES

		Variance components		
	All chromosomes	Big biarmed	Small biarmed	Telocentric
Between cells within clones (σ_γ^2)	12·2	1·7	2·1	7·9
Between clones within lines (σ_β^2)	7·4	0·9	3·6	3·5
Between lines (σ_α^2)	2·1*	3·6	1·0*	13·6
Overall mean number of chromosomes	58·5	19·9	5·4	33·2

Big and small biarmed are respectively the biarmed chromosomes which are larger or smaller than the largest telocentric.

* These are the only components not significant at the 1 per cent level.

The analysis is based on karyotypes made from six to nine cells from each clone. The clones were derived from the five lines 1W1 (three clones), 2W1 (two clones), 3W4 (six clones), 4W7 (one clone) and 4W10 (seven clones) as described in the text. The hierarchical analysis of variance model is based on fitting the observed linear model

$$y_{ijk} = \mu + \alpha_i + \beta_{ij} + \gamma_{ijk}$$

where y_{ijk} is the observed number of chromosomes in the kth cell of the jth clone of the ith line.

α_i is the effect of the ith line.
β_{ij} is the effect of the jth clone of the ith line.
γ_{ijk} is the effect of the kth cell of the jth clone of the ith line.

The variance components σ_γ^2, σ_β^2, σ_α^2 are respectively the estimated variances of the cell effects (γ_{ijk}), the clone effects (β_{ij}) and the line effects (α_i). (For a description of a hierarchical analysis of variance see, for example, Graybill[19].)

The variance between cells within clones is a measure of the amount of variation in chromosome number which accumulates during the cloning process up to the time of karyotype analysis, namely, about twenty generations. The component between clones within lines is a measure of the amount of variation that has accumulated within the line from the time of its formation after fusion, namely, about 230 days or 150 generations. This includes variation resulting from early rapid breakdown of the initial presumed nucleus containing a diploid human complement as well as subsequent changes. The differences between lines reflect overall differences between the initial hybrid colonies.

Because the between clone components are highly significant, clonal segregation within lines for karyotypic constitution which is analogous to that observed for the LDH patterns and reactions to anti-human serum must occur. The data are, unfortunately, not adequate to show any significant associations between the numbers of chromosomes in the various categories and the presence of human G6PD, LDH or antigens. The significant variation between lines in the number of big biarmed and telocentric chromosomes suggests that early events following fusion contribute to chromosomal variation and lead to important average differences, between the lines, of the type we have already seen in the LDH patterns. It is interesting that there is no significant variation in total chromosome numbers between lines. The mean total is very close to that for 1R. There is, in fact, a significant negative correlation for the lines, between the number of big biarmed chromosomes and the combined number of telocentric and small biarmed chromosomes ($r = -0·9$, 5 per cent $> P > 2$ per cent). The same correlation for clones within lines is -0.28, and is not significant, while that for cells within clones is only -0.006.

As Green *et al.*[3,4] have emphasized, the human–mouse hybrid seems fortunately to have the appropriate balance between karyotypic stability and chromosomal loss to provide the basis for the assignment of genes to linkage groups and identifiable chromosomes. The loss of human chromosomes after fusion produces a variety of genetically segregant clones from any given hybridization, which, after initial rapid chromosomal evolution, remain stable enough for a detailed analysis of their phenotypes.

The karyotypic evolution of a hybrid clone might conceivably provide a model for tumour evolution and progression. If the major changes take place very early, perhaps even during the first one or two divisions after nuclear fusion, the subsequent evolution may depend on strong selective pressures. These may lead to co-adapted chromosome combinations within any cell as suggested by the correlation between the number of big biarmed and short chromosomes in the different lines. They may also result in associations between different types of cells within the lines. This "population structure" of the hybrid lines emphasizes the importance of clonal analysis. Selective interaction could result in spurious statistical linkages between markers on different chromosomes. The production of "reduced" lines, containing only one human chromosome[4], would circumvent this problem, but we believe that useful information can be obtained from the joint segregation of genetic markers in the hybrids.

We thank Bruna Miggiano for her technical assistance and Dr Eric Shooter for his advice about electrophoretic systems. We also thank Professor H. Harris for providing us with Sendai virus. This work was supported in part by a US Public Health Service research career programme award to W. B. and by US Public Health Service, US National Science Foundation and US Atomic Energy Commission grants.

Received March 7; revised April 23, 1969.

[1] Barski, G., Sorieul, S., and Cornefert, F. R., *CR Acad. Sci.*, **251**, 1825 (1060).

[2] Ephrussi, B., and Sorieul, S., *Approaches to the Genetic Analysis of Mammalian Cells*, 81 (Univ. of Michigan Press, 1962).

[3] Weiss, M. C., and Green, H., *Proc. US Nat. Acad. Sci.*, **58**, 1104 (1967).

[4] Matsuya, Y., Green, H., and Basilico, C. *Nature*, **220**, 1199 (1968).

[5] Harris, H., Watkins, J. F., Ford, C. E., and Schoefl, G. I., *J. Cell Sci.*, **1**, 1 (1966).

[6] Miggiano, V., Nabholz, M., and Bodmer, W., *Wistar Institute Symposium* (in the press).

[7] Littlefield, J. W., *Proc. US Nat. Acad. Sci.*, **50**, 568 (1963).

[8] Littlefield, J. W., *Cold Spring Harbor Symp. Quant. Biol.*, **29**, 161 (1964).

[9] Szybalski, W., Szybalski, E. H., and Brockman, R. W., *Proc. Amer. Assoc. Cancer Res.*, **3**, 272 (1961).

[10] Davidson, J. D., Bradley, T. R., Roosa, R. H., and Law, L. W., *J. Nat. Cancer Inst.*, **29**, 789 (1962).

[11] Lesch, M., and Nyhan, W. L., *Amer. J. Med.*, **36**, 561 (1964).

[12] Seegmiller, J. E., Rosenbloom, F. M., and Kelley, W. M., *Science*, **55**, 1682 (1968).

[13] Friedmann, T., Seegmiller, T. E., and Subak-Sharpe, J. H., *Nature*, **220**, 272 (1968).

[14] Markert, C. L., *Science*, **146**, 1329 (1964).

[15] Weiss, M. C., and Ephrussi, B., *Genetics*, **54**, 1095 (1966).

[16] Migeon, B. R., *Biochem. Genet.*, **1**, 305 (1968).

[17] Davis, B. J., and Ornstein, L., *Ann. NY Acad. Sci.*, **121**, 321, 404 (1964).

[18] Fritz, P. J., and Jacobson, K. B., *Science*, **140**, 64 (1963).

[19] Graybill, F. A., *An Introduction to Linear Statistical Models*, 1 (McGraw-Hill, New York, 1961).

[20] Kraus, A., and Neely, jun., C., *Science*, **145**, 595 (1964).

19

LINKAGE BETWEEN HUMAN LACTATE DEHYDROGENASE A AND B AND PEPTIDASE B

F. H. Ruddle, V. M. Chapman, T. R. Chen, and R. J. Klebe

METHODS for establishing human gene linkage relationships, using somatic cell hybrids between man and mouse, have recently been developed[1,2]. They are based on electrophoretic differences between homologous enzymes, on differences between chromosomes, and on the predominant diminution of only human chromosomes from the hybrids. The characterization of enzyme phenotypes in clonally derived cell populations containing different numbers and combinations of human chromosomes can be used to infer human gene/gene and gene/chromosome linkage relationships.

It has been suggested, on the basis of clonal differences in the activity of total LDH B, that the genes coding for the A and B subunits of lactate dehydrogenase (LDH) are unlinked[3,4]. Because the electrophoretic mobilities of the mouse and human B subunits are identical, the conclusion was only tentative, however, and other possibilities exist. For example, LDH A and LDH B structural genes may be linked, but other unlinked gene(s) may modulate the synthetic rate of mouse and human LDH B subunits. This suggestion is supported by genetic evidence for genes which control the expression of LDH B in mouse erythrocytes[5].

We argued that the problem might be resolved using either a mouse or human structural variant of the LDH B subunit, and describe here a human LDH B variant, LDH B (var), and its segregation in clonal populations of mouse/human hybrids. This provides strong evidence for the absence of linkage between human LDH A and LDH B genes. Besides scoring the LDH phenotypes, we have ascertained the expressions of a total of fifteen different human enzyme phenotypes in a total of twenty-six clones and sixty-six subclones. These data suggest that the human genes for LDH B and peptidase B (Pep B) are linked, and this is supported by the accompanying report of Santachiara *et al.*[6]. Moreover, negative linkage of many other genes can be inferred.

Somatic Cell Hybrids and Clones

Peripheral leucocytes were collected from the proposita, J, and her mother, C. The leucocytes were separated from defibrinated whole blood by dextran sedimentation[7], and then hybridized to mouse RAG cells using β-propiolactone-inactivated Sendai virus[8,9]. The parental cells were mixed at a ratio of 1:1 at a total density of $4·0 \times 10^6$ per ml. in suspension with 1,000 h.a.u. of Sendai virus. The hybrids were selected in Dulbecco-Vogt modified Eagle's medium[10] with 10 per cent newborn calf serum lacking γ-globulin and supplemented with hypoxanthine $(10^{-4}$ M), aminopterin $(4 \times 10^{-7}$ M), and thymidine $(1·6 \times 10^{-5}$ M) (HAT medium[11]). The RAG parent is sensitive to this medium because it lacks hypoxanthine-guanine phosphoribosyl transferase (HG-PRT) activity[8]. Leucocytes do not proliferate *in vitro*. The hybrids proliferate in HAT because suppression of proliferation is recessive, and the HG-PRT deficiency in RAG is complemented by the unimpaired human HG-PRT gene. The selection system has been described in detail elsewhere[2,8,9,11]. The mouse RAG cell line was derived from a renal adenocarcinoma of a BALB/cd mouse. It is particularly appropriate for hybridization because of its relatively unmodified karyotype[9].

Two independent series of hybrids were produced, and termed C and J after the respective human donors (Fig. 1). Mass cultures of C and J hybrids were recovered one month after hybridization. These possessed very weak activity for human LDH A and LDH B (var) phenotypes, indicating that only a few cells in the mass heterogeneous populations carried the LDH A and B (var) genes. The C and J populations were cloned to give the C and J series of independently derived clones. The LDH phenotypes of these clones were then examined to isolate hybrids possessing the LDH B (var) phenotype. Derivative subclones were produced from clones in the J series as shown in Fig. 1.

Linkage Relationships of LDH A and LDH B Genes

Previous studies have shown that lactate dehydrogenase is a tetramer[12]. Two separate and distinct gene loci, termed A and B, code for subunit polypeptides which are distinct in their amino-acid composition[13]. The A and B loci are generally expressed in most somatic cell types, but in differing activity ratios[14]. The a and b subunits interact to give rise to five molecular forms of the tetrameric enzyme, namely, a_4, a_3b, a_2b_2, ab_3 and b_4. These molecular forms possess different electrophoretic mobilities, and they correspond respectively to the LDH isozymes LDH-5, 4, 3, 2 and 1. The isozymes can be visualized by previously described electrophoretic procedures[5].

Allelic variants at the LDH A and B loci have been reported in man[15-18] and other mammals[13]. Analysis of the heritability of these variants indicates that the A and B loci are on autosomes. Because the variants are rare, no families which segregate variants for both A and B have been found, and the linkage of LDH A and B has not yet been determined in man. Genetically determined variants for LDH isozymes have been reported in fishes[19]. The number and complexity of LDH isozymes are greater in fishes than mammals, but A and B forms have been designated and homologized to the A and B forms in mammals[19]. Evidence that A and B variant forms segregate non-randomly[20-22] suggests that the loci are possibly linked. In mammals and birds an additional, unique form of LDH, LDH X, is expressed in testicular tissues and sperm[23]. LDH X is coded by the LDH C locus. Recently, evidence has been presented that the LDH C and LDH B loci are very closely linked in pigeons[24]. No evidence which demonstrates linkage relationships between LDH A and B in birds or mammals has been reported from genetic analysis *in vivo*.

LDH-5 homopolymers in mouse and man differ considerably in their electrophoretic mobilities and this provides good qualitative marker characteristics for mouse and human LDH A genes (Fig. 2: channels 8, 9 and 10). The LDH-1 homopolymers in mouse and man are not detectably different in their electrophoretic mobilities. The LDH B wild type mouse and human genes cannot therefore be distinguished in hybrid cells on a qualitative

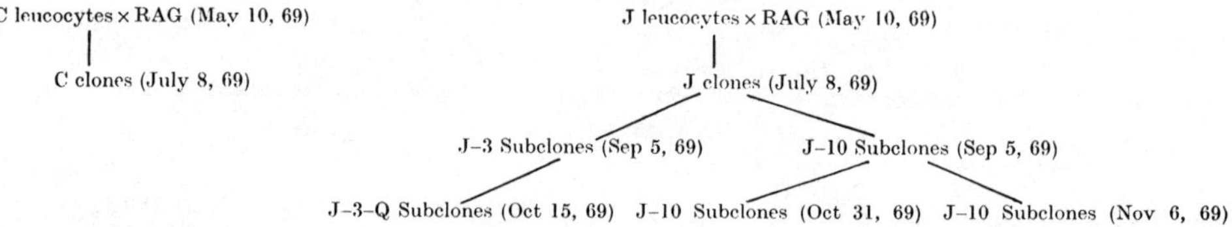

Fig. 1. Relationships of primary clones and derived subclones. Dates are given for the initiation of hybrids and the dates on which primary clones and derived subclones were picked.

Table 1. SEGREGATION OF HUMAN LDH A AND LDH B (VAR) PHENOTYPES IN PRIMARY, INDEPENDENT CLONES OF C HYBRIDS

C	1	2	3	6	7	8	9	11	13	15	16	17	18	19	21
LDH A	−	−	−	−	−	−	−	−	−	−	−	−	−	+	−
LDH B (var)	−	−	−	−	−	−	−	−	−	−	−	−	−	+	+

(−), No enzyme activity; (+), activity.

basis (Fig. 2: channels 8 and 10). Quantitative differences in the expression of mouse and human "b" subunits occurs, however, in the tissue cell lines used in the formation of the hybrids described here. Mouse LDH-1, 2 and 3 possess very reduced activities in the RAG cell line, and this is also true for other mouse lines such as A9 (Fig. 2: channel 9). Human "b" subunits are produced at a significantly higher rate in cell line WI-38 and other human cell lines such as KB, resulting in appreciable activity in LDH-1, 2 and 3 (Fig. 2: channel 8). Differences in activity of LDH-1 have been used to determine whether the human LDH B gene is present in hybrid cells between mouse and man[3]. It is possible, however, that regulatory phenomena could mimic such effects, and this makes it very desirable to use qualitative mutant forms of either the mouse or human LDH B gene for marker purposes.

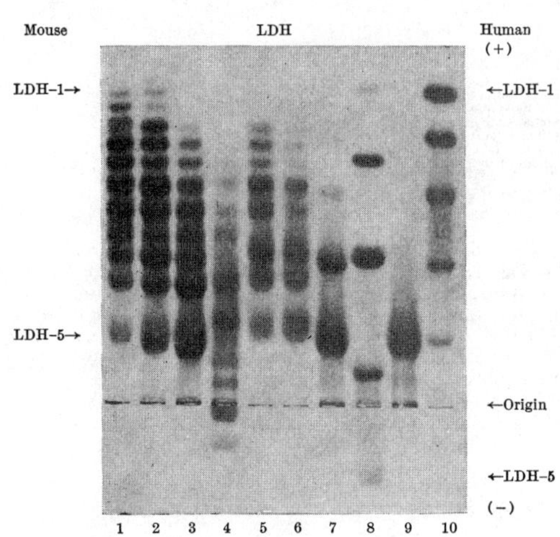

Fig. 2. LDH zymogram showing various mouse/man hybrid phenotypes. The channels were as follows: 1–7, mouse/man hybrid extracts; 8, human tissue culture extract (KB); 9, mouse tissue culture extract (A9); 10, mouse kidney extract. The human LDH phenotypes were scored as follows: (1) LDH A⁻/LDH B (var)⁺; (2) LDH A⁻/LDH B (var)⁺; (3) LDH A⁻/LDH B (var)⁺; (4) LDH A⁺/LDH B (var)⁺; (5) LDH A⁻/LDH B (var)⁺; (6) LDH A⁻/LDH B (var)⁺; (7) LDH A⁻/LDH B⁻. Gel electrophoresis was performed in a Tris-EDTA-borate gel at pH 8·6 (see ref. 5).

An LDH B (var) allele was detected in the course of a survey of newborn babies in the New Haven population[25]. Only one variant form of LDH B was detected in 860 individuals sampled. The homopolymer LDH B (var) migrated slightly faster than wild type human LDH-2, and at approximately the same position as mouse LDH-2 (Fig. 3: channels 2 and 4). The family of the proposita was studied, and the inheritance of the variant allele determined. If one assumes codominance, the variant segregates as an autosomal gene, in agreement with previous family studies pertaining to LDH inheritance[15–18]. The variant was expressed as a heterozygote in all individuals who possessed it.

Table 2. SEGREGATION OF HUMAN LDH A AND LDH B (VAR) PHENOTYPES IN PRIMARY, INDEPENDENT CLONES OF J HYBRIDS

J	1	2	3	4	5	6	7	8	9	10	11
LDH A	−	−	−	−	−	−	−	−	−	+	−
LDH B (var)	−	−	+	−	−	−	−	−	−	+	−

(−), No enzyme activity; (+), activity.

The patterns of LDH expression in the J and C clones provide strong evidence that LDH A and LDH B are not linked in man: clones C-21 and J-3, which both express the human LDH phenotype LDH A⁻/B (var)⁺, provide the critical evidence (Tables 1 and 2, Figs. 2 and 3). This configuration is more informative than the phenotype LDH A⁺/B⁻, because of the heterozygous nature of the human somatic cell parent, and the fact that the non-variant form of human LDH B is electrophoretically identical with mouse LDH B. The configuration LDH A⁺/B⁻ is less informative than LDH A⁻/B (var)⁺, because the human LDH B non-variant gene could in fact be present, but incorrectly scored because of its co-migration with mouse, or because of a regulatory phenomenon. The argument for non-linkage is greatly strengthened by the occurrence of phenotypic configurations in two completely independent series of clones derived from different donors (Tables 1 and 2). The independence of the observations weakens counter arguments based on the possibility of chromosome fragmentation. The conclusion that LDH A and B are not linked is further supported by segregation of these markers in derivative secondary series of clones (Tables 3–7, and Figs. 2 and 3).

Table 3. SEGREGATION OF HUMAN ENZYME PHENOTYPES IN SECONDARY CLONES OF THE J-3 CLONE

J:	3	A	D	E	G	I	J	L	M	N	O	P	Q	S	T
LDH A	−	−	−	−	−	−	−	−	−	−	−	−	−	−	−
LDH B (var)	+	−	+	+	+	−	+	+	+	+	+	+	+	+	+
LDH B (act)	+	+	+	+	+	−	+	+	+	+	+	+	+	+	+
Pep B	+	+	+	+	+	−	+	+	+	+	+	+	+	+	+
ADA	−	−	−	−	−	−	−	−	−	−	−	−	−	−	−
GOT	+	+	+	+	+	−	+	+	+	+	+	+	+	+	+
GPI	−	−	−	−	−	−	−	−	−	−	−	−	−	−	−
G6PD	+	+	+	+	+	+	+	+	+	+	+	+	+	+	+
MOR	−	−	−	−	−	−	−	−	−	−	−	−	−	−	−

(−), No enzyme activity; (+), activity.

It could also be argued that this interpretation might be incorrect because of the possible existence of genes which regulate the expression of LDH A. For example, human LDH A and B could indeed be linked, but a third unlinked gene might either specifically suppress or activate the expression of human LDH A. Two observations contradict this. First, the action of the repressor or activator would need to be species specific, for mouse LDH A is always expressed at uniform activity levels. Second, derivative subclones from J-3 (Table 3) and J-3-Q (Table 4) do not provide an example of a transition from human LDH A (−) to LDH A (+) as might be expected if an unlinked regulator gene were segregating. These arguments are not in themselves definitive, but they strengthen the conclusion that LDH A and B are not linked.

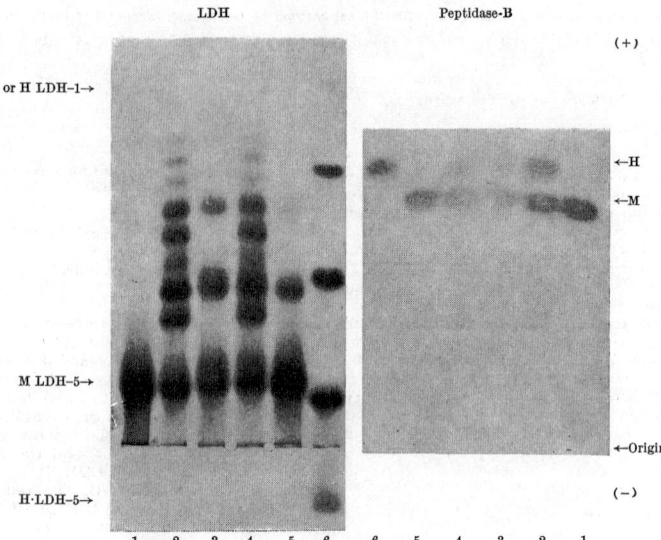

Fig. 3. Zymograms stained for LDH and Pep B activity. The slices were cut from the same gel. Channels having the same numbers represent identical samples. (1) Mouse control (A9); (2) mouse/man hybrid with an LDH A-/LDH B (var)+ human phenotype; (3) mouse/man hybrid with an LDH A-/LDH B (act)+ human phenotype; (4) mouse/man hybrid with an LDH A-/LDH B (var)+ human phenotype; (5) mouse/man hybrid with an LDH A-/LDH B- phenotype; and (6) human control (KB). The channels on the Pep B zymogram were similarly loaded and the human phenotypes were scored as follows: (1) mouse (control), (2) Pep B+, (3) Pep B+, (4) Pep B+, (5) Pep B-, and (6) human (control). The Pep B stain uses leucyl-glycyl-glycine as substrate[27].

Linkage of Human LDH B and Pep B

In all our sixty-six derivative subclones there was a positive correlation between the expression of Pep B, and either the expression of human LDH B (var) or increased LDH B activity, which is presumed to indicate the presence of the human LDH B wild type allele (Tables 3–7). Appreciable activity of isozymes LDH-1 and 2 was used as a criterion for the presence of the non-variant human LDH B gene (Fig. 3: channels 3 and 5). The analysis of this presumed linkage relationship is unfortunately complicated by the apparent distribution of the LDH B non-variant allele to all of the subclone series. Nevertheless, clones which express LDH B (var) activity invariably also possess Pep B activity.

Table 4. SEGREGATION OF HUMAN ENZYME PHENOTYPES IN SECONDARY CLONES OF THE J-3-Q CLONE

	J-3-Q	1	2	3a	3b	4	7	8	9	10	11	12	13	14	15	16	19	20
LDH A	−	−	−	−	−	−	−	−	−	−	−	−	−	−	−	−	−	−
LDH B (var)	+	+	−	−	+	+	+	−	+	+	+	+	+	+	+	+	+	+
LDH B (act)	+	+	+	−	+	+	+	+	+	+	+	+	+	+	+	+	+	+
Pep B	+	+	+	−	+	+	+	+	+	+	+	+	+	+	+	+	+	+
Pep A	+	+	+	+	+	+	+	+	+	+	+	+	+	+	+	+	+	+
ADA	−	−	−	−		−	−	−	−	−	−	−	−	−	−	−	−	−
GOT	+	+	+	+	+	+	+	+	+	+	+	+	+	+	+	+	+	+
GPI	−	−	−	−	−	−	−	−	−	−	−	−	−	−	−	−	=	=
G6PD	+	+	+	+	+	+	+	+	+	+	+	+	+	+	+	+	+	+
IPO	±								+				+		+			+
MOD	±	−	−	−	−	−	−	−	−	−	−	−	−	−	−	−	−	−
MOR	±	−	−	−	±	−	−	−	−	−	±	±	−	−	−	−	±	−
PGM₁		+	+	+	+		+		±		+	+	±	+	±	+	+	±

(−), No enzyme activity; (+), activity; (±), an equivocal result; elsewhere no determination was made.

Table 5. SEGREGATION OF HUMAN ENZYME PHENOTYPES IN SECONDARY CLONES OF THE J-10 CLONE

J:	10	A	B	C	D	E	F	G	H	I	J	M	N	O	P	Q	S	T
LDH A	+	+	+	+	+	±	+	+	+	+	+	+	+	+	+	+	+	+
LDH B (var)	+	−	−	−	+	−	+	−	−	+	−	−	+	−	+	−	−	−
LDH B (act)	+	+	+	+	+	+	+	+	+	+	+	+	+	+	+	+	+	+
Pep B	+	+	+	+	+	+	+	+	+	+	+	+	+	+	+	+	±	+
ADA	−	−			−		−	−	−	−	−		−	−				
GOT	+	+		+		+		+	+	+	+	+		+	+			
GPI	−	−			−		−	−	−	−	−		−	−				
G6PD	+	+		+		+		+	+	+	+	+		+	+			
IDH	−				−		−											
IPO	+				+		+											
MOD	+	+		+		+		+	+	+	+	+		+				
MOR	±	−		−		−		+	−	−	+							

(−), No enzyme activity; (+), activity; (±), an equivocal result: elsewhere no determination was made.

I *LDH B regulator gene is linked to Pep B*

Possible configuration	Expected phenotypes
(a) Pep B / LDH B / reg	Pep B⁺/LDH B⁺
(b) Pep B / reg	Pep B⁺/LDH B⁻ (species specificity)
	Pep B⁺/LDH B⁺ (species-non-specificity)
(c) LDH B	Pep B⁻/LDH B⁺

II *Pep B regulator gene is linked to LDH B*

Possible configuration	Expected phenotypes
(a) LDH B / Pep B / reg	Pep B⁺/LDH B⁺
(b) LDH B / reg	Pep B⁻/LDH B⁺
(c) Pep B	Pep B⁺/LDH B⁻

III *Pep B and LDH B linkage units segregate concomitantly*

Possible configuration	Expected phenotypes
(a) LDH B Pep B	LDH B⁺/Pep B⁺
(b) 2 LDH B Pep B	LDH B⁺⁺/Pep B⁺
(c) 2 LDH B 2 Pep B	LDH B⁺⁺/Pep B⁺⁺
(d) LDH B 2 Pep B	LDH B⁺/Pep B⁺⁺

Fig. 4. Hypotheses for the segregation of linked regulatory genes and non-random segregation of chromosomes as explanations for observed associations of Pep B and LDH B phenotypes. We assume that the presence of the regulator gene is required for the expression of the specified structural gene; that is, the regulation is positive. It is further assumed that only the non-variant form of human LDH B is used. Species specificity implies that the regulator affects the activity of genes within its own genome. According to hypothesis III, the numeral "2" signifies two chromosomes bearing the designated gene. (−) signifies no enzyme activity; (+) signifies activity.

Other interpretations of the data are possible (Fig. 4). For example, the Pep B structural gene could be linked to a regulator gene which is required for the expression of mouse and human LDH B activity (hypothesis I; Fig. 4; I). In that case, one would expect configurations of the type Pep B⁺/LDH B⁻ (Fig. 4; Ib). We have not observed such a combination in the J and C series of clones using activity levels of LDH B, or the presence of the LDH B (var), as criteria for the presence of human LDH B alleles. To be certain, our populations would have to be completely devoid of the wild type human LDH B allele. One observation consistent with the linkage of Pep B and LDH B structural genes is that clones which possessed LDH B (var) activity frequently showed a concomitant increase in Pep B activity (Fig. 3), supporting a linkage involving the structural genes of Pep B and LDH B. Such a dosage relationship would not be predicted by hypothesis I (Fig. 4).

Another possibility is that a regulator gene controlling Pep B activity is linked to the structural gene for LDH B (hypothesis II: see Fig. 4; II). This model predicts a segregant class which would express human LDH but not human Pep B (Fig. 4; IIb). In our observations, the LDH B (var) expression was always associated with the Pep B expression. This result therefore contradicts the alternative hypothesis II (Fig. 4).

A third alternative to the linkage of human Pep B and LDH B structural genes is that they reside on separate chromosomes which are either retained or lost from cells always concomitantly (hypothesis III: see Fig. 4; III). The large number of clones in our study and in that of Santachiara's *et al.*[6] rules out a chance association of this type. It could be argued, however, that the simultaneous presence or absence of the chromosomes was co-adaptive in terms of affecting cellular viability, but this is made less acceptable by the dosage relationship between LDH B (var) and Pep B (Fig. 3). The co-adaptation argument is more difficult to accept if a 1:1 multiplicity relationship must prevail between the homologues carrying the human

Pep B and LDH B genes, as is the case if the model is to agree with observations which indicate a dosage relationship.

In conclusion, the hybrid clone data most simply suggest that there is a linkage between the structural genes of LDH B and Pep B in man. More definitive evidence will come from a cytological search for association between the enzyme phenotypes and a particular chromosome(s). Chromosome studies on the J series clones are now being performed.

Linkage of Additional Isozyme Markers

We have also determined the expression of the following human isozymes in many of the subclones: Pep A, B, C[26,27]; adenosine deaminase (ADA)[28]; glutamate-oxalo-acetate transaminase (GOT)[29,30]; glucose phosphate isomerase (GPI)[31]; glucose-6-phosphate dehydrogenase (G6PD)[32]; isocitrate dehydrogenase (IDH)[33]; indophenol oxidase (IPO)[34]; malate : NADP oxidoreductase (decarboxylating enzyme) (MOD or NADP-MDH)[35,36]; malate : NAD oxidoreductase (MOR or NAD-MDH) (ref. 37 and unpublished work of T. B. Shows, V. M. C. and F. H. R.); and phosphoglucomutase 1 (PGM₁)[39,40]. There was no suggestion of positive linkage between any of the additional markers (Tables 3–6). Unfortunately, no positive statement can be made regarding negative linkages, because the data are based on segregation patterns in subclones, and not all clones were examined for all markers. Taking into account these limitations of subclone analysis[2], a summary of the linkage relationships is presented in Fig. 5.

Human G6PD activity was observed in every subclone. The *X* chromosome to which G6PD is linked would be

Table 6. SEGREGATION OF HUMAN ENZYME PHENOTYPES IN SECONDARY CLONES OF THE J-10-M CLONE

J-10:	M	3	4b	6	7	9	10	11b
LDH A	+	+	+	+	+	+	+	+
LDH B (var)	−	−	−	−	−	−	−	−
LDH B (act)	+	−	+	+	−	+	+	+
Pep B	+	−	+	+	−	+	+	+
Pep A	+	+	+	+	+	+	+	+
Pep C		+	+	+	±	+	+	+
GOT	+	+	+	+	+	+	+	+
GPI	−	−	−	−	−	−	−	−
G6PD	+	+	+	+	+	+	+	+
IPO	+	−	−	±	−	+	−	+
MOD	+	±	−	−	±	+	−	+
MOR	+	−	−	−	−	−	+	−
PGM₁		+	+	+	+	+	+	+

(−), No enzyme activity; (+), activity; (±), an equivocal result; elsewhere no determination was made.

Table 7. SEGREGATION OF HUMAN ENZYME PHENOTYPES IN SECONDARY CLONES OF THE J-10-H CLONE

J-10:	H	1	4	4a	5	6	7	9	10	11	12	13
LDH A	+	+	+	+	+	+	+	+	+	+	+	+
LDH B (var)	−	−	−	−	−	−	−	−	−	−	−	−
LDH B (act)	+	+	+	+	+	+	−	−	+	+	+	+
Pep B	+	+	+	+	+	+	−	−	+	+	+	+
Pep A	+	+	+	+	+	+	+	+	+	+	+	+
Pep C		+	+	+	+	+	+	+	+	+	+	+
GOT	+	+	+	+	+	+	+	+	+	+	+	+
GPI	−	−	−	−	−	−	−	−	−	−	−	−
G6PD	+	+	+	+	+	+	+	+	+	+	+	+
IPO	+	+	+	−	+	+	+	−	+	+	+	+
MOD	+	+	+	−	−	+	+	−	+	+	+	+
MOR	−	−	−	−	−	−	−	−	−	−	−	−
PGM₁		+	+	+	+	+	+	+	+	+	+	+

(−), No enzyme activity; (+), activity; (±), an equivocal result; elsewhere no determination was made

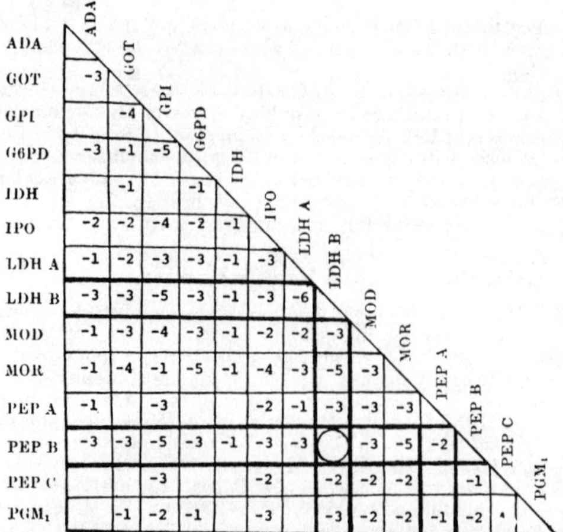

Fig. 5. Phenotype association in two independent and seven derivative families of clones. Summary of segregation data shown in primary clones C and J, and secondary clones J-3, J-3-Q, J-10, J-10-M, J-10-H. Negative associations between pairs of phenotypes are measured in terms of numbers of clone series (not individual clones within a series) which show negative correlations. This was done because clones within a series may not represent independent segregation events. Zero scores indicate possible positive linkages. Many of the zero scores such as between PGM_1 and IDH are probably not significant because of absence of segregation and/or paucity of phenotype testing.

	ADA	GOT	GPI	G6PD	IDH	IPO	LDH A	LDH B	MOD	MOR	PEP A	PEP B	PEP C
GOT	-3												
GPI		-4											
G6PD	-3	-1	-5										
IDH		-1		-1									
IPO	-2	-2	-4	-2	-1								
LDH A	-1	-2	-3	-3	-1	-3							
LDH B	-3	-3	-5	-3	-1	-3	-6						
MOD	-1	-3	-4	-3	-1	-2	-2	-3					
MOR	-1	-4	-1	-5	-1	-4	-3	-5	-3				
PEP A	-1		-3			-2	-1	-3	-3	-3			
PEP B	-3	-3	-5	-3	-1	-3	-3	○	-3	-5	-2		
PEP C			-3			-2		-2	-2	-2		-1	
PGM_1		-1	-2	-1		-3		-3	-2	-2	-1	-2	*

expected to be retained in all instances because it also carries the gene for HG-PRT. If we assume that human HG-PRT is present in all clones, this would indicate that X chromosome rearrangements which would segregate HG-PRT and G6PD do not occur. The absence of such rearrangements in sixty-six subclones suggests either close linkage of the two genes or a low rate of X chromosome rearrangement or both.

Chromosome Constitution of the Hybrids

Preliminary analysis of the chromosome constitution of the hybrid clones has been performed. Human chromosomes were identified on the basis of arm length measurements, and their failure to anneal with mouse satellite DNA or its complementary RNA, using *in situ* cytological annealing (unpublished procedure of M. Pardue and J. Gall and ref. 40) (Figs. 5 and 6). The modal number of human chromosome numbers varies between ten and sixteen in most clones. All classes of chromosomes are present, and identification of human D and G chromosomes can be obtained by the *in situ* annealing technique. Because the subclone series differ with respect to the presence or absence of LDH A, it was possible to confirm our previous linkage assignment of LDH A to a C group chromosome (unpublished results of C. Boone and F. H. R.). In most clones, it was possible to identify a human C group chromosome with arm length measurements which were consistent with those of the human X chromosome.

In conclusion, we agree with Santachiara *et al.*[6] that the strategy for future linkage studies will involve the

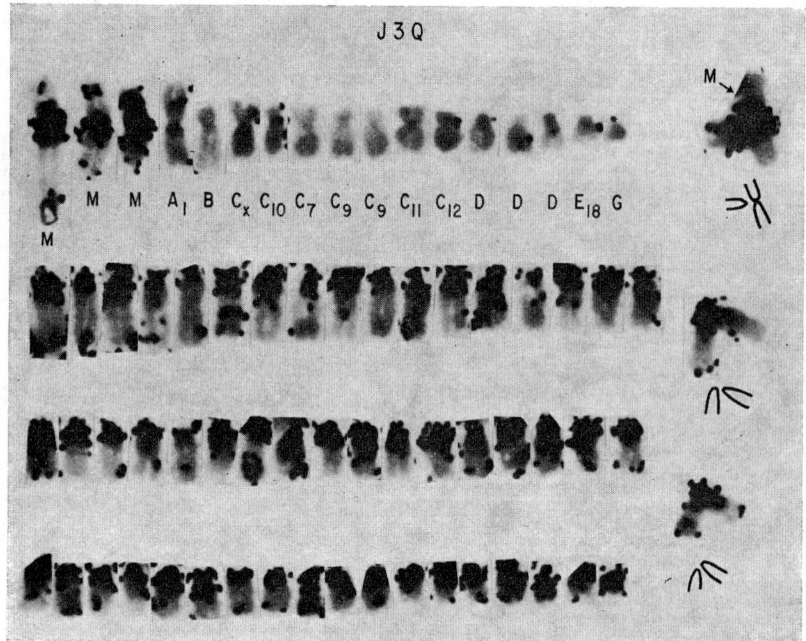

Fig. 6. Verification of human chromosome identity by means of nucleic acid annealing. Mouse bi-arm chromosomes are identified by the symbol "M"; the various types of human chromosomes by their group designations as stipulated by the London Conference[42]. Determination of human chromosome identity was made on the basis of absolute length and centromere position. Preparations were made by the air-dry method of Moorhead *et al.*[43]. The fixed, air-dried metaphase preparations were incubated with isotopically labelled RNA complementary to mouse satellite DNA. This material specifically anneals to mouse satellite DNA which is located in the centromere regions of the intact chromosomes. In autoradiograms, the chromosomes of mouse origin are isotopically labelled over their centromeres. Human chromosomes are unlabelled. The method especially permits the more certain identification of human D and G chromosomes.

isolation of large numbers of primary, independently derived hybrid clones which can be analysed for a large number of unit markers such as antigens, isozymes, as well as chromosomes. In instances where ambiguities exist, subcloning can be used to resolve them. We believe the method will contribute powerfully to our knowledge of human gene linkage relationships.

We thank Miss E. Nichols, Miss G. Yun, Mrs S. Speer, Mrs I. Kleinman, Mrs G. Fazzalaro and Mrs M. Reger for assistance; Dr J. G. Gall and Dr M. Pardue, of Yale University, for assistance with the *in situ* nucleic acid annealing of chromosome preparation; and Dr R. Davison and Dr J. A. Cortner, Division of Human Genetics, Children's Hospital of Buffalo, for help with ascertaining the LDH B variant.

We learn that Dr Westerveld and Dr Bootsma of Rijswijk, The Netherlands, using human–Chinese hamster somatic cell hybrids, have recently obtained data (unpublished) which confirm our observation on the independent segregation of human LDH A and LDH B loci.

Received March 22; revised April 6, 1970.

[1] Ruddle, F. H., in *Park City International Symposia on Problems in Biology I. RNA in Development* (edit. by Hanley, W.) (in the press).
[2] Ruddle, F. H., *Symp. of the Intern. Soc. Cell Biol.* (Academic Press, in the press).
[3] Nabholz, M., Miggiano, V., and Bodmer, W., *Nature*, **223**, 358 (1969).
[4] Boone, C., and Ruddle, F. H., *Biochem. Genet.*, **3**, 119 (1969).
[5] Shows, T. B., and Ruddle, F. H., *Proc. US Nat. Acad. Sci.*, **614**, 574 (1968).
[6] Santachiara, A. Silvana, Nabholz, M., Miggiano, V., Darlington, A. J., and Bodmer, W., *Nature*, **227**, 248 (1970) (preceding article).
[7] Bodmer, W. F., Tripp, M. B., and Bodmer, J. A., *Histocompatibility Testing* (edit. by Curtoni, E. S., Malting, P. L., and Tisi, R. M.), 341 (1967).
[8] Miggiano, V., Nabholz, M., and Bodmer, W., in *Heterospecific Genome Interaction*, The Wistar Institute Symp. Monograph No. 9, 61 (1969).
[9] Klebe, R. J., Chen, T. R., and Ruddle, F. H., *J. Cell Biol.* (in the press).
[10] Vogt, M., and Dulbecco, R., *Proc. US Nat. Acad. Sci.*, **46**, 365 (1960).
[11] Littlefield, J., *Science*, **145**, 709 (1964).
[12] Appella, E., and Markert, C. L., *Biochem. Biophys. Res. Commun.*, **6**, 171 (1961).
[13] Shaw, C., and Barto, E., *Proc. US Nat. Acad. Sci.*, **50**, 211 (1963)
[14] Markert, C. L., and Moller, F., *Proc. US Nat. Acad. Sci.*, **45**, 753 (1959).
[15] Nance, W. E., Claflin, A., and Smithies, O., *Science*, **142**, 1075 (1963).
[16] Boyer, S. H., Falner, D. C., and Watson-Williams, E. J., *Science*, **141**, 642 (1963).
[17] Kraus, A. P., and Neely, C. L., *Science*, **145**, 595 (1964).
[18] Davidson, R. G., Fildes, R. A., Glen-Bott, A. M., Harris, H., and Robson, E., *Ann. Hum. Genet.*, **29**, 5 (1965).
[19] Whitt, G. S., *Science*, **166**, 1156 (1969).
[20] Morrison, W. J., and Wright, J. E., *J. Exp. Zool.*, **163**, 259 (1966).
[21] Allison, W. S., *Ann. NY Acad. Sci.*, **151**, 180 (1968).
[22] Morrison, W. J., *Trans. Amer. Fisheries Soc.*, **99**, 193 (1970).
[23] Blanco, A., and Zinkham, W. H., *Science*, **139**, 601 (1963).
[24] Zinkham, W. H., Isensee, H., and Renwick, J. H., *Science*, **164**, 185 (1969).
[25] Lubs, H. A., and Ruddle, F. H., *Edinburgh Meetings on Cytogenetics (1969)*. In: *Human Population Cytogenetics*, P. A. Jacobs, W. H. Price, and P. Law, Eds., Edinburgh Univ. Press, 1970, pp 119–142.
[26] Lewis, W. H. P., and Harris, H., *Ann. Hum. Genet.*, **32**, 317 (1969).
[27] Lewis, W. H. P., and Harris, H., *Ann. Hum. Genet.*, **33**, 301 (1970).
[28] Spencer, N., Hopkinson, D. A., and Harris, H., *Ann. Hum. Genet.*, **32**, 9 (1968).
[29] Detter, J. C., Ways, P. O., Giblett, E. R., Baughan, M. A., Hopkinson, D. A., Povey, S., and Harris, H., *Ann. Hum. Genet.*, **31**, 329 (1968).
[30] DeLorenzo, R. J., and Ruddle, F. H., *Biochem. Genet.* (in the press).
[31] DeLorenzo, R. J., and Ruddle, F. H., *Biochem. Genet.*, **3**, 151 (1969).
[32] Shows, T. B., Tashian, R. E., Brewer, G. J., and Dern, R. J., *Science*, **145**, 1056 (1964).
[33] Henderson, N. S., *J. Expt. Zool.*, **158**, 263 (1965).
[34] Brewer, G. J., *Amer. J. Hum. Genet.*, **19**, 674 (1967).
[35] Henderson, N. S., *Arch. Biochem. Biophys.*, **117**, 28 (1966).
[36] Shows, T. B., and Ruddle, F. H., *Science*, **160**, 1356 (1968).
[37] Davidson, R. G., and Cortner, J. A., *Nature*, **215**, 761 (1967).
[38] Parrington, J. M., Cruckshank, G., Hopkinson, D. A., Robson, E. B., and Harris, H., *Ann. Hum. Genet.*, **32**, 27 (1968).
[39] Shows, T. B., Ruddle, F. H., and Roderick, T. H., *Biochem. Genet.*, **3**, 25 (1969).
[40] Gall, J., and Pardue, M., in *Methods in Enzymology, 12C* (edit. by Moldave, K., and Grossmar, L.) (in the press).
[41] Ressler, E. O., Thompson, G., and Joseph, R., *Nature*, **210**, 695 (1966).
[42] London Conference Study Group, *Cytogenetics*, **2**, 264 (1963).
[43] Moorhead, P. S., Nowell, P. C., Mellman, W. J., Battips, D. M., and Hungerford, D. A., *Exp. Cell Res.*, **20**, 613 (1961).

139

20

Reprinted from *Natl. Acad. Sci. (USA) Proc.* **68**:116–120 (1971)

Mitotic Separation of Two Human X-Linked Genes in Man–Mouse Somatic Cell Hybrids

O. J. MILLER*, P. R. COOK‡, P. MEERA KHAN§, S. SHIN§, AND M. SINISCALCO§

* Departments of Human Genetics and Development, and of Obstetrics and Gynecology, College of Physicians and Surgeons, Columbia University, New York; ‡ Sir William Dunn School of Pathology, University of Oxford; and § Department of Human Genetics, University of Leiden

Communicated by S. Spiegelman, October 21, 1970

ABSTRACT Six interspecific somatic hybrid cell lines were derived from a mouse line deficient in hypoxanthine: guanine phosphoribosyltransferase (HGPRT) and human diploid cells with normal enzyme activity. Human HGPRT was present in all six hybrids and the clones derived from them. However, in two of the six, and in some clones from another two, human glucose-6-phosphate dehydrogenase (G6PD) was absent. Since the structural loci for both these enzymes are X-linked in man, these findings suggest that these two loci have separated quite frequently through chromosome breakage and that they must be rather far apart on the X chromosome.

The potential importance of man–mouse somatic cell hybrids for the genetic analysis of man became apparent after the discovery by Weiss and Green (1) that this type of interspecific hybrid undergoes a rapid loss of human chromosomes. They attempted to determine the chromosomal location of the human structural gene for thymidine kinase (TK) by obtaining cell hybrids between normal human diploid cells and mouse cells that were deficient in TK and therefore unable to grow in HAT selective medium (2). The appearance of hybrid clones that grew in HAT medium suggested that human *tk* locus was being retained and expressed in the hybrid cells. Presumably, the human chromosome carrying the *tk* locus was selectively retained. Further studies (3, 4) led to the conclusion that the chromosome in question belongs to the human E-group, since a small submetacentric chromosome was present in all of the hybrid cells grown in HAT medium, even though other human chromosomes were lost. This conclusion was strengthened by the results of a back selection experiment: when these cells were grown in 5-bromodeoxyuridine (which selects for TK deficiency) this chromosome was lost. Recently, Migeon *et al.* have shown (5) that the electrophoretic mobility of the TK produced by the hybrid cells grown in HAT medium resembles that of human enzyme. An alternative hypothesis—of a reverse mutation at the mouse *tk* locus—was therefore unlikely.

Abbreviations: HGPRT, hypoxanthine: guanine phosphoribosyltransferase (EC 2.4.2.8); Hgprt, phenotype denoting ability or otherwise to synthesize this enzyme; *hgprt*, gene directing this synthesis; G6PD, G6pd, and *g6pd*, similarly for glucose-6-phosphate dehydrogenase (EC 1.1.1.49); TK, thymidine kinase (EC 2.7.1.21); HAT, hypoxanthine–aminopterin–thymidine selective growth medium.

Requests for reprints should be addressed to Instituut voor Anthropogenetica, Rijksuniversiteit Leiden, Wassenaarseweg 62, Leiden, The Netherlands.

Using similar methods, Nabholz *et al.* (6) showed that the hybrid clones derived from a fusion between normal human lymphocytes and a mouse line deficient in hypoxanthine: guanine phosphoribosyl transferase (HGPRT) always retained a genetic marker of the human X-chromosome, glucose-6-phosphate dehydrogenase (G6PD), and that G6PD was always lost by hybrid cells that survived back-selection (for (HGPRT) deficiency) in 8-azaguanine. Since the X-linkage of the human loci for *hgprt* and *g6pd* has been established beyond doubt by pedigree analysis (7, 8) and other studies (9, 10), these findings provide indirect support for the hypothesis that the survival of the hybrid cells in HAT selective medium depends on the retention and expression of at least part of the human X-chromosome. Direct support for this conclusion has been provided by Shin *et al.* (11), who showed that the HGPRT produced by these hybrids is electrophoretically indistinguishable from the human HGPRT, but clearly different from the mouse enzyme.

The purpose of this paper is to report the first evidence of mitotic separation of linked loci in somatic cell hybrids. A preliminary account of some of this work has been given elsewhere (12, 13).

METHODS

Human diploid cells were obtained from three sources: blood of an erythroblastotic newborn male infant (RBC), with red-cell precursors making up 50% of the nucleated cells; peripheral blood lymphocytes from three adults (two female, JW and ADC, and one male, CF) grown for 3 days in the presence of phytohemagglutinin; and fibroblasts from an embryonic female (HF$_2$) and an adult with the sex chromosomal complement of XXXXY (4XY). The mouse cells were of the A$_9$ line, which is deficient in HGPRT, and the B$_{82}$ line, deficient in TK. Both lines were originally derived from the L-cell line by Littlefield (14, 15). Neither cell line will grow in HAT medium (2, 14). Murine and human cells were fused by the use of UV-inactivated Sendai virus (16). Six viable hybrid lines free of the parental A$_9$ cells and one free of the parental B$_{82}$ cells were obtained by growth of cultures containing the fused cells in HAT medium. Colonies of hybrid cells appeared in 2–5 weeks. Mass cultures were obtained by transferring individual colonies to glass culture vessels and passing them in HAT medium at approximately weekly intervals.

The chromosome complement and enzyme content of each hybrid cell line were determined at intervals. G6PD activity was determined by the method of Motulsky *et al.* (17), and

TABLE 1. *Presence and activity of enzymes in mouse–human hybrids*

| Cell type | Identi-fication number[*] | No. of metaphase cells analyzed | Modal no. chromosomes | Enzyme studies | | | | | |
| | | | | Human | | Mouse | | Activity | |
				HGPRT	G6PD	HGPRT	G6PD	HGPRT[†]	G6PD[‡]
A$_9$	...	21	56,57	0	0	0	+	0.02	0.156
L$_{929}$	3	...	...	0	0	+	+	101±20§	0.154
A$_9$–RBC2	5	36	60,61	+	+	0	+	15–39	0.176
A$_9$–CF	6	28	61	+	+	0	+	18–27	0.220
A$_9$–ADC	7	24	56	+	+	0	+	13–38	0.257
A$_9$–4XY	8	20	57,58	+	+	0	+	13–40	0.257
A$_9$–RBC1	9	45	57	+	0	0	+	34–50	0.149
A$_9$–JW	10	31	55	+	0	0	+	4–96	0.126
B$_{82}$–HF2	11	23	64	0	0	+	+	118	0.134

* Corresponds to channel number in Figs. 1 and 2.

† Nanomoles of IMP formed per hr per mg protein. The values are not directly comparable as they increased with time after hybridization.

‡ International units per mg of soluble protein. Values usually increased slightly with time after hybridization.

§ SD, $n = 7$.

expressed in international units per mg of soluble proteins. The type of G6PD, whether human or mouse, was determined electrophoretically by the method of Rattazzi *et al.* (18). The activity of HGPRT was determined by the method of Harris and Cook (19), and the enzyme was characterized electrophoretically as described by Shin *et al.* (11). Control enzyme preparations were obtained from normal human lymphocytes, normal diploid fibroblastic cells, or mouse L-cells. Usually, sonicates of 1–$3 \times 10_6$ cells were required for each set of quantitative and qualitative enzyme assays. Growth of L-cells in medium 199 + 10% fetal calf serum or Dulbecco-modified Eagle's medium + 10% fetal calf serum with or without HAT did not affect the electrophoretic mobility of either enzyme.

Hybrid cell lines were cloned in HAT or 8-azaguanine in microtest tissue-culture plates (Falcon). All the hybrid lines and the clones derived from them have been stored at $-90°$C for future use.

RESULTS

Chromosome studies

The A$_9$ mouse cells used in these experiments had modal numbers of 56 and 57 chromosomes, including several biarmed chromosomes indistinguishable from the human X. The mouse–man hybrid cells had modal numbers of 55–61 chromosomes. Because several mouse and human chromosomes were morphologically similar, the number of mouse or human chromosomes in any hybrid could not be determined, although most of them were clearly of mouse origin.

Enzyme studies: HGPRT

The A$_9$ mouse cell had no detectable HGPRT activity, but considerable enzyme activity was present in the other parental cell types and in every hybrid cell line (Table 1). The electrophoretic mobility of mouse HGPRT is clearly different from that of human HGPRT (Fig. 1). The HGPRT activity in all of the A$_9$–human hybrid cell lines has the same mobility as the human enzyme, while the B$_{82}$–human hybrid cells have HGPRT with the mobility of the mouse enzyme.

These results suggest that every A$_9$–human hybrid of the present series owes its survival in HAT medium to the production of HGPRT of human type. The B$_{82}$–human hybrid cells, for which a condition for survival in HAT is the retention of the human *tk* locus and not necessarily of human HGPRT, produce mouse HGPRT but not human HGPRT, probably because they have lost the human X-chromosome. This conclusion is supported by the absence of human G6PD in these cells.

Enzyme studies: G6PD

All the parental cell lines used for the hybridization experiments and all hybrid lines had G6PD activity (Table 1). However, electrophoretic studies performed on cell sonicates of the hybrid lines and on controls showed differences in the G6PD patterns (see Fig. 2). Human G6PD and mouse G6PD migrate as single bands (channels 1 and 3). A mixture of the two does not lead to the formation of additional bands (channel 2). Four of the six A$_9$–human hybrids (channels 5–8) showed an enzyme band with intermediate mobility in addition to both human and mouse G6PD. This probably represents heteropolymeric G6PD molecules made up of murine and human G6PD subunits (6). The other two A$_9$–human hybrids (channels 9 and 10), as well as the B$_{82}$–human hybrid (channel 11), contained only murine G6PD. The relative intensities of the murine, heteropolymeric, and human G6PD bands in the four A$_9$–human hybrids (channels 5–8) were different for each hybrid; in two instances (channels 5 and 6) the murine G6PD was clearly in excess of the human G6PD (M > H); in two other hybrids (channels 7 and 8) the amount of human G6PD was equal to, if not greater than, the amount of murine G6PD (M < H). Under our assay conditions, the banding pattern of G6PD of similar man–mouse hybrid clones, which were kindly provided by Drs. Ruddle (20) and Bodmer (6), was always found to be of the type M < H (see channel 4).

Cloning studies

The variable banding pattern of G6PD in our series of A$_9$–human hybrids might reflect mixed populations of cells with

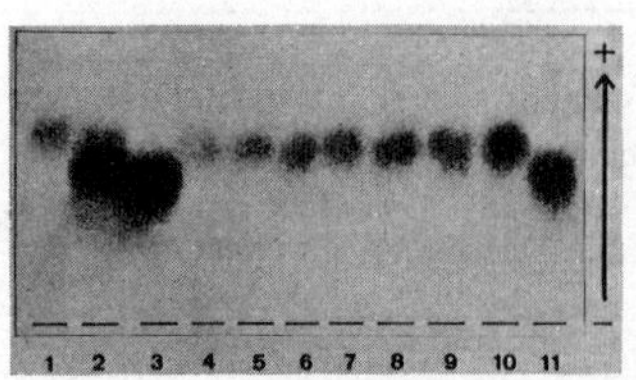 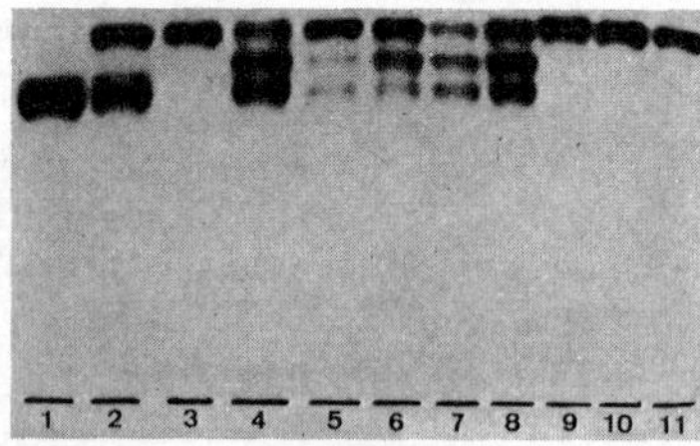 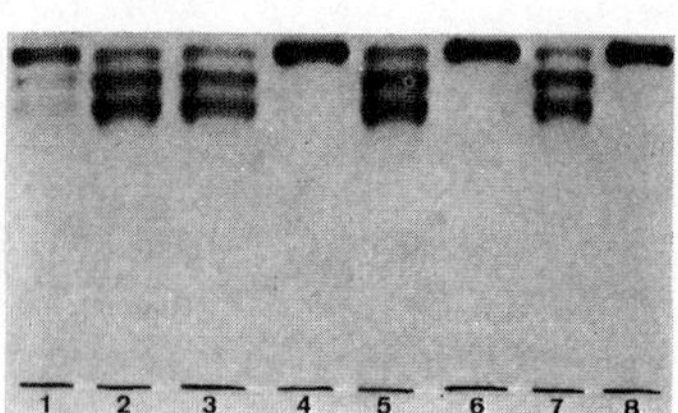

(*Left*) Fig. 1.   Electrophoretic mobility of murine and human HGPRT in man–mouse somatic cell hybrids and controls.

Channel *1*, human fibroblasts; *2*, mixture of human and mouse cells; *3*, mouse cells (line L-929); *4*, man–mouse hybrid line (3W4) from Nabholz *et al.* (6) derived from a HGPRT-deficient mouse line and human lymphocytes; *5–10*, the man–mouse hybrid lines of the present series, derived from the A_9 mouse line (HGPRT-deficient) and human diploid cells of different types; *11*, the man–mouse hybrid line derived from the B_{82} mouse line (TK-deficient) with human diploid fibroblasts.

The murine HGPRT moves more slowly than the human enzyme. All the man–mouse hybrids isolated in the HAT medium and derived from a HGPRT-deficient mouse line produce a "human-like" HGPRT (channels 4–10). The man–mouse hybrid derived from the TK-deficient mouse line (channel 11) produced a "murine-like" HGPRT.

(*Center*) Fig. 2.   Electrophoretic mobility and intensity of murine (fast), human (slow), and heteropolymeric (intermediate) G6PD in hybrids and controls. Channels are the same as for Fig. 1.

The normal banding pattern is the one of channel 4 (the man–mouse hybrid of Nabholz *et al.*), in which the intensity of the human G6PD is greater than that of the murine G6PD.

The hybrids of channels 9 and 10 possess only murine G6PD (though they still possess human HGPRT, Fig. 1). The hybrid of channel 11 has lost human G6PD and human HGPRT (Fig. 1). The variable banding pattern of the hybrids of channels 5–8 is due to mixture of cells having the "normal" banding pattern (as in channel 4) with cells having only murine G6PD.

(*Right*) Fig. 3.   G6PD patterns in a mass culture (channel 1, A_9–RBC2) and in representative clones derived from it (channels 3, 4, 5, 6, 7, and 8). Channel 2 shows the control hybrid (3W4) provided by Nabholz *et al.* Three of the man–mouse hybrid clones have only murine G6PD. The banding pattern of G6PD in the clones of channels 3, 5, and 7 is now identical to the normal banding pattern of channel 2, where the human G6PD is in excess of the murine G6PD.

the intensity pattern of cloned hybrids (M < H) (channel 4) together with hybrids possessing only murine G6PD. In order to test this possibility, we cloned the six A_9–human hybrid lines and the B_{82}–human hybrid line in HAT medium and studied each clone for its HGPRT and G6PD enzyme patterns. Table 2 summarizes the results for a total of 121 such clones. Studies of HGPRT showed that all the 105 hybrid clones derived from the six A_9–human mass cultures possessed human HGPRT, while the 16 hybrid clones derived from the B_{82}–human cell line possessed murine HGPRT. Of these first 105 hybrid clones, which were derived from six independent fusion experiments between A_9 and human cells, only 58 possessed human G6PD and a triple-banded pattern. In 50 of these, the pattern of murine, heteropolymeric and human G6PD was of the type M < H. The remaining 8 "clones" still had a triple-banded pattern of the type M > H, which suggests that they still may be mixtures of cells, some of which possess only murine G6PD. Of the two A_9–human hybrids that contained a mouse G6PD band of greater intensity than the human band (M > H), one (A_9–RBC2, see Fig. 3, channel 1) yielded clones of two types—some clones possessing the triple-banded pattern M < H (Fig. 3, channels 3, 5, and 7) like that of clones of the hybrids produced by Nabholz *et al.* (Fig. 3, channel 2), and other clones possessing only murine G6PD (Fig. 3, channels 4, 6, and 8). The 16 clones derived from the mass culture of B_{82}–human hybrid had only murine G6PD.

DISCUSSION

In somatic cell hybrids derived from normal human diploid cells and HGPRT-deficient mouse cells, survival in HAT medium depends on the retention and expression of the human structural locus for HGPRT. Since the loci for both HGPRT

and G6PD are located on the X-chromosome, failure to find human G6PD in 47 out of 105 clones derived from four of the six hybrid lines studied is unexpected.

The simplest explanation of these findings is that the X-chromosome has broken between the loci for *hgprt* and *g6pd*, with obligatory retention of the *hgprt* locus and occasional loss of the *g6pd* locus. Such a loss of loci may provide a mechanism for "mitotic segregation" in somatic hybrid cells. The two human X-linked loci in question must necessarily be far apart from one another for breakage to have occurred so frequently between them.

Alternative explanations for the absence of human G6PD in the A_9–human hybrid lines that possess human HGPRT include (*i*) use of human cell donors who have a G6PD-negative allele and therefore are deficient in G6PD activity, (*ii*) fresh somatic mutation to deficiency, (*iii*) repression of the human *g6pd* locus, and (*iv*) back mutation or virus-dependent enzyme induction at the mouse *hgprt* locus.

The first possibility has been ruled out by the direct examination of the donors. All were found to be of normal phenotype, generally called Gd(+); B (21).

Fresh somatic mutation cannot be excluded, but it is known to be a very rare event. It is unlikely that fresh mutation could account for the high frequencies of exceptions observed (47 in 105), expecially in the absence of a selective pressure for cells lacking human G6PD.

Repression of the synthesis or inhibition of the activity of human G6PD also seems unlikely in view of the normal production of mouse G6PD. In addition, after cloning, the cells never underwent further phenotypic changes even after numerous cell divisions, and those that had been derived from the mass cultures with murine G6PD alone never produced human G6PD. Back mutation or virus-dependent enzyme

TABLE 2. *Characteristics (human or murine) of enzymes in hybrids and clones*

		Mass cultures			Clones		
						G6PD†	
					MHyH	MHyH	
Hybrid	No.*	HGPRT	G6PD	HGPRT	M > H	M < H	M
A_9–RBC2	5	H	MHyH (M>H)	H	1	18	3
A_9–CF	6	H	MHyH (M>H)	H	5	12	
A_9–ADC	7	H	MHyH (M $\leqslant$ H)	H	2	6	5
A_9–4XY	8	H	MHyH (M $\leqslant$ H)	H		14	
A_9–RBC1	9	H	M	H			20
A_9–JW	10	H	M	H			19
B_{82}–HF2	11	M	M	M			16

H, human; Hy, hybrid; M, murine. M > H means more murine than human enzyme present.

* Corresponds to channel number in Figs. 1 and 2.

† Numbers refer to the number of clones in each category.

induction at the mouse *hgprt* locus, resulting in the production of HGPRT with an electrophoretic mobility indistinguishable from that of human HGPRT, can be excluded with reasonable confidence because the reversion to Hgprt$^+$ was never observed in the control fusions within the A_9 mouse parental line. Furthermore, the HGPRT produced by all the A_9–human hybrids, with or without the human G6PD, is always of the same type and is electrophoretically indistinguishable from the human HGPRT. Even if the A_9 cell could revert to Hgprt+, the possibility that the new HGPRT activity would be always "human-like" is rather small.

Loss of the structural locus is therefore the most likely explanation for the absence of the human G6PD. This could be the result of a chromosomal breakage followed by the loss or suppression of that part of the human X-chromosome carrying the *g6pd* locus.

On the basis of the frequent separation of the two loci observed in our series of A_9–human hybrids, it can be concluded that the genes for HGPRT and G6PD are not closely linked. Pedigree analysis of the only family so far reported as segregating at both these loci (22) suggests that they are indeed "loosely linked". This is in accord with our conclusion.

In contrast to our results, Nabholz *et al.* (6) did not observe loss of human G6PD in 42 clones derived from five man–mouse somatic cell hybrids of the same type. Ruddle (personal communication), however, found cytological evidence of X-chromosome breakage in one of the hybrid clones derived from his series of man–mouse hybrids, and Migeon and Miller (4) suggested the occurrence of similar chromosome instability in hybrids derived from human diploid cells and mouse cells deficient in TK. They noted a steady increase with time in the number of cells able to grow in HAT medium despite the absence of detectable human E-group chromosomes and suggested that the human *tk* locus may have been translocated to a mouse chromosome. It seems probable, therefore, that we have observed the separation of linked loci at such high frequencies because our hybrids have all been grown in the HAT selective medium for an extended period of time (over a year).

Live Sendai virus can produce chromosome breaks in cultured cells (23). Chromosome breakage and rejoining have also been observed in mammalian interspecific somatic cell hybrids produced by fusion with inactivated virus (24).

Saksela *et al.* (25) have shown that chromosome breakage may even be an unavoidable consequence of virus-induced fusion. This could be a serious source of error when one is determining new linkage groups by means of the man–mouse somatic cell hybrid technique. On the other hand, it is an ideal tool, if not a necessary requirement, for estimating the genetic distance between linked loci, as demonstrated in the present series of experiments with the two X-linked markers.

The analysis of interspecific somatic cell hybrids may also be utilized for mapping the chromosomes of other animals, since preferential loss of the chromosomes of one of the species in interspecific hybrids is not limited to man–mouse hybrids (24, 26). Linkage groups can best be determined under conditions of maximum chromosome stability, while map distances and gene order can perhaps be determined most efficiently by exposure of the test cells to a chromosome-breaking agent under standard conditions prior to hybridization.

Cell fusions and some of the subsequent work were carried out in Professor Henry Harris' laboratory, Sir William Dunn School of Pathology, University of Oxford, while O. J. M. was a U.S. National Science Foundation Senior Postdoctoral fellow. This work was supported by the U.S. Public Health Service, research grants HD 00516 and GM 13415 and the Health Research Council of the City of New York, grant 1–192.

The authors thank Drs. W. Bodmer and M. Nabholz for samples of some of their hybrid cell lines and Mrs. C. M. Kooij-Veldkamp for her valuable collaboration in the cloning experiments. The expert technical assistance of Miss A. J. Hasenoot and Mr. A. M. Bogaart is gratefully acknowledged.

1. Weiss, M. C., and H. Green, *Proc. Nat. Acad. Sci. USA*, **58**, 1104 (1967).
2. Szybalska, E. H., and W. Szybalski, *Proc. Nat. Acad. Sci. USA*, **48**, 2026 (1962).
3. Matsuya, Y., H. Green, and C. Basilico, *Nature*, **220**, 1199 (1968).
4. Migeon, B. R., and C. S. Miller, *Science*, **162**, 1005 (1968).
5. Migeon, B. R., S. W. Smith, and C. L. Leddy, *Biochem. Genet.* **3**, 583 (1969).
6. Nabholz, M., V. Miggiano, and W. Bodmer, *Nature*, **223**, 358 (1969).
7. Seegmiller, J. G., F. M. Rosenbloom, and W. N. Kelley, *Science*, **155**, 1682 (1967).
8. Marks, P. A., and R. T. Gross, *J. Clin. Invest.*, **38**, 2253 (1959).
9. Migeon, B. R., V. M. Der Kaloustian, W. L. Nyhan, W. J. Young, and B. Childs, *Science*, **160**, 425 (1968).

10.   Davidson, R. G., Nitowsky, H. M., and Childs, B., *Proc. Nat. Acad. Sci. USA*, **50**, 481 (1963).

11.   Shin, S., P. Meera Khan, and P. R. Cook, *Biochem. Genet.* in press.

12.   Siniscalco, M., in *Control Mechanism in Expression of Cellular Phenotypes*, ed. M. A. Padykula (Academic Press, New York), in press.

13.   Siniscalco, M., *Proc. 3rd Int. Congr. Congenital Malformations* (The Hague, 1969). *Excerpta Medica* (1970), pp. 72–83.

14.   Littlefield, J. W., *Nature*, **203**, 1142 (1964).

15.   Littlefield, J. W., *Exp. Cell Res.*, **41**, 190 (1966).

16.   Harris, H., and J. F. Watkins, *Nature*, **205**, 640 (1965).

17.   Motulsky, A. G., J. Vandepitte, and G. R. Fraser, *Amer. J. Hum. Genet.*, **18**, 514 (1966).

18.   Rattazzi, M. C., L. F. Bernini, G. Fiorelli, and P. M. Mannucci, *Nature*, **213**, 79 (1967).

19.   Harris, H., and P. R. Cook, *J. Cell Sci.*, **5**, 121 (1969).

20.   Ruddle, F. H., in *Control Mechanisms in Expression of Cellular Phenotypes*, ed. M. A. Padykula (Academic Press, New York), in press.

21.   World Health Organization, 1967. Standardization of procedures for the study of Glucose-6-phosphate dehydrogenase, *WHO Techn. Rep. Ser.* 366, p. 53.

22.   Nyhan, W. L., B. Bakay, J. D. Connor, J. F. Marks, and D. K. Keele, *Proc. Nat. Acad. Sci. USA*, **65**, 214 (1970).

23.   Nichols, W. W., A. Levan, and B. A. Kihlman, in *Cytogenetics of Cells in Culture*, ed. R. J. Harris (Academic Press, New York), 1964, p. 255.

24.   Siniscalco, M., *Atti Ass. Genet. Ital.*, **15**, 3 (1970).

25.   Saksela, E., P. Aula, and K. Cantell, *Ann. Med. Exp. Fenn.*, **43**, 132 (1965).

26.   Scaletta, L. J., N. B. Rushforth, and B. Ephrussi, *Genetics*, **57**, 107 (1967).

21

Reprinted from *Cell* **12**:205–218 (1977)

LOCALIZATION OF THE HUMAN α-GLOBIN STRUCTURAL GENE TO CHROMOSOME 16 IN SOMATIC CELL HYBRIDS BY MOLECULAR HYBRIDIZATION ASSAY

A. Deisseroth, A. Nienhuis, P. Turner, R. Velez,
W. F. Anderson, F. Ruddle, J. Lawrence,
R. Creagan, and R. Kucherlapati

Summary

We have used 16 human × mouse somatic cell hybrids containing a variable number of human chromosomes to demonstrate that the human α-globin gene is on chromosome 16. Globin gene sequences were detected by annealing purified human α-globin complementary DNA to DNA extracted from hybrid cells. Human and mouse chromosomes were distinguished by Hoechst fluorescent centromeric banding, and the individual human chromosomes were identified in the same spreads by Giemsa trypsin banding. Isozyme markers for 17 different human chromosomes were also tested in the 16 clones which have been characterized. The absence of chromosomal translocation in all hybrid clones strongly positive for the α-globin gene was established by differential staining of mouse and human chromosomes with Giemsa 11 staining. The presence of human chromosomes in hybrid cell clones which were devoid of human α-globin genes served to exclude all human chromosomes except 6, 9, 14 and 16. Among the clones negative for human α-globin sequences, one contained chromosome 2 (JFA 14a 5), three contained chromosome 4 (AHA 16E, AHA 3D and WAV R4D) and two contained chromosome 5 (AHA 16E and JFA14a 13 5) in >10% of metaphase spreads. These data excluded human chromosomes 2, 4 and 5 which had been suggested by other investigators to contain human globin genes. Only chromosome 16 was present in each one of the three hybrid cell clones found to be strongly positive for the human α-globin gene. Two clones (WAIV A and WAV) positive for the human α-globin gene and chromosome 16 were counter-selected in medium which kills cells retaining chromosome 16. In each case, the resulting hybrid populations lacked both human chromosome 16 and the α-globin gene. These studies establish the localization of the human α-globin gene to chromosome 16 and represent the first assignment of a nonexpressed unique gene by direct detection of its DNA sequences in somatic cell hybrids.

Introduction

Methods previously used for the chromosomal localization of unique gene sequences on human autosomes have been based on family linkage analysis or on the use of somatic cell hybrids (McKusick and Ruddle, 1977). The latter approach has depended on the existence of interspecific allelic variation of the gene to be mapped as well as on the constitutive expression of these genes in mouse × human fibroblast hybrid cells (Creagan and Ruddle, 1974). The chromosomal assignments of markers for differentiated cells have been more difficult to establish since these genes are usually not expressed in the human × mouse fibroblast hybrid cells. To overcome this limitation, we have used a direct approach to detect unique gene sequences in several hybrid cell clones (Deisseroth and Nienhuis, 1976). Hybrid cell DNA was annealed to purified human α-globin complementary DNA. The presence or absence of human α-globin sequences was correlated with the presence or absence of the individual human chromosomes in 16 different somatic cell hybrid lines.

Previous mapping studies of the human α- and β-globin genes began with the characterization of a family in which variants of both globins existed (Smith and Torbert, 1958). While these studies indicated that the human α- and β-globin structural genes were not linked, formal proof of the asyntenic relationship between these two genes was only recently obtained using somatic cell hybrids and a molecular hybridization assay specific for each gene (Deisseroth, Velez and Nienhuis, 1976a). A number of linkage analyses based on family studies have failed to establish conclusively the chromosomal localization of the genes, possibly due to the small number of individuals in each family and to the absence of detectable, tightly linked polymorphic genes (Neel, Schull and Shapiro, 1952; Maynard-Smith, Penrose and Smith, 1961; Nance et al., 1970). Price and his co-workers (Price, Conover and Hirschhorn, 1972; Price and Hirschhorn, 1975a; Price and Hirschhorn, 1975b) have reported that the human globin genes are present on human chromosome 2 and on one of the B group chromosomes. They used in situ molecular hybridization of rabbit globin mRNA to metaphase spreads of hu-

146

man chromosomes. Theoretical objections to the design of these experiments have cast doubt on their validity (Bishop and Jones, 1972).

We have chosen to reexamine the chromosomal localization of the human globin genes using a more direct approach. The DNA–cDNA hybridization assay we used is more specific and sensitive than the in situ methods used by Price et al. (1972). The cDNA probes for the human globin genes do not react with mouse globin DNA sequences, and are specific for either the human α– or the β–globin genes under the stringent conditions of hybridization we have used (Deisseroth et al., 1976a). We also characterized the chromosomal composition of each clone by treating metaphase spreads with Giemsa trypsin, which generates specific banding patterns in each chromosome. The same metaphase spreads were counter-stained with Hoechst 33258, which is a species-specific centromeric stain used to distinguish human and mouse chromosomes. The alkaline Giemsa stain (Giemsa 11) was also used to allow identification of the human chromosomes. Comparison of the chromosomal composition of the hybrid clones with the presence or absence of the human α–globin gene permitted identification of human chromosome (H.C.) 16 as the one which contains the human α–globin locus. In a final test of the association of H.C. 16 and the α–globin structural gene, hybrid clones were grown in selective media designed to promote the retention or elimination of this chromosome. The analysis of these clones confirmed that the presence of the human α–globin gene correlates with the presence of H.C. 16. The technique that we used to localize the human α–globin gene is a general one. We believe that it is the method of choice for the localization of unique gene sequences for which a cDNA exists. It is especially useful for the localization of genes not constitutively expressed in hybrid cells, and we expect that its use will lead to the mapping of genes which code for specialized products in differentiated cells.

Results

Determination of Chromosomal Composition and Globin Gene Sequence Content of Human × Mouse Hybrid Cell Clones

A series of 16 human X mouse hybrid cell clones was chosen on the basis of the complementarity of the overlapping subsets of human chromosomes which they contained. The cell lines were expanded to provide a sufficient number of cells to yield 3–6 mg of DNA. Cells were incubated in vinblastine sulfate, and metaphase spreads were generated at the conclusion of the cell expansion. The human chromosomes present in each cell line were identified by a variety of staining techniques: Giemsa 11 differential staining of human and mouse chromosomes (Bobrow, Madan and Pearson; 1972; Kucherlapati et al., 1975; Friend, Chen and Ruddle, 1976), sequential staining of a single metaphase spread by Giemsa-trypsin treatment and the Hoechst 33258 fluorescent stain (Yoshida, Ikouchi and Sasaki, 1975; Kozak, Lawrence and Ruddle, 1977), and, in some cases, quinacrine fluorescence banding (Caspersson, Zech and Johansson, 1970). Differentiation of human and mouse chromosomes by the Giemsa 11 staining technique is shown in a metaphase spread of hybrid cell line WAV R4D in Figure 1. Characterization of the specific chromosomes studied by the Giemsa-trypsin-Hoechst 33258 technique is presented in Figure 2. An average of 81 metaphase spreads (range 45–167) was studied for each hybrid clone. Analysis of this number of metaphase spreads for each hybrid clone makes the probability of detecting a chromosome present in only 10% of the cells of a hybrid population 97%. The chromosomal composition of the hybrid clones studied in this manner is presented in Table 1. The fraction of metaphase spreads containing each chromosome is also listed for each hybrid clone. These data were derived by use of the Giemsa-trypsin-Hoechst combination staining technique and, in some cases, Giemsa 11 staining of human chromosomes.

Isozymal markers for 17 different human chromosomes were also used to characterize the composition of human chromosomes in each human X mouse hybrid clone (Nichols and Ruddle, 1973). The sensitivity of each isozymal assay was established by study of protein extracts of artificial mixtures of human and mouse fibroblast cell lysates (E. A. Nichols and F. H. Ruddle, manuscript in preparation). This established relative sensitivities of the isozymal assay for human chromosomes relative to the sensitivity of the DNA-cDNA assay for the presence of the human globin genes. The results of the isozymal analysis are also presented in Table 1.

The species specificity of the DNA-cDNA molecular hybridization assay for the human globin genes was established by incubating human spleen or mouse embryo DNA with human α– or β–globin cDNA. As shown by the data presented in Figure 3, the conditions of hybridization were sufficiently stringent to permit detection of the human genes without significant cross-reaction of the human cDNAs with mouse globin gene sequences. The sensitivity of the DNA-cDNA assay was estimated by reacting the human cDNAs with artificial mixtures of human and mouse DNA. The results obtained with a 1:3 mixture (human:mouse) and a 1:4 mixture (human:mouse) with the α and β probes, respectively, are shown in Figure 3. We have con-

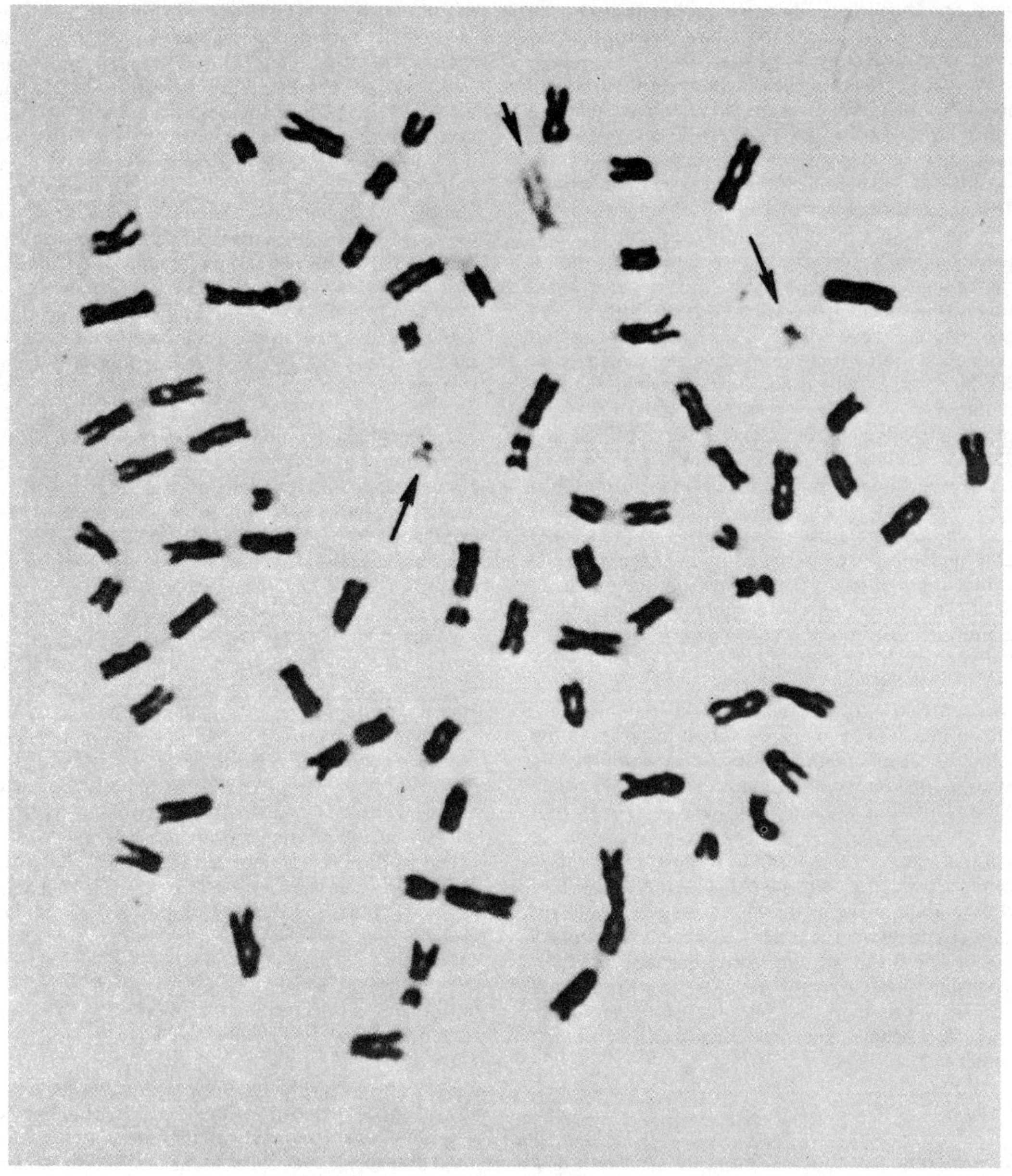

Figure 1. Examples of Human Chromosomes Detected by the Giemsa 11 Staining Technique in a Human × Mouse Hybrid Cell Clone, WAV R4D

This particular subclone of WAV was found to contain a B group chromosome and two G group chromosome which are indicated by the arrows. The human chromosomes (arrows) exhibit a light grey, even staining color in this black and white reproduction, while all the mouse chromosomes exhibit a darker stain on the arms and a lighter staining area in the centromeric region. This cell line, WAV R4D, was shown to contain human chromosomes 4, 21 and 22 by Giemsa-trypsin-Hoechst analysis, and to be devoid of human α–globin genes (see Table 1).

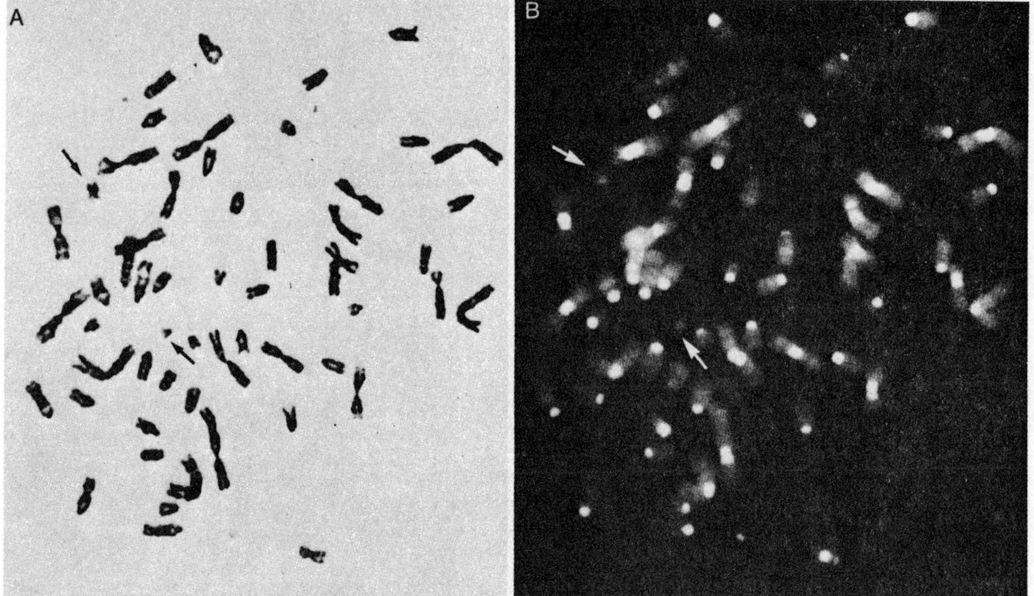

Figure 2. Giemsa-Trypsin Banding and Hoechst Fluorescent Staining of a Single Metaphase Spread of a Human × Mouse Hybrid Clone WAIV A

(A) Giemsa trypsin banding with the human chromosomes indicated by arrows. This hybrid clone was strongly positive for the human α-globin gene by nucleic acid hybridization assay, and contained primarily H.C. 16 and 22. (B) Hoechst 33258 staining of the same metaphase spread shown in (A) illustrating the presence of intense centromeric fluorescence in most mouse chromosomes and its absence on all human chromosomes.

cluded, based on these data and on analysis of DNA mixtures containing a lower fraction of human DNA (data not shown), that we could detect a globin gene on a chromosome which occurs in only 10% of the cells of a hybrid population.

Three different methods were used to prepare a complementary DNA probe enriched in α-globin sequences. Method I utilized human globin mRNA isolated from reticulocytes of a patient with β-thalassemia major. This mRNA contains approximately 70% α-, 20% γ- and 10% human β-mRNA sequences, and the probe synthesized with reverse transcriptase contains approximately the same proportions of the three species of cDNA. Complementary DNA prepared by Method I was used for analysis of DNA from hybrid cell clones AIM 23a X-1, AIM 23a old, JFA 14a 5, JFA 14a 13 5, J10H7 and IL II 5. The second method used for preparing an α-enriched cDNA (Method II) depended on purification of α-globin mRNA by polyacrylamide electrophoresis in 98% formamide (Forget et al., 1975). RNA was prepared from reticulocytes of patients with sickle cell anemia. It lacked significant contamination with γ-mRNA sequences. The resulting purified α-globin mRNA fraction therefore contained no γ-mRNA and only 10-15% human β-mRNA. This α-enriched probe (85-90% human α-globin cDNA) was used for analysis of DNA for hybrid cell clones AHA 3D, AHA 16E, IL II 5 4D

WAIV A (grown in nonselective media), WAV, WAV R4D and WAV R4D A19. The third method involved removal of the contaminating β-cDNA sequences from the human α-enriched probe prepared by Method II and was accomplished as follows. Method II human α-cDNA was annealed to a 4 fold excess of β-enriched (Hb H) mRNA (80% β:20% α) for 2 hr at 62°C. This time interval was selected so that 30-40% of the total cDNA was incorporated into duplexes as determined in a preliminary experiment. In this mRNA-driven annealing reaction, the β-cDNA contaminating sequences should be among the first incorporated into mRNA-cDNA hybrid molecules. These duplexes were removed by chromatography on hydroxyapatite (Benz et al., 1977) leaving a probe which we estimate was >95% α-cDNA sequences. The α-enriched probe prepared by Method III was used for analysis of DNA hybrid cell lines AHA 16D, WAIV A (grown in adenine-alanosine medium) and WAIV A (grown in diaminopurine medium). The results of the DNA-cDNA analysis for the presence of human α-globin genes in each hybrid cell clone are summarized in Table 1.

Synteny and Chromosomal Testing of Human × Mouse Hybrid Cell Clones

The data presented in Table 1 were used to determine which human chromosomes were present in

Table 1.

CELL LINE	1	2	3	4	5	6	7	8	9	10	11	12
AHA 3D	0/45 -PGM -PEP C	2/45 -IDH -MDH	17/45	10/45	2/45	0/45 -MOD	0/45	5/45	0/45 -AK_1	8/45* -GOT	31/45 +LDH A	14/45 +LDH B
AHA 16D	2/60 -PEP C	10/60 +MDH +IDH	40/60	4/60	1/60	2/60 -MOD	10/60	1/60	0/60	12/60 +GOT	23/60 +LDH A	0/60 -LDH B
AHA 16E	13/70	0/70 -MDH -IDH	20/70	8/70	13/70	0/70 -MOD	0/70	3/70	0/70 -AK_1	17/70	45/70 +LDH A	26/70 +LDH B
AIM 23 OLD HAT MEDIUM	30/68 +PEP C	0/68 -MDH -IDH	29/68	2/68	1/68	0/68 -MOD	0/68	0/68	0/68	21/68 +ALD K +GOT	0/68 -LDH A	30/68 +LDH B
AIM 23 X-1 HAT MEDIUM	28/70 +PEP C	0/70 -MDH -IDH	41/70	6/70	8/70	0/70 -MOD	0/70	3/70	0/70	22/70 +ALD K +GOT	3/70 -LDH A	20/70 +LDH B
IL II-5	0/66 -PGM -PEP C	0/66 -MDH -IDH	0/66	0/66	0/66	0/66 -MOD	18/66	0/66	0/66 -AK_1	0/66 -GOT	0/66 -LDH A	23/66 +PEP B +LDH B
IL II-54D	0/70 -PGM -PEP C	0/70 -MDH -IDH	0/70	0/70	0/70	0/70 -MOD	14/70	0/70	0/70 -AK_1	0/70 -GOT	0/70 -LDH A	15/70 +PEP B +LDH B
JFA 14a 5	0/160	55/160 +MDH +IDH	0/160	0/160	0/160	0/160 -MOD	0/160	0/160	0/160	0/160 -ALD K	0/160 -LDH A	0/160 -PEP B
JFA 14a 13-5	0/167 -PEP C	0/167 -MDH -IDH	0/167	0/167	10/167	0/167 -MOD	0/167	0/167	0/167	0/167 -ALD K	0/167 -LDH A	0/167 -PEP B -LDH B
J10H7	10/69 +PEP C	1/69 -MDH -IDH	40/69	0/69	0/69	27/69 +MOD	34/69	0/69	0/69	29/69 +GOT	52/69 +LDH A	0/69 -PEP B -LDH B
WAV	0/84 -PGM -PEP C	0/84 -MDH -IDH	0/84	23/84	0/84	0/84 -MOD	0/84	0/84	0/84 -AK_1	0/84 -GOT	0/84 -LDH A	0/84 -LDH B
WAV R4D DAP MEDIUM	0/107 -PGM -PEP C	0/107 -MDH -IDH	0/107	27/107	0/107	0/107 -MOD	0/107	0/107	0/107 -AK_1	0/107 -GOT	0/107 -LDH A	0/107 -LDH B
WAV R4D A19 DAP MEDIUM	0/70 -PGM -PEP C	0/70 -MDH -IDH	0/70	0/70	0/70	0/70 -MOD	0/70	0/70	0/70 -AK_1	0/70 -GOT	0/70 -LDH A	0/70 -LDH B
WAIV A	0/95 -PGM -PEP C	0/95 -MDH -IDH	0/95	0/95	0/95	0/95 -MOD	0/95	0/95	0/95 -AK_1	0/95 -GOT	0/95 -LDH A	0/95 -PEP B -LDH B
WAIV A-AA	0/46 -PGM	0/46 -MDH -IDH	0/46	0/46	0/46	0/46 -MOD	0/46	0/46	0/46	0/46	0/46	0/46
WAIV A-DAP	0/50 -PGM	0/50 -MDH -IDH	0/50	0/50	0/50	0/50 -MOD	0/50	0/50	0/50	0/50	0/50	0/50

CELL LINE	13	14	15	16	17	18	19	20	21	22	X	HUMAN ALPHA GLOBIN GENE
AHA 3D	0/45 -ESD	3/45 +NP	0/45 -HEX A -MPI	0/45 -APRT	0/45	7/45 +PEP A	0/45* +GPI	0/45 -ADA	0/45 -SOD	0/45	0/45* +PGK +G6PD	−
AHA 16D	2/60 -ESD	0/60 -NP	0/60 -MPI	9/60	1/60 -GK	15/60 +PEP A	13/60 +GPI	14/60 +ADA	17/35 +SOD	17/35	35/60 +G6PD	+
AHA 16E	9/70	4/70	0/70	0/70 -APRT	12/70	14/70	11/70 +GPI	8/70	23/70	30/70	32/70	−
AIM 23 OLD HAT MEDIUM	0/68 -ESD	1/68	5/68 +HEX A +MPI	0/68	0/68 -GK	1/68 -PEP A	0/68 -GPI	0/68 -ADA	34/68	1/68	29/68 +G6PD	−
AIM 23 X-1 HAT MEDIUM	0/70 -ESD	0/70	13/70 +HEX A +MPI	3/70	0/70	3/70 -PEP A	0/70 -GPI	0/70 -ADA	10/70	0/70	37/70 +G6PD	±
IL II-5	0/66 -ESD	0/66 -NP	11/66 +HEX A +MPI	0/66 -APRT	16/66	0/66 -PEP A	13/66 +GPI	0/66 -ADA	4/41 +SOD	3/41	0/66 -PGK -G6PD	−
IL II-54D	0/70 -ESD	0/70 -NP	0/70 -HEX A -MPI	0/70 -APRT	42/70	0/70 -PEP A	7/70 +GPI	0/70 -ADA	7/70 +SOD	0/70	0/70 -PGK -G6PD	−
JFA 14a 5	0/160 -ESD	0/160 -NP	0/160 -HEX A	0/160 -APRT	0/160 +GK	0/160 -PEP A	0/160 -GPI	0/160	0/160	0/160	0/160 -G6PD	−
JFA 14a 13-5	0/167 -ESD	0/167 -NP	0/167 -PK -HEX A -MPI	0/167 -APRT	0/167 -GK	106/167 +PEP A	0/167 -GPI	0/167	0/167	0/167	0/167 -G6PD	−
J10H7	0/69	0/69 -NP	6/69 +HEX A +PK	4/69	37/69	27/69 +PEP A	0/69 -GPI	0/69 -ADA	12/69	17/69	35/69 +G6PD	?+
WAV	0/84 -ESD	0/84 -NP	0/84 -HEX A -MPI	54/84 +APRT	0/84	0/84 -PEP A	0/84 -GPI	9/84* -ADA	47/59 +SOD	43/59	0/84 -PGK -G6PD	+++
WAV R4D DAP MEDIUM	0/107 -ESD	0/107 -ND	0/107 -HEX A -MPI	0/107 -APRT	0/107	0/107 -PEP A	0/107 -GPI	0/107	25/37 +SOD	24/37	0/107 -PGK -G6PD	−
WAV R4D A19 DAP MEDIUM	0/70 -ESD	0/70 -ND	0/70 -HEX A -MPI	0/70 -APRT	0/70	0/70 -PEP A	0/70 -GPI	0/70	15/41 +SOD	15/41	0/70 -PGK -G6PD	−
WAIV A	0/95 -ESD	0/95 -NP	0/95 -HEX A -MPI	56/95 +APRT	0/95	0/95 -PEP A	0/95 -GPI	0/95 -ADA	0/95 -SOD	27/95	0/95 -PGK -G6PD	+++
WAIV A-AA	0/46 -ESD	0/46	0/46 -MPI	32/46 +APRT	0/46	0/46	0/46 -GPI	0/46	0/46 -SOD	0/46	0/46 -PGK	++++
WAIV A-DAP	0/50 -ESD	0/50	0/50 -MPI	0/50 Weakly +APRT	0/50	0/50	0/50 -GPI	0/50	0/50 -SOD	24/50	0/50	±

(*) The isozyme analysis in these four cases does not agree with the chromosomal data. In the remaining 199 entries, however, both the isozymal and chromosomal analyses were in agreement.

(±) indicates percentage of hybridization in the Cot analysis which is no greater than 10% of background.

(?+) indicates that hybridization was attributed to β- and γ-cDNA (see text).

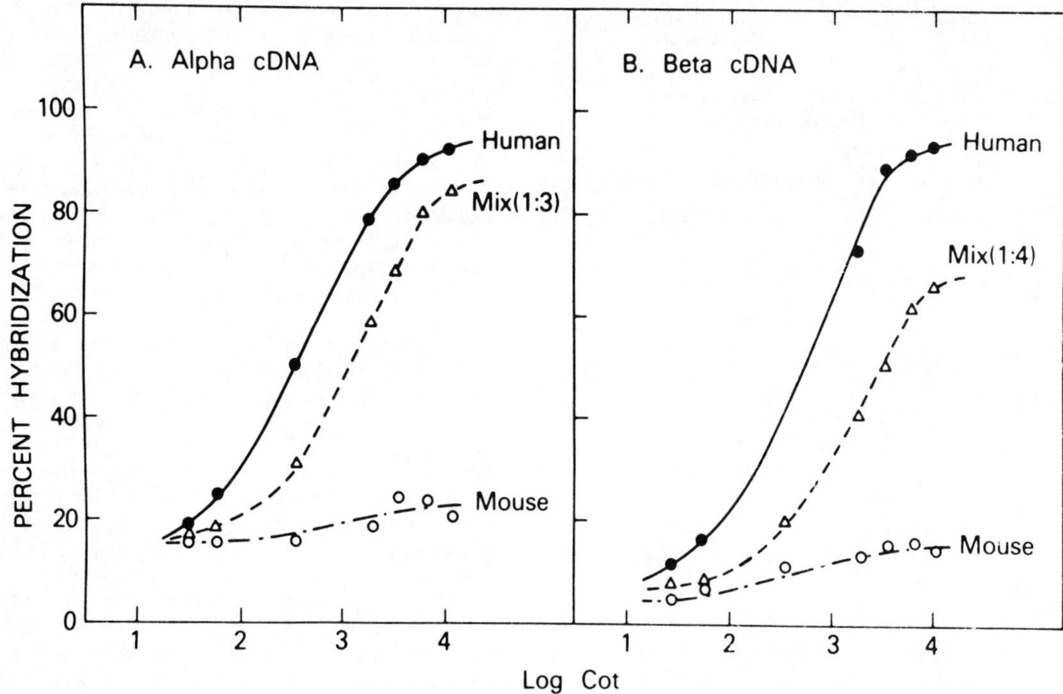

Figure 3. Species Specificity and Sensitivity of the DNA-cDNA Assay for Human Globin Genes

Reaction mixtures contained either human α-(A) or human β-(B) globin cDNA and DNA from human spleen (●), from mouse embryo (○) or a mixture of the two (△). Human α- and β-globin cDNA used in these experiments was prepared by Method I.

human X mouse hybrid clones found to be devoid of detectable levels of human α-globin genes. We believe that the presence of an intact chromosome in >10% of the metaphase spreads of a hybrid clone devoid of globin genes is the strongest test of discordancy of the chromosome and the α-globin gene. Table 2 summarizes the clones we studied which contained human chromosomes in >10% of the cells of the population and which were devoid of human α-globin genes. All the human chromosomes were found in clones devoid of human α-globin genes except for H.C. 6, 9, 14 and 16.

Exclusion of Human Chromosomes 2, 4 and 5
Each of the three human chromosomes 2, 4 and 5 proposed as possible sites of human globin structural genes (Price et al., 1972) were found in cell lines which lacked human α-globin genes. H.C. 2 was shown to be present in clone JFA 14a 5 in 55 of the 160 metaphase spreads studied by the two banding techniques. This cell line was devoid of detectable levels of human α-globin genes. In addition, two isozyme markers for H.C. 2, isocitrate dehydrogenase and malate dehydrogenase, were present in protein extracts of this clone. H.C. 4 was present in 11–25% of the cells of the hybrid clones AHA 3D, AHA 16 E and WAV R4D, all of which were found to be devoid of human α-globin genes (see

Tables 1 and 2). H.C. 5 was present in two clones (AHA 16E and JFA 14a 13-5) which were devoid of detectable levels of human α-globin genes.

Human chromosomes 2, 4 and 5 could not be found in many of the hybrid clones most strongly positive for the human α-globin gene. H.C. 2 and H.C. 5 were absent from all metaphase spreads of each of the three hybrid clones (WAV, WAIV A and WAIV A-AA) which contained the human α-globin gene. H.C. 4 was absent from all metaphase spreads analyzed of two of the three hybrid cell lines (WAIV A and WAIV A-AA) which were strongly positive for the human α-globin gene. These data show that there is no correlation between the presence or absence of chromosomes 2, 4 or 5 with the presence or absence of the human α-globin gene. We conclude that the previous tentative assignments of human α-globin genes to H.C. 2, 4 or 5 based on in situ molecular hybridization assays are not in agreement with our present data which are based on direct detection of the globin genes in somatic cell hybrids.

Evidence Implicating Human Chromosome 16 as Bearing the α-Globin Gene
Among the chromosomes (6, 9, 14 and 16) which were not excluded on the basis of the data presented in Table 2, only H.C. 16 or its enzymatic

Table 2. Hybrid Cell Lines Which Are α–Globin-Negative with Chromosome Present in >10% of Metaphase Spreads

Chromosome	Cell Clone
1	AHA 16 E (13/70), AIM 23 OLD (30/68)
2	JFA 14a5 (55/160)
3	AHA 3D (17/45), AHA 16 E (20/70), AIM 23 OLD (29/68)
4	AHA 3D (11/45), AHA 16 E (8/70), WAV R4D (27/107)
5	AHA 16 E (13/70), JFA 14a 13-5 (10/167)
6	
7	IL-II-5 (18/66), IL II-54D (14/70)
8	AHA 3D (5/45)
9	
10	AHA 3D (8/45), AHA16 E (16/70), AIM 23 OLD (21/68)
11	AHA 3D (31/45), AHA16 E (45/70)
12	AHA 3D (14/45), AHA16 E (26/70), AIM 23 OLD (30/68), IL II-5 (23/66), IL II-544D (15/70)
13	AHA 16E (9/70)
14	
15	IL II 5 (11/68)
16	
17	IL II-5 (16/66), IL II-5 (42/70), AHA 16E (12/70)
18	AHA16 E (14/70), JFA 14a 13-5 (106/167)
19	AHA16 E (11/70), IL II-5 (13/66)
20	AHA 16 E (8/70)
21	AHA16 E (23/70), AIM 23 OLD (34/68), WAV R4D (25/37), WAV R4D A19 (15/41)
22	AHA16 E (30/70), WAV R4D (24/37), WAV R4D A19 (15/41)
X	AHA16 E (32/70), AIM 23 OLD (29/68)

marker, adenosyl phosphoribosyl transferase (APRT), was found to be present in the hybrid clones containing the human α–globin gene. The three clones most strongly positive for the α gene (WAV, WAIV A and WAIV A-AA) all contained H.C. 16 in >50% of the metaphase spreads studied in each cell line. All these cell lines contained human APRT as determined by isozyme assay. All the cell lines for which a borderline but positive result for the presence of the human α–globin gene was obtained (AHA 16 D, AIM 23 X-1 and WAIV A-DAP) contained either a low level of H.C. 16 or a weakly positive result for the presence of APRT (see Table 1). No other human chromosome was common to the three clones which were the most strongly positive for the human α–globin gene. As shown by the data in Tables 1 and 2, no cell line devoid of the human α–globin gene was found to contain H.C. 16 in any of the metaphase spreads studied, and none of the clones negative for the α–globin gene was positive for APRT. We observed a very good correlation between the presence or absence of human chromosome 16 and the presence or absence of the human α–globin gene. On the basis of all these data, the human α–globin gene was tentatively assigned to H.C. 16.

The studies conducted on one hybrid clone, J10H7, were at variance with this assignment of the human α–globin gene to H.C. 16. This clone contains H.C. 16 in only 4 out of 69 of its metaphase spreads. Nonetheless, the incubation of the J10H7 DNA with the α-enriched cDNA (Method I) resulted in a hybridization of 30% over background. As noted above and in Experimental Procedures, this α-enriched cDNA (Method I) contains approximately 70% α sequences. The remainder are γ (20%) and β (10%). The DNA of clone J10H7 reacted strongly with human β–cDNA and thus contained high levels of human β–globin gene sequences. Since the human β– and γ–globin genes are known to be linked on a single chromosome, high levels of human γ–globin gene sequences were also present in the DNA of J10H7. We believe that the positive hybridization observed between the DNA of J10H7 and the human γ–globin cDNA represents a false positive, arising from the annealing of the β– and γ–cDNA sequences which contaminate this cDNA (Method I) with the β– and γ–globin gene sequences present in the J10H7 DNA. The other clones found to be strongly positive for the human α–globin genes were studied using human α–globin cDNA (Methods II and III) in which only very low levels or no human β– and no human γ–cDNA sequences were present. Furthermore, none of the clones strongly positive for α–globin genes (WAV, WAIV A, WAIV A-AA) were found to contain detectable levels of human β–globin genes. Thus all the positive hybridization reactions listed in Table 1 between human α–globin cDNA and DNA from each of the hybrid clones except for J10H7 do represent annealing of the α–globin component of the α-enriched cDNA with human α–globin genes.

Use of Selection and Counter Selection to Test the Linkage between Human Chromosome 16 and the α–Globin Gene

To test further for the presence of the human α–globin gene on H.C. 16, we chose to study the human X mouse clones, WAV and WAIV A. Both clones contained H.C. 16 in >50% of the metaphase spreads, and most other human chromosomes were absent. As shown by the data in Figure 4, the DNA of these cell lines reacted strongly with human α–globin cDNA (Method II) without showing any significant reaction with human β–globin cDNA.

The presence of the enzyme APRT on human H.C. 16 and its absence in the mouse fibroblast parent (A9) (Tischfield and Ruddle, 1974) of cell hybrids WAV and WAIV A allowed us to test the assignment of the human α–globin gene to H.C. 16

by counter selection. Growth of human × mouse hybrid cells in medium supplemented with a substituted purine, diaminopurine (DAP), results in the death of hybrid cells containing APRT and therefore selects for hybrid cells which have lost H.C. 16 (Long and Green, 1971; Rappaport and DeMars, 1973, Tischfield and Ruddle, 1974; Kahan, Held and Demars, 1974).

Hybrid clone WAV was cloned in DAP, and the clones WAV R4D and WAV R4D A19 were obtained. The composition of human chromosomes in WAV, WAV R4D and WAV R4D A19 are presented in Table 1. H.C. 16 was present in WAV, and DNA from this line contained the human α-globin gene (Figure 4), whereas the DNA derived from subclones WAV R4D and WAV R4D A19 (which had lost H.C. 16) lacked the human α-globin gene (see Table 1 and Figure 4). Thus the presence or absence of H.C. 16 correlated very well with the presence or absence of the human α-globin gene in these experiments.

A second series of experiments conducted with clone WAIV A was designed to test the α-globin gene content of cells derived from the same hybrid clone which were selected to lose or retain H.C. 16. As shown in Table 1 and Figure 4, the DNA of WAIV A is rich in human α-globin gene sequences but devoid of human β-globin genes. WAIV A cells were grown in medium supplemented with DAP in concentrations (20 μg/ml) which permit selective growth of cells lacking H.C. 16. In contrast to the DAP-resistant subclones of WAV discussed above, WAIV A cells resistant to DAP (and therefore H.C. 16-negative) were selected from mass populations without cloning. In addition, WAIV A cells were grown in medium supplemented with heat-inactivated fetal calf serum, adenine and the antibiotic alanosine. This latter medium permits the growth

of cells which contain APRT and selects against those cells lacking this enzyme (Gale and Schmidt, 1968; Tischfield and Ruddle, 1974). Since the mouse parent (A9) of WAIV A was negative for APRT, only those hybrid cells retaining H.C. 16 grow in the alanosine-adenine-supplemented medium (AA).

As shown by the data presented in Table 1, WAIV A cells expanded in AA medium (WAIV A–AA) exhibited H.C. 16 in 70% of metaphase spreads, whereas WAIV A cells expanded in medium supplemented with DAP (WAIV A–DAP) had H.C. 16 in none of the 50 metaphase spreads studied (Table

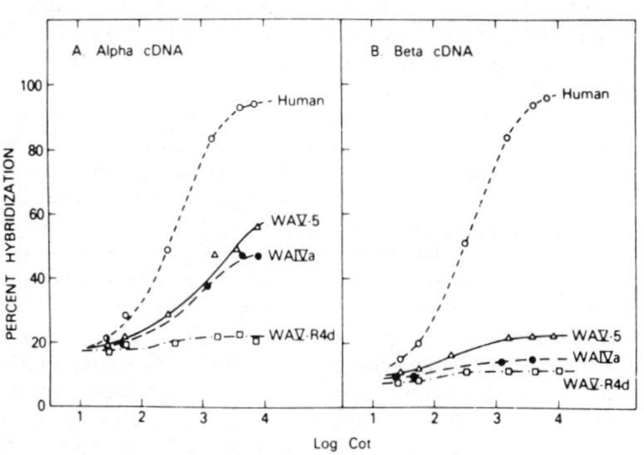

Figure 4. Detection of Human α-Globin Genes in Somatic Cell Hybrids

Human α-(A) or β-(B) globin cDNA reacted with DNA purified from human spleen (O), hybrid cell WAV-5 ($\triangle$), hybrid cell WAIV a ($\bullet$) or hybrid cell WAV-R4d ($\square$). Human α- and β-globin cDNA used in these experiments was prepared by Method II.

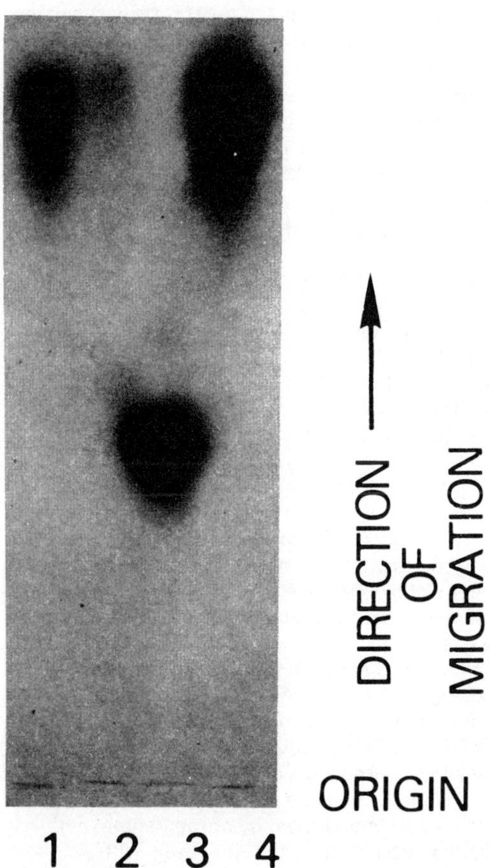

Figure 5. Electrophoresis of Protein Extracts of Hybrid Clone WAIV A Expanded in AA and DAP Selective Medium (See Text) Stained with a Reaction Mixture Specific for the Enzyme Adenosine Phosphoribosyltransferase, a Marker for H.C. 16

Channel 1 contains extract from WAIV A expanded in AA medium; Channel 2 contains extract from WAIV A expanded in DAP medium; Channel 3 contains extract from mouse fibroblasts; Channel 4 contains extract from human fibroblasts.

1). The data in Figures 5 and 6 indicate that WAIV A-AA was strongly positive for APRT and the human α-globin gene, while WAIV A-DAP was only weakly positive for the human α-globin gene and also for APRT. As stated above, the WAIV A-DAP cells were obtained without cloning. It is not unusual to have a small population of cells remaining which possess H.C. 16 and APRT when selection of cells resistant to DAP is conducted from mass populations without cloning.

The human α-globin cDNA (Method III) used in these experiments (Figure 6) was prepared by further purification of the human α-globin cDNA obtained by Method II to remove the small amount of contaminating human β-globin cDNA (10%) present in this probe (see Experimental Procedures). The thermal stability of duplexes formed by incubation of this highly purified α–cDNA (Method III) with the DNA purified from human spleen or hybrid cell (WAIV A-AA) DNA are compared in Figure 6B. The melting transition occurs at essentially the same temperature with each DNA, indicating that we are detecting authentic α gene sequences in the DNA from the hybrid cell WAIV A expanded in AA medium.

The data in Figure 6A were mathematically transformed to yield a linear relationship, the slope of which is proportional to the α gene sequence content of the individual DNAs (Figure 6C). By this method, we estimate that the α gene sequence content of WAIV A-AA DNA is one fourth that of human fibroblast DNA. This value is consistent with the fact that H.C. 16 is monosomic in the hybrid cells, that the mouse chromosome complement is hyperdiploid and that we identified H.C. 16 in only 70% of the metaphase spreads of WAIV A-AA. In contrast, the α-globin gene content of WAIV A-DAP was only $^1/_{15}$ that of human DNA. Although we identified H.C. 16 in none of the 50 metaphase spreads of this hybrid clone, a chromosome absent in 50 metaphase spreads could be present in up to 7% of the cells in the population (90% confidence; $N = 50$). Since the APRT assay was very weakly positive for WAIV A-DAP (see Figure 5), and since the isozyme assay has been shown by E. A. Nichols and F. H. Ruddle (manuscript in preparation) to be very sensitive (positive at a dilution of 1/64), it can be concluded that the low frequency of H.C. 16 which was demonstrated by chromosomal and isozymal analysis correlates very well with the low level of the human α-globin gene sequences present in this line (see Figure 6). Thus the use of positive and negative selection to obtain separate populations which contained a high or a low level of H.C. 16 from the same clone of hybrid cells (WAIV A) allowed us to compare the human α-globin gene content of each. The presence or absence of H.C. 16 in these two populations corre-

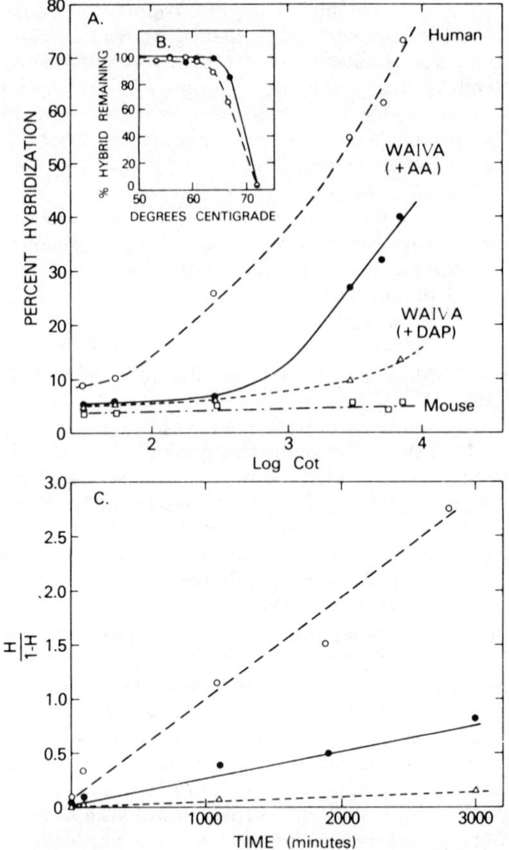

Figure 6. Analysis of DNA Extracted from WAIV A Cells Expanded either in Adenine and Alanosine (●) or Diaminopurine (△)

The α–cDNA used in these experiments was synthesized using α-globin mRNA purified by polyacrylamide gel electrophoresis. The α–cDNA was hybridized to β-enriched mRNA and the β-β duplexes were removed, yielding a probe which is essentially pure α-globin cDNA (see Experiment Procedures). (A) The reaction mixtures contained human α-globin cDNA and human fibroblast DNA (○), DNA from WAIV A expanded in AA medium (●), DNA from WAIV A expanded in DAP medium (△) or mouse embryo DNA (□). 1 mg of each DNA sample was mixed with 0.015 ng of α-cDNA, and a Cot analysis was performed as described in Experimental Procedures. The fraction of cDNA incorporated into hybrid in each of the individual Cot points was determined by incubation with S1 nuclease (Benz et al., 1977). These hybridization reactions do not go to completion, and the $Cot_{1/2}$ is higher than that obtained by hydroxylapatite analysis (Figure 3), as anticipated from the results obtained by Britton and Davidson (1976). (B) The thermal stability of duplexes formed between the α-enriched cDNA and either human spleen (○) or WAIV A (expanded in AA medium) DNA (●) was measured as previously described (Benz et al., 1977). 1 mg of DNA was mixed with 0.015 ng of cDNA in 70 μl and annealed to a Cot of 5 × 10³. Individual capillary tubes were incubated at the indicated temperatures for 5 min, and the fraction of double-stranded cDNA was determined by S1 analysis. (C) The data presented in Figure 5A were transformed to yield a linear relationship, the slope of which is directly proportional to the α-globin gene content of the individual DNA samples. Human α-globin cDNA used in these experiments was prepared by Method III.

lated very well with the presence or absence of the human α–globin gene.

Discussion

The evidence which we have presented in this report, based on direct measurement of the human globin genes in a series of 16 human X mouse somatic cell hybrid cell clones, establishes the location of human α–globin gene to human chromosome 16. DNA-cDNA hybridization was used to determine the presence or absence of the human α–globin gene. We correlated these results with the human chromosomal composition determined by the use of a variety of techniques on a large number of metaphase spreads in each clone. All human chromosomes, except for H.C. 6, 9, 14 and 16, were found to be present in clones devoid of human α–globin genes. Furthermore, we found a positive correlation between H.C. 16 and the α–globin gene. The use of counter selection with diaminopurine allowed us to correlate the loss of H.C. 16 with the loss of the human α–globin gene in two independently derived human X mouse hybrid clones, WAV and WAIV A: cloning in DAP was used with the former, and selection from mass populations in DAP was used with the latter These selection experiments ruled out the presence of undetectable chromosomal translocations or other unrecognized human chromosomal material containing the human α–globin gene and strongly support the assignment of the human α–globin gene to chromosome 16.

Several factors appear to be of importance for the successful use of the DNA-cDNA molecular hybridization assay for the chromosomal localization of structural genes encoding markers of human differentiated cells in somatic cell hybrids. The sensitivity of the DNA–cDNA hybridization assay must be similar to the sensitivity of methods used for characterization of the human chromosomes in the hybrid clones. This avoids detection of a gene carried by a chromosome present in such a low fraction of the cells of a hybrid population so as to be undetectable by the methods used for the chromosomal analysis. Because loss of human chromosomes (which occurs during serial passage) alters the composition of the human chromosomes of a hybrid clone, the number of cell generations during expansions of clones must be minimized. Moreover, cells used for chromosomal analysis must be obtained from the same population of cells at the end of the expansion which is used for DNA-cDNA hybridization studies.

A number of independent chromosomal staining methods was used to characterize the human chromosomes present in each human X mouse hybrid clone to avoid the bias and pitfalls present in the use of any one technique. The Giemsa-11 stain was useful for the detection of translocations and the differentiation of human and mouse chromosomes, while the combination of Hoechst 33258 staining and Giemsa trypsin treatment on the same metaphase spread facilitated identification of human chromosomes in the human X mouse hybrid cells. At least two observers independently studied all metaphase spreads to avoid individual bias. The most useful test to exclude a chromosome from candidacy was the presence of an intact human chromosome at a high frequency in a clone in which there were no detectable human α–globin genes present. This approach, emphasizing chromosome exclusion on the basis of its presence in a cell line devoid of the gene, also minimizes the risk that an unrecognized translocation between mouse and human chromosomes will lead to incorrect conclusions. Definitive chromosomal localization of a gene requires a positive correlation between the presence of the chromosome and the gene. The localization of the human α–globin gene in our studies to H.C. 16 was facilitated by the presence of APRT on H.C. 16, which permitted us to select for human X mouse hybrid cells which had retained or lost this chromosome.

At the start of our work on the chromosomal localization of the human globin genes (Deisseroth and Nienhuis, 1976; Anderson et al, 1976), we initially studied a series of nine human X mouse hybrid cell clones (AIM 23-1, JFA 14a 13, J10H12, WAII A, AIM 3A, AIM 11A, AIM 23a-2, JFA 14B and WAIV A). In these early experiments, only 15–20 metaphase spreads of each cell line were characterized by quinacrine fluorescent banding. Cells taken for chromosomal analysis were not always derived from specimens pooled from multiple flasks harvested at the end of the cell expansion. Although there was a good correlation between APRT and the presence of the human α–globin gene, we realized that these results could not be considered conclusive because the experiments on which they were based were not conducted following the guidelines outlined in the previous section. We therefore chose to analyze an entirely new series of hybrid clones (the subject of this report) following all the strategems summarized here. Application of these principles of chromosomal localization to those new hybrid clones provided a decisive confirmation of our initial impression that the human α–globin gene was on H.C. 16.

The purity of the cDNA is of critical importance, as illustrated by our experience with the clone (J10H7). DNA from these cells gave a ''positive'' reaction with the α probe, although H.C. 16 was not present after expansion. We recognized that the ''positive'' reaction most probably represents a reaction of β and γ gene sequences known to be in the hybrid cell DNA with the corresponding cDNA sequences which contaminated the α probe (Method I). In subsequent experiments, we used methods to prepare human α–globin cDNA (Methods II and III) which resulted in probes of greater purity. The specificity of the annealing of a purified probe to its complementary sequences is also an important factor. In our experiments, increasing the temperature at which hybridization occurred virtually eliminated cross-reaction of the mouse globin genes with the human cDNA probe. We have found that a minimal nucleotide sequence difference of 2–3% is required to achieve a specific and quantitative annealing reaction (Benz et al., 1977). The thermal stability of purified duplexes (as shown in Figure 6B) may be used to verify that only completely complementary gene sequences have reacted with the cDNA.

Several hybrid clones which contained H.C. 2, 4 or 5 at a substantial frequency in the absence of most other human chromosomes were used to test the previous chromosomal assignments of the human globin genes reported by Price et al. (1972) which were based on in situ molecular hybridization. These hybrid clones (containing H.C. 2, 4 or 5) were devoid of human α-globin genes when tested by the DNA-cDNA molecular hybridization assay. This assay is much more sensitive than the in situ method and was conducted under conditions which ensured detection of globin genes present on chromosomes in hybrid populations at very low frequencies. On the basis of our studies, we have concluded that the previous assignments were most probably incorrect. Bishop and Jones (1972) have already discussed the reasons for the insensitivity of the methods used in these studies which may explain their inaccuracy. Atwood et al. (1975) attempted to test the accuracy of the in situ hybridization technique by applying it to metaphase spreads derived from mouse fibroblasts carrying translocations involving the mouse globin genes. These investigators stated that they were able to identify correctly the mouse chromosomes containing the mouse globin genes (Atwood et al., 1975) and indicated that they were applying this technique to the study of the human globin genes (Atwood et al., 1976). Although statistical analysis of preliminary experiments of these studies have been reported (Atwood et al., 1976). a final report which contains primary data supporting the previous assignments of the human globin genes to H.C. 2, 4 or 5 has not yet been published by these investigators.

Cheung et al. (1976), using a variation on the in situ technique involving fluorescent beads, failed to detect localization of human globin mRNA to H.C. 2, and instead localized both α– and β–globin genes to H.C. 4 and H.C. 5. Initial reports of family studies suggesting a linkage between the human β-globin gene and the MN locus reported by Weitkamp, Adams and Rowley (1972) were not confirmed in later studies by these same investigators using larger numbers of families (L. Weitkamp and G. Stammatoyannopoulous, manuscript submitted). These latter studies also did not support tight linkage between human globin genes and known markers of H.C. 2, 4 or 5. Other published studies (Gandini et al., 1977) utilized measurement of the α/β-globin biosynthetic ratios, rather than direct detection of globin gene sequences to develop data, suggesting the presence of human α–globin genes on H.C. 4. These studies may not bear on the structural gene localization but possibly on the factors which govern their expression, since the data were derived from the relative level of synthesis of α- and β-globin in the peripheral blood of individuals in whom a variety of abnormalities of H.C. 4 had been defined. These differences found among the proposed chromosomal assignments for the human globin genes based on Mendelian genetic studies, or on in situ molecular hybridization studies discussed above, suggest that these methods may be inappropriate ones for the solution of the human globin structural gene mapping problem.

The method we have used for the chromosomal localization of the human α-globin gene was chosen because it is based on direct detection of the globin genes by a sensitive species-specific molecular hybridization assay. These techniques are particularly suited for the chromosomal localization of structural genes which are not constitutively expressed in hybrid cells or skin fibroblasts. We are currently applying this method to localize the human β-globin gene and expect that these methods will soon be applied to the chromosome localization of other human structural genes which serve as markers of cellular differentiation in man.

Experimental Procedures

Cell Hybrids

Several series of human X mouse hybrid cell populations which have been discussed in previous publications were used in these studies. Subclones J10H7 and J10H12 of the hybrid cell J10H, which was produced by fusion of the RAG mouse line (deficient in hypoxanthine phosphoribosyltransferase) with human peripheral leukocytes (Ruddle et al., 1970), were isolated in medium supplemented with hypoxanthine, aminopterin and thymidine (HAT medium) as outlined by Littlefield (1964). Hybrid cell lines JFA 14a 5 and JFA 14a 13-5 are subclones of the hybrid line JFA, which was derived by fusion of the mouse line A9 with a human skin fibroblast line [carrier of the fusion product 14/22 as reported by Tischfield and Ruddle (1974)], and were isolated in alanosine-adenine selective media (Kusano et al., 1971; Tischfield and Ruddle, 1974). Clones WAIV A and WAV, which were generated by the fusion of the human fibroblast line WI–38 with the mouse cell line A–9 (Tischfield and Ruddle, 1974), were also isolated in the alanosine-adenine selective medium. Subclones of WAV, WAV R4D and WAV R4D A19 were isolated from WAV by cloning WAV in diaminopurine as outlined by Tischfield and Ruddle (1974). WAIV A-AA and WAIV A-DAP were selected from WAIV A by expansion of mass populations in AA and DAP medium, respectively. The AIM 23a lines were generated by the fusion of the mouse line A-9 with the human fibroblast GM-17 which contains a 15/18 translocation (Elsevier et al., 1974) and were isolated in HAT medium. The clones AHA 13D, AHA 16D and AHA 16E were subclones of the hybrid population derived by R. Kucherlapati by fusion of a human fibroblast (carrying a 6/21 fusion product) with the mouse cell A9 (Ruddle et al., 1970). Clones IL-II-5 and IL-II-3 R4D were derived by fusion of mouse LM(TK−) cells with IMR 32 cells (human neuroblastoma) (McMorris et al., 1973). All cell lines used for these studies were tested for mycoplasma infection by Flow Laboratories (Rockville, Maryland) and indpendently by the use of a culture technique as described by Barile (1973). All 16 hybrid cell clones studied were found to be mycoplasma-free except for JFA 14a 5, JFA 14a 13 5 and WAV. Each of these latter lines were originally mycoplasma-free at the time of their isolation, and no change in their chromosomal composition attended the emergence of this infection by mycoplasma orale or arginini. Giemsa 11 staining of metaphase spreads of these two lines did not reveal any chromosomal fragmentation among the human chromosomes, and isozyme analysis in these lines showed no disruption of normal linkage relationships among known markers of human chromosomes. Clone AHA 16D also contained a mycoplasma infection and a single translocation not involving H.C. 16. Analysis of the data presented in Table 1 shows that there is no relationship between the presence or absence of mycoplasma infection and the presence or absence of the human α-globin genes in the various hybrid clones studied.

Cell lines were prepared for DNA extraction and chromosomal analysis by expansion in monolayer in T-150 flasks to yield 4–8 × 10^9 cells. At the end of the expansion, the cells from all the flasks were pooled. An aliquot of cells was taken from a pool of trypsinized cells from several different flasks and explanted in monolayer culture for chromosomal analysis. After 1–3 passages, these cells were cultured in vinblastine sulfate for 2 hr, and metaphase spreads were generated in a large number of slides as described below. Another aliquot of the cell suspension was used to generate protein extracts for isozymal analysis. The remainder was allocated for purification of DNA. We combined all the cells of each hybrid cell line at the end of an expansion to generate identical populations of cells for the chromosomal studies, isozymal analysis and DNA-cDNA hybridization assays.

Isozymal Analysis

The following 24 isozymal markers were used to screen the hybrid clones for the presence of the human X chromosome and 16 different human autosomes by starch gel electrophoresis as outlined previously (Nichols and Ruddle, 1973): chromosome 1: phosphoglucomutase (PGM, E.C.2.7.5.1.) and peptidase C (PEP C, E.C.2.7.2.3); chromosome 2: isocitrate dehydrogenase (IDH, E.C.1.1.1.42) and (NAD)-malate dehydrogenase, cytoplasmic (MDH, E.C.1.1.1.37); chromosome 6: (NADP)-malic enzyme, cytoplasmic (MOD, E.C.1.1.1.40); chromosome 9: adenylate kinase (AK₁, E.C.2.7.4.3); chromosome 10: glutamate oxaloacetate transaminase (GOT, E.C.2.6.1.1.) and adenosine kinase (ALD K.E.R.2.7.1.20); chromosome 11: lactate dehydrogenase-A (LDH A, E.C.1.1.1.); chromosome 12: lactate dehydrogenase B (LDH B, E.C.1.1.1.27) and peptidase-B (PEP-B, E.C.2.7.2.3); chromosome 13: esterase D (ESD, E.C.3.1.1.1); chromosome 14: purine nucleoside phosphorylase (NP, E.C.2.3.2.1); chromosome 15: hexosaminidase A (HEX A, E.C.3.2.1.30), mannosephosphate isomerase (MPI, E.C.5.3.1.8) and pyruvate kinase-3 (PK, E.C.2.7.1.40); chromosome 16: adenine phosphoribosyl transferase (APRT, E.C.2.4.2.7); chromosome 17: galactokinase (GK, E.C.2.7.1.6); chromosome 18: peptidase-A (PEP A, E.C.2.7.2.3); chromosome 19: glucose phosphate isomerase (GPI, E.C.5.3.1.9); chromosome 20: adenosine deaminase (ADA, E.C.3.5.4.2); chromosome 21: cytoplasmic indophenol oxidase (SOD, E.C.1.6.4.3); chromosome X: glucose-6-phosphate dehydrogenase (G6PD, E.C.1.1.1.49) and phosphoglycerate kinase (PGK, E.C.2.7.2.3).

Chromosomal Analysis

Log-phase cultures were treated sequentially with vinblastine sulfate, hypotonic solutions and methanol:acetic acid fixative. One drop of the cell suspension was used to generate metaphase spreads on cleaned glass slides as described previously (Friend et al., 1976). A variety of specific chromosomal staining techniques were used to identify the human chromosomes present in each hybrid clone. Quinacrine mustard treatment was used (Caspersson et al., 1970) to generate banding patterns on mouse and human chromosomes. Differential color staining of human and mouse chromosomes with Giemsa (pH 11), as described by Friend et al. (1976), allowed identification of human chromosomes of the A, B, D and G groups, as well as detection of any interspecific chromosomal translocations. A combination of Giemsa-trypsin (Kozak et al., 1977) and Hoechst 33258 (Yoshida et al., 1975) sequential staining of the same metaphase spread as reported by Kozak, et al. (1977) permitted assortment of mouse and human chromosomes and identification of specific human chromosomes. Metaphase spreads were first stained by the Giemsa-trypsin technique and photographed. After destaining by immersion in methanol:glacial acetic acid (3:1 by volume), the slides were exposed to Hoeschst 33258. The intense centromeric fluorescence which develops in most of the mouse telocentric and all the mouse biarmed, but none of the human chromosomes, permits identification of the chromosomes of the mouse and human parent species. All metaphase spreads treated as described above were photographed with the Zeiss photomicroscope III or the Leitz photomicroscope, and printed enlargements were made for analysis.

Extraction of DNA

3–12 ml of packed hybrid cell volume (approximately 2–6 × 10^9 cells) were used for preparation of DNA as outlined previously (Marmur, 1961; Deisseroth et al., 1976a; Deisseroth et al., 1976b). The cell pellets were added directly to 10–15 vol of 0.35 M NaCl, 1.0 M sodium perchlorate and 0.1 M EDTA (pH 8.0), and then homogenized 3 times at full speed with a tight-fitting teflon glass homogenizer. The resulting solution was extracted with an equal volume of phenol:chloroform (1:1) by shaking for 30 min at room temperature. The phases were separated by centrifugation and the aqueous phase was removed. 2 vol of cold (−20°C) ethanol were gently poured over the top of the aqueous layer, and the DNA was removed by spooling onto a glass pipette. The DNA was dissolved into 1 × SSC [0.15 M NaCl and 0.015 M sodium citrate (pH 7.0)] by gently stirring overnight. RNAase (previously boiled for 10 min to activate DNAase) was added to a final concentration of 100 μg/ml, and the solution was incubated at 37°C for 2 hr. Pronase (predigested for 2 hr at 37°C) was added to a final concentration of 100 μg/ml, and the incubation was continued for 2 hr. Extraction with an equal volume of chloroform was performed twice. The concentration of nucleic acid in the final aqueous solution was adjusted to 1 mg/ml, and then 10 ml aliquots were sonicated with a Sonifer Cell Disrupter with a microprobe (Model W185 Heat Systems-Ultrasonics Incorporated, Plainville, New York). This procedure results in an average DNA fragment size of 300–400 nucleotides. After sonication, 2 vol of cold ethanol were added, and the DNA was recovered (after 8 hr at −20°C) by centrifugation. The DNA was then dissolved in 2–4 ml of water and desalted by centrifugation through a dry pad of coarse Sephadex G25. Human and mouse DNA were extracted from human spleen and mouse embryos, respectively, by identical techniques, except that the tissue was cut up, placed in the solution of NaCl, sodium perchlorate and EDTA buffer, and ground in a Waring Blendor before homogenization with the teflon glass homogenizer.

Preparation of Complementary DNA

Human α-globin cDNA (Method I), used for analysis of the DNA of hybrid clones AIM 23a-X-1, AIM 23a old, JFA 14a 5, JFA 14a 13 5, J10H7 and IL II 5, was prepared by incubating mRNA extracted directly from abnormal human reticulocytes with the viral enzyme, RNA-directed DNA polymerase (Verma et al., 1972; Kacian et al.,

1972; Ross et al., 1972; Weiss et al., 1976). Patients with homozygous β thalassemia have a quantitative deficiency of β-globin mRNA (Benz and Forget, 1975). Thus RNA prepared from their reticulocytes contains mainly α-globin mRNA and may be used to synthesize an α-enriched cDNA. Human α-(20%) and human β-(10%) globin mRNA is also present in these fractions. Lysates of peripheral blood from these patients were prepared, crude RNA was obtained by phenol extraction and a 7–12S fraction was recovered by sucrose gradient fractionation as previously described (Nienhuis, Falvey and Anderson, 1974; Velez et al., 1975). Mouse globin mRNA was prepared from the reticulocytes of mice injected with phenylhydrazine. The polysomes were spun out of a membrane-free lysate, and the 10S messenger RNA was recovered by sucrose gradient fractionation (Nienhuis, Falvey and Anderson, 1974).

Synthesis of cDNA (Methods I, II, III) was by incubation of mRNA with RNA-directed DNA polymerase (reverse transcriptase) exactly as previously described (Wilson et al., 1975; Benz et al., 1977). The concentration of nucleotides was 100 μM for dATP, dGTP and dTTP, and 50 μM for dCTP. ^{32}P-dCTP of specific activity 150–250 Ci/mmole (supplied by Amersham-Searle or New England Nuclear) was used in the reaction to give a probe whose final specific activity ranged between 150,00–250,000 cpm/ng. A fraction of the cDNA ranging in size from 500–600 nucleotides was obtained after alkaline sucrose gradient fractionation (Deisseroth et al., 1976b; Benz et al., 1977). α- and β-enriched cDNAs were freshly prepared each time a set of hybrid cell DNAs were available for analysis and used within 10 days of synthesis.

For certain experiments (hybrid clones WAV, WAV R4D, WAV R4D A19 WAIV A, AHA 3D, AHA 16E and IL II 54D), we used an α-enriched cDNA (Method II) prepared from mRNA isolated from reticulocytes of patients with sickle cell anemia. An 8–12S human mRNA fraction obtained by sucrose gradient fractionation (Nienhuis et al., 1974) was further purified by oligo (dT)-cellulose chromatography (Aviv and Leder, 1972). The retained RNA was eluted with 10 mM Tris-HCl (pH 7.5) and concentrated by ethanol precipitation. 10 μg were applied to tube gels in 90% formamide containing 20 mM barbital and 20 mM NaCl as previously described (Forget et al., 1975; Kazazian et al., 1975; Benz et al., 1977). Individual gels were then stained with methylene blue, and the slices containing the faster migrating of the two 10S bands were cut out, homogenized in SDS buffer, extracted with phenol, recovered by ethanol precipitation, further purified by oligo(dT)-cellulose chromatography and used to make α-enriched cDNA.

The α cDNA (Method III) used in the experiments illustrated in Figure 6 for study of DNA from clones WAIV A–AA, WAIV A–DAP and AHA 16D was further purified as follows. Immediately after preparation of the cDNA, a Crot analysis was performed by annealing the α-enriched cDNA to a 4 fold excess of a partially purified mRNA from reticulocytes of a patient with Hb H disease. This RNA contained β and α sequences in a ratio of approximately 80:20 and was approximately 10% globin mRNA. Annealing reactions were performed in 10 μl solutions sealed in glass capillaries at 62°C in 50% formamide exactly as described previously (Wilson et al., 1975). Individual capillaries were removed at specified times, frozen and stored until the incubation period was complete. Analysis with S1 nuclease to determine the percentage of duplex cDNA was performed (Benz et al., 1977).

A two-transition Crot curve was obtained. The first transition occurred with a Crot$_{1/2}$ of annealing at 2.5×10^{-2} M/l X sec, whereas the second occurred at a Crot$_{1/2}$ of 1.3×10^{-1}. Since this was an RNA excess reaction, the fast reacting component represented annealing of β-mRNA with the β-cDNA component contaminating the α probe, whereas the slower reacting component represented annealing of α-mRNA sequences with the α-cDNA.

For the preparative reaction, 35 ng of ^{32}P-cDNA (spec. act. 280,000 cpm/ng) were annealed with 1200 ng of the partially purified Hb H mRNA in a 750 μl reaction. This was incubated for 2 hr (Crot = 1×10^{-1}) at 62°C and then diluted 10 fold in 0.05 M phosphate buffer (pH 6.8). The single-stranded fraction was purified by hydroxylapatite chromatography as previously described

(Benz et al., 1977). It was desalted by passage over a 1.6 X 60 cm column of coarse G-50 Sephadex and concentrated by ethanol precipitation. The sample was then applied to an alkaline sucrose gradient to ensure that molecules of 500–600 nucleotides were used in subsequent reactions and also to free the purified probe from any remaining mRNA sequences.

DNA-cDNA Hybridization Reactions

1 or 2 mg of hybrid cell, control mouse or human DNA were used in each analysis. A total reaction mix of 70 or 140 μl contained 0.016–0.055 ng of cDNA (spec. act. = 150,000–280,000 cpm/ng) in 3 X SSC and 50% formamide. Individual 12 or 20 μl aliquots were sealed in glass capillary tubes and incubated for times ranging from 10 min to 48 hr to generate a Cot curve. The globin gene sequence to cDNA ratio for human DNA ranged from 10–25 in these experiments, whereas the corresponding sequence ratio for the hybrid cell line was clearly contingent upon its human globin gene sequence content. When a chromosome bearing a human globin gene was present in a low fraction of the cells, an excess of globin gene sequence over cDNA did not exist. For most of the experiments, the fraction of single- and double-stranded cDNA in the individual Cot points was determined by batch chromatography on hydroxylapatite (Deisseroth et al., 1976a). The data in Figure 6 were obtained by incubation with S1 nuclease under conditions in which single-stranded cDNA was digested (Benz et al., 1977).

Acknowledgments

Part of this work was completed while A.D. was the recipient of a National Research Fellowship. The support of an NIH grant awarded to F.R. is also acknowledged. We wish to acknowledge gratefully the skill and support of Elizabeth Nichols without whose expertise the analysis of isozymal markers for human and mouse chromosomes could not have been performed. We wish to thank Drs. J. and D. Beard for providing RNA-directed DNA polymerase through the Office of Program Resources and Logistics, Viral Oncology, National Cancer Institute. Thanks are also given to Dr. Edward Benz of the Clinical Hematology Branch for assistance in preparation of the human α-globin mRNA by formamide gel electrophoresis, and to Stephen Meador for help in expansion of some of the hybrid clones. We also acknowledge the technical assistance of Sylvan Von Der Pool, Adam Messer, Francie Lawyer, Ann Cunningham, Teresa Caryk and Marcia Willing. We appreciate the assistance of Exa Murray in the preparation of the manuscript.

Received May 13, 1977

References

Anderson, W. F., Deisseroth, A. B., Velez, R., Nienhuis, A. W., Ruddle, F. H. and Kucherlapati, R. S. (1976). In Thirs International Workshop on Human Gene Mapping, Birth Defects: Original Article Series, *XII*, No. 7, (New York: The National Foundation) pp. 367–371.

Atwood, K. C., Henderson, A. S., Kacian, D. and Eicher, E. (1975). Cytogenet. Cell Genet. *14*, 59–61.

Atwood, K. C., Yu, M. T., Eicher, E. and Henderson, A. S. (1976). In Third International Workshop on Human Gene Mapping, Birth Defects: Original Article Series, *XII*, No. 7, (New York: The National Foundation), pp. 372–375.

Aviv, H. and Leder, P. (1972). Proc. Natl. Acad. Sci. USA *68*, 1408–1412.

Barile, M. F. (1973). In Contamination in Tissue Culture, J. Fogh, ed. (New York: Academic Press), pp. 132–162.

Benz, E. J. and Forget, B. G. (1975). Prog. Hematol. *9*, 107–155.

Benz, E. J., Geist, C. E., Steggles, A. W., Barker, J. E. and Nienhuis, A. W. (1977). J. Biol. Chem. *252*, 1908–1916.

Bishop, J. O. and Jones, K. W. (1972). Nature *240*, 149–150.

Bobrow, M., Madan, K. and Pearson, P. L. (1972). Nature New Biol. *238*, 122–124.

Britten, R. J. and Davidson, E. H. (1976). Proc. Nat. Acad. Sci. USA *73*, 415–419.

Caspersson, T., Zech, L. and Johansson, C. (1970). Exp. Cell Res. *62*, 490–492.

Cheung, S. W., Tishler, P. V., Atkins, L., Senguptu, S., Madert, E. and Forget, B. (1976). J. Cell Biol. *70*, 221a.

Creagan, R. F. and Ruddle, F. H. (1974). In Second International Workshop on Human Gene Mapping, Birth Defects: Original Article Series, *II*, No. 3, (New York: The National Foundation), pp. 112–116.

Deisseroth, A. and Nienhuis, A. (1976). In Vitro *12*, 734–742.

Deisseroth, A., Velez, R. and Nienhuis, A. (1976a). Science *191*, 1262–1264.

Deisseroth, A., Velez, R., Burke, R., Minna, J., Anderson, F. W. and Nienhuis, A. (1976b). Somatic Cell Genet. *2*, 373–384.

Elsevier, S. M., Kuchalapati, R. F., Nichols, E. A., Willike, K., Creagan, R. P., Giles, R. E., McDougal, J. K. and Ruddle, F. H. (1974). Nature *251*, 633.

Forget, B. G., Housman, D., Benz, E. J., Jr. and McCaffrey, R. P. (1975). Proc. Nat. Acad. Sci. USA *72*, 984–988.

Friend, K. K., Chen, S. and Ruddle, F. H. (1976). Somatic Cell Genet. *2*, 153.

Gale, G. R. and Schmidt, G. B. (1968). Biochem. Pharmacol. *17*, 363–368.

Gandini, E., Dallapiocola, B., Laurent, C., Suerinc, E. F., Farabasco, A., Conconi, F. and DelSenno, L. (1977). Nature *265*, 65–66.

Kacian, D. L., Spiegelman, S., Bank, A., Jerada, M., Metafora, S., Dow, L., and Marks, P. A. (1972). Nature New Biol. *235*, 167–169.

Kahan, B., Held, K. R. and DeMars, R. (1974). Genetics *78*, 1143–1156.

Kazazian, H. H., Cinder, C. P., Snyder, P. C., Van Benedez, R. J. and Woodhead, A. P. (1975). Proc. Nat. Acad. Sci. USA *72*, 567–571.

Kozak, C. A., Lawrence, J. B. and Ruddle, F. H. (1977). Exp. Cell Res. *105*, 109–117.

Kucherlapati, R., Hilwig, I., Gropp, A. and Ruddle, F. H. (1975). Human Genet. *27*, 9–14.

Kusano, T., Long, C. and Green, H. (1971). Proc. Nat. Acad. Sci. USA *68*, 82–86.

Littlefield, J. (1964). Science *145*, 709.

Marmur, J. (1961). J. Mol. Biol. *3*, 208–218.

Maynard-Smith, S., Penrose, L. S. and Smith, C. A. B. (1961). Mathematical Tables for Research Workers in Human Genetics. (London: Churchill).

McKusick, V. A. and Ruddle, F. H. (1977). Science *196*, 390–405.

McMorris, F. A., Chen, T. R., Riccuiti, F., Tischfield, J., Creagan, R. F. and Ruddle, F. H. (1973). Science *179*, 1129.

Nance, W. E., Conneally, M., Kang, K. W., Reed, T., Shroder, T. and Rose, S. (1970). Am. J. Human Genet. *22*, 453–459.

Neel, J. V., Schull, W. J. and Shapiro, H. S. (1952). Am. J. Human Genet. *4*, 204–208.

Nichols, E. A. and Ruddle, F. H. (1973). J. Histol. Chem. Cytochem. *21*, 1066–1081.

Nienhuis, A. W., Falvey, A. and Anderson, W. F. (1974). Methods in Enzymology, *30* (New York: Academic Press), pp. 621–630.

Price, P. M. and Hirschhorn, K. (1975a). Fed. Proc. *34*, 2227–2232.

Price, P. M. and Hirschhorn, K. (1975b). Cytogenet. Cell Genet. *14*, 225–231.

Price, P. M., Conover, J. H. and Hirschhorn, K. (1972). Nature *237*, 340–342.

Rappaport, H. and Demars, R. (1973). Genetics *75*, 335–345.

Ross, J., Aviv, H., Scolnick, E. and Leder, P. (1972). Proc. Nat. Acad. Sci. USA *69*, 264–268.

Ruddle, F. H., Chapman, V. N., Chen, T. R. and Klebe, R. J. (1970). Nature *227*, 251–257.

Smith, E. W. and Torbert, J. V. (1958). Bull. Johns Hopkins Hosp. *102*, 38–41.

Sumner, A. T., Evans, H. J. and Buckland, R. A. (1971). Nature New Biol. *232*, 21–32.

Tischfield, J. A. and Ruddle, F. H. (1974). Proc. Natl. Acad. Sci. USA *71*, 45–48.

Velez, R., Kantor, J. A., Picciano, D. J., Anderson, W. F. and Nienhuis, A. W. (1975). J. Biol. Chem. *250*, 3193–3198.

Verma, I. M., Temple, G. F., Fan, H. and Baltimore, D. (1972). Nature New Biol. *235*, 163–167.

Weiss, G. B., Wilson, G. N., Steggles, A. W. and Anderson, W. F. (1976). J. Biol. Chem. *251*, 3425–3431.

Weitkamp, L. R., Adams, M. S. and Rowley, P. T. (1972). Human Hered. *22*, 566–572.

Wilson, G. N., Steggles, A. W., Kantor, J. A., Nienhuis A. W. and Anderson, W. F. (1975). J. Biol. Chem. *250*, 8604–8613.

Yoshida, M. C., Ikeuchi, T. and Sasaki, M. (1975). Proc. Jap. Acad. *51*, 184–187.

Editor's Comments
on Papers 22, 23, and 24

22 CARTER
Effects of Cytochalasins on Mammalian Cells

23 BUNN, WALLACE, and EISENSTADT
Cytoplasmic Inheritance of Chloramphenicol Resistance in Mouse Tissue Culture Cells

24 EGE and RINGERTZ
Preparation of Microcells by Enucleation of Micronucleate Cells

FUSION OF SUBCELLULAR FRAGMENTS

The cell hybridization experiments described earlier involved the fusion of intact cells. Experiments based on the fusion of subcellular fragments also have been carried out and have provided information on a broad range of topics, including virus activation, cytoplasmic inheritance, and gene mapping. These studies were made possible by the demonstration by Carter (Paper 22) that the fungal metabolite cytochalasin B can cause enucleation of cultured mouse cells. When cultured mouse cells were treated with cytochalasin B, the nuclei in a high percentage of the cells were extruded, remaining attached to the remainder of the cell only by a thin strand of cytoplasm. This cytoplasmic strand could be broken mechanically, resulting in the complete enucleation of the cells. Many cells exposed to cytochalasin B for twenty-four hours lost their nuclei even without mechanical treatment. The enucleated cells resulting from cytochalasin treatment appeared to remain viable for twenty-four hours after enucleation. These studies raised the possibility of using chemically-induced enucleation to analyze the functional relationship between the nucleus and cytoplasm in cultured mammalian cells.

Although spontaneous enucleation was observed in cytochalasin B treated cells, the frequency with which it occurred was found to be variable and often too low to be useful. Improved procedures for enucleating cultured cells have been developed, combining cytochalasin B treatment with centrifugation to rupture the connection of the extruded nuclei to the cells (Prescott et al., 1972; Wright and Hayflick, 1972). Through the use of these procedures, enucleation of close to 100% of treated cells can be obtained.

Early studies demonstrated that subcellular fragments resulting from cytochalasin B treatment could be fused with other cells or fragments. In the first such study of this type, enucleated cells (cytoplasts) were fused with intact cells, and fusion products were identified as nucleated cells that exhibited surface characteristics of the cytoplast donor (Poste and Reeve, 1971). The fusion products resulting from the fusion of cytoplasts with whole cells are now generally referred to as *cybrids* (see Paper 23).

In another study, cytoplasts from SV40-permissive cells were fused with SV40-transformed cells. Following fusion with the cytoplasts, the virus in the transformed cells was reactivated. These results indicated that the cytoplasm of permissive cells contains all the factors necessary for virus reactivation and that the nucleus of the permissive cell is not required for reactivation (Croce and Koprowski, 1973).

Cytoplasts have been fused not only with intact cells but also with the nuclei resulting from cytochalasin treatment. Such fusion between cytoplasts and nuclei leads to the formation of *reconstituted* cells (Veomett et al., 1974; Fge et al., 1974), which are capable of undergoing mitosis. Subsequent studies from several laboratories showed that reconstituted cells are capable of long-term prolification.

In the first experiments on the fusion of cytoplasts with intact cells, the experimental design precluded the identification and isolation of pure populations of proliferating cybrids. The markers for the cytoplast contribution in those experiments were surface characteristics passively donated to the cybrids, and these markers would be lost upon proliferation. In contrast, the identification and isolation of proliferating cybrids required cytoplasmic markers that could be retained over long periods of time. The existence of such markers was demonstrated by Eisenstadt and co-workers in 1974 (Paper 23), in experiments that provided the first evidence for cytoplasmic inheritance in cultured mammalian cells.

These studies were based on the isolation of cells resistant to the drug chloramphenicol (CAP), which inhibits mitochondrial protein synthesis in cultured cells. (CAP resistance in yeast was shown previously to be determined by mutations in mitochondrial DNA.) Mouse cells resistant to CAP were enucleated using cytochalasin B, and the cytoplasts were fused with intact CAP sensitive cells. Upon selection with CAP, cells that exhibited the CAP resistance of the cytoplast donor and the nuclear characteristics of the CAP sensitive parental cell were isolated. These results demonstrated that CAP resistance in mouse cells is cytoplasmically inherited. Similar results were obtained for the cytoplasmic transfer of CAP resistance in human cells (Wallace et al., 1975). These experiments established the foundation for the study of cytoplasmic inheritance in mammalian cells.

Another type of subcellular fragment has been used to transfer only small numbers of chromosomes from one cell to another. When cells are cultured in the presence of mitotic inhibitors, cell division is inhibited and the nuclear membrane can reform around small groups of chromosomes. Thus, treated cells may develop a large number of micronuclei, in extreme cases even approaching the number of chromosomes in a cell (Stubblefield, 1964).

Ege and Ringertz (Paper 24) treated micronucleated cells with cytochalasin B and were able to isolate a new type of subcellular fragment, which was referred to as a *microcell*. These microcells contained a micronucleus apparently surrounded by an intact plasma membrane. Cytochemical DNA determinations indicated that the smallest microcells contained as few as one or two chromosomes. It was also shown that microcells could be fused with intact cells. These studies provided the first practical way of transferring only a few chromosomes at a time from one cell to another. It was recognized that the transfer of only a small number of chromosomes into hybrids could be of value for gene mapping, but the proliferative capacity of the fused cells could not be demonstrated at the time. As mentioned earlier, proliferating microcell hybrids resulting from the fusion of microcells with intact cells have been isolated and shown to be useful for gene mapping (Fournier and Ruddle, 1977). Microcell hybrids also appear to be potentially useful for studies on gene regulation or cell-virus interactions, where the transfer of only a few chromosomes from one cell to another would be advantageous.

REFERENCES

Croce, C. M., and H. Koprowski, 1973, Enucleation of Cells Made Simple and Rescue of SV40 by Enucleated Cells Made Even Simpler, *Virology* **51:**227–229.

Ege, T., U. Krondahl, and N. R. Ringertz, 1974, Introduction of Nuclei and Micronuclei into Cells and Enucleated Cytoplasms by Sendai Virus Induced Fusion, *Exp. Cell Res.* **88:**428–432.

Fournier, R. E., and F. H. Ruddle, 1977, Microcell-Mediated Transfer of Murine Chromosomes into Mouse, Chinese Hamster, and Human Somatic Cells, *Natl. Acad. Sci. (USA) Proc.* **74:**319–323.

Poste, G., and P. Reeve, 1971, Formation of Hybrid Cells and Heterokaryons by Fusion of Enucleated and Nucleated Cells, *Nature New Biol.* **229:**123–125.

Prescott, D. M., D. Myerson, and J. Wallace, 1972, Enucleation of Cells with Cytochalasin B, *Exp. Cell Res.* **71:**480–485.

Stubblefield, E., 1964, DNA Synthesis and Chromosome Morphology of Chinese Hamster Cells Cultured in Media Containing N-Deacetyl-N-Methylcolchicine (Colcemid), in *Cytogenetics of Cells in Culture*, R. J. Harris, ed., International Society of Cell Biology Symposium, vol. 3, Academic Press, New York, pp. 223–298.

Veomett, G., D. M. Prescott, J. Shay, and K. R. Porter, 1974, Reconstruction of Mammalian Cells from Nuclear and Cytoplasmic Components Separated by Treatment with Cytochalasin B., *Natl. Acad. Sci. (USA) Proc.* **71:**1999–2002.

Wallace, D. C., C. L. Bunn, and J. M. Eisenstadt, 1975, Cytoplasmic Inheritance of Chloramphenicol Resistance in Human Tissue Culture Cells, *J. Cell Biol.* **67:**174–188.

Wright, W. E., and L. Hayflick, 1972, Formation of Anucleate and Multinucleate Cells in Normal and SV40 Transformed WI-38 by Cytochalasin B., *Exp. Cell Res.* **74:**187–194.

22

Effects of Cytochalasins on Mammalian Cells

by
S. B. CARTER
Research Department, Pharmaceuticals Division,
Imperial Chemical Industries Limited,
Alderley Park, Macclesfield, Cheshire

The cytochalasins—a group of mould metabolites—inhibit movement and cytoplasmic cleavage in cultured cells. At higher doses they cause nuclear extrusion which may lead to total enucleation

SEVERAL species of moulds have been found to produce a number of chemically related metabolites which show unusual biological activity. The name "cytochalasin" (Greek *cytos*, a cell; and *chalasis*, relaxation) is suggested for this new class of compounds. The name is intended as a general description of the effects which are characteristic of these substances. It is not meant to imply a particular mode of action.

Four cytochalasins have been isolated in these laboratories by Dr. W. B. Turner. These have been designated *A*, *B*, *C*, and *D*. Cytochalasins *A* and *B* were obtained from culture filtrates of *Helminthosporium dematioideum*. Cytochalasins *C* and *D* were isolated from *Metarrhizium anisopliae*. Cytochalasin-like activity seems to be widely distributed. Filtrates from nine other species (representing four orders of fungi) have been found to produce similar effects. The four cytochalasins so far isolated show essentially similar activity but differ in potency. Cytochalasins *C* and *D* are about ten times as active as cytochalasins *A* and *B*.

The experiments described here were carried out with cytochalasin *B*. This compound has a novel macrolide

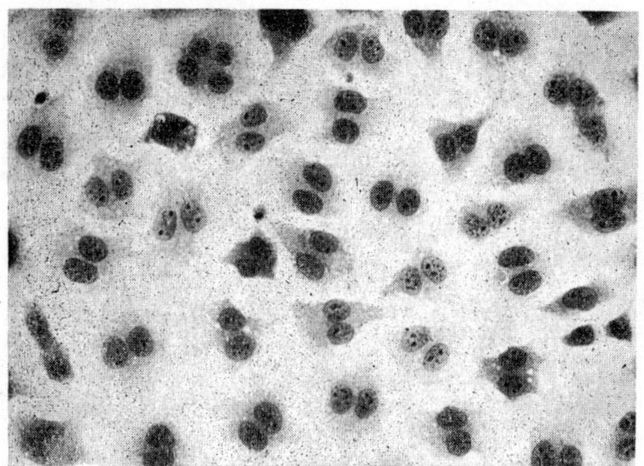

Fig. 1. Binucleated "*L*" cells produced by exposure to cytochalasin *B* at 1·0 μg/ml. for 24 h (× 250).

structure in which the lactone ring is joined to a bicyclic lactam system. Its molecular structure is reported elsewhere[1].

Earle's "*L*" strain of mouse fibroblast was used in most of these experiments. The cells were cultured on glass cover slips using Eagle's medium[2] with 8 per cent inactivated calf serum. For observations on living cells, oil-sealed culture chambers were used[3]. The compound is poorly soluble in water, and appropriate dilutions in culture medium were made from a stock solution in dimethyl sulphoxide (0·1 per cent).

Prevention of cytoplasmic cleavage. Cytochalasin *B* prevents cytoplasmic cleavage at a concentration of 0·5 μg/ml. After 24 h exposure to this concentration, many cells are found to contain two nuclei. At 1 μg/ml., inhibition of cytoplasmic cleavage is virtually complete (Fig. 1). Nuclear division proceeds normally, and following this a deep cleavage furrow may develop. This furrow fails to separate the cells completely, however, and subsequently retreats. If the compound is removed by perfusion with normal medium, the binucleated cells produced undergo multiple cytoplasmic cleavage following the next nuclear division. No compound has previously been described which specifically and consistently prevents cytoplasmic cleavage without interfering with nuclear division. Wolpert has pointed out the potential value of such a compound for investigating the cleavage mechanism[4].

Cells remain viable for many days in the continuous presence of cytochalasin *B*. Although the mitotic rate is reduced, many consecutive nuclear divisions may take place without cleavage. Very large multinucleated cells are produced which are clearly visible to the naked eye. All the nuclei in a particular cell appear to enter mitosis synchronously. Synchrony of nuclear division in multinucleated cells is regarded as the rule in a very wide range of biological material. If all the nuclei divide together, it would be expected that the number in each cell would double after each mitotic cycle. In fact this does not happen. More than a hundred individual cells have been followed over a period of 7 days under continuous treatment with cytochalasin *B*, to investigate the progressive increase in nuclearity which occurs. The usual pattern is for the nuclear complement of each cell to increase one at a time. This is a surprising finding which seems to conflict with an established cytological principle. It may be that synchronous nuclear division in multinucleated cells cannot safely be inferred from fixed preparations in which all the nuclei appear to be in the same mitotic phase. The simplest explanation for the production of only one nucleus per cycle is that a multinucleated cell may enter mitosis when only one of its nuclei is fully prepared and competent to divide. This implies that the other nuclei may be

induced to go through a "pseudomitotic" cycle which closely imitates mitosis, but involves no overall change in the nuclei taking part. Such a cycle could entail chromatin condensation and "chromosome" formation, followed by complete reconstruction of each nucleus in its original form. It is difficult to obtain direct evidence for this explanation because of the large number of chromosomes involved. If it can be shown to occur, however, pseudomitosis would have important implications for the initiation of nuclear division and the mechanism of chromosome formation.

Inhibition of cell motility. Time lapse cinematography reveals that "*L*" cells stop moving when the culture is perfused with cytochalasin *B* at 0·5 μg/ml. The effect is immediate and is reversed with equal rapidity when normal medium is restored. Treated cells remain in the same position for many days. This greatly facilitates the long term investigation of nuclear changes by time lapse techniques.

Ruffling of the cell margin is also inhibited. Ruffle formation is resumed as soon as the compound is removed.

No effect has been observed on the motility of a number of fresh water ciliates and flagellates even at doses as high as 50 μg/ml. Similarly the motility of mouse spermatozoa was unimpaired at this concentration.

Extrusion of nuclei. Nuclear extrusion is occasionally observed in "*L*" cells exposed to cytochalasin *B* at 1 μg/ml. At a dose of 10 μg/ml., however, a high proportion of cells extrude their nuclei (Figs. 2 and 3).

The initial stages of this process are dramatically rapid. Within minutes the nucleus has formed a prominent bulge in the cell membrane. This bulge progresses rapidly

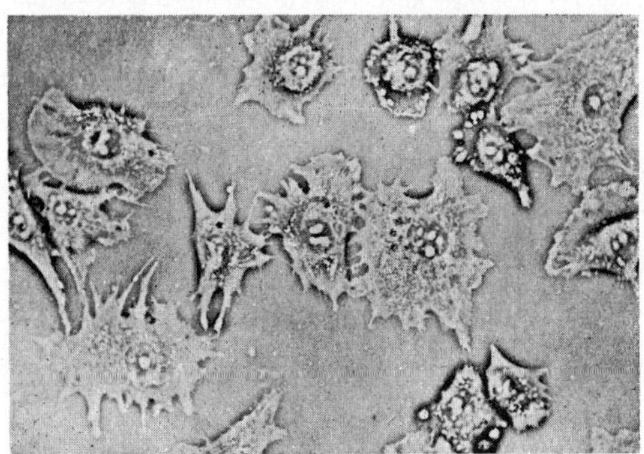

Fig. 2. Living "*L*" cells before exposure to cytochalasin *B*. (Phase contrast × 350.)

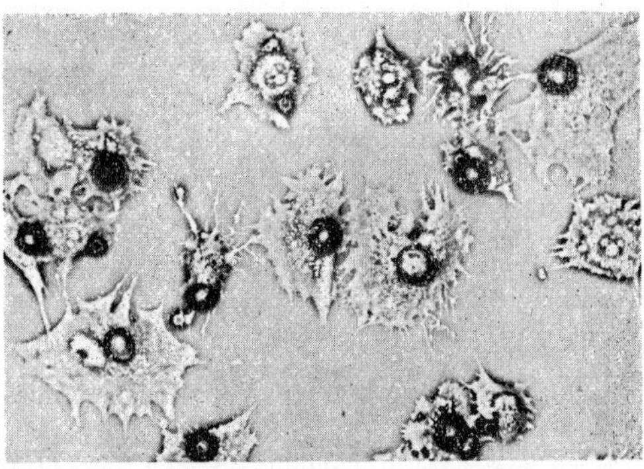

Fig. 3. The same field as Fig. 2 showing nuclear extrusion 1 h after exposure to cytochalasin *B* at 10 μg/ml. (Phase contrast × 350.)

until the nucleus appears to be entirely outside the cell, but still attached by a delicate cytoplasmic thread. At this stage the process can be reversed by washing with fresh medium. The nucleus is drawn back into the cell and is functionally unimpaired. If the nucleus is allowed to remain fully extruded for several hours, irreversible damage may be done and the nucleus becomes pycnotic.

The thin thread connecting nucleus and cytoplasm becomes progressively longer and more attenuated. It can easily be severed with a micromanipulator needle. Spontaneous enucleation also occurs, and many cells can be found without nuclei after exposure to the compound for 24 h. The mechanism later proposed for nuclear extrusion does not account for the complete expulsion of the nucleus. When this occurs it may be dependent on loss of nuclear viability or accidental rupture of the connecting thread. Cells which have lost their nuclei in this way have been found to be viable 24 h later, but no detailed investigation of their ultimate fate has been attempted. Such cells are capable of limited movement if placed in normal medium.

Mechanism of action. Cells which have been exposed to cytochalasin *B* at 0·5 µg/ml. are flatter than untreated cells, and each cell occupies a proportionately larger area of glass surface. The cell outline tends to be relatively simple and approximates to a circular shape. This appearance suggests that adhesion between the cell and the glass may be increased. This could indicate that the compound is adsorbed at the cell/glass interface, lowering the interfacial tension between them. Cell adhesion to glass may be regarded as a three way competition for mutual contact between cell, glass and culture medium. It should be kept in mind, therefore, that a reduction in cell/medium boundary tension, or an increase in glass/medium boundary tension, could have the same effect. These possibilities are not mutually exclusive.

Cells which have been dosed with cytochalasin *B* in suspension adhere to a glass surface very much more rapidly than untreated cells. After attachment, however, their rate of spreading is less than that of control cultures. This could mean that adsorption of the compound on the cell surface also increases its surface viscosity.

Cytochalasins *C* and *D* completely prevent cell cleavage at 0·05 µg/ml. A mould filtrate at present under investigation appears to contain an even more active metabolite. Crude extracts of this filtrate have shown cytochalasin-like activity at a dilution of 1/20,000,000. The high potency of these compounds is associated with very rapid action and ready reversibility. Such features fit well with the suggestion that they act as highly specific surface active agents which are preferentially adsorbed on biological membranes.

Several theories have been put forward to account for cytoplasmic cleavage (see review by Wolpert[4]). Whatever the underlying mechanism, it seems likely that if a cleaving cell is attached to glass, the process cannot be properly completed unless the cell membrane is detached from the glass surface along the line of the cleavage furrow. If cytochalasins increase the adhesion between cell and glass, they could slow down or prevent this detachment. The development of a cleavage furrow might also be retarded by an increase in surface viscosity produced by adsorption of cytochalasin. The cleavage mechanism is closely involved with the presence of the mitotic apparatus. The orientation and position of this apparatus determine the subsequent position of the line of cleavage. In order to prevent cleavage, therefore, it may only be necessary to *slow down* the process of furrowing. The mitotic apparatus is dismantled shortly after nuclear division, and if furrowing is not completed by this time, the relevant mechanism may no longer operate, and the furrow will retreat.

The mechanism of cell motility is poorly understood and many theories have been advanced. The theory of cell movement I have previously put forward[5,6] is particularly amenable to the interpretation of cytochalasin effects in terms of surface activity. According to this theory, cell movement is basically the result of cell extension by "passive" spread over a surface which can be wetted by the cell. On a heterogeneous surface, the cell moves in a direction which represents an increase in cell to substrate adhesion. This can be shown experimentally using surfaces of graded adhesiveness. Cells move up such an adhesion gradient with improved efficiency and in a highly directional manner. I have suggested the name "haptotaxis" to refer to cell movement in the direction of increased cell to substrate adhesion. Haptotaxis is proposed as the key principle in determining the motile behaviour of tissue cells. The specific application of this idea to the interpretation of contact inhibition, cancer invasion and chemotaxis has previously been discussed[5]. Cell motility on a homogeneous surface can also be interpreted as a result of haptotactic movement. In this case the difference in substrate adhesiveness which determines cell movement is considered to be created by the cell itself. The leading edge of a cell moving on glass is breaking new ground on a clean area of substrate. This could allow the leading edge to maintain a permanent advantage over the trailing edge where the glass surface may be rendered less adhesive through contamination by previous contact with the cell. The theory may be applied to an idealized cell of circular shape moving on a uniform flat surface. The theory then requires that the boundary tension forces tending to move the cell are proportional to its circumference, whereas the frictional forces tending to resist movement are proportional to its area. Area increases as the square of the circumference, so if cell to substrate adhesion is progressively increased, a point will be reached at which the sum of the restraining forces exceeds the forces tending to move the cell. At this point the cell can no longer move. Such a mechanism may underlie the effect of cytochalasin *B* on cell motility. It is interesting to note that "epithelial" cells in culture—usually so called because they tend to be flattened and relatively well spread on glass—are relatively immobile. Cultured cells which normally show rapid movement on glass sometimes produce polyploid giant cells. These cells tend to occupy a disproportionately large area of glass and are generally incapable of movement.

Cytochalasins also inhibit the ruffling movements at the margins of certain motile cells. I have interpreted ruffle formation as the result of a local retreat of the cell margin, causing a piling up of cell material. This retreat is considered to occur in the face of advancing culture medium which is successfully competing for contact with the glass at these points. An increase in the adhesion of the cell to glass due to the adsorption of cytochalasin at this interface would be expected to inhibit ruffle formation if this interpretation of ruffling is correct. If cytochalasins have an effect on surface viscosity as previously suggested, this could also be expected to impede cell movement and ruffle formation.

The explanation for nuclear extrusion follows a similar pattern. In this case, however, it is necessary to postulate adsorption of cytochalasin at an internal cell interface. This might account for the fact that nuclear extrusion requires a higher concentration of compound than the other effects. If cytochalasin is adsorbed on the internal surface of the cell membrane, or on the external surface of the nuclear membrane, it might lower the boundary tension between them. Increased adhesion between these structures would increase the area of any region of chance contact, and tend to make the cell membrane wrap itself closely round the nucleus until it was virtually excluded from the cell. Although it could be almost completely pinched off in this way, the nucleus would remain attached so long as the cell membrane remained intact. The essentially competitive nature of adhesion was emphasized in relation to the adhesion of cells to glass in the presence of culture medium. Similarly, adhesion between nucleus and cell membrane cannot be fully treated without reference to cytoplasmic fluid acting as a third phase. Again it should be made clear that adhesion between nucleus and cell membrane could involve surface effects at more than one of the three interfaces concerned.

A curious and puzzling observation was made at an early stage in these investigations. Nuclear extrusion occurred more frequently in cells cultured in perfusion chambers than when they were grown in Petri dishes. The significant difference proved to be that cells in the perfusion chambers were attached to the under surface of the cover glass. It seemed unlikely that gravity could play a significant part in nuclear extrusion, as gravitational forces are exceedingly small at the cellular level. The mechanism proposed for nuclear extrusion provides a ready explanation. When cells are growing on the upper surface of a cover glass, the nucleus may fall under gravity within the cell and make contact with the lower cell membrane. In this region the cell membrane is not free to wrap round the nucleus because it is more strongly adherent to the glass substrate.

Potential value of cytochalasins.   The cytochalasins produce a remarkable range of interesting biological effects. By interfering with specific cell activities such as cytoplasmic cleavage and cell movement, they should prove useful as research tools for investigating these important aspects of cell biology. Their potency, reversibility, and lack of general toxicity are particularly valuable features in this connexion. Their action in preventing cytoplasmic cleavage makes it possible to produce polyploid cells at will. This could be of value, for example, in determining the influence of different, known degrees of polyploidy on specific aspects of cell behaviour

and function. The possible significance of apparent mitotic synchrony in multinucleated cells without nuclear doubling has already been mentioned. The elucidation of this problem could have important consequences for the physiology of mitosis. The use of these compounds for the non-surgical removal of cell nuclei seems to offer new opportunities for investigating the functional relationship of nucleus and cytoplasm. Complete enucleation only occurs in a proportion of cells, but refinements in technique may make it possible to enucleate entire cell populations. When fully extruded, the connecting thread between nucleus and cytoplasm is extremely tenuous. This must restrict any functional interchange between them and may prevent it altogether. If so, it may not be necessary to enucleate the cell completely in order to study some aspects of the interdependence of nucleus and cytoplasm. Because nuclear extrusion is reversible, short-term experiments of this kind may have the additional advantage of allowing the restoration of the nucleus at a later stage.

Many experimental uses for these compounds can be envisaged. Whether or not their usefulness is likely to extend beyond the laboratory will only become apparent when they have been investigated more widely.

I thank Miss Elaine Bowker for technical assistance.

Received November 28, 1966.

[1] Aldridge, D. C., Armstrong, J. J., Speake, R. N., and Turner, W. B., *Chem. Commun.* (in the press).
[2] Eagle, H., *Science*, **130**, 432 (1959).
[3] Carter, S. B., *Exp. Cell Res.*, **42**, 395 (1966).
[4] Wolpert, L., *Intern. Rev. Cytol.*, **10**, 163 (1960).
[5] Carter, S. B., *Nature*, **208**, 1183 (1965).
[6] Carter, S. B., *Nature* (p. 256 of this issue).

23

Reprinted from *Natl. Acad. Sci. (USA) Proc.* **71**:1681–1685 (1974)

Cytoplasmic Inheritance of Chloramphenicol Resistance in Mouse Tissue Culture Cells

(enucleation/5-bromodeoxyuridine resistance/cell fusion/mitochondria/somatic cell genetics)

C. L. BUNN, DOUGLAS C. WALLACE, AND J. M. EISENSTADT

Departments of Human Genetics and Microbiology, Yale University, School of Medicine, 310 Cedar Street, New Haven, Connecticut 06510

Communicated by Edward A. Adelberg, January 28, 1974

ABSTRACT   A chloramphenicol-resistant mutant, isolated from mouse A9 cells, was enucleated and fused with a nucleated chloramphenicol-sensitive mouse cell line. Resultant fusion products, cytoplasmic hybrids (or "cybrids"), were selected as resistant to chloramphenicol, and had the nuclear markers and chromosome complement of the chloramphenicol-sensitive parent. These cybrids appeared at the high frequency of 2–8 per 10^4 cells plated. Neither parent produced any colonies when plated under identical selective conditions. Fusion between enucleated chloramphenicol-sensitive cell fragments and the chloramphenicol-sensitive cell produced no resistant colonies, suggesting that chloramphenicol resistance is not due to an increase in the ratio of cytoplasm to nucleus. Furthermore, fusions between resistant and sensitive nucleated cells produced resistant hybrids at a frequency 100 times less than that of resistant cybrids. Thus, these stable chloramphenicol-resistant cybrids result from the fusion of a chloramphenicol-resistant cytoplasm with a chloramphenicol-sensitive cell. It is proposed, therefore, that chloramphenicol resistance is a cytoplasmically inherited characteristic in this mouse cell line.

It has been established that mitochondria contain, in addition to their own DNA, the biochemical apparatus for translation and transcription of this genetic information. Studies on the genetics of mitochondria are most advanced in yeast, where several antibiotic resistance and respiratory deficiency mutations are known to be coded in mitochondrial DNA (mt-DNA) and cytoplasmic genetic recombination has been demonstrated (1, 2). Mammalian cell mtDNA is smaller in size than yeast mtDNA. Hybridization studies have shown that mammalian mtDNA codes for ribosomal RNA and 12 distinct 4S RNAs (3).

A genetic approach to mtDNA function in mammalian cells requires the isolation of mutants with altered mitochondrial properties, and the demonstration of the cytoplasmic inheritance of such properties. Chloramphenicol (CAP) inhibits mitochondrial protein synthesis in human HeLa cells (4), and this laboratory has recently described a HeLa mutant whose mitochondrial protein synthesis is resistant to CAP (5, 6). Further, CAP resistance in yeast is coded by mtDNA (7). However, no method has been described as yet to demonstrate cytoplasmic inheritance in mammalian cells.

A CAP-resistant mutant of the mouse line A9 has been isolated in this laboratory in a manner similar to that for the HeLa CAP-resistant mutant. This paper describes the experimental evidence that CAP resistance in mouse cells is cytoplasmically inherited.

MATERIALS AND METHODS

Strains and Culture Conditions. Strains A9 and LMTK⁻ are subclones of mouse L-cells, a line of aneuploid fibroblasts. Strain A9 is deficient in hypoxanthine phosphoribosyltransferase (HPRT; EC 2.4.2.8) activity and resistant to 8-azaguanine; LMTK⁻ is deficient in thymidine kinase (TK; EC 2.7.1.75) activity and resistant to 5-bromodeoxyuridine (BrdU) (8, 9). Both lines are sensitive to 50 μg/ml of CAP and do not grow in hypoxanthine–aminopterin–thymidine (HAT) medium (10).

Strain 501-1 is a CAP-resistant mutant subclone isolated from A9 by Spolsky in this laboratory. Cultures of A9 inoculated at 2×10^6 cells per flask (75 cm²) were treated with 2.5 μM ethidium bromide for 18 hr. The cells were washed, fed with fresh medium, and incubated for 8 hr, when 50 μg of CAP per ml of medium was added to each flask. Mutant colonies appeared in $2\frac{1}{2}$ months, and were then cloned. CAP resistance is a stable characteristic of the mutant 501-1.

Growth media, cloning conditions, and the methods of determining growth curves and sensitivity curves have all been described by Spolsky and Eisenstadt (5). A CAP-sensitive strain in the presence of CAP undergoes approximately four cell divisions before growth ceases. In such cases the mean cell division time is reported.

Enucleation. Strains 501-1 and A9 were enucleated by a modification of the technique described by Croce and Koprowski (11). Cells were treated with 20 μg of cytochalasin B per ml of medium for 3 hr at 37°, and then centrifuged in the same medium at 10,000 rpm in an SW27 rotor for 45 min. The percentage of enucleated cells in the resultant cell population was estimated by staining an aliquot of the cell suspension with lactoacetic-orcein (12) and counting at least 200 cells.

We have chosen to use the nomenclature suggested by Shay *et al.* (13) in referring to the resultant enucleated cell fragments as *cytoplasts* and the nuclear fragments as *karyoplasts*.

Cell Fusion. Five million cells of each parent were mixed in suspension with 1000 HAU of β-propiolactone-inactivated Sendai virus (pH 7.8) and incubated at 4° for 10 min. Nonselective medium was added and the cells were incubated for 30 min at 37°, and then distributed at various concentrations into flasks containing selective media.

The product of the fusion of a cytoplast and a cell will be referred to as a *cytoplasmic hybrid* (*cybrid*) as opposed to *hybrid*, which refers to the fusion of two nucleated cells.

Abbreviations: TK, thymidine kinase; BrdU, 5-bromodeoxyuridine; HPRT, hypoxanthine phosphoribosyltransferase; CAP, chloramphenicol; HAT, hypoxanthine–aminopterin–thymidine; mtDNA, mitochondrial DNA.

168

TABLE 1. *Transfer of chloramphenicol resistance by cytoplasts*

Cells	No. of cells plated	Average no. of colonies per flask*	Colonies per 10^6 cells
en501-1 × LMTK⁻	2×10^6	Confluent (2)	
	1×10^6	Confluent (2)	
	5×10^5	Confluent (3)	
	2.5×10^5	70 (3)	280
	1×10^5	18 (3)	180
	8×10^4	14 (1)	175
	5×10^4	5 (2)	100
en501-1	2×10^6	0	0
LMTK⁻	2×10^6	0	0

Cells were fused and plated in selective medium containing BrdU (30 μg/ml of medium) and CAP (50 μg/ml of medium) as described in *Methods*. The prefix "en" denotes a culture previously treated with cytochalasin B to induce enucleation (see *Methods*). This treatment produced 93% enucleation of 501-1. Cells (5×10^6) of LMTK⁻ and 5×10^6 cells of the enucleated preparation of 501-1 were fused. Colonies growing on flasks in the selective medium were counted when clearly visible (see *text*).
* Numbers in parentheses represent the number of flasks counted at that particular cell number.

Chromosome Analysis. Cells were prepared for chromosome examination by treatment in late exponential phase with colchicine. They were then swollen by incubation for 15 min in hypotonic buffer, fixed with methanol–glacial acetic acid (3:1), and stained with lacto-acetic orcein.

RESULTS

Transfer of Chloramphenicol Resistance by Cytoplasts. In order to demonstrate cytoplasmic inheritance of any characteristic, one must be able to distinguish between nuclear gene inheritance and the postulated cytoplasmic gene inheritance. One way of achieving this distinction in mammalian cells is to separate physically the nucleus from the cytoplasm, and to demonstrate the transfer and stable inheritance of a given characteristic by the cytoplasm in the absence of the nucleus. This has recently become possible following the demonstration of cytochalasin B-induced enucleation in mammalian tissue culture cells (11, 13, 14). Such an experimental system would entail the enucleation of the postulated cytoplasmic mutant and its Sendai virus-induced fusion with a nucleated "wild-type" cell, followed by growth under conditions that select for the product of fusion between mutant cytoplast (enucleated cell) and wild-type cell. These growth conditions should also select against both parents, nuclear–nuclear hybrids, and any mutant cells not enucleated by cytochalasin B treatment.

Accordingly, the CAP-resistant mutant, 501-1, was enucleated as described in *Methods* and fused with the CAP-sensitive LMTK⁻. The fusion mixture was then maintained in the presence of 30 μg of BrdU per ml of medium to select against nucleated 501-1 cells and any hybrids resulting from the fusion of nucleated 501-1 and LMTK⁻. Fifty micrograms of CAP per ml of medium was also added to select against LMTK⁻. Hence, the only cells capable of growing in this selective medium would be BrdU-resistant and CAP-resistant, and would presumably result from LMTK⁻ cells having received CAP resistance from the 501-1 cytoplast.

Two fusion experiments between enucleated 501-1 (en-501-1) and LMTK⁻ were performed, and the results of one are shown in Table 1. BrdU-resistant, CAP-resistant colonies appeared in flasks in both experiments at different but high frequencies relative to normal mutation frequencies. Visible colonies appeared in flasks 9 days after fusion, and were counted after 12 days. The number of such colonies was approximately proportional to the number of cells inoculated into each flask (Table 1).

No colonies appeared in any flasks containing one or the other parent under identical conditions of Sendai virus treatment and BrdU and CAP selection medium. The en501-1 cells in the presence of BrdU become large, flat, and gray within 3 days and continue growing at a decreasing rate. If

TABLE 2. *Comparison of transfer of chloramphenicol resistance in cell–cell and cell–cytoplast fusions*

Fusion/parents	% Enucleation (501-1 or A9)	Ratio of parents	Selective medium	No. of cells plated $\times 10^6$ (0 time)	No. of colonies per 10^6 cells (12–14 days)
en501-1 × LMTK⁻	93	1:1	BrdU + CAP	9.7	184
en501-1 × LMTK⁻	57	1.25:1	BrdU + CAP	3.5	832
enA9 × LMTK⁻	84	1:1	BrdU + CAP	8.6	0
501-1 × LMTK⁻	0	1:1	BrdU + CAP	8.7	0*
501-1 × LMTK⁻	0	0.07:1	BrdU + CAP	4.8	0
501-1 × LMTK⁻	0	1:1	HAT + CAP	9.0	176
en501-1	93	—	BrdU + CAP	2.0	0
en501-1	57	—	BrdU + CAP	3.6	0
501-1	0	—	BrdU + CAP	4.4	0
501-1	0	—	HAT + CAP	5.0	0
enA9	84	—	BrdU + CAP	4.4	0
LMTK⁻	0	—	BrdU + CAP	8.7	0
LMTK⁻	0	—	HAT + CAP	5.0	0

The methods of enucleation, cell fusion, and selective techniques were as described for Table 1 and in *Methods*.
* Two colonies appeared after prolonged incubation (see *text*).

169

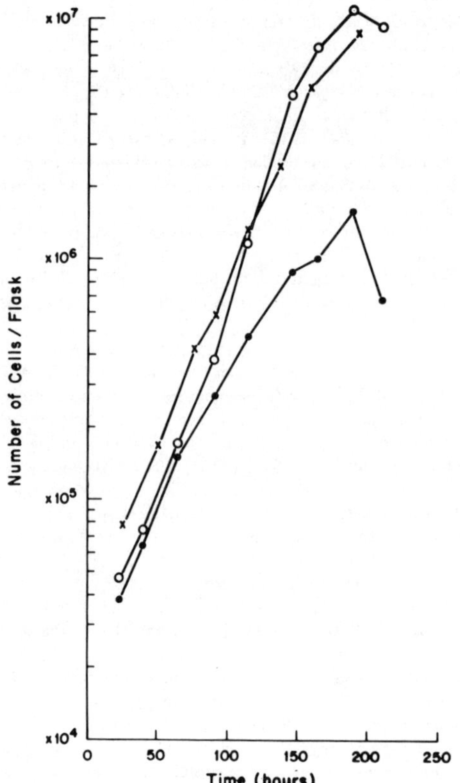

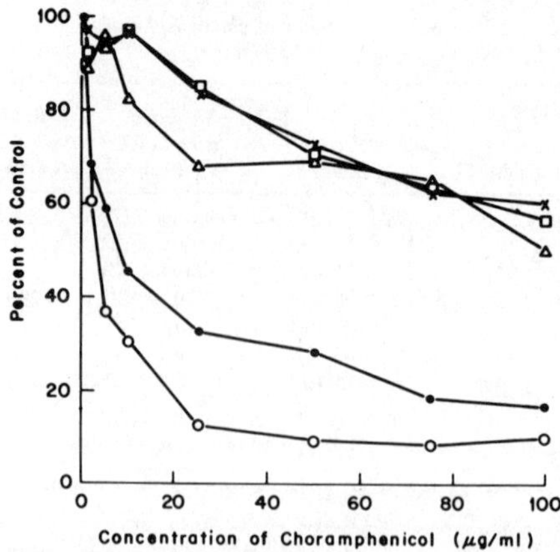

Fig. 2. Sensitivity of LEA clones, 501-1, A9, and LMTK⁻, to chloramphenicol. 5 × 10⁴ cells of each culture were inoculated into flasks. Twenty-four hours later CAP was added in the concentrations indicated to duplicate flasks. Six days later, the cells were harvested and counted as described in *Methods*. Clones LEA-10a and LEA-2a were derived from the fusion of an enucleated culture of 501-1 and LMTK⁻ as described in *Results*. The data are reported as percentages of the number of cells in flasks in the absence of CAP. (×) LEA-10a; (□) LEA-2a; (△) 501-1; (●) LMTK⁻; (○) A9.

Fig. 1. Chloramphenicol resistance of a LEA clone. Cells (5 × 10⁴) of LEA clone 2a, derived from an en501-1 × LMTK⁻ fusion and LMTK⁻, were inoculated into flasks, and cell counts were taken daily as described in *Methods*. (●) LMTK⁻ in the presence of 50 μg of CAP per ml of medium; (○) LMTK⁻ in the absence of CAP; (×) LEA-2a in the presence of 50 μg of CAP per ml of medium.

such cells reach confluence, they stop growing and die slowly. Otherwise, they become vacuolated and die within 2 weeks. In the experiment described in Table 1, there were occasionally some nonviable large, flat, vacuolated 501-1 cells remaining attached to the flask. Strain LMTK⁻ in the presence of CAP undergoes approximately four successively slower cell divisions, becoming granular and vacuolated in appearance, and then dies. These LMTK⁻ cultures have no viable cells after 10 days in the presence of CAP. The absence of any colonies when parents alone are plated demonstrates that these BrdU-resistant, CAP-resistant colonies have not appeared as a result of mutations of LMTK⁻ cells to CAP resistance, nor of 501-1 cells to BrdU resistance.

The results of further control experiments are shown in Table 2. The fusion of the enucleated CAP-sensitive strain, A9, and LMTK⁻ failed to produce any CAP-resistant clones (Table 2, line 3). Hence CAP resistance depends upon the presence of the enucleated CAP-resistant mutant, and is not due merely to an increase in the amount of cytoplasm relative to the nucleus in the "cybrid".

A crucial question was whether CAP resistance might have been transferred to LMTK⁻ by fusion with some of the remaining nucleated 501-1 cells. To test this possibility, fusions were performed under identical BrdU and CAP selection conditions with nucleated 501-1 and LMTK⁻ cells. These cells were combined in the ratios of 1:1 and 0.07:1, the latter ratio designed to mimic the 7% nucleated 501-1 cells present in the enucleated 501-1 cell preparation used in Table 1. Only two colonies in 10⁶ cells resulted from the 1:1 parent fusion, and none from the 0.07:1 fusion (Table 2, lines 4 and 5). These two colonies appeared after prolonged incubation in selective medium, one after 18 days and the other after 26 days. Both these colonies contained cells resembling LMTK⁻ in morphology. They may have arisen by mutation of LMTK⁻, or from the fusion of a G₁ or S phase 501-1 nucleated cell with a mitotic LMTK⁻ cell, resulting in premature chromosome condensation. Premature chromosome condensation has been described in a variety of animal species (15, 16), and might lead to loss of 501-1 chromosomes as observed for HeLa, Chinese hamster, and chick cells (16, 17). However, either event would be expected to occur at low frequency and could not account for the results presented in Table 1, where BrdU-resistant, CAP-resistant colonies appeared at least 100-fold more frequently. Colonies of large, gray, flat cells also appeared in these nucleated fusion experiments after 14 days, presumably containing nuclear complements of both LMTK⁻ and 501-1, but all of these died by 26 days.

CAP-resistant binucleate cell hybrids were specifically selected in HAT medium after fusion of 501-1 and LMTK⁻ for comparison with the CAP-resistant cybrid clones. The selection of HAT-resistant hybrids from the fusion of H-PRT⁻ and TK⁻ cells has been described by Littlefield (10). These hybrids appeared at a frequency of approximately 2 per 10⁴ cells plated (Table 2, line 6), a rate similar to that of

TABLE 3. *Properties of parents, hybrids, and cybrids*

Strains	Cell division time (hr)*		Chloramphenicol sensitivity† (μg/ml)	Mean‡ chromosome no.	5-Bromo-§ deoxyuridine	8-Azaguanine¶
	−CAP	+CAP				
501-1	19.6	21.4	100	47.2 (3.8, 21)‡	S	R
LMTK⁻	19.2	40.8	8	44.6 (3.8, 33)	R	S
LEA 1a	N.T.	19.4	100	46.6 (5.1, 11)	R	S
LEA 2a	N.T.	19.4	>100	48.1 (3.3, 24)	R	S
LEA 10a	N.T.	19.4	>100	47.6 (2.2, 22)	R	S
ALM 1	N.T.	N.T.	>50	91.2 (14.1, 15)	S	S
ALM 2	N.T.	N.T.	>50	97.3 (10.9, 10)	S	S
ALM 10	N.T.	N.T.	>50	97.8 (15.0, 14)	S	S

LEA cybrid clones were obtained from the fusion of enucleated 501-1 cells with LMTK⁻ cells followed by selection in BrdU and CAP (see *Results*); ALM clones were similarly derived from 501-1 × LMTK⁻ fusions and selection in HAT + CAP medium.

* Cell division time was determined from growth curves of strains in the presence and absence of 50 μg/ml of CAP (see *Methods*). LMTK⁻ does not maintain growth in 50 μg/ml of CAP.

† CAP sensitivity is defined as that concentration of drug which limits cell growth in 6 days to 50% of the control level in the absence of CAP (see *Methods*).

‡ Chromosome staining and counting were performed as described in *Methods*. Numbers in *parentheses* after each mean chromosome number are the standard deviations and number of cells counted, respectively. The following anomalous counts were omitted from calculations of means and standard deviations: 84 from LEA 1a; 78, 97, and 98 from LEA 10a; and 37 from ALM 10. The high chromosome numbers may be due to aberrant chromosome duplication without cell division. The low number would presumably be a nonviable cell.

§ Sensitivity (S) to 5-bromodeoxyuridine is defined as the inability of cells to grow continuously in the presence of 30 μg of drug per ml of medium for 14 days. Resistant (R) cells are unaffected by this concentration of BrdU.

¶ Sensitivity (S) to 8-azaguanine is defined as the inability of cells to grow continuously in the presence of 12 μg of drug per ml of medium for 14 days. Resistant (R) cells are unaffected by this concentration of 8-azaguanine.

N.T. indicates not tested.

cybrid colonies (Table 1, and line 1, Table 2). Further characterization of these hybrids will be described in Table 3. No colonies appeared in parental cultures under any of the selective conditions described in Table 2.

Cybrids from the fusion of enucleated 501-1 (derived from strain A9) and LMTK⁻ (derived from strain L) are referred to as LEA strains (L × enucleated A9). Hybrids from nucleated 501-1 × LMTK⁻ fusions are referred to as ALM strains (A9 × LM, both nucleated). Sixteen LEA strains, one from each flask, were then cloned in soft agar in the presence of 50 μg of CAP per ml of medium.

Analysis of the Products of Cell–Cell and Cell–Cytoplast Fusions. If LEA cells are the product of the fusion between LMTK⁻ and a CAP-resistant cytoplasm of 501-1, they should have the following characteristics: (*1*) CAP resistance similar to 501-1; (*2*) a chromosome complement similar to LMTK⁻; (*3*) the BrdU resistance and 8-azaguanine sensitivity of LMTK⁻; and (*4*) a morphology similar to LMTK⁻. The following experiments indicate that cloned LEA cells do indeed possess all these characteristics.

Fig. 1 illustrates that the growth rate of a representative LEA clone, LEA-2a, in the presence of 50 μg/ml of CAP is almost identical with that of LMTK⁻ in the absence of CAP. By contrast, LMTK⁻ in the presence of CAP undergoes four successively slower cell divisions and then dies. Fourteen other LEA clones, and mixed cultures from three flasks each containing more than 50 BrdU and CAP-resistant colonies (Table 1), have growth characteristics identical to LEA-2a in the presence of CAP.

The resistance of LEA clones 2a and 10a to a range of CAP concentrations (0–100 μg/ml of medium), as indicated by extent of growth after 6 days, was compared to that of 501-1, A9, and LMTK⁻. The results are shown in Fig. 2. A third LEA clone gave identical results to LEA-2a and LEA-10a. The LEA clones are clearly as resistant as 501-1 over a wide range of CAP concentrations, and in fact show a slightly increased degree of resistance in the range 10–50 μg of CAP per ml. This may be due to the presence of BrdU (30 μg/ml) in the medium in which LEA clones are routinely maintained. Other experiments in this laboratory suggest that BrdU may slightly increase antibiotic resistance in TK⁻ strains.

Exposure of LMTK⁻ or A9 cells to each level of CAP for periods longer than 6 days would cause total cell death. Hence, the curve in Fig. 2 does not indicate that 17% of LMTK⁻ cells or 10% of A9 cells are CAP-resistant. Longer exposure of LEA clones and 501-1 to 75 or 100 μg of CAP per ml may also cause inhibition greater than that shown in Fig. 2. However, LEA clones and 501-1 can be grown indefinitely in 50 μg of CAP per ml, and show no decrease in viability or alteration in division time.

The CAP sensitivities of LEA clones, ALM strains, and the parent strains 501-1 and LMTK⁻ are compared in Table 3, in terms of cell division time and the CAP concentration that will inhibit to 50% of the control level in 6 days. Clearly, LEA clones are as CAP-resistant as 501-1.

Table 3 also shows that the LEA clones possess a mean chromosome number similar to that of LMTK⁻ and 501-1. This is, however, about half the number of chromosomes of the CAP-resistant binucleate ALM strains that were selected from 501-1 × LMTK⁻ fusions in HAT and CAP. Therefore, LEA clones have not arisen from the fusion of two nucleated cells without chromosome loss. Nor are LEA clones the result of a simple mutation to TK⁻ in 501-1, cells as they are sensitive to 8-azaguanine while 501-1 cells are resistant (Table 3). Finally all LEA clones had morphology similar to LMTK⁻, which was quite different from either 501-1 cells or ALM cells (not shown).

DISCUSSION

The isolation and characterization of CAP-resistant cells resulting from the fusion between enucleated CAP-resistant mouse cells and nucleated, CAP-sensitive mouse cells has been described. These fusion products, or cybrids, appear at high frequency and are stable with respect to CAP resistance. They resemble the nucleated parent cell in terms of chromosome complement and nuclear markers and the enucleated parent cell in terms of CAP resistance. No CAP- and BrdU-resistant cells were obtained from either parent under identical selection conditions. Hence, such cybrid cells do not arise from mutations in either parent. Furthermore, control fusions between enucleated and nucleated CAP-sensitive cells produced no CAP-resistant cells. Fusions between nucleated, CAP-resistant and CAP-sensitive cells under the same selective conditions produced CAP-resistant hybrids at a frequency 100 times lower than the rate of appearance of CAP-resistant cybrids. Consequently, the appearance of these CAP-resistant cybrids at such frequencies could not be explained by the fusion of two nucleated cells, nor by the increase in the amount of cytoplasm per nucleus.

Similar results showing the transfer of CAP resistance have also been obtained in this laboratory with the CAP-resistant human HeLa mutant described by Spolsky and Eisenstadt (5).

Thus, the genetic information for CAP resistance resides in the cytoplasm of mouse cells. This information may be encoded in mtDNA, or possibly in one of the other types of cytoplasmic DNAs reported in mammalian cells. These are spcDNA (18), microsome-associated DNA (19, 20), informational or I-DNA (21), and membrane-associated cmDNA (22). All are suggested to be of nuclear origin. If these classes of cytoplasmic DNA are dependent on the nucleus for function and replication, it is unlikely that they are capable of transferring permanent CAP resistance in the absence of the nucleus.

CAP resistance has been shown to be expressed at the level of mitochondrial protein synthesis in HeLa cells (6). Preliminary experiments indicate that this is also true for the CAP-resistant mouse mutant 501-1 described in this paper. Mitochondrial DNA has been shown to code for CAP resistance in yeast (8). It appears likely that the mutation to CAP resistance in mouse cells may also have occurred in mtDNA.

The mechanism of the transfer process at the cellular or molecular level is unclear, as is the fate of the CAP-sensitive cytoplasm. Fusions between human and mouse cells have shown that human mtDNA is lost along with human chromosomes (23, 24), but may be retained when reverse segregation occurs, that is, when mouse chromosomes are eliminated (25). No selective pressure for either species of mitochondria was used in these fusions (23–25). It is now possible to select for the retention of human mitochondria in interspecific fusions where human chromosomes are lost.

We thank Dr. Spolsky for providing the mouse mutant line, and A. Eisenstadt, K. Beattie, and D. Molony for their technical assistance. This work was supported in part by USPHS Research Grants GM-18186, GM-1948, and GM-20124 (to J.M.E.), and USPHS Training Grant GM-00275 (to D.C.W.).

1. Linnane, A. W., Haslam, J. M., Lukins, H. B. & Nagley, P. (1972) *Annu. Rev. Microbiol.* **26**, 163–198.
2. Borst, P. (1972) *Annu. Rev. Biochem.* **41**, 333–376.
3. Wu, M. Davidson, N., Attardi, G. & Aloni, Y. (1972) *J. Mol. Biol.* **71**, 81–93.
4. Firkin, F. C. & Linnane, A. W. (1968) *Biochem. Biophys. Res. Commun.* **32**, 398–402.
5. Spolsky, C. M. & Eisenstadt, J. M. (1972) *FEBS Lett.* **25**, 319–324.
6. Kislev, N., Spolsky, C. M. & Eisenstadt, J. M. (1973) *J. Cell Biol.* **57**, 571–579.
7. Coen, D., Deutsch, J., Netter, P., Petrochilo, E. & Slonimski, P. P. (1970) *Soc. Exp. Biol. Symp.* **24**, 449–496.
8. Littlefield, J. W. (1963) *Proc. Nat. Acad. Sci. USA* **50**, 568–576.
9. Kit, S., Dubbs, D. R., Piekarski, L. J. & Hsu, T. C. (1963) *Exp. Cell Res.* **31**, 297–312.
10. Littlefield, J. W. (1964) *Science* **145**, 709–710.
11. Croce, C. M. & Koprowski, J. (1973) *Virology* **51**, 227–229.
12. Welshons, W. J., Gibson, B. H. & Scandlyn, B. J. (1962) *Stain Tech.* **37**, 1–5.
13. Shay, J. W., Porter, K. R. & Prescott, D. M. (1973) *J. Cell Biol.* **59**, Abstr. 623.
14. Carter, S. B. (1967) *Nature* **213**, 261–264.
15. Johnson, R. T., Rao, P. N. & Hughes, H. D. (1970) *J. Cell Physiol.* **76**, 151–158.
16. Rao, P. N. & Johnson, R. T. (1972) *J. Cell. Sci.* **10**, 495–513.
17. Schwartz, A. G., Cook, P. R. & Harris, H. (1972) *Nature New Biol.* **230**, 5–8.
18. Smith, C. A. & Vinograd, J. (1972) *J. Mol. Biol.* **69**, 163–178.
19. Schneider, W. C. & Kuff, E. L. (1969) *J. Biol. Chem.* **244**, 4843–4851.
20. Bond, H. E., Cooper, J. A., Courington, D. P. & Wood, J. S. (1969) *Science* **165**, 705–706.
21. Bell, E. (1969) *Nature* **224**, 326–328.
22. Meinke, W., Hall, M. R., Goldstein, D. A., Kohne, D. E. & Lerner, R. A. (1973) *J. Mol. Biol.* **78**, 43–56.
23. Clayton, D. A., Teplitz, R. L., Nabholz, M., Dovey, H. & Bodmer, W. (1971) *Nature* **234**, 560–562.
24. Attardi, B. & Attardi, G. (1972) *Proc. Nat. Acad. Sci. USA* **69**, 129–133.
25. Coon, H. G., Horak, I. & Dawid, I. B. (1973) *J. Mol. Biol.* **81**, 285–298.

Preparation of microcells by enucleation of micronucleate cells

T. EGE and N. R. RINGERTZ, *Institute for Medical Cell Research and Genetics, Karolinska Institutet, Medical Nobel Institute, S-104 01 Stockholm 60, Sweden*

Much of the usefulness of the cell hybridization technique in chromosome mapping and the analysis of gene regulation derives from the fact that interspecific hybrids, i.e. hybrids formed by the fusion of cells from two different species, undergo chromosome segregation and preferentially lose chromosomes from one of the two parental cells (for reviews see [1, 2, 3]). With closely related species the chromosome segregation, however, is very slow and unpredictable. As an alternative approach to chromosome segregation we have explored the possibility of introducing at the time of cell fusion only a portion of a diploid genome equivalent to a few chromosomes into another cell by means of Sendai virus-induced cell fusion [4]. The technique developed is based on the preparation of new types of cells, which we have named *microcells*. These cells were produced by first inducing micronucleation in cells adhering to plastic discs and then drawing out the micronuclei by centrifugation in the presence of cytochalasin B. We present here a characterization of microcells which demonstrates, in particular, that they are suitable for somatic cell hybridization in being able to undergo Sendai virus-induced fusion with other cells. The routine used for preparing microcells from rat L6 myoblasts [5] was as follows:

Material and Methods

L6 cells (2×10^6) were inoculated into plastic dishes (90 mm) containing plastic discs ($\varnothing$ 25 mm) which had been punched out of the bottom of Falcon tissue culture dishes. After 12 h of growth in Eagle's MEM supplemented with 10 % calf serum in an atmosphere of 5 % CO_2 at 37°C, colchicine was added to give a final conc. of 2 μg/ml. The cultures were then incubated for another 48 h in order to allow micronuclei to develop. Among the interphase cells present on the plastic discs at the end of the colchicine treatment >80 % contained 2–30 micronuclei of varying size (fig. 1 b), whereas the rest of the cells contained a single nucleus. Enucleation of micronucleated cells was carried out in two stages. First, loosely attached cells were removed by centrifuging the preparations in round-bottomed plastic tubes containing 3–4 ml phosphate-buffered saline (PBS) supplemented with 10 % calf serum. The discs were positioned with the cell side facing the bottom of the centrifuge tube and a cylindrical plastic plug which exactly fitted into the tube was added on top of the plastic slide in order to fix its position. Centrifugation was carried out at 10 000 rpm (12 000 g) for 10 min in a Sorvall centrifuge (SS 34 rotor) prewarmed to 37°C. In order to enucleate the preparations were transferred to new centrifuge tubes containing PBS, 10 % calf serum and 10 μg/ml of cytochalasin and centrifuged for 20 min at 18 000 rpm (39 000 g) at 37°C.

This resulted in enucleation of 90–100 % of the L6 cells. After enucleation, the plastic plugs and discs were removed. The pellet material was stirred up in the medium remaining in the tubes and the resulting suspension was centrifuged in small plastic centrifuge tubes (250 g for 5 min at 22°C). The supernatant was then removed and the pellet washed once in Eagle's MEM without serum and stored at $+4$°C until used. For fusion experiments the washed pellet was suspended in Earle's salt solution. The number of intact and damaged cells was examined by diluting the pellet suspension 1 : 1 in a 0.4 % trypan blue solution in PBS and then scoring the percentage of stained (damaged) and non-stained cells. For microscopy and cytochemistry samples of the pellet suspension were transferred to glass slides where the cells were allowed to attach for 30 min at 22°C. Alternatively, smears were prepared. The preparations were then washed in saline and fixed in ethanol/acetone (1 : 1) for at least 30 min before staining with May-Grünwald-Giemsa. Other preparations were stained with acridine orange for microfluorimetric measurements or DNA at 530 nm [6], Feulgen for microspectrophotometric

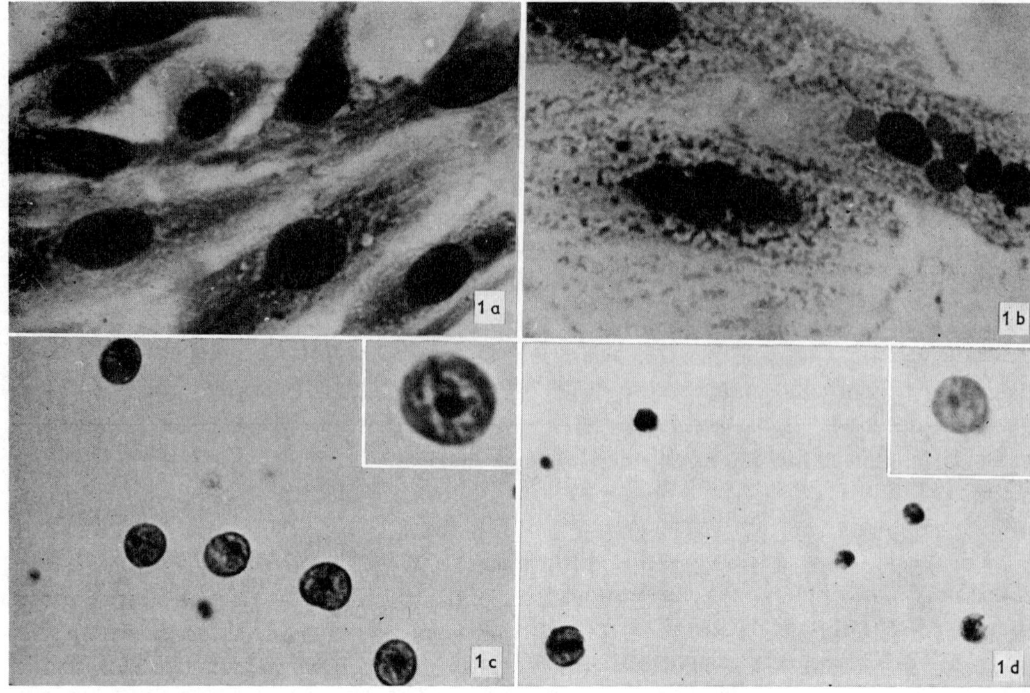

Fig. 1. (*a*) Intact L6 rat myoblasts; (*b*) micronucleated L6 myoblasts; (*c*) minicells obtained from normal L6 cells; (*d*) microcells obtained by enucleating micronucleated L6 cells. Insets show a minicell (*c*) and a microcell (*d*) at a higher magnification.

DNA determination or a combination of Feulgen and Naphthol Yellow for DNA and protein determination [7, 8].

Results and Discussion

Data reported in the preceding paper [9] show that enucleation of normal mononucleate L6 cells results in pellet preparations containing *minicells*. The minicells (fig. 1*c*) were found to have an intact nucleus but had lost 80–90 % of the original cytoplasm leaving only a narrow rim of cytoplasm between the plasma membrane and the nuclear membrane [9]. In the present study it was found that enucleation of micronucleate cells resulted in the formation of even smaller, cell-like structures (fig. 1*d*) which we will refer to as *microcells*. Each of these cells contained a micronucleus which gave a

positive Feulgen reaction. The mean nuclear diameter, DNA content and amount of cell protein is indicated in table 1 which also shows similar data for normal cells and minicells. In acridine orange-stained preparations the nuclei of the microcells were seen to contain a dispersed, interphase type of chromatin which gave a yellowish green fluorescence. Occasionally small red nucleoli were seen. The nuclei were surrounded by a small amount of red cytoplasm. The observation that 80–90 % of the unfixed microcells excluded trypan blue suggested that, like the minicells [9], the microcells were surrounded by an intact plasma membrane. Measurements of the nuclear diameter of minicells prepared from normal cells and of microcells prepared from micronucleate cells showed that although the majority of the

Table 1. *Properties of microcells, minicells and intact L6 cells*

	Nuclear area (μm^2)		DNA^a (arbitrary units)		Total protein^b (arbitrary units)		Protein^a/DNA^b relative ratios (arbitrary units)	
	$\bar{x}\pm$S.E.M.	n	$\bar{x}\pm$S.E.M.	n	$\bar{x}\pm$S.E.M.	n	$\bar{x}$	n
Intact cells	—	—	319 ± 24	62	$1\,398\pm69$	62	4.8	62
Minicells	171 ± 6	50	377 ± 14	61	415 ± 22	61	1.1	61
Microcells	70 ± 7	50	153 ± 13	69	267 ± 27	69	1.9	69
Microcells (corrected for max. minicell contamination)	48 ± 3	40	111 ± 8	57	187 ± 15	57	1.9	57

^{a} DNA determinations based on Feulgen microspectrophotometry.
^{b} Total quantity of Naphthol Yellow staining cell protein.

nucleate structures in the microcell preparation had nuclei smaller than those of G1 minicells, there were also some cells which

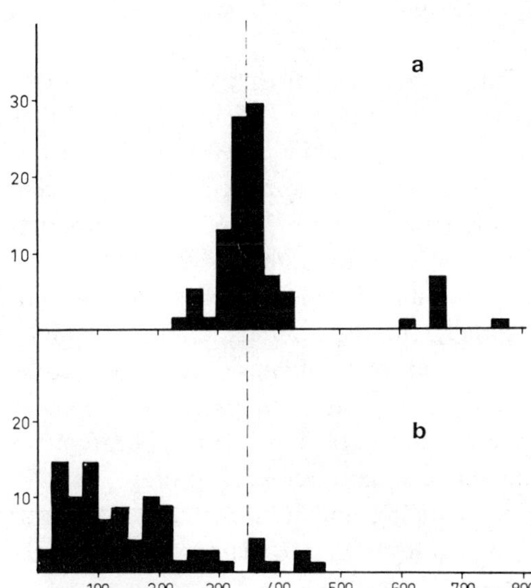

Fig. 2. *Abscissa:* DNA content in arbitrary Feulgen units; *ordinate:* % of cells.
Microspectrophotometric DNA measurements of Feulgen-stained minicells (*a*) obtained by enucleation of normal L6 cells and microcells (*b*) obtained by enucleation of L6 populations, where approx. 70 % of the cells contained 2–30 micronuclei. The microcells have DNA contents below the normal G1 value (---).

had nuclei of the same size as those of normal minicells. Presumably these cells were minicells derived from mononucleate cells which had failed to undergo micronucleation. As with the minicell preparations [9], there were also some cytoplasmic fragments lacking nuclei.

Cytochemical DNA determinations by acridine orange microfluorimetry and Feulgen microspectrophotometry showed that the microcells contained DNA quantities below the normal G1 level (fig. 2). The smallest of the micronuclei in the microcells contained DNA quantities equivalent to approx. 1/30 of the G1 level of the intact nuclei. This would be equivalent to 1–2 chromosomes since the L6 myoblasts have a near diploid chromosome number of 40–41 chromosomes (Zech, personal communication).

In order to assess the practical usefulness of the microcells in cell fusion experiments, attempts were made to fuse microcells with intact mononucleate L6 cells. In order to facilitate recognition of the micronuclei the microcells were prepared from L6 cells prelabelled for 48 h with ^{3}H-thymidine. For cell fusion a suspension of labelled microcells

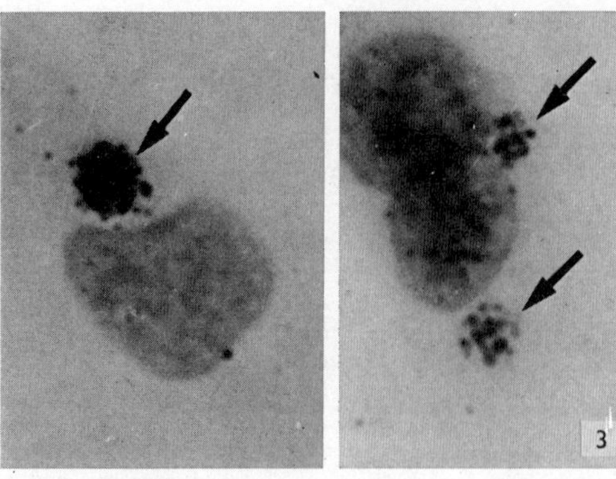

Fig. 3. Autoradiograms showing ³H-thymidine labelled L6 micronuclei (*arrows*) introduced into unlabelled L 6 cells by Sendai virus induced cell fusion.

(also containing some minicells, see fig. 2) and UV-inactivated Sendai virus was added to a monolayer of unlabelled L6 cells attached to a plastic disc (technique, see [10]). In order to confine the microcell preparation to a small area on the disc a plastic ring was mounted on the monolayer before the microcells were added. This ring was removed after incubation with the Sendai virus. The addition of Sendai virus caused agglutination and fusion of both microcells and minicells to the monolayer. Unfused micro- and minicells were removed by twice washing the preparation with tissue culture medium (without serum), the first time 1 h after fusion, the second immediately before fixation. Preparations fixed 20 h after fusion contained a large number of cells with one or more micronuclei in addition to the main nucleus (fig. 3). The micronuclei had dispersed chromatin, were found in the same optical plane as the main nucleus, and appeared to be well integrated into the cytoplasm. There were also some pycnotic micronuclei with condensed chromatin. These nuclei were surrounded by an empty zone as if they had been phagocytosed. Autoradiography showed that practically all micronuclei observed were ³H-

thymidine labelled and thus originated from the microcells. No properly integrated micronuclei were seen if virus was omitted from the fusion mixture.

The results obtained show that by combining the ability of mitotic inhibitors to cause micronucleation [11] and the enucleation technique developed by Prescott et al. [12], it is possible to generate a new type of cell, the microcell, which contains only a small part of a normal diploid genome. Furthermore, most of these microcells appear to be surrounded by an intact plasma membrane since they fail to stain with trypan blue and apparently carry receptors for Sendai virus on their surface. In this respect the microcell resembles the minicell obtained by enucleating mononucleate near-diploid L6 cells (see ref. [9]). These structures as well as the small cytoplasmic fragments extracted from the cytoplasm during enucleation are surrounded by a plasma membrane clearly visible in the electron microscope.

The demonstration that microcells can be fused with normal cells indicates that they offer a new possibility of introducing a few chromosomes of one cell into another cell. The long-term fate of these fusion products is presently being examined. Whether or not the introduced chromosomes become integrated into the host cell's mitotic spindle and carried into daughter cells remains to be established. If this is the case, the microcells have a number of important applications in somatic cell genetics and virology, e.g. in gene mapping and in the study of tumour viruses integrated on mammalian chromosomes.

The authors wish to thank U Krondahl, G. Blomgren, G. Jacobson and M. Lundström for valuable assistance and Dr G. Auer and Dr A. Zetterberg for help with the Naphthol Yellow staining method.

The investigation was supported by grants from the Swedish Cancer Society.

References

1. Ephrussi, B, Hybridization of somatic cells. Princeton Univ Press (1972).
2. Ruddle, F H, Adv human genetics 30 (1972) 173.
3. Ringertz, N R & Savage, R E, Cell hybrids. Academic Press (1975).
4. Harris, H & Watkins, J F, Nature 205 (1965) 640.
5. Yaffe, D, Proc natl acad sci US 61 (1968) 474.
6. Rigler, R, Acta physiol Scand 67 (1966) suppl. 267.
7. Deitsch, A, Lab invest 4 (1955) 324.
8. Gaub, J, Auer, G & Zetterberg A. In preparation.
9. Ege, T, Hamberg, H, Krondahl, U, Ericsson, J & Ringertz, N R, Exptl cell res 87 (1974) 365.
10. Carlsson, S-A, Savage, R E & Ringertz, N R, Nature 228 (1970) 869.
11. Phillips, S G & Phillips, D M, J cell biol 40 (1969) 248.
12. Prescott, D M, Myerson, D & Wallace, J, Exptl cell res 71 (1972) 480.

Part II

GENE TRANSFER

Editor's Comments
on Papers 25 Through 29

25   **SZYBALSKA and SZYBALSKI**
*Genetics of Human Cell Lines, IV. DNA-Mediated Heritable
Transformation of a Biochemical Trait*

26   **MUNYON et al.**
*Transfer of Thymidine Kinase to Thymidine Kinaseless L Cells
by Infection with Ultraviolet-Irradiated Herpes Simplex Virus*

27   **WIGLER et al.**
*Transfer of Purified Herpes Virus Thymidine Kinase Gene to
Cultured Mouse Cells*

28   **WIGLER et al.**
*Transformation of Mammalian Cells with Genes from
Procaryotes and Eucaryotes*

29   **CAPECCHI**
*High Efficiency Transformation by Direct Microinjection of
DNA into Cultured Mammalian Cells*

DNA-MEDIATED TRANSFORMATION

Much important work in microbial genetics, and, indeed, the
demonstration that DNA is the genetic material, was based upon
DNA-mediated transformation. (Unless otherwise indicated, the term
transformation will be used in this section to mean the introduction
of foreign DNA into cells and the acquisition of heritable characteristics by the recipient cells. Unfortunately, *transformation* also has
been used to refer to changes in the growth characteristics and
malignant potential of cultured cells.) In light of the obvious potential
of gene transfer systems for studies with higher organisms, it is not
surprising that the earliest attempts to develop mammalian transformation systems date back more than two decades. These attempts
appeared feasible, as mammalian cells were known to take up nucleic
acids and to be susceptible to infection by naked viral nucleic acids.
However, early attempts at transforming mammalian cells did not

provide convincing results, at least in part because of the absence of appropriate selectable markers.

The first apparently successful demonstration of DNA-mediated transformation of mammalian cells was provided by Szybalska and Szybalski in 1962 (Paper 25). These experiments utilized the then recently developed HAT selective system (see also Papers 34 and 36), and the selectable marker was the expression of inosinic acid pyrophosphorylase (now known as hypoxanthine-guanine phosphoribosyltransferase, HGPRT). HGPRT deficient human cells (see Paper 35) were incubated with DNA isolated from HGPRT positive human cells, and colonies were selected in HAT medium, in which only HGPRT positive cells should survive. The cells selected in HAT medium expressed HGPRT activity, and they appeared to be transformants. The development of HGPRT positive cells was DNA dependent, and the frequency of cells surviving in HAT medium was proportional to the concentration of exogenous DNA. However, because cells of the same species were used as both donor and recipient, it was not possible to prove that the HGPRT activity in the presumptive transformants was due to the expression of transferred genetic information. In addition, the results were very difficult to reproduce. Nevertheless, these experiments stimulated much activity on the transformation of mammalian cells. Progress was not rapid, and only a decade later were reliable transformation systems for mammalian cells developed.

The foundation for much of the recent work on DNA-mediated transformation was established by Munyon and co-workers in 1971 (Paper 26). Rather than cellular DNA as the transforming vector, a DNA virus, herpes simplex virus (HSV), was used as the transforming agent. HSV carries a gene for thymidine kinase (TK), an enzyme that is normally expressed by cultured cells. However, the form of TK produced by the virus is easily distinguishable from the form produced by mammalian cells. TK deficient mouse cells were treated with HSV (ultraviolet irradiated to prevent lytic infection), and TK positive colonies were selected in HAT medium (see Paper 3). The identification of the type of TK produced by these cells was of critical importance, and it was shown that the TK in the selected cells corresponded to the viral (donor) and not the cellular (recipient) form. This study thus provided the first unequivocal demonstration of the genetic tranformation of an enzymatic trait in mammalian cells. The majority of gene transfer studies performed since that time have involved the HSV-TK system.

The occurrence of genetic transformation in the HSV-TK system was confirmed by Davidson and co-workers (Davidson et al., 1973). In addition, these studies demonstrated the instability of expression of the transferred genetic information in transformed cells. Such instability has since been observed as a common feature of gene transfer systems. In the HSV-transformed cells, the expression of the viral TK gene could be alternately suppressed and reactivated at very high frequencies (greater than 10%). The viral TK gene in the transformants could also enter a latent state in which it was retained in the cells for long periods in the absence of expression and from which it could be reactivated only at very low frequencies (approximately 0.0001%). Recent evidence has suggested that changes in viral TK expression in the transformed cells may involve alterations in the methylation of the viral TK gene (Clough et al., 1982).

Two technical developments that proved to be very important for the transformation of mammalian cells by isolated DNA were provided by Graham and co-workers in studies with adenoviruses. The naked DNA of adenovirus 5 is able to infect human cells, and it is also capable of causing morphological transformation. The infectivity of the viral DNA was found to be greatly increased by coprecipitating the DNA with calcium phosphate (Graham and van der Eb, 1973). In subsequent studies, the viral DNA was cleaved with restriction endonucleases to generate a variety of specific fragments, and the fragments were tested in a bioassay to identify which fragments had transforming activity. When untransformed cells were exposed to purified DNA fragments (coprecipated with calcium phosphate), a specific fragment of viral DNA capable of causing morphological transformation was identified (Graham et al., 1975).

This approach of combining restriction endonuclease digestion with a bioassay for the transforming activity of purified DNA fragments was used by Wigler, Silverstein, Axel, and co-workers (Paper 27) to isolate a small fragment of the HSV-1 genome carrying the TK gene. HSV DNA was cleaved with a variety of restriction endonucleases, and purified DNA fragments (coprecipated with calcium phosphate) were tested for the ability to transfer TK activity to TK deficient mouse cells. The results identified a 3.4 kilobase Bam fragment that was capable of efficiently transforming TK deficient cells. Thus, the viral TK gene was localized to a small fragment of the HSV-1 genome. Similar results were obtained in studies with HSV-2 (Maitland and McDougall, 1977). The purified HSV-TK gene has become an important vector for transferring other genes as well.

The approach taken in the studies to purify the HSV-TK gene has been used to isolate a number of cellular genes that specify enzymes conferring the ability to grow under selective conditions. Similarly, cellular oncogenes, which cause malignant transformation, were isolated by Weinberg and co-workers (Shih et al., 1979) and by Cooper and co-workers (Cooper et al., 1980). In these experiments, the selected trait was the altered growth characteristics of the transformants.

Although the transformation bioassay method made it possible to transfer and purify a variety of mammalian cell genes, this approach was limited to genes that confer selectable properties. The range of genes that could be used in gene transfer studies was greatly expanded when Wigler, Silverstein, Axel, and co-workers (Paper 28) developed a co-transformation system in which nonselectable markers also are transferred. In these studies, TK deficient cells were exposed to a DNA mixture containing the HSV-TK gene as well as another well-defined DNA sequence, for example, a cloned globin gene. TK positive transformants were selected and tested for the nonselected marker, and it was found that the majority of the TK transformants also had stably incorporated the nonselected DNA sequence. Ligation of the nonselected DNA sequence to the TK gene was not required for co-transformation to occur. These studies provided a method by which essentially any gene could be introduced into cultured cells. The isolation of transformants carrying the nonselected marker did not depend on a selective advantage conferred by the gene, nor even on the expression of the gene in the transformants. Subsequent studies in several laboratories showed that genes introduced by co-transformation could be transcribed in the recipient cells.

The majority of the co-transformation studies to date have utilized the HSV-TK gene as the selectable marker. However, any selectable marker can be used in a co-transformation system. Selectable markers that do not require the introduction of a mutation into the recipient cells prior to selection would be especially useful for co-transformation studies, as any cell could then serve as the recipient. One such marker appears to be the bacterial gene for xanthine-guanine phosphoribosyl transferase (Mulligan and Berg, 1981). Selection for this enzyme is possible even in the presence of the cellular HGPRT activity.

Most studies on DNA-mediated transformation in mammalian cells have used the calcium phosphate precipitation technique to introduce DNA into cells. However, it is possible to microinject DNA directly into the nucleus of cultured cells, using glass micropipettes (see Diacumakos et al., 1970). Recent studies by Capecchi (Paper 29) and by Anderson, Diacumakos, and co-workers (Anderson et al., 1980) have shown that microinjection of DNA can result in high efficiency transformation of mammalian cells. In the studies by Capecchi, the HSV-TK gene incorporated into an E coli plasmid was microinjected into the nucleus of TK deficient mouse cells. At least

50% of the microinjected cells transiently expressed TK activity. The frequency of stable transformants, however, was much lower, being on the order of one transformant per 500–1000 cells microinjected. The frequency of transformation was increased greatly when specific SV40 DNA sequences were introduced into the plasmid with the HSV-TK gene. When these plasmids were used, the frequency of stable TK transformants was approximately 20% of the cells microinjected. With transformation frequencies of this order of magnitude, it should be possible to isolate transformants without any selection pressure. Thus, any gene conceivably could be transferred into any cell.

High efficiency transformation also was observed in the experiments of Anderson et al. (1980). In these experiments, TK deficient mouse cells were microinjected with a mixture of two recombinant plasmids, one containing the HSV-TK gene and the other containing the human β globin gene. Approximately 5% of the cells microinjected gave rise to stable TK transformants (selected in HAT medium). All the transformants selected for TK expression were found to contain human β globin sequences, for which no selective pressure was applied. Thus, co-transformation occurs with microinjection of DNA, as well as with exposure of cells to calcium phosphate precipitated DNA (see Paper 28).

REFERENCES

Anderson, W. F., L. Killos, L. Sanders-Haigh, P. J. Kretschmer, and E. G. Diacumakos, 1980, Replication and Expression of Thymidine Kinase and Human Globin Genes Microinjected into Mouse Fibroblasts, *Natl. Acad. Sci. (USA) Proc.* **77:**5399–5403.

Clough, D. W., L. M. Kunkel, and R. L. Davidson, 1982, 5-Azacytidine-Induced Reactivation of a Herpes Simplex Thymidine Kinase Gene, *Science* **216:**70–73.

Cooper, G. M., S. Okenquist, and L. Silverman, 1980, Transforming Activity of DNA of Chemically Transformed and Normal Cells, *Nature* **284:**418–421.

Davidson, R. L., S. J. Adelstein, and M. N. Oxman, 1973, Herpes Simplex Virus as a Source of Thymidine Kinase for Thymidine Kinase-Deficient Mouse Cells: Suppression and Reactivation of the Viral Enzyme, *Natl. Acad. Sci. (USA) Proc.* **70:**1912–1916.

Diacumakos, E. G., S. Holland, and P. Pecora, 1970, A Microsurgical Method for Human Cells In Vitro: Evolution and Applications, *Natl. Acad. Sci. (USA) Proc.* **65:**911–918.

Graham, F. L., and A. J. van der Eb, 1973, A New Technique for the Assay of Infectivity of Human Adenovirus 5 DNA, *Virology* **52:**456–467.

Graham, F. L., P. J. Abrahams, C. Mulder, H. L. Heijneker, S. O. Warnaar, F. A. de Vries, W. Fiers, and A. J. van der Eb, 1975, Studies on In

Vitro Transformation by DNA and DNA Fragments of Human Adeno-
viruses and Simian Virus 40, in *Tumor Viruses,* Cold Spring Harbor
Symposium on Quantitative Biology, vol. 39, Cold Spring Harbor
Laboratory, Cold Spring Harbor, N.Y., pp. 637–650.

Maitland, N. J., and J. K. McDougall, 1977, Biochemical Transformation
of Mouse Cells by Fragments of Herpes Simplex Virus DNA, *Cell*
11:233–241.

Mulligan, R. C., and P. Berg, 1981, Selection for Animal Cells that Express
the *Eschericha Coli* Gene Coding for Xanthine-Guanine Phosphoribosyltransferase, *Natl. Acad. Sci. (USA) Proc.* **78:**2072–2076.

Shih, C., B.-Z. Shilo, M. P. Goldfarb, A. Dannenberg, and R. A. Weinberg,
1979, Passage of Phenotypes of Chemically Transformed Cells via
Transfection of DNA and Chromatin, *Natl. Acad. Sci. (USA) Proc.*
76:5714–5718.

25

Reprinted from *Natl. Acad. Sci. (USA) Proc.* **48**:2026–2034 (1962)

GENETICS OF HUMAN CELL LINES, IV.   DNA-MEDIATED HERITABLE TRANSFORMATION OF A BIOCHEMICAL TRAIT

BY ELIZABETH HUNTER SZYBALSKA AND WACLAW SZYBALSKI

MCARDLE MEMORIAL LABORATORY, UNIVERSITY OF WISCONSIN

Communicated by R. Alexander Brink, October 15, 1962

The phenomenon of DNA-mediated genetic transformation,[1] first recorded for Pneumococcus,[2] and later extended to several other bacterial species and to infectious viral DNA was only recently demonstrated to occur in mammalian cells.[3, 4] That the phenomenon may be more generalized in mammalian systems is suggested by the related observations: the uptake of nucleic acids by mammalian cells,[5–15] and the infectivity of naked nucleic acids isolated from mammalian viruses.[16–18]

Numerous attempts have been made to demonstrate genetic transformation in mammalian cells,* but the systems employed apparently were not sufficiently selective.   In the majority of cases the experiments were performed partially or entirely

186

with intact animals, relying on morphological traits,[19] immunological characters,[20] or drug resistance[21] as markers. Success was claimed only for the last markers,[21] but the systems did not permit adequate substantiation. In an attempt to transform tissue culture cells to aminopterin resistance, no transformants were detected above the background of spontaneous mutants.[15] As for other systems in higher organisms, marginal mention should be made of the claimed genetic transformation of morphological characters in ducks, which still awaits substantiation.[22]

There exist also in the literature reports of various nonspecific actions of DNA on cells of higher organisms, including effects on resistance to viruses,[23] mutagenesis,[24] teratological changes,[25] cytological effects,[26] tumorigenesis,[27] and partial reversal of UV killing.[28] The synthesis of modified hemoglobins by immature human erythrocytes *in vitro* has been interpreted as a specific response to both DNA[29] and RNA.[30]

Reported interactions between mammalian cells include cell fusion[31] and formation of chromosomal hybrids by joint cultivation of two cell lines with morphological chromosomal markers.[32]

With the discovery of highly selective genetic markers in the human cell line D98S,[3, 4, 33] a means became available for detecting genetic transformants with a high degree of resolution, i.e., as few as one per 10^7 cells. A mutational system, which involves the selection of inosinic acid pyrophosphorylase (IMPPase)-positive cells from an IMPPase-negative population, was described earlier.[3] In preliminary experiments utilizing this selective system, genetic transformation was demonstrated for D98 cells.[3, 4] The present communication documents in detail and extends these observations.

Materials and Methods.—Cell lines and culture media: The D98S human cell line, the wild-type strain employed in these studies, was originally derived by single clone isolation from the Detroit 98 normal sternal bone marrow line.[34] The mutant lines, the isolation and properties of which were described in earlier publications,[3, 4, 33] originated as follows: D98/AG is an 8-azaguanine-resistant derivative of the D98S line; D98/AH and D98/AH-2 are 8-azahypoxanthine-resistant derivatives of the D98/AG line; D98/APt is an aminopterin-resistant derivative of the D98S line. HeLa cells and rat ML-2 cells were kindly provided by Dr. G. C. Mueller; Dr. L. Siminovitch contributed mouse embryo and L60 cells.

The standard culture medium was a modified Eagle's basal medium supplemented with 10 per cent horse serum (E_{90} medium), used in conjunction with a 5 per cent CO_2 atmosphere.[33]

Isolation of donor DNA: For the isolation of DNA, the donor cells were suspended in 10 volumes of standard saline-citrate (SSC) (0.15 M NaCl + 0.02 M Na$_3$·citrate), and lysed with 2% sodium lauryl sulfate. After adjusting the NaCl concentration to 1 M, exhaustive deproteinization was carried out by repeated shaking with a 4:1 chloroform-butanol mixture and centrifugation. The nucleic acids were precipitated by addition of two volumes of 95% ethanol, spooled on a glass rod, and redissolved in 5 ml of 0.015 M NaCl + 0.02 M Na$_3$·citrate, after which the NaCl concentration was readjusted to 0.15 M. Some preparations were rendered RNA-free by treatment for 2 hr (37°C) with 50 μg/ml RNase (heated to 100°C for 10 min), followed by deproteinization and two more precipitations with ethanol.

Procedure for transformation (adopted on the basis of experiments described in subsequent sections): For the preparation of the recipient cell suspension, 4- to 5-day old, heavily seeded cultures were grown up in 4-oz prescription bottles. The cell sheet was rinsed twice with balanced salt solution (BSS) (8.0 gm NaCl; 0.4 gm KCl; 0.35 gm NaHCO$_3$; 1.0 gm glucose; per liter of water) and exposed 4–6 min to 0.25% pancreatin in BSS. The cells were detached from the glass by knocking the bottle against a hard surface and suspended in 3 ml (per bottle) of phosphate-buffered saline (PBS) (7.0 gm NaCl; 0.4 gm KCl; 2.75 gm Na$_2$HPO$_4$; 0.25 gm NaH$_2$PO$_4$·H$_2$O; 1.0 gm glucose; per liter of water) supplemented with 50 μg/ml spermine·HCl. The cell concentration of the combined suspension was then adjusted to 500,000 cells/ml by diluting with PBS.

To 1.5 ml aliquots of the recipient cell suspension the appropriate amounts of DNA solution and/or PBS were added, bringing the total volume to 2.0 ml/tube. After a 15-min period (37°C), the contents of each tube were distributed between five 60 mm plastic petri dishes containing 5 ml of HAT medium (E_{90} medium + 5 μg/ml hypoxanthine, 0.1 μg/ml aminopterin, and 5 μg/ml thymidine). Following 12–14 days of incubation, with medium changes every 2 to 3 days, the plates were rinsed with PBS, and the colonies were fixed, stained, and counted.[33] Cell viability (plating efficiency) during the DNA treatment was determined at intervals by plating 0.02 ml samples of the reaction mixture in E_{90} medium and scoring the colonies after 7 days incubation.

Results and Discussion.—Conditions for transformation: In developing the standard procedure for assaying transformation in D98 cells, guidance was provided by the experience of other investigators in related fields, including bacterial transformation, and infectivity and uptake of nucleic acids in mammalian systems. Consideration was given to the following five aspects:

(1) *The selective system* employs the IMPPase-positive D98S or D98/AG line as DNA donor and the IMPPase-negative D98/AH-2 line as recipient.[3, 4] Highly selective conditions for the scoring of IMPPase-positive transformants are provided by a medium containing 0.1 μg/ml aminopterin, as inhibitor of *de novo* purine synthesis, 5 μg/ml thymidine, to counteract the accompanying block in thymidylate synthesis, and 5 μg/ml hypoxanthine, as sole purine source (HAT medium). Under these conditions the IMPPase-positive donors are able to form colonies with normal plating efficiency and growth rate, whereas the IMPPase-negative recipient is completely inhibited, being unable to utilize hypoxanthine even at 100 times higher concentrations. Another feature of this selective system, contributing to its high resolution, is the nonmeasurable rate of spontaneous mutation (less than 10^{-7}) towards capacity to grow on the selective HAT medium exhibited by the receptor strain, D98/AH-2, which was chosen on this basis.

(2) *The isolation of transforming DNA* was conducted under conditions similar to those routinely applied in this laboratory for the preparation of *Bacillus subtilis* DNA of high specific transforming activity. The procedure is outined in *Materials and Methods.*

(3) *The competence of the recipient cells*, as measured by the proportion transformed (transformant frequency) at saturating DNA concentrations, was reasonably constant under the conditions outlined in *Materials and Methods.* Therefore, no systematic evaluation of the effect of physiological state on competence was carried out, especially since the routinely prepared recipient cells satisfied the requirement for high plating efficiency.

(4) *The exposure of recipient cells to donor DNA* was carried out under conditions designed to be optimum for both cell survival and DNA uptake. The medium satisfied the following requirements: (*a*) freedom from serum, a source of nucleases; (*b*) physiological osmotic environment and effective buffering capacity, as provided by PBS; (*c*) the presence of a polybasic component, known to enhance the cellular uptake and inhibit the enzymatic degradation of nucleic acids.[9, 14, 35] Both glass-attached and suspended cells were employed as recipients, but since treatment of cells in suspension proved more practical and permitted greater accuracy it became the standard procedure.

During periods of exposure not exceeding 30 min, the survival of the recipient cells suspended in PBS was 90 to 100%. Therefore, protective additives, such as glycerol, were found unnecessary. When DNA was present in concentrations exceeding 10 µg/ml, viable cell losses of 10 to 15% at 15 min and 20 to 30% at 30 min were encountered. Toxic effects of DNA have been reported by others.[24, 36] The optimum pH range was determined to be 7.0 to 7.5. Below pH 6.5 the transformant yield was sharply reduced.

The addition of 50 µg/ml spermine·HCl was found not to have any adverse effect on cell viability and appeared to be essential to the transformation process (see Table 1). Protamine was tried as an alternative, but at the concentrations em-

TABLE 1

DNA-MEDIATED GENETIC TRANSFORMATION IN D98 CELLS: TRANSFER OF CAPACITY TO UTILIZE HYPOXANTHINE FROM IMPPASE-POSITIVE DONORS TO IMPPASE-NEGATIVE D98/AH-2 RECIPIENT CELLS

DNA donor	IMPPase activity of donor	Treatment of donor DNA	No. of Transformants/ml* Donor DNA concentration (µg/ml)			
			0	1	10	100
D98S	+	...	0	1	32	128
D98/AG	+	...	0	2	19	62
D98/AH	−	...	0	..	0	0
D98/AH-2	−	...	0	..	0	0
D98/AG	+	RNase†	..	..	..	65
D98/AG	+	DNase‡	..	..	..	0
D98/AG	+	spermine omitted	..	..	0	0

* Assayed under standard conditions, as described in *Materials and Methods*, in presence of 50 µg/ml spermine·HCl except where indicated.
† Treated with heated (10 min, 100°C) RNase preparation (50 µg/ml) for 30 min at 37°C.
‡ Treated with DNase preparation (2 µg/ml) for 30 min at 37°C in presence 10^{-2} M Mg++.

ployed it caused precipitation of the DNA. Application of osmotic shock, for enhancing the uptake of the nucleic acid by the recipient cells,[8, 37] was found impractical, since exposure of D98 cells to 1 M NaCl even for brief periods proved detrimental to cell viability.

Recipient cells were allowed to react with donor DNA for 5–60 min prior to plating. The maximum ultimate yield of transformants was reached between 5 and 10 min; beyond that time no further increase occurred. It has been reported that the absorption of DNA by mammalian cells, as with bacteria, is a very rapid process.[7, 13]

(5) *The phenotypic expression period* in the transformation process was analyzed by plating and preincubating DNA-treated recipient cells for periods of up to 4 days in nonselective medium prior to imposition of the selective conditions. Since the yield of transformants was independent of the length of the preincubation, and since large populations of D98/AH-2 cells were observed to survive attached in HAT medium for 3–4 days, it was concluded that integration of the genetic determinant and its phenotypic expression was accomplished well within this period.

Specific role of donor DNA in transformation: The specific role of donor DNA in the transformation process was established by the following observations: (1) Transformants are produced only in response to a specific DNA, isolated from IMPPase-positive strains, while DNA isolated from IMPPase-negative cells is inactive (Tables 1 and 2). (2) DNase completely destroys the transforming activity, while RNase action is without effect (Table 1). (3) The yield of transformants is a

TABLE 2

CHARACTERISTICS OF TWO TRANSFORMANT LINES AS COMPARED WITH RECEPTOR STRAIN, DNA DONORS, AND SPONTANEOUS "REVERTANT"

	Recipient D98/AH-2	DNA donors		Spontaneous "revertant" D98/AH-R	Transformants	
		D98S	D98/AG		D98/AH-TS (Donor: D98S)	D98/AH-TAG (Donor: D98/AG)
Sensitivity to purine analogs:*						
AG	400	0.12	12	100	0.2	0.15
AH	1800	0.12	0.28	80	0.13	0.09
TG	12	0.006	0.12	8	0.01	0.008
Capacity to utilize hypoxanthine†	> 500	0.08	0.5	0.1	0.15	0.07
IMPPase activity‡	–	+++	+++	+	+++	...
Transforming activity of isolated DNA:§						
4 µg/ml	0	8	9	...	12	...
40 µg/ml	0	102	48	...	92	...

* Concentration (µg/ml) permitting 50% survival (colony formation); AG = 8-azaguanine, AH = 8-azahypoxanthine, TG = 6-thioguanine.

† Concentration of hypoxanthine (µg/ml) permitting 50% survival in presence of 0.02 µg/ml aminopterin and 5 µg/ml thymidine.

‡ Assayed with cell-free enzyme preparations in presence of hypoxanthine-C14 and PRPP.

§ Transformants/ml, assayed under standard conditions, as described in *Materials and Methods.*

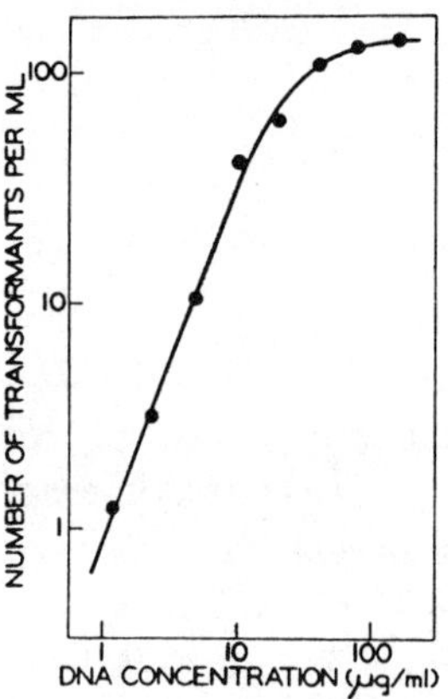

FIG. 1.—Number of IMPPase-positive transformants per ml of "reaction mixture," assayed in HAT medium, following treatment of IMPPase-negative D98/AH-2 recipient cells with increasing concentrations of DNA isolated from IMPPase-positive D98S donor cells, under conditions described in *Materials and Methods.*

function of the DNA concentration, as shown in Table 1 and Figure 1. The response is roughly linear up to about 20 µg DNA/ml, with the curve leveling off at a yield of approximately 150 transformants per ml (4 × 10⁻⁴ transformants per recipient cell). (4) DNA and transforming activity sediment in similar fashion in the CsCl equilibrium density gradient, as described in the following section.

Fractionation of transforming principle in the CsCl density gradient: In addition to providing further evidence that DNA and the transforming principle are indeed identical, it was hoped that CsCl gradient centrifugation[38] would permit separation of active fractions from the competing and inactive DNA. Since the specific transforming activity of D98 DNA is low, it was necessary to handle relatively larger amounts than in the analogous bacterial experiments.[39] On the other hand, overloading the centrifuge tube with DNA prevents sharp fractionation. The procedure outlined below is necessarily a compromise between the two requirements.

Approximately 50 µg of DNA in 3 ml of CsCl solution (density 1.69 gm/cm³), overlayered with 2 ml of paraffin oil, was spun for 72 hr in the S39L swinging bucket rotor of the Spinco Model L preparative centrifuge at 35,000 rpm. Five-drop fractions (approximately 50 μl) were collected as described by Szybalski.[40] OD_{260} was determined for each undiluted fraction, using 20 μl microcuvettes. The fractions were then dialyzed against SSC, since CsCl was found to be toxic in concentrations above 0.05 M. The transforming activities (number of transformations per fraction), assayed as described in *Materials and Methods*, are plotted in Figure 2,

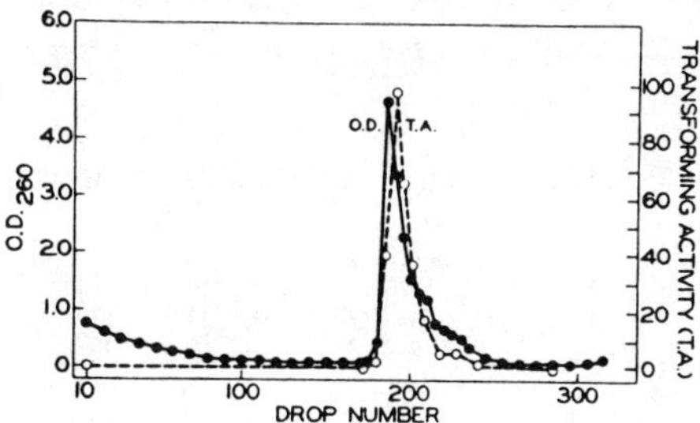

Fig. 2.—Distribution of D98S DNA (O.D.), measured as UV absorbance at 260 mμ and of its transforming activity (T.A.), expressed as the number of IMPPase-positive transformant colonies per fraction, among 50 μl fractions collected after CsCl equilibrium density gradient centrifugation (35,000 rpm; 20°C; 72 hr).

along with the OD_{260} readings. It is apparent that transforming activity sediments in the same buoyant density range as DNA. However, it bands more sharply than the total DNA, and the peak is shifted by one fraction from the peak OD_{260}, indicating partial purification of the active transforming material.

Properties of the transformants: Of several transformant clones isolated and characterized all closely resembled the D98S line, as regards their ability to utilize hypoxanthine and their sensitivity to various purine analogs (see Table 2). Enzymatic assay (kindly performed by Dr. R. W. Brockman and P. Stutts of the Southern Research Institute, Birmingham, Ala.) on one of the transformant lines, D98/AH-TS, revealed normal IMPPase activity, comparable to that of the donor lines, D98/AG and D98S. By the same test, employing C^{14}-labeled hypoxanthine or guanine as substrate in the presence of PRPP, no IMPPase activity could be detected in the receptor strain.[41] The aminopterin sensitivity of the transformants did not differ from that of the receptor and donor lines. During repeated subculture, the transformant lines bred true, with respect to the above properties, indicating the heritable nature of the change. DNA was isolated from the transformant line, D98/AH-TS, and its transforming activity proved to be comparable to that of the other IMPPase-positive donors, D98S and D98/AG (see Table 2).

The transformants could readily be distinguished from the spontaneously occurring "revertant," D98/AH-R, which utilizes hypoxanthine efficiently but shows intermediate degrees of resistance to hypoxanthine and guanine analogs and only a very low IMPPase activity (see Table 2). This "revertant," described earlier,[3, 4] can be isolated from D98/AH populations, although not from the D98/AH-2 line, employed as recipient in the transformation studies.

Transforming activity of heterologous DNA's and interference by nontransforming DNA's: The observation that IMPPase-positive transformants are produced only in response to DNA isolated from IMPPase-positive D98 donors lead to two further questions. (1) Would heterologous DNA's isolated from other human cell lines and from unrelated mammalian sources known to possess an active IMPPase be effective as the transforming agent; and (2) would genetically inactive DNA's interfere with the transformation process, in analogy with the interference phenomenon observed in bacterial transformation?

(1) Highly polymerized DNA was isolated from the HeLa cell line of human malignant origin, from the ML-2 rat cell line, and from mouse embryo and L60 cells. As can be seen in Table 3, HeLa DNA appeared to have a very low transforming activity, while the mouse and rat DNA's were inactive.

(2) The effect of nontransforming homologous (D98/AH) DNA and of heterologous human (HeLa) and mouse (L60) DNA on transformations initiated by D98S donor DNA at near saturation level was determined. As indicated in Table 3,

TABLE 3

TRANSFORMING ACTIVITY OF HETEROLOGOUS DNA'S AND INTERFERENCE BY NONTRANSFORMING DNA'S

D98S DNA concentration	—	No. of Transformants/ml*				Mouse L60 DNA	Mouse Embryo DNA	Rat ML-2 DNA
		D98/AH DNA		HeLa DNA				
		10 μg/ml	50 μg/ml	10 μg/ml	50 μg/ml	50 μg/ml	50 μg/ml	50 μg/ml
0 μg/ml	0	0	0	0	2	0	0	0
10 μg/ml	52	26	15	35	27	30	..	..

* Assayed under standard conditions, as described in *Materials and Methods*.

transformation was markedly inhibited by D98/AH DNA at a concentration equal to that of donor DNA, and the effect was magnified when the ratio of interfering DNA to donor DNA concentration was increased. Similar effects were observed for HeLa DNA and L60 DNA.

Attempted transformation in other systems: The experiments described above were based on selection of IMPPase-positive transformants from an IMPPase-negative recipient population in HAT medium (D98/AH → D98S). Other selective systems available for D98 cells include: (1) from sensitivity to 100-fold resistance to 8-azaguanine (D98S → D98/AG); and (2) from sensitivity to over 5,000-fold resistance to 8-azahypoxanthine (D98S, D98/AG, or D98/AH-TS → D98/AH).[3, 4, 33] The first system was handicapped by a high spontaneous background of resistant cells (10^{-3} to 10^{-4}) and was therefore abandoned. In the second system no transformations were detected, even though periods as long as 5 days were allowed to elapse between plating of the DNA treated cells and addition of the selective agent.

Attempted genetic recombination between intact cells: A preliminary test for genetic interaction between intact cells was carried out, employing two singly marked mutants, D98/AH, resistant to more than 1,000 μg/ml 8-azahypoxanthine, and D98/APt, resistant to 0.1 μg/ml aminopterin. A 4-oz prescription bottle was seeded with a mixed inoculum of approximately 10^6 D98/AH and D98/APt cells. This culture was carried in E_{90} medium for a total of 14 days with one transfer after 7 days. On the 7th and 14th days, 10^6 cells of this mixed culture were plated in a doubly selective medium containing 10 μg/ml 8-azahypoxanthine and 0.02 μg/ml aminopterin, concentrations completely inhibitory to lines D98/APt and D98/AH respectively. No recombinants were produced, since no colonies appeared in this medium.

Summary.—The treatment of the IMP-pyrophosphorylase (IMPPase)-deficient human cell line, D98/AH-2, with DNA isolated from IMPPase-positive D98S or D98/AG cells resulted in the appearance of IMPPase-positive genetically transformed cells, which were detectable under highly selective conditions. DNAse but not RNAse treatment abolished the transforming activity. The yield of transformants was a linear function of the transforming DNA concentration up to the level of about 20 μg DNA/ml, reaching a plateau of approximately 4×10^{-4} transformations per recipient cell at 150 μg DNA/ml. Transforming activity banded in the CsCl equilibrium gradient more sharply than the total DNA mass. DNA isolated from the D98/AH-2 recipient cells or from the cells of other mammals did not transform *per se*, but interfered with the active transformation process. No recombinations between intact cells were detected, using aminopterin and 8-azahypoxanthine resistance as selective markers.

We wish to express our gratitude to Mrs. Sandra Joannes Antoine for her able assistance with some phases of these experiments. This research was supported by grant no. CY-5215 from the National Cancer Institute.

The following abbreviations are used in this paper: DNA, deoxyribonucleic acid; RNA, ribonucleic acid; DNase, deoxyribonuclease; RNase, ribonuclease; IMP, inosinic acid; IMPPase, inosinic acid pyrophosphorylase; PRPP, α-5-phosphoribosyl-pyrophosphate; UV, ultraviolet light; OD_{260}, optical density at 260 mμ (1 cm light path).

* The term genetic transformation is used here with the same operational meaning implied as in bacterial genetics, i.e., a heritable change in the recipient cell produced by highly polymerized DNA extracted from donor cells bearing the new character. This phenomenon bears no obvious relationship to the term "transformation" which has been used widely in the tissue culture literature and has been applied to heritable changes in the growth pattern of fresh cell explants and similar changes caused by viral infection; cf. Swim, H. E., *Ann. Rev. Microbiol.*, **13**, 141 (1959), and Shein, H. M., and J. F. Enders, these PROCEEDINGS, **48**, 1164 (1962).

[1] Avery, O. T., C. M. MacLeod, and M. McCarty, *J. Exptl. Med.*, **79**, 137 (1944).

[2] Griffith, F., *J. Hyg.*, **27**, 113 (1928).

[3] Szybalski, W., E. H. Szybalska, and G. Ragni, National Cancer Institute Monograph No. 7, 75 (1962).

[4] Szybalski, W., and E. H. Szybalska, *Univ. of Mich. Med. Bull.*, **28**, 277 (1962).

[5] Hudnik-Plevnik, T. A., V. R. Glisin, and M. M. Simic, *Nature*, **184**, 1818 (1959); Kay, E. R. M., *Nature*, **191**, 387 (1961); Rieke, W. O., *J. Cell Biol.*, **13**, 205 (1962); Schmizu, T., S. Koyama, and M. Iwafuchi, *Biochim. Biophys. Acta*, **55**, 795 (1962).

[6] Gartler, S. M., *Nature*, **184**, 1505 (1959); *Biochem. and Biophys. Res. Comm.*, **3**, 127 (1960).

[7] Sirotnak, F. M., and D. J. Hutchison, *Biochim. Biophys. Acta* **36**, 246 (1959).

[8] Chorazy, M. R., H. H. Baldwin, and R. K. Boutwell, *Fed. Proc.*, **19**, 307 (1960); Chorazy, M. R., *Acta Union Internat. Contre le Cancer*, **18**, 244 (1962).

[9] Amos, H., *Biochem. Biophys. Res. Comm.*, **5**, 1 (1961).

[10] Bensch, K. G., and D. W. King, *Science*, **133**, 381 (1961).

[11] Borenfreund, E., and A. Bendich, *J. Biophys. and Biochem. Cytol.*, **9**, 81 (1961).

[12] Hill, M., *Nature*, **189**, 916 (1961); Hill, M., and J. Jakubickova, *Exp. Cell Research*, **26**, 541 (1962).

[13] Wilczok, T., *Neoplasma*, **8**, 5 (1961), and **9**, 369 (1962).

[14] Cocito, C., A. Prinzie, and P. DeSomer, *Experientia*, **18**, 218 (1962).

[15] Mathias, A. P., and G. A. Fischer, *Biochem. Pharmacol.*, **11**, 69 (1962).

[16] Herriot, R. M., *Science*, **134**, 256 (1961); Colter, J. S., and K. A. O. Ellem, *Fed. Proc.*, **20**, 650 (1961).

[17] DiMayorca, G. A., B. E. Eddy, S. E. Stewart, W. S. Hunter, C. Friend, and A. Bendich, these PROCEEDINGS, **45**, 1805 (1959); Weil, R., *Virology*, **14**, 46 (1961); Atanasiu, P., G. Orth, J. P. Rebiere, M. Boiron, and C. Paoletti, *C.R. Acad. Sci. (Paris)*, **254**, 4228 (1962); Gerber, P., *Virology*, **16**, 46 (1961).

[18] Ito, Y., *Virology*, **12**, 596 (1960), and these PROCEEDINGS, **47**, 1897 (1961).

[19] Perry, T. L., and D. Walker, *Proc. Soc. Exp. Biol. Med.*, **99**, 717 (1958); Bearn, J. G., and K. S. Kirby, *Exp. Cell Research*, **17**, 547 (1959); Tigyi, A., I. Benedeczky, and K. Lissak, *Acta Biol. Acad. Sci. Hung.*, **10**, 197 (1959).

[20] Medawar, P. B., *Nature*, **182**, 62 (1958); Holoubek, V., and L. Hnilica, *Canad. J. Biochem. Physiol.*, **39**, 1478 (1961); Floersheim, G. L., *Nature*, **193**, 1266 (1962).

[21] Kurita, S., C. Takemura, A. Hoshino, and K. Kimura, *Gann*, **49** Suppl., 66 (1958); Kurita, S., *J. Nagoya Med. Assoc.*, **81**, 1156 (1961), *Abst. Jap. Med.*, **1**, #3828 (1961); Blumenthal, G. H., and D. M. Greenberg, Abstracts of Papers, VIII Int. Cancer Congress, Moscow, p. 139 (1962); Podgayetskaya, D. Ja., V. M. Bresler, and Y. M. Olenov, Abstracts of Papers, VIII Int. Cancer Congress, Moscow, p. 139 (1962), and *Cytologia*, **4**, 59 (1962).

[22] Benoit, J., P. Leroy, C. Vendrely, and R. Vendrely, *C.R. Acad. Sci. (Paris)*, **244**, 2321 (1957); *Trans. N.Y. Acad. Sci.*, **22**, 494 (1960).

[23] Kantoch, M., and F. B. Bang, these PROCEEDINGS, **48**, 1553 (1962).

[24] Fahmy, O. G., and M. J. Fahmy, *Nature*, **191**, 776 (1961).

[25] Martinovitch, P. N., D. T. Kanazir, Z. A. Knezevitch, and M. M. Simitch, *J. Embryol. Exp. Morph.*, **10**, 167 (1962).

[26] Leuchtenberger, C., R. Leuchtenberger, and E. Uyeki, these PROCEEDINGS, **44**, 700 (1958); Frederic, J., and J. Corin-Frederic, *C.R. Soc. Biol.*, **156**, 742 (1962).

[27] Meek, E. S., and T. F. Hewer, *Brit. J. Cancer.*, **13**, 121 (1959); Stolk, A., *Naturwissenschaften*, **47**, 88 (1960).

[28] Djordjevic, O., L. Kostic, and D. Kanazir, *Nature*, **195**, 614 (1962).

[29] Kraus, L. M., *Nature*, **192**, 1055 (1961).

[30] Weisberger, A. S., these PROCEEDINGS, **48**, 68 (1962).

[31] Okada, Y., T. Suzuki, and Y. Hosaka, *Med. J. Osaka Univ.*, **7**, 709 (1957); Klein, R., *C.R. Soc. Biol.*, **155**, 1917 (1961).

[32] Barski, G., S. Sorieul, and F. Cornefert, *C.R. Acad. Sci. (Paris)*, **251**, 1825 (1960); Sorieul, S., and B. Ephrussi, *Nature*, **190**, 653 (1961).

[33] Szybalski, W., and M. J. Smith, *Proc. Soc. Exp. Biol. Med.*, **101**, 662 (1959).

[34] Berman, L., and C. S. Stulberg, *Proc. Soc. Exp. Biol. Med.*, **93**, 730 (1956).

[35] Smull, C. E., M. F. Mallette, and E. H. Ludwig, *Biochem. Biophys. Res. Comm.*, **5**, 247 (1961); Herbst, E. J., and B. P. Doctor, *J. Biol. Chem.*, **234**, 1497 (1959); Keister, D. L., *Fed. Proc.*, **17**, 84 (1958).

[36] Smith, A. G., and H. R. Cress, *Lab. Invest.*, **10**, 898 (1961).

[37] Koch, G., S. Koenig, and H. Alexander, *Virology*, **10**, 329 (1960).

[38] Meselson, M., F. W. Stahl, and J. Vinograd, these PROCEEDINGS, **43**, 581 (1957).

[39] Szybalski, W., Z. Opara-Kubinska, Z. Lorkiewicz, E. Ephrati-Elizur, and S. Zamenhof, *Nature*, **188**, 743 (1960).

[40] Szybalski, W., *Experientia*, **16**, 164 (1960).

[41] Szybalski, W., E. H. Szybalska, and R. N. Brockman, *Proc. Amer. Assoc. Cancer Research*, **3**, 272 (1961).

Reprinted from *J. Virology* **7**:813–820 (1971)

Transfer of Thymidine Kinase to Thymidine Kinaseless L Cells by Infection with Ultraviolet-Irradiated Herpes Simplex Virus

WILLIAM MUNYON, EDMUNDO KRAISELBURD,[1] DANIEL DAVIS,[2] AND JUDITH MANN

L cells lacking thymidine kinase (TK) activity (Ltk⁻ cells) have been stably transformed to a TK-positive phenotype by infection with ultraviolet-irradiated herpes simplex virus (HSV-UV). The highest frequency of the Ltk⁻ to Ltk⁺ transformation observed in these experiments was approximately 10^{-3}, whereas no measurable transformation was observed (less than 10^{-8}) in the absence of HSV-UV infection. Cell lines of HSV-transformed Ltk⁺ cell lines contain 7 to 24 times as much TK activity as do the parental Ltk⁻ cells, and they have been maintained in culture for a period exceeding 8 months. The kinetics of thermal inactivation of the TK activity derived from an Ltk⁺ HSV-transformed cell line and the TK activity from Ltk⁻ cells lytically infected with infectious HSV are similar. Both of these TK activities are much more thermolabile than the TK activity present in wild-type L cells. A mutant strain of HSV which does not induce TK activity during lytic infection does not cause the Ltk⁻ to Ltk⁺ transformation. These data suggest that either an HSV TK gene has been transferred to Ltk⁻ cells or that an HSV gene product has caused the expression of a previously repressed cellular enzyme.

There has been much interest recently concerning the development of methods by which external genetic information could be stably introduced into an eukariotic cell line. As a first step, we proposed to look for a carrier mechanism which would enable the desired information to enter the cell and be established in an inheritable manner. An obvious choice was a mammalian virus. Our requirements for such a suitable virus were (i) that there be some basis for belief that the virus is capable of a nonlytic relationship with cells, (ii) that the virus be a nuclear deoxyribonucleic acid (DNA) virus, and (iii) that it contain a gene(s) that could be selected for in a tissue culture system.

In view of these criteria, herpes simplex virus (HSV) was chosen. HSV is able to establish persistent infections in man such as herpes labiallis. Evidence has been obtained (18) indicating that Epstein-Barr (EB) virus, a herpesvirus, is vertically transmitted in cultured cell lines derived from Burkitt tumors and is able to transform human leukocytes in vitro in regard to their growth characteristics (8, 21). Zur Hausen and Schulte-Holthausen (24) have shown that the tumor tissue from Burkitt tumors contains DNA homologous to EB virus DNA. Marek's disease of chickens (4) and lymphomas in rabbits (H. Hinze, Bacteriol. Proc., p. 157, 1969) have been shown to be associated with viral agents having an HSV morphology. There is, therefore, much evidence to indicate that HSV and other viruses with similar morphology are able to establish nonlytic relations with animal cells.

HSV is a DNA-containing virus (1), and its DNA replicates in the nucleus of infected cells (20). A nuclear DNA virus was chosen for study because it seemed that it would offer maximum opportunity for possible genetic interchange between the genomes of the virus and the cell.

Kit et al. (10) have reported that by cultivating LM cells (Ltk⁺ cells) in the presence of successively increasing concentration of bromodeoxyuridine (BUDR), an altered cell strain was produced which has very low amounts of thymidine kinase (TK) activity (Ltk⁻ cells). HSV is able to induce TK activity in Ltk⁻ cells during the course of a lytic infection (9). This HSV-induced TK

[1] Department of Biology, State University of New York at Buffalo, Buffalo, N.Y.

[2] Department of Biophysics, State University of New York at Buffalo, Buffalo, N.Y.

activity has been shown to be immunologically different from the Ltk$^+$ TK activity (12).

The existence of a virus which is potentially a carrier of a TK gene and a cell line which is deficient in this activity present the opportunity to test the possibility that HSV-infected Ltk$^-$ cells are able to acquire permanently the virus TK activity.

Since HSV has a strong cytopathic effect in Ltk$^-$ cells, ultraviolet-irradiated HSV (HSV-UV) were used for infection. UV irradiation of transducing particles prior to infection increases the frequency of stable transduction in bacterial systems (7). It has been shown that vaccinia that had been inactivated by UV irradiation is able to initiate TK synthesis in Ltk$^-$ cells, indicating that a UV-inactivated virus may still contain a functional TK gene and that the gene can be expressed by a UV-killed virus (17).

Littlefield (13) has shown that TK-containing cells (Ltk$^+$ cells) can be selected from an Ltk$^-$ population by including aminopterin, thymidine, hypoxanthene, and glycine in the tissue culture medium (HAT medium). By using a similar system, we have infected Ltk$^-$ cells with HSV-UV and then selected for cells which have acquired TK activity (Ltk$^+$ cells).

MATERIALS AND METHODS

Virus. HSV type 1 was obtained from David Yohn in 1966. Except where noted otherwise in this paper, stock virus was prepared in primary rabbit kidney (PRK) monolayers as follows. Cell layers that had been infected 20 to 24 hr previously were frozen and thawed twice in growth medium. This lysate was centrifuged at $500 \times g$ for 10 min, cultured for sterility, and stored at -70 C. Virus stocks prepared in this way had 10^5 to 10^7 plaque-forming units (PFU)/ml and 10^9 to 5×10^{10} particles/ml.

HSV (B2006), a strain of HSV, was obtained from Saul Kit and Del Dubbs and has been described by them (5). This is a mutant HSV that does not induce TK activity in Ltk$^-$ cells during a lytic infection.

The cell line on which a particular virus stock had been grown and the number of consecutive passages in that cell line are indicated by an abbreviation for the cell line followed by the passage number. For example, HSV-PRK p = 7 refers to herpes simplex virus that has been serially passaged in PRK cells seven times.

Virus preparations were not sonically treated before or after UV irradiation because sonic treatment did not increase the titer of infectious virus preparations.

UV irradiation of virus. A GE Germicidal Lamp G8T5 was used to irradiate suspensions of stock virus. Viral suspensions in growth medium were irradiated at a distance of 13 inches (33.3 cm) from the lamp. Irradiation at this distance produced a UV intensity of 23 ergs per mm^2 per sec. Samples (2 ml) of stock virus suspensions were irradiated in 6-cm plastic petri dishes. During irradiation the petri dishes were rotated

at 100 rev/min on a rotary platform shaker and were exposed to atmospheric oxygen at room temperature.

Media. Growth medium contained Eagle's medium supplemented with the nonessential amino acids and 5% calf serum (6). TK plus selective medium was growth medium with the following additions: methotrexate (6.0×10^{-7} M), thymidine (1.6×10^{-5} M), adenosine and guanosine (each at 5×10^{-5} M), and glycine (10^{-4} M).

Cell lines. Ltk$^-$ cells were obtained from Kit et al. (10). These cells were cloned once and were serially cultivated in growth medium containing 20 μg of BUDR per ml. The phenotypic characterization of this cell line is that (i) it is resistant to growth inhibition by BUDR at 20 μg/ml, (ii) it incorporates very little tritiated thymidine into its DNA, and (iii) high speed ($100,000 \times g \times 1$ hr) supernatant fluid extracts of Ltk$^-$ cells contain very low levels of TK activity.

A9 cells were obtained from Littlefield (13). This cell line was used as the L cell type that contained TK activity (Ltk$^+$ cells). These cells were grown in growth medium without 8-azaguanine, except in the experiments involving phosphoribosyl transferase activity.

KB cells were obtained from the cell service unit at Roswell Park Memorial Institute. They were grown in monolayer cultures in growth medium.

Cell lines derived from Ltk$^-$ cells infected with HSV-UV and grown in TK plus selective medium are referred to in this paper as Ltk$^+$-transformed cells. Specific Ltk$^+$-transformed cell lines will be identified as Ltk$^+$ HSV-. The number following the hyphen after HSV identifies the particular clone.

PRK cells were obtained from 10-day-old rabbits, and tissue cultures were prepared from them as described by Merchant et al. (19).

Human embryonic lung cells were obtained from Jun Minowada at Roswell Park Memorial Institute. This cell line was grown in Eagle's medium containing 10% fetal calf serum.

Virus plaque titrations. Plaque titrations were performed as described by Roizman and Roane (22), except that monolayers of PRK cells were used.

Virus particle counts. Virus particle counts were performed as described by electron microscopy (23) and standardized by using latex particles.

Colony counting. In the experiments in which colony counts were made, four to six replicate petri dishes were counted for each determination. The Ltk$^+$ HSV-transformed colonies varied greatly in size. Colonies of less than 50 cells were not counted. Data are expressed as the average number of Ltk$^+$-transformed colonies per dish.

Tritiated thymidine uptake. This was performed as described by Breslow and Goldsby (3). The medium used for all cell types was TK plus selective medium without methotrexate.

Thymidine kinase assay. The TK assay was as described by Kit et al. (11) with the following modifications. Pellets of cells were suspended in a sonically treated buffer (five times their volume) containing 0.01 M maleate tris(hydroxymethyl)aminomethane (Tris; pH 6.5), 0.15 M KCl, 0.02 M MgCl$_2$, 0.001 M mercaptoethanol, 2×10^{-4} M adenosine triphosphate (ATP), 10^{-5} M thymidine and were sonically treated.

The standard assay mixture contained in a total of 120 μliters the following final concentration of solutes: 0.1 M maleate Tris (*pH* 6.5), 0.01 M ATP, 0.02 M MgCl₂·6H₂O, 8.3 × 10⁻⁵ M ³H-thymidine (specific activity approximately 1 mCi/mmole), and between 100 and 700 μg of protein. Under these conditions, enzyme activity was proportional to the amount of cell extract added and to the time of incubation, up to 30 min at 36.5 C.

The amount of phosphylated thymidine produced was measured as described by Breitman (2). Protein was determined by the method of Lowry et al. (15).

Thermal denaturation experiments. Cell extracts were prepared in a manner similar to that described for the TK assay. Each sample, containing 4.4 mg of protein in a total volume of 2 ml, was incubated at 52 C in a water bath. Samples of 0.2 ml were removed at indicated time intervals, and portions were assayed for TK activity.

Preparation of HSV antiserum. New Zealand rabbits were immunized by three successive injections of purified HSV-KB p = 10. The first injection was given intravenously and contained 5 × 10¹⁰ virus particles. Two subsequent intramuscular injections of 10¹⁰ HSV particles followed the first injection at monthly intervals. The HSV used for immunization was purified by two cycles of rate zonal sedimentation in 5 to 40% Ficol (Pharmacia Ltd.) gradients.

RESULTS

Reversion frequencies of Ltk⁻ cells to Ltk⁺ cells. In these experiments, spontaneous reversion of Ltk⁻ cells to the Ltk⁺ phenotype is less than 1 in 10⁸. In the course of 20 experiments (measured by the formation of colonies by Ltk⁻ cells in TK plus selective medium), 10⁶ cells were seeded in each of a total of 100 petri dishes containing selective medium. No colony formation was observed. Similar results have been reported by Littlefield (14).

Transformation of Ltk⁻ to Ltk⁺ via infection with HSV-UV. Formation of colonies in TK plus selective medium was observed when Ltk⁻ cells were infected with HSV-UV. These experiments were carried out as follows. Sparse monolayers of Ltk⁻ cells, growing in 6-cm diameter plastic petri dishes, were inoculated with 0.2 ml of HSV-UV. After an adsorption period of 1 hr, the inoculum was aspirated, and 5 ml of growth medium containing 0.2% pooled human gamma globulin (Hyland) was added to each dish. The infected cells were then incubated for 24 hr to allow for gene establishment and expression. The growth medium then was removed and replaced with an equal volume of TK plus selective medium. TK plus selective medium caused between 50 and 80% of the cells to die and detach from the surface of the petri dishes. At 5 days after infection, the suspension of dead cells and medium was aspirated and replaced with fresh TK plus

selective medium. The dishes were then incubated for a total of 16 to 20 days at 37 C in a humidified incubator with a CO₂-air atmosphere at 10 and 90%, respectively. Colonies that had formed were then either stained with crystal violet-formaldehyde at 0.1 and 3.7%, respectively, or used for clone isolation. Under these circumstances colony formation was regularly observed. Colony size and morphology varied widely. Approximately one-fourth to one-third of the colonies grew to a size of 2 mm or larger in diameter. A continuous spectrum of smaller colonies also formed. Many of the small colonies did not increase in size, even if incubation was continued for more than 19 days. Often the cells of these small nongrowing clones showed cytological evidence of thymidine starvation and probably represented cells in which TK activity was lost after a few divisions.

No colony formation was observed when cells were infected with unirradiated HSV. In this case a massive cytopathic response, typical of herpes infection and not characteristic of the effect of the TK plus selective medium, was observed.

Infection of Ltk⁻ cells with UV-irradiated vaccinia did not result in the formation of Ltk⁺ colonies.

Cells containing phosphoribosyl transferase (PRT) activity can be selected for in a tissue culture system by using HAT medium as described by Littlefield (13). A9 cells are deficient in PRT activity. Infection with HSV-UV did not enable A9 cells to form colonies in HAT medium.

Factors affecting the frequency of Ltk⁻ to Ltk⁺ transformation. The following factors have been found to alter the frequency of the HSV-irradiated Ltk⁻ to Ltk⁺ transformation: UV dose, virus dose, time of incubation in growth medium after infection, and cell density.

The frequency of colony formation versus the dose of UV received by HSV is shown in Fig. 1. Optimal colony formation was observed when HSV was irradiated for 9 to 12 min. Note that virus that had been irradiated until no infectivity could be detected was still effective in causing the Ltk⁻ to Ltk⁺ transformation. The number of colonies formed varied linearly with relative virus dose in the experiment shown in Fig. 2. In several experiments, however, this relation was not linear, comparing undiluted virus versus the lowest virus dilution. This effect is observed in Fig. 3. Dilution of the undiluted virus (plotted as a relative dose of 25) by a factor of 5 did not reduce the number of colonies produced by a factor of 5. The subsequent dilution of these virus preparations by a factor of 5 did reduce the number of colonies by approximately one-fifth. The use of high multiplicities alone, however, cannot account

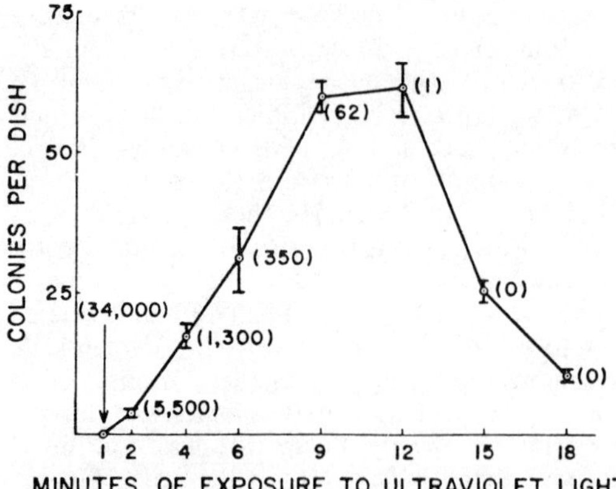

FIG. 1. *An HSV-PRK p = 7 preparation, containing 6.9 × 10⁹ particles and 6.3 × 10⁶ plaque-forming units (PFU) per 0.2 ml before irradiation, was exposed to UV irradiation for the intervals indicated on the abscissa. The number of PFU per 0.2 ml surviving a given dose of UV is given by the number in parentheses by each point. A 0.2-ml amount of this preparation was added to petri dishes, each of which contained 3.8 × 10⁵ Ltk⁻ cells. Other details are as described in Materials and Methods. Vertical lines through the points of this and the other figures represent the magnitudes of the standard deviations of the graphed values.*

for this nonlinearity, since the multiplicity used in the experiment shown in Fig. 3 is lower than in Fig. 2. This effect was independent of the amount of UV irradiation given to the virus.

The HSV-UV infected cells were incubated in growth medium for variable periods prior to the addition of TK plus selective medium (Table 1). The optimum period of incubation in growth medium was found to be 24 hr.

The number of colonies formed depended also on the number of cells present in the petri dish at the time of infection. Optimal cell density was between 2.5 × 10⁵ and 6.3 × 10⁵ cells per petri dish. Suppression of colony formation was observed when higher cell densities were used (Fig. 4).

Treatment of HSV-UV with HSV immune rabbit serum prevented colony formation (Table 2). Heating of HSV-UV at 50 C inactivated the transforming activity of the virus (Table 3). Incubation of HSV-UV with deoxyribonuclease had no significant effect (Table 3).

To demonstrate the formation of Ltk⁺ HSV transformation, it was necessary to use high multiplicities of virus particles per cell. Other experiments indicated that the efficiency of HSV particle uptake by Ltk⁻ cells is very low, so that the number of virus particles adsorbed is much lower than the number of virus particles added. For example, after being exposed to a monolayer of Ltk⁻ cells for 1 hr, HSV-UV inocula were re-

moved and used to infect a second set of Ltk⁻ monolayers. Ltk⁺ HSV-transformed colonies (70 to 90%) formed on the second set. The proportion of HSV-UV particles adsorbed by Ltk⁻ monolayers under the same conditions used in the transformation experiments was determined by counting virus particles in solution before and after being exposed to Ltk⁻ cells. In one experiment, no decrease in particle count was observed.

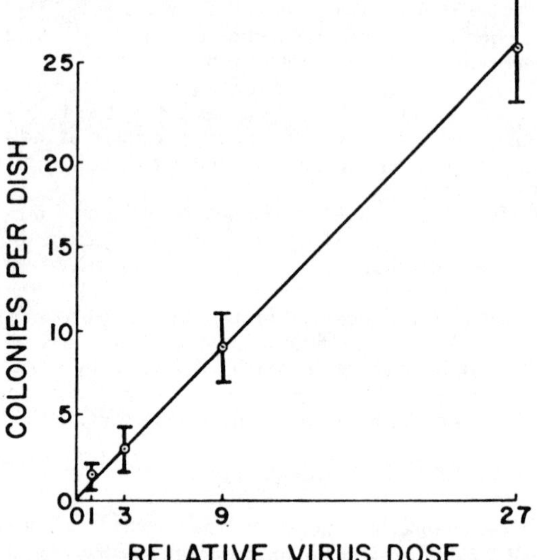

FIG. 2. *Petri dishes containing 1.4 × 10⁶ Ltk⁻ cells/dish were infected with 0.2 ml of HSV-Ltk⁻ p = 1 that had been UV irradiated for 9 min. The particle count of the undiluted preparation was 3.1 × 10⁹ particles/0.2 ml. A relative virus dose of 27 corresponds to the undiluted virus suspension.*

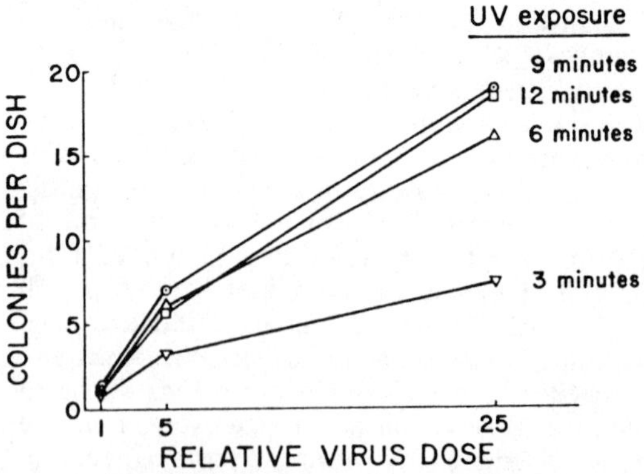

FIG. 3. *Petri dishes containing 1.7 × 10⁶ Ltk⁻ cells/dish were infected with 0.2 ml of HSV-PRK p = 3 that had been irradiated for the intervals indicated. The particle count of the undiluted virus suspension was 1.2 × 10⁹/0.2 ml. Relative dose of 25 represents the undiluted inoculum.*

In a second experiment, a 15% decrease was observed. It is therefore concluded that in these experiments only a relatively small fraction of the total HSV particles of an inoculum are taken up

TABLE 1. *Effect on colony formation of incubation of infected cells in growth medium prior to the addition of selective medium*

Period of incubation in growth medium (hr)[a]	Avg no. of colonies	SD[b]
0[c]	1.7	2.1
12	7.2	1.3
24	46.8	9.7
36	—[d]	
48	—[d]	

[a] Ltk$^-$ cells (8.8 × 10^5 per dish) were exposed to Ltk$^-$ HSV p = 1 (ca. 8.8 × 10^8 particles) that had been irradiated with UV light for 8 min. Virus was allowed to adsorb to the cells for 1 hr at room temperature. After adsorption 5 ml of growth medium containing 0.2% pooled human gamma globulin was added to each petri dish, and the dishes were incubated at 36.5 C for variable intervals as indicated. This interval of incubation was terminated by adding 0.05 ml of a solution containing methotrexate (6.0 × 10^{-5} M), thymidine (1.6 × 10^{-3} M), adenosine and guanosine (each at 5 × 10^{-3} M), and glycine (10^{-2} M). Cells were then incubated for 18 days, counting from the time of infection. Colonies were fixed, stained, and counted as previously described.

[b] Standard deviation.

[c] At the time of infection, each petri dish contained approximately 2.3 × 10^6 Ltk$^-$ cells.

[d] Cell multiplication during incubation in growth medium for periods exceeding 24 hr produced relatively dense monolayers. Under these conditions colonies were small and difficult to count.

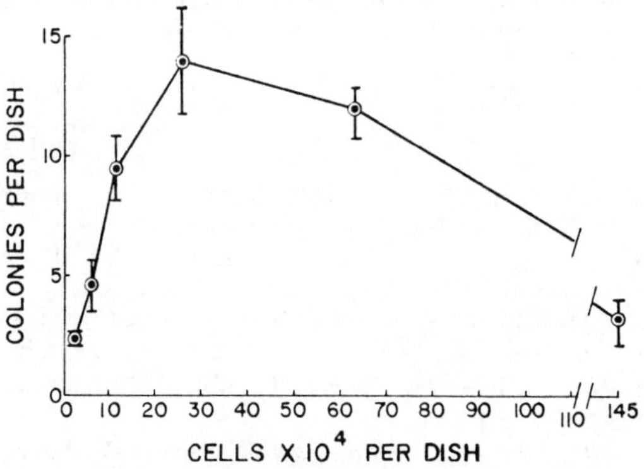

FIG. 4. *Petri dishes containing the indicated number of Ltk$^-$ cells were inoculated with 0.2 ml of a suspension of HSV-PRK p = 7 that had been irradiated for 9 min. The virus preparation contained 6.9 × 10^9 particles (6.3 × 10^6 plaque-forming units before irradiation) per 0.2 ml.*

TABLE 2. *Neutralization of colony-forming activity of HSV-UV by herpes antiserum*

Serum treatment[a]		Avg no. of colony	SD[b]
Type	Dilution		
Immune	1/4	0.0	
Preimmune	1/4	7.4	2.6
Immune	1/40	0.5	0.55
Preimmune	1/40	11.3	2.2
Immune	1/320	8.9	3.1
Preimmune	1/320	6.0	4.0
Balanced salt solution		6.3	2.7

[a] HSV-PRK p = 1 (ca. 2 × 10^9 particles/0.2 ml) was irradiated with UV light for 9 min. Irradiated virus (1.2 ml) was combined with 0.4 ml of serum or serum dilutions and incubated at 37 C for 1 hr in plastic test tubes. The sera used were heat-inactivated at 56 C for 1 hr. Samples (0.2 ml) of virus-serum mixtures were added to petri dishes containing 1.4 × 10^6 cells per dish. Addition of growth medium and thymidine kinase plus selective medium was carried out as described in Materials and Methods. Colonies were stained and counted 17 days after infection.

[b] Standard deviation.

TABLE 3. *Effects of heat and deoxyribonuclease digestion on colony-forming activity of HSV-UV[a]*

Treatment[b]	Avg no. of colonies	SD[c]
0 C	6.3	2.7
56 C for 5 min	No colonies formed	
56 C for 30 min	No colonies formed	
Deoxyribonuclease digestion	8.0	3.2

[a] HSV-PRK p = 1 (ca. 2 × 10^9 particles/0.2 ml) was irradiated with UV light for 9 min.

[b] HSV-UV was either heated as indicated in the table or incubated at 37 C for 15 min in the presence of 400 μg of bovine pancreatic deoxyribonuclease per ml. Samples (0.2 ml) of these preparations were inoculated on 1.4 × 10^6 Ltk$^-$ cells. Other procedures are as described in Materials and Methods.

[c] Standard deviation.

by the Ltk$^-$ cells after 1 hr of adsorption. For this reason, the number of Ltk$^-$ to Ltk$^+$-transforming events per infecting particle has not been computed.

Effect of BUDR on colony formation. The omission of thymidine from TK plus selective medium suppressed colony formation (*see* Table 4). The substitution of thymidine with BUDR or iododeoxuridine in TK plus selective medium completely suppressed colony formation.

Properties of the Ltk$^+$-transformed clones. A

series of 10 Ltk$^+$ HSV-transformed cell lines, derived from single clones, have been subcultured in TK plus selective medium twice a week for the past 8 months. The ability of these cell lines to incorporate tritiated thymidine into cellular DNA and the level of TK activity in supernatant fluid extracts of these cells are shown in Table 5. Relative to the Ltk$^-$ cell line, the Ltk$^+$ HSV-transformed cell lines were between 460 and 1,900 times more active in taking up tritiated thymidine into cellular DNA and contained 7 to 24 times as much TK activity.

Identity of the origin of the TK activity in Ltk$^+$ transformed cells. A mutant strain of HSV, HSV(B2006), does not induce TK activity in Ltk$^-$ cells during lytic infection. Infection of Ltk$^-$ cells with HSV-UV (B2006) did not cause the formation of colonies in TK plus selective medium (Table 6). This suggests that the infecting HSV must contain a functional TK gene to be able to cause the production of Ltk$^+$ cells.

HSV strains, which had been grown on Ltk$^-$ cells and then inactivated with UV, were able to cause the production of Ltk$^+$-transformed cells (Table 6). This suggests that the production of Ltk$^+$-transformed cells by HSV-UV is not due to the encapsidization of a fragment of the host cell DNA in the HSV virion. HSV which had been passed a greater number of times on Ltk$^-$ cells was not used because this strain of HSV replicates poorly on Ltk$^-$ cells.

Studies on the kinetics of thermal inactivation of TK activity were carried out with cell extracts derived from (i) Ltk$^+$ cells (A9 cells), (ii) Ltk$^-$ cells lytically infected with HSV, and (iii) Ltk$^+$ HSV-transformed cells (Fig. 5). The TK activity of the Ltk$^+$ HSV-transformed cells was similar

TABLE 4. *Effects of thymidine, bromodeoxyuridine (BUDR), and iododeoxyuridine (IUDR) on colony formation of HSV-UV infected Ltk$^-$ cells[a]*

Ingredient added to medium[b]	Avg no. of colonies	SD[c]
No addition.........	2.6	1.1
Thymidine..........	12	5.4
BUDR (20 µg/ml)...	No colonies formed	
IUDR (20 µg/ml)...	No colonies formed	

[a] HSV-Ltk$^-$ p = 1 containing 3.1 × 10^9 particles/0.2 ml was irradiated with UV light for 6 min. This virus (0.2 ml) was inoculated into petri dishes containing 3 × 10^6 Ltk$^-$ cells. Other procedures are as described in Materials and Methods.

[b] Growth medium containing methotrexate (6.0 × 10^{-7} M), adenosine and guanosine (each at 5.0 × 10^{-5} M), and glycine (10^{-4} M).

[c] Standard deviation.

TABLE 5. *Rate of incorporation of tritiated thymidine into DNA and thymidine kinase activity of Ltk$^+$ HSV-transformed cell lines*

Cell lines	Incorporation of ^{3}H-thymidine into DNA[b] (counts per min per cell division per hr)	Thymidine kinase activity[c]
Ltk$^+$ (A9)	1.42	0.81
Ltk$^-$	0.003	0.03
Ltk$^+$ HSV-3[a]	2.4	0.46
Ltk$^+$ HSV-11	3.2	0.44
Ltk$^+$ HSV-20	2.7	0.85
Ltk$^+$ HSV-21	2.8	0.72
Ltk$^+$ HSV-22	2.0	0.63
Ltk$^+$ HSV-23	1.4	0.65
Ltk$^+$ HSV-28		0.60
Ltk$^+$ HSV-33	1.6	0.33
Ltk$^+$ HSV-39	5.7	0.59
Ltk$^+$ HSV-40	2.9	0.26

[a] Ltk$^+$ HSV-transformed cell lines were isolated from clones derived from Ltk$^-$ cells that had been infected with HSV-UV as described in Materials and Methods. The number after the HSV is an arbitrary clone designation.

[b] Monolayer cultures of the indicated cell lines were exposed to growth medium containing adenosine and guanosine (each at 5 × 10^{-5} M), glycine (10^{-4} M), and ^{3}H-thymidine (10^{-7} M), (specific activity of 10^4 µCi/µmole) for 1 hr. The rate of cell division was determined by counting replicate monolayer cultures 8 hr before and 21 hr after the cells were exposed to the tritiated thymidine. The experimental error was approximately 15%.

[c] Picomoles of thymidine monophosphate per µg of protein in the uncentrifuged cell sonically treated material per 20 min of incubation.

TABLE 6. *Ltk$^-$ to Ltk$^+$ transformation by using a thymidine kinase-deficient HSV*

Virus type	MOI[a]	Frequency of Ltk$^+$ HSV transformation[b]
HSV-HEL p = 3	18,000	8 × 10^{-4}
HSV(B2006)[c]-PRK p = 3	16,400	None
HSV-Ltk$^-$ p = 1	14,700	1.9 × 10^{-3}

[a] Multiplicity of infection (particles of HSV-UV added per cell). All virus preparations were irradiated for 9 min. The cell count per dish was 8.5 × 10^4 cells.

[b] This data was computed as follows: Ltk$^+$ HSV-transformed colonies per Ltk$^-$ cells per dish at time of infection.

[c] HSV(B2006) is an HSV mutant that does not induce thymidine kinase activity during lytic infection.

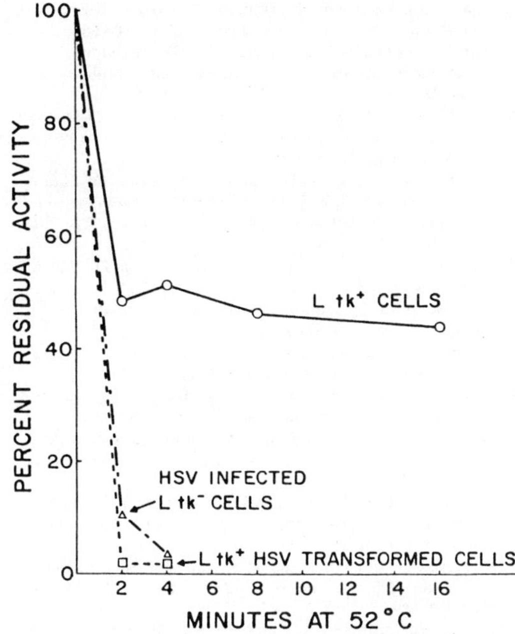

FIG. 5. *Extracts were prepared from Ltk+ cells, (A9 cells) HSV-infected Ltk- cells (8 hr postinfection), and Ltk+ HSV-20 clone of transformed cells. Before heating, these extracts were all diluted to a protein concentration of 2.2 mg/ml. The extracts were quickly brought to 52 C, and samples were removed and quickly cooled at the times indicated in the figure.*

(in respect to the kinetics of thermal inactivation) to the activity present in cells lytically infected with HSV and was much more thermolabile than the TK activity found in wild-type Ltk+ cells.

DISCUSSION

The Ltk- cells used in these experiments had no measurable reversion to Ltk+ cells (less than 10^{-8}). When Ltk- cells were infected with HSV-UV, a transformation frequency to the Ltk+ genotype of 10^{-5} to 10^{-3} was observed. The number of colonies formed was proportional to the relative virus dose used and depended also on the amount of UV received by the virus. Colony formation was neutralized by HSV antiserum, was thermolabile, and was not affected by deoxyribonuclease digestion. These data indicate that the appearance of Ltk+-transformed cells was caused by the infection of Ltk- cells with HSV-UV.

The failure of a TK-minus mutant of HSV (one that does not induce TK activity during lytic infection) to cause the transformation of Ltk- to Ltk+ cells provides genetic evidence to suggest that the new TK activity in the Ltk+ HSV-transformed cells is either derived from a structural gene in the HSV genome or a previously

repressed product of the cells was derepressed by an HSV gene product. The kinetics of thermal denaturation of the TK activity of the Ltk+-transformed cells is similar to that of the TK activity induced by HSV infection and different from the wild-type TK enzyme activity. These observations suggest that the TK activity present in Ltk+ HSV-transformed cells is associated with a different protein than the TK activity of wild-type L cells.

Experiments now in progress indicate that the major TK activities of Ltk+ and Ltk+ HSV-transformed cells migrate at different rates in polyacrylamide gel electrophoresis. These results will be reported separately.

The clones that arose in TK plus selective medium differed from the parental Ltk- cells by the following criteria. (i) They did not grow in BUDR-containing media; (ii) they grew continuously in TK plus selective medium; (iii) they contained 7 to 24 times more TK activity than did Ltk- cells, and (iv) they incorporated 960 to 1,900 times more tritiated thymidine into cellular DNA than did Ltk- cells.

The addition of TK plus selective medium prior to 24 hr after HSV-UV infection strongly inhibits colony formation. The requirements for a period of incubation in growth medium after infection suggests that gene expression takes approximately 24 hr, i.e., there is a lag between infection and the occurrence of TK activity in colony-forming cells. In support of this, the inclusion of BUDR in the growth medium added to the cells after infection (and then replaced with TK plus selective medium 24 hr later) does not significantly inhibit colony formation. These results are compatible with the following hypothetical sequence of events resulting in the formation of Ltk+ transformed cells. (i) DNA synthesis is required for the "establishment" of the TK gene activity in the potentially Ltk+ HSV-transformed cells and (ii) expression of the TK activity in this cell.

The significant decrease in colony number when cell density is greater than 6.3×10^5 cells per dish may be due to contact inhibition of the newly transformed cell or the formation of diffusible inhibitory macromolecules (e.g., interferon). A similar process may be operative when HSV-UV infected cells are incubated in growth medium for more than 24 hr prior to the addition of the TK plus selective medium. In this circumstance, a dense monolayer may form due to cell growth during this period.

To our knowledge, this is the first instance in which a gene of a known function has been stably acquired by an eukariotic cell line due to virus infection. This may be a first step in establishing a system in which viruses are used as carrier mechanisms to enable desired information to be transferred to a mammalian cell and to be established in an inheritable fashion.

In bacterial systems UV irradiation increases the frequency of transduction (7), presumably by increasing the probability of recombination. Casto (Bacteriol. Proc., p. 196, 1970) and Lytle et al. (16) have reported that the exposure of cultured cells to UV light increases the frequency of transformation by DNA-containing viruses. A similar effect may be operating in the Ltk$^-$ cell to Ltk$^+$ cell transformation caused by HSV-UV.

Preliminary experiments have indicated that Ltk$^+$-transformed cell lines have increased resistance to UV irradiation as compared to the parental strain. This would be consistent with the possibility that HSV-UV can confer UV repair capability. It may therefore be possible to insert this information into cell lines lacking UV repair mechanisms, e.g., cell lines derived from patients with xeroderma pigmentosa.

ACKNOWLEDGMENTS

These studies were supported by the John A. Hartford Foundation, Inc., Public Health Service grant CA-07745, and institutional grant IN-54-J-23 from the American Cancer Society.

LITERATURE CITED

1. Becker, Y., H. Dym, and I. Sarov. 1968. Herpes simplex virus DNA. Virology 36:184–192.
2. Breitman, T. 1963. The feedback inhibition of thymidine kinase. Biochim. Biophys. Acta 67:153–155.
3. Breslow, R., and R. Goldsby. 1969. Isolation and characterization of thymidine transport mutants of Chinese hamster cells. Exp. Cell Res. 55:338–346.
4. Churchill, A. E., and P. M. Biggs. 1967. Agent of Marek's disease in tissue culture. Nature (London) 215:528–530.
5. Dubbs, D., and S. Kit. 1964. Mutant strains of herpes simplex deficient in thymidine kinase inducing activity. Virology 22:493–502.
6. Eagle, H. 1959. Amino acid metabolism in mammalian cell culture. Science 130:432–437.
7. Garen, A., and N. Zinder. 1955. Radiological evidence for partial genetic homology between bacteriophage and host bacteria. Virology 1:347–376.
8. Henle, W., V. Diehl, and G. Kahn. 1967. Herpes type virus and chromosome marker in normal leukocytes after growth with irradiated Burkitt cells. Science 157:1064–1065.
9. Kit, S., and D. Dubbs. 1963. Acquisition of thymidine kinase activity by herpes simplex infected mouse fibroblast cells. Biochem. Biophys. Res. Commun. 11:55–59.
10. Kit, S., D. Dubbs, L. Piekarski, and T. Hsu. 1963. Deletion of thymidine kinase activity from L cells resistant to bromodeoxyuridine. Exp. Cell Res. 31:297–312.
11. Kit, S., L. Piekarski, and D. Dubbs. 1963. Effects of 5-fluorouracil, actinomycin D and mitomycin C on the induction of thymidine kinase by vaccinia infected L-cells. J. Mol. Biol. 7:497–510.
12. Klemperer, H., G. Haynes, W. Shedden, and D. Watson. 1967. A virus-specific thymidine kinase in BHK 21 cells infected with herpes simplex virus. Virology 31:120–128.
13. Littlefield, J. 1963. The inosinic acid pyrophosphorylase activity of mouse fibroblasts partially resistant to 8-azoguanine. Proc. Nat. Acad. Sci. U.S.A. 50:568–573.
14. Littlefield, J. 1965. Studies on thymidine kinase in cultured mouse fibroblasts. Biochim. Biophys. Acta 95:14–22
15. Lowry, O. H., N. J. Rosebrough, A. L. Farr, and R. J. Randall. 1951. Protein measurement with the Folin phenol reagent. J. Biol. Chem. 193:265–275.
16. Lytle, C. D., K. B. Hellman, and N. C. Telles. 1970. Enhancement of viral transformation by ultra-violet light. Int. J. Radiat. Biol. 18: 297–300.
17. McAuslan, B. 1963. The induction and repression of thymidine kinase in the pox virus infected HeLa cell. Virology 21:383–389.
18. Maurer, B., T. Imamura, and S. Wilbert. 1970. Incidence of EB virus-containing cells in primary and seconadry clones of several Burkitt lymphoma cell lines. Cancer Res. 30: 2870–2875.
19. Merchant, D., R. Kahn, and W. Murphy. 1965. Handbook of cell and organ culture, p. 28. Burgess Publishing Co., Minneapolis.
20. Morgan, C., H. Rose, M. Holden, and E. Jones. 1959. Electron microscopic observations on the development of herpes simplex virus. J. Exp. Med. 110:643–656.
21. Pope, J. H., M. K. Horne, and W. Scott. 1968. Transformation of foetal human leukocytes in vitro by filtrates of a human leukemic cell line containing herpes-like virus. J. Cancer 3:857–866.
22. Roizman, B., and D. Roane. 1961. A physical difference between two strains of herpes simplex virus apparent on sedimentation in cesium chloride. Virology 15:75–79.
23. Watson, D. H., W. C. Russell, and P. Wildy. 1963. Electron microscopic particle counts on herpes virus using the phosphotungstate negative staining technique. Virology 19:250–260.
24. Zur Hausen, H., and H. Schulte-Holthausen. 1970. Presence of EB virus nucleic acid homology in a "virus free" line of Burkitt tumor cells. Nature (London) 227:245–248.

27

Reprinted from *Cell* **11**:223–232 (1977)

TRANSFER OF PURIFIED HERPES VIRUS THYMIDINE KINASE GENE TO CULTURED MOUSE CELLS

M. Wigler, S. Silverstein, L.-S. Lee, A. Pellicer, Y.-c. Cheng, and R. Axel

Summary

Treatment of Ltk⁻, mouse L cells deficient in thymidine kinase (tk), with Bam I restriction endonuclease cleaved DNA from herpes simplex virus-1 (HSV-1) produced tk⁺ clones with a frequency of $10^{-6}/2$ μg of HSV-1 DNA. Untreated cells or cells treated with Eco RI restriction endonuclease fragments produced no tk+ clones under the same conditions. The thymidine kinase activities of four independently derived clones were characterized by biochemical and serological techniques. By these criteria, the tk activities were found to be identical to HSV-1 tk and different from host wild-type tk. The tk⁺ phenotype was stable over several hundred cell generations, although the rate of reversion to the tk⁻ phenotype, as judged by cloning efficiency in the presence of bromodeoxyuridine, was high (1–5×10^{-3}). HSV-1 DNA Bam restriction fragments were separated by gel electrophoresis, and virtually all activity, as assayed by transfection, was found to reside in a 3.4 kb fragment. Transformation efficiency with the isolated fragment is 20 fold higher per gene equivalent than with the unfractionated total Bam digest. These results prove the usefulness of transfection assays as a means for the bioassay and isolation of restriction fragments carrying specific genetic information. Cells expressing HSV-1 tk may also provide a useful model system for the detailed analysis of eucaryotic and viral gene regulation.

Introduction

The isolation of specific fragments of eucaryotic DNA has permitted an analysis of the structural organization of specific genes and may ultimately provide information on the mechanism regulating the expression of these genes. Cleavage of the genome with restriction endonucleases followed by molecular cloning of these fragments in bacterial plasmids has permitted the isolation and amplification of specific genes (Cohen and Chang, 1974). Identification of eucaryotic genes within recombinant plasmid DNAs, however, requires molecular hybridization with purified RNA or DNA probes capable of reacting specifically with a single gene. Analysis of the biological activity of isolated DNA fragments by transfection provides an alternate means of identifying specific eucaryotic genes. This approach further demonstrates that the DNA contained within a given fragment includes the information required to code for the entire structural gene, in a form recognizable by the transcriptional and translational machinery of the host cell.

This experimental design has been used to identify specific fragments of the SV40 and adenovirus genomes containing the genes for malignant transformation (Graham et al., 1975). Another gene amenable to isolation by restriction endonuclease cleavage in concert with transfection assays is the thymidine kinase (tk) gene of herpes simplex virus (HSV-1). Infection of cells with ultraviolet-irradiated herpes virus results in the introduction and stable expression of multiple viral gene functions (Macnab and Timbury, 1976). HSV-1 thymidine kinase activity has been transferred to tk-deficient mouse L cells (Ltk⁻) by infection with inactivated virions (Munyon et al., 1971).

We have therefore attempted to isolate a specific DNA fragment containing the thymidine kinase gene from the HSV-1 genome using transfection of this gene function as a bioassay. The choice of this system was dictated by several considerations. First, the viral genome is orders of magnitude less complex than the eucaryotic genome. This greatly enhances the prospects for successful transfection and allows the possibility of purification of active restriction fragments by size alone. Second, the tk⁺ phenotype can be efficiently selected over a tk⁻ phenotypic background by utilizing growth conditions in which the salvage pathway enzyme thymidine kinase is necessary for survival. There exist cell lines deficient in tk with low rates of spontaneous reversion to tk⁺ which can be used as recipients. Third, the tk gene is an ideal subject for mutational analysis because the tk⁺ or the tk⁻ phenotype can be selected under various conditions. Fourth, the gene product, thymidine kinase, is a well characterized viral protein of known function.

In this report, we demonstrate the stable transfer of HSV-1 tk activity to mouse L cells (Ltk⁻) by transfection with HSV-1 DNA cleaved by restriction endonuclease Bam I. The tk gene can be transfected using an electrophoretically pure fragment 3.4 kb in length. The transformed mouse cells with restored tk activity synthesize an enzyme with antigenic and electrophoretic properties identical to that coded for by the herpes virus genome. Furthermore, this gene function is stably expressed under selective pressure for several hundred generations.

Results

Transfection of tk Activity with Fragments of HSV DNA

The isolation of a specific fragment of the HSV-1 genome containing the thymidine kinase gene requires that we identify a restriction endonuclease capable of digesting HSV DNA, which makes no internal cleavages within the tk gene. Identification of such a DNA fragment in addition requires a cell line that will stably express the tk function upon competent transfection. Ltk⁻ clone d, a clone of mouse cells resistant to bromo-deoxyuridine (BdUrd) and deficient in cytoplasmic thymidine kinase (Kit et al., 1963) was therefore chosen for transfection experiments. Ltk⁻ cells are unable to grow in medium containing HAT (hypoxanthine, aminopterin and thymidine), in which survival depends upon the presence of both salvage pathway enzymes thymidine kinase and hypoxanthine–guanosine phosphoribosyl transferase (Littlefield, 1963). The cells have a very low rate of spontaneous reversion to the tk⁺ phenotype, as judged by ability to form colonies in HAT-containing medium, and were used as host recipients to demonstrate that ultraviolet-inactivated HSV-1 virions could infect and stably confer HSV tk activity (Munyon et al., 1971).

Viral DNA for transfection was extracted from virions grown in Vero cells and purified free of contaminating host sequences by velocity sedimentation or CsCl equilibrium density centrifugation. Purity was monitored by isopycnic centrifugation in CsCl. This DNA was then cleaved with a series of restriction endonucleases. The DNA products of these digestions were separated by electrophoresis on 0.5% agarose slab gels (Figure 1). Although all the enzymes used for cleavage require the recognition of a unique hexanucleotide pair for activity, significant differences in the number of cleavage sites for the different enzymes is apparent. The gel profiles shown in Figure 1 reflect complete digestion by the endonuclease since first, incubation for an additional 3 hr results in no change in the band pattern; second, the addition of a second dose of enzyme at 3 hr followed by a second incubation did not alter the digestion profile; and third, adenovirus 2 DNA was completely digested under our reaction conditions.

In initial experiments, Bam I- and Eco RI-digested HSV DNA (Figure 1, slots E and G) were used in transfection assays. Cells were plated at a density of 6 × 10⁵/100 mm petri dish in growth medium. Culture medium was removed 24 hr later, and cell monolayers were overlaid with restriction-cleaved

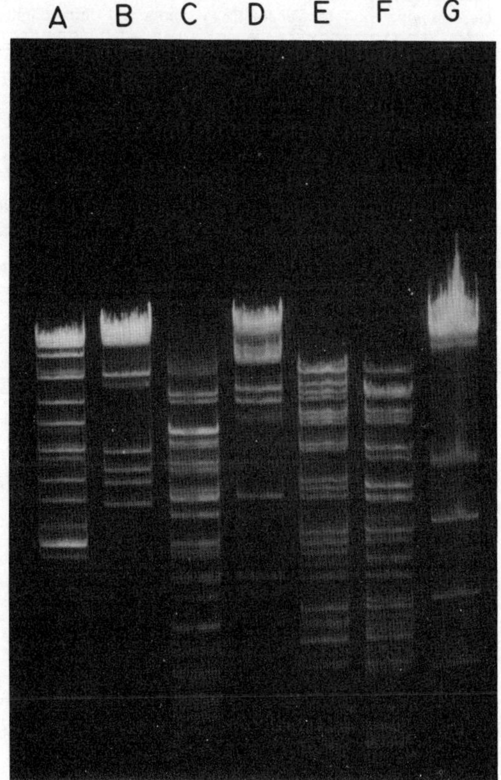

Figure 1. Digestion of HSV-1 DNA with DNA Restriction Endonucleases

1.0 µg of HSV-1 DNA was incubated with 3 U of various restriction enzymes for 3 hr at 37°C. The resultant DNA fragments were analyzed by electrophoresis on a 17 cm 0.5% agarose slab gel. Gels were stained with ethidium bromide and photographed under short-wave ultraviolet illumination. (A) Hpa I; (B) Bgl II; (C) Sal I; (D) Hind III; (E) Bam I; (F) Bam I + Eco RI; (G) Eco RI.

HSV-1 DNA co-precipitated with calcium phosphate (Graham and van der Eb, 1973). Salmon sperm DNA was used as carrier to yield a total DNA concentration of 20 µg/ml. After a 30 min exposure to DNA, cells were fed growth medium. 24 hr later, cultures were refed growth medium containing HAT and subsequently fed HAT medium every 2–3 days. After 2 weeks, surviving colonies were counted. Table 1 summarizes the results from four experiments. In all four experiments, untreated cultures, or cultures treated with either Salmon sperm DNA alone or with Eco R1-digested HSV-1, exhibited no surviving colonies in HAT. We estimate from these and other experiments that the reversion rate of these cells to the tk⁺ phenotype is <10⁻⁸. By contrast, cultures treated with Bam 1-digested HSV-1 DNA consistently displayed surviving colonies at a frequency of approximately 1 colony per 10⁶ cells per 2 µg HSV-1 DNA.

Table 1. Summary of Transfection Experiments with Unfractionated Restriction Endonuclease Digests of HSV-1 DNA

	μg HSV-1 DNA per Dish[a]	Experiment 1		Experiment 2		Experiment 3		Experiment 4	
		Σ/Σ[b]	Specific Activity[c]	Σ/Σ	Specific Activity	Σ/Σ	Specific Activity	Σ/Σ	Specific Activity
Eco RI-Digested									
HSV-1 DNA	4.00	0/2	0.00	0/2	0.00				
	2.00	0/2		0/2				0/5	0.00
	1.00	0/2		0/2					
	0.50	0/2		0/2					
	0.25	0/2		0/2					
Bam I-Digested									
HSV-1 DNA	4.00			4/2	0.50	7/2	0.90		
	2.00			1/2	0.25	6/2	1.50		
	1.00			0/2	0.00	2/2	1.00	3/5	0.60
	0.50			0/2	0.00	0/2	0.00		
	0.25			1/2	2.00	3/4	3.00		
Bam I-, Eco RI- Digested HSV-1 DNA	2.00							0/5	0.00
Salmon Sperm DNA		0/5		0/5		0/5		0/5	
Untreated[a]		0/5		0/5					

[a] All dishes received 10 μg DNA in 0.5 ml using salmon sperm DNA as carrier, as described in Experimental Procedures, except "untreated" cultures, which did not receive even salmon sperm DNA.
[b] Σ/Σ = total number of colonies per number of replicate dishes.
[c] Specific activity = HAT-resistant colonies per 10^6 cells exposed per μg of HSV DNA.

These data suggest that Bam I cleavage of HSV-1 DNA generates at least one DNA fragment containing information for the entire tk structural gene. Eco RI fragments display no activity in transfection assay, presumably because cleavage occurs within the gene. To verify this, the infectivity of HSV DNA, which was digested with both Bam I and Eco RI, was assayed by transfection. This preparation of doubly cleaved fragments was not capable of generating HAT-resistant colonies.

Transformed Cells Express HSV tk Activity
Electrophoretic Mobility of tk in Transformed Cells
Proof that transformation of the mouse Ltk⁻ phenotype results from the introduction and expression of viral DNA fragments requires us to demonstrate the viral origin of the tk expressed in the transfected clones. To this end, colonies were picked from different culture dishes, grown into mass cultures and further analyzed. Four clones were chosen, and their tk activity was characterized by electrophoretic mobility, immunologic neutralization and substrate specificity.

Electrophoretic profiles of the cytosol fractions derived from Ltk⁻, L1210 (a mouse leukemic line), Vero and Ltk⁻HSV⁺ (Ltk⁻ 12 hr post-infection with HSV-1 at 10 pfu/cell) are presented in Figure 2. As expected, Ltk⁻ showed a small peak of activity of mitochondrial tk at an R_f value of 0.9. In contrast, Ltk⁻HSV⁺, the infected cell homogenate, has an additional peak migrating with an R_f of 0.45. This value is in agreement with previous studies of the HSV-1-induced tk (Cheng and Ostrander, 1976). The L1210 cell homogenate showed the normal pattern of mouse cytoplasmic and mitochondrial tk with activity migrating at an R_f of 0.2 and 0.9, and with no detectable activity with an R_f of 0.45. Vero cells, in which the HSV-1 was grown, similarly revealed no tk activity at an R_f of 0.45. The second peak of activity at $R_f = 0.55$ in the electrophoretic pattern of Vero could be due to the mitochondrial tk and has an electrophoretic mobility similar to the human mitochondrial tk (Lee and Cheng, 1976).

Studies were performed on the electrophoretic mobility of four herpes-transformed cell lines: LH1A2-1, LH2-1, LH5-1 derived from cultures exposed to 4, 2 and 0.25 μg DNA (Table 1, experiment 2), and LH5C2-2 derived from cultures exposed to 0.5 μg DNA (experiment 3). These lines were maintained in continuous culture in medium containing HAT for approximately 30 cell doublings. The main tk activity was consistently found at the position $R_f = 0.45$, in agreement with that of Ltk⁻HSV⁺ (Figure 3). Although mouse mitochondrial tk is present at $R_f = 0.9$, no mouse cytoplasmic tk is found in these lines.

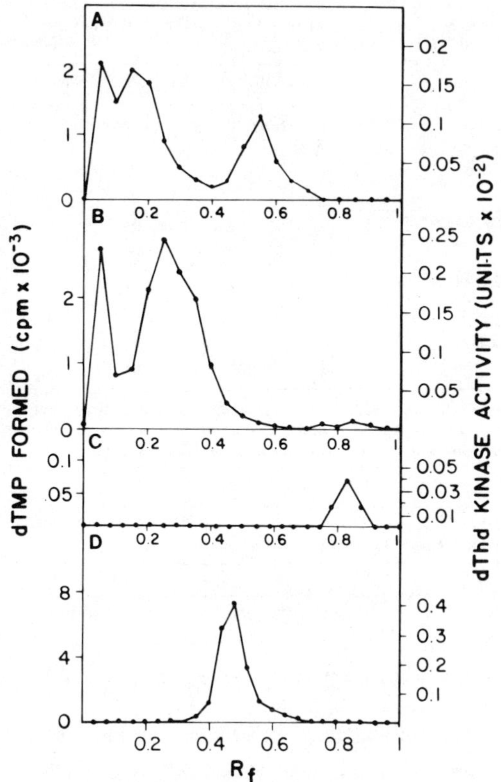

Figure 2. Electrophoretic Pattern of Thymidine Kinase Activities from Cytoplasmic Fractions of Various Cell Lines

30,000 × g supernatants of homogenates were applied to 5% polyacrylamide disc gels. Gels were sliced, and each slice was assayed for thymidine kinase activity as described previously (Lee and Cheng, 1976). Specific activities of samples are as indicated in Table 2. Electrophoretic mobilities (R_f values) were calculated with reference to the electrophoretic mobility of bromphenol blue. (A) Vero; (B) L1210; (C) Ltk⁻; (D) Ltk⁻HSV⁺.

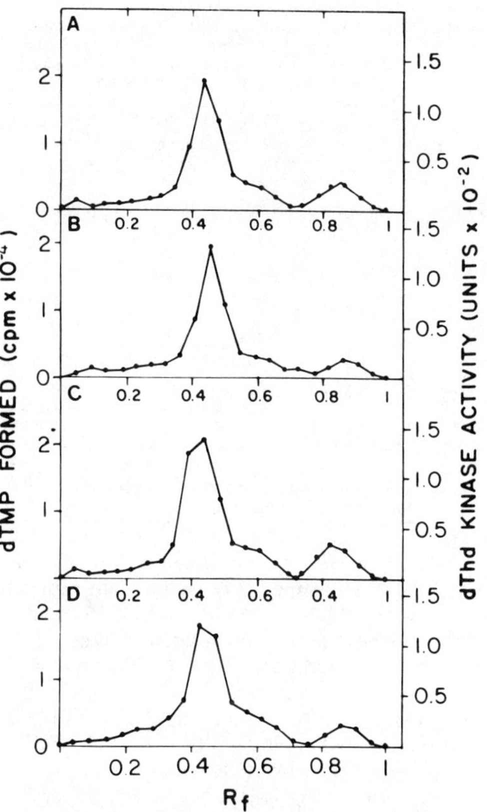

Figure 3. Electrophoretic Pattern of Thymidine Kinase Activities from Cytoplasmic Fractions of Four HSV-1 DNA-Transformed Cell Lines

The 30,000 × g supernatants of four transfected lines of mouse Ltk⁻ cells were assayed for thymidine kinase activity following gel electrophoresis as described in Figure 2. Specific activities of the samples are indicated in Table 2. (A) LH5C2-2; (B) LH5-1; (C) LH2-1; (D) LH1A2-1.

Antigenic Identity of tk in Transformed Cells

Antisera raised against purified HSV tk effectively neutralizes the enzymatic activity of viral, but not cellular, tk (Klemperer et al., 1967). We would therefore predict that the tk activity of our transfected clones should be completely neutralized by these antisera. The experiments shown in Table 2 indicate that the tk activity of four transformed lines, LH2-1, LH5-1, LH5C2-2 and LH1A2-1, can be neutralized by antisera to purified HSV-1 tk. The tk activities of Vero, Ltk⁻ and A9, a mouse L cell with wild-type tk⁺, are not neutralized by the same antisera. These data demonstrate that the tk activity present in the transformed cell lines is antigenically related to purified HSV-1 tk. The residual activity remaining after neutralization of the transformed cell extracts may represent mouse mitochondrial tk activity. It is also apparent from these data that the

four transformed lines have at least 20 times the tk activity of the parental Ltk⁻ cell.

Substrate Specificity of tk in Transformed Cells

As a final criterion of identification, the substrate specificity of the tk activity found in the four transformed cell lines was analyzed. Two known inhibitors of herpes-specific tk were used, 5–ethyl deoxyuridine (5–ethyl dUrd) and 5–allyl dUrd (Cheng et al., 1976b). They inhibited phosphorylation of thymidine by 80 and 60%, respectively, in all four cell lines. These same drugs inhibited the tk activity in Ltk⁻HSV⁺ cells, but had no effect on extracts from Vero or A9 cells (Table 3).

Stability of the tk⁺ Phenotype

Fourteen colonies were picked from experiments 2 and 3 (Table 1) and grown in HAT medium. Of these fourteen colonies, two displayed multiple abortive

Table 2. Specific Neutralization of HSV-1 Thymidine Kinase

| | Activity with Preimmune Serum | | Activity with Antiserum | |
Source of tk	Units per ml	Units per mg	Units per mg	% Residual Activity
Vero	1.0	0.25	0.37	150
A9	1.67	1.50	1.50	100
Ltk$^-$	0.06	0.007	0.007	100
Ltk$^-$HSV$^+$	20.0	2.80	0.13	5
LH1A2-1	1.1	0.19	0.017	9
LH2-1	1.15	0.15	0.019	12
LH5-1	0.9	0.14	0.015	10
LH5C2-2	1.0	0.15	0.015	10

30,000 × g supernatants of homogenates (S-30) from various cell lines were mixed with preimmune sera or antisera to purified HSV-1 tk, and tk activity was assayed as described in Experimental Procedures. Activity is expressed both as units per ml of S-30 and units per mg protein within the S-30 fraction.

Table 3. Effect of Pyrimidine Analogs on Thymidine Kinase Activity Derived from Various Sources

| | % Activity in the Presence of Analogs | |
Source of tk	5–Ethyl dUrd	5–Allyl dUrd
Vero	105	85
A9	100	100
Ltk$^-$	86	36
Ltk$^-$HSV$^+$	20	39
LH1A2-1	20	42
LH2-1	28	48
LH5-1	26	39
LH5C2-2	24	44

Assay of the effect of analogs (100 μM) was carried out as described in Experimental Procedures. % activity is calculated with respect to the activity in the absence of an analog.

Table 4. Cloning Efficiency of Various Cell Lines in a Variety of Selective and Nonselective Media

Cell Line	MEM[a]	HAT[b]	BdUrd[c]
A9	3.6×10^{-1}	ND	$<1.0 \times 10^{-5}$
Ltk$^-$	3.0×10^{-1}	$<10^{-7}$	2.7×10^{-1}
LH2-1	2.9×10^{-1}	2.2×10^{-1}	1.0×10^{-3}
LH5-1	2.0×10^{-1}	2.3×10^{-1}	2.0×10^{-3}
LH5C2-2	2.9×10^{-1}	3.8×10^{-1}	4.5×10^{-3}

[a] Nonselective medium (MEM, 5% fetal calf serum).
[b] tk$^+$ selective medium (MEM, 5% fetal calf serum, 15 μg/ml hypoxanthine, 1 μg/ml aminopterin, 5 μg/ml thymidine, 15 μg/ml glycine).
[c] tk$^-$ selective medium (MEM, 5% fetal calf serum, 30 μg/ml BdUrd).
Cells were plated in replicate, and colonies were stained and counted after 2 weeks as described in Experimental Procedures.

colonies (that is, formed small colonies which never grew larger than fifty cells) when replated, but twelve could be grown for at least 25 cell doublings under continuous selective pressure. Of these twelve, four were chosen for further study and have now been carried in HAT medium for over 6 months. Similarly, eleven colonies were picked from later experiments (see below, Table 6). One colony formed abortive colonies on replating, and ten lines have now been maintained for several months of continuous passage in HAT medium. We conclude that the acquisition of the HAT resistance phenotype and, presumably, the tk$^+$ phenotype is stable under selective conditions.

To assess the stability of the tk phenotype under other conditions, transformed cells, after approximately 50 doublings in continuous culture in HAT medium, were plated at low density into one of three media, and their cloning efficiency in these media was examined. The media were unsupplemented growth medium, growth medium supplemented with HAT, which selects for the tk$^+$ phenotype, or growth medium supplemented with BdUrd, which selects for the tk$^-$ phenotype. The data presented in Table 4 are summarized as follows; the cloning efficiency of the tk$^+$-transformed cells was the same in either (HAT) or nonselective media, and resembled that of the parental Ltk$^-$ line in nonselective media (30%). The cloning efficiency of these lines in the presence of BdUrd was reduced by two orders of magnitude to between 0.1 and 0.45%. It is of interest that this level of cloning efficiency is some 100 fold higher than that of A9, a mouse L cell derivative, or other tk$^+$ mouse cells. In this respect, these cells are similar to those tk$^+$ cells produced by infection of tk$^-$ cells with ultraviolet-inactivated virus (Davidson, Adelstein and Oxman, 1973).

Identification and Isolation of tk Active Bam Fragment

The observation that the DNA products of Bam I cleavage of HSV DNA can stably transfect tk activity suggests the use of this assay to identify the specific DNA fragment containing the thymidine kinase gene. The experimental design we have chosen involves the electrophoretic separation of specific groups of DNA and ultimately of individual DNA fragments. The fragment in which the tk gene resides is then readily identified by transfection with these fractionated populations of DNA. To this end, a Bam I digest of HSV DNA (Figure 1) was fractionated by electrophoresis on a 45 cm, 1% agarose slab gel. These DNA fragments were divided into five size classes and extracted from the agarose slab. DNA was purified free of agarose by hydroxylapatite and Sephadex G-50 chromatography, and was again analyzed by agarose gel electrophoresis (Figure 4).

The isolated size classes seen in Figure 4 were

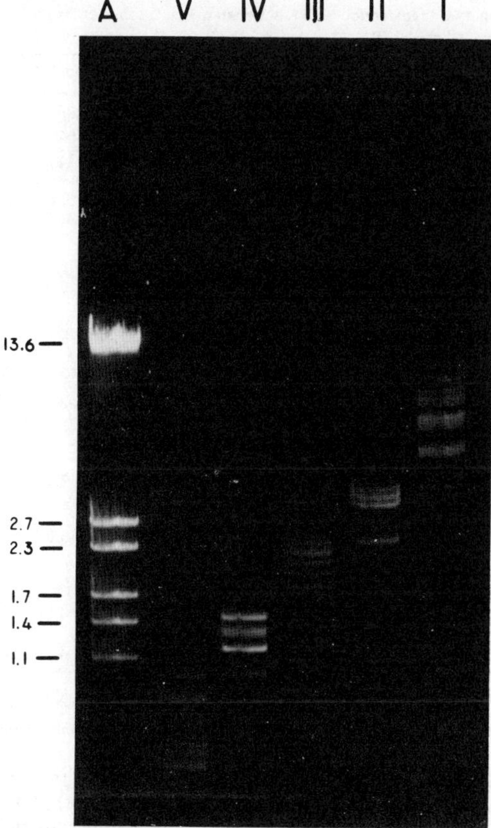

Figure 4. Fractionation of Bam 1-Cleaved HSV-1 DNA

HSV-1 DNA was digested with Bam I endonuclease, and the resultant fragments were separated on a 45 cm 1% agarose slab gel. This preparative gel was sliced, and the DNA corresponding to five discrete size classes was extracted from the gel and electrophoresed on a 0.5% agarose slab. The five size classes of DNA are shown in slots I–V. Slot A contains a preparation of Eco RI-digested Ad2 DNA as a size marker (Pettersson et al., 1973).

then used in transfection experiments following the same protocol as was used for unfractionated total digest. The results of this experiment are summarized in Table 5. Transfection activity is restricted to size class III. The small amount of activity seen in size class II probably results from the contamination of that class with size class III, as can be seen in Figure 4. Of particular interest is the observation that the specific activity of size class III is about 10 colonies per 10^6 cells per μg genome equivalent, which is about 20 times the specific activity of the unfractionated genome. These data indicate that the tk gene is located on one of five well resolved fragments ranging in molecular weight from 2.5–3.7 kb. This size class was further fractionated into its five discrete fragments (Figure 5). From the molar yield of these fragments after Bam I digestion of

Table 5. Transfection with Size Classes of Bam I-Cleaved HSV-1 DNA

Size Class	μg Equivalents of HSV-1 DNA per Dish	Σ/Σ[a]	Specific Activity[b]
I	4.0	0/5	0.0
II	4.0	3/3	0.25
III	4.0	93/3	7.75
IV	4.0	0/5	0.0
V	4.0	0/5	0.0

[a] See Table 1.
[b] Specific activity calculated as before, but based on μg equivalents of HSV-1 DNA.

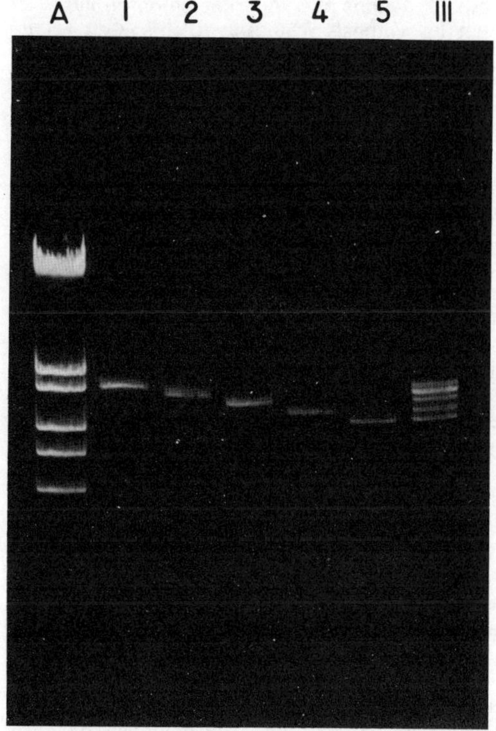

Figure 5. Isolation of the Bam I Fragment of HSV-1 DNA Containing the tk Gene

The DNA bands present in size class III (Figure 4) of a Bam I digest of HSV-1 DNA were fractionated into five fragments on 45 cm agarose slab gels. The isolated fragments were analyzed by electrophoresis on a 1% agarose slab gel. Slot A contains Eco RI-cleaved Ad2 DNA as molecular weight markers. Slots 1–5 contain the isolated fragments of size class III (see Figure 4). Slot III contains the unfractionated DNA of size class III.

total HSV DNA (Figure 1), it is probable that each of these fragments represents a homogeneous species of DNA. These individual fragments obtained from two separate preparations of HSV DNA were then assayed for their ability to transfect the tk gene (Table 6). Experiments 1 and 2 both indicate that significant transfection activity resides only in fragment 2 of size class III. The other purified fragments of class III as well as class II DNA have little or no activity. The structural gene for tk is therefore contained within a single DNA fragment 3.4 kb in length. Again it should be noted that the specific activity of band 2 is about 20 fold higher per genome equivalent than unfractionated total digest, with colonies appearing in cultures treated with as little as 0.05 μg HSV-1 DNA equivalents or about 1 ng of DNA.

Discussion

The genome of herpes simplex virus consists of about 10^8 daltons of DNA, sufficient information to direct the synthesis of 50–100 proteins (Kieff, Bachenheimer and Roizman, 1971). The enzyme thymidine kinase consistently appears shortly after infection, and considerable evidence indicates that it is coded for by the viral genome (Honess and Watson, 1974; Summers, Wagner and Summers, 1975). Considering the relatively small size of the viral genome, we would expect that a specific fragment could be obtained which could transfer viral tk activity to tk⁻ cells. In this report, we describe the isolation of a 3.4 kb fragment of HSV-1 DNA which contains the structural gene for tk and which can stably transfect cells lacking this activity. The ease with which this gene can be identified in transfection experiments suggests that this approach may be applied to the localization of any gene for which selection criteria are available.

Our studies indicate that treatment of Ltk⁻ cells with Bam I-digested HSV-1 DNA results in the stable transformation of a population of cells now expressing a tk⁺ phenotype and capable of forming colonies in HAT selection media. Eco RI-treated HSV-1 DNA does not show this activity, presumably because this enzyme cleaves within the tk gene. It is possible, however, that Eco RI does not fragment the tk gene but liberates a series of fragments capable of transfecting lytic functions of HSV. This would result in cell death and obviously obscure the transfer of tk activity. If this were the case, we would expect that digestion of HSV DNA with both Bam I and Eco RI should leave the tk structural gene intact and result in fragmentation of the DNA coding for lytic functions. Preliminary results from our laboratory indicate that this doubly digested DNA is not competent in transfection assays, suggesting that Eco RI indeed cleaves within the tk gene. More definitive data will be obtained by transfection with the Eco RI cleavage productions of the 3.4 kb Bam I fragment.

It is of obvious importance to demonstrate that tk activity transfected by Bam I-cleaved HSV DNA is of viral origin. The spontaneous rate of reversion of

Table 6. Transfection with Fractionated Size Class III of Bam I-Cleaved HSV-1 DNA

| | μg Equivalents per Dish | Experiment 1 | | Experiment 2 | |
		Σ/Σ[a]	Specific Activity[b]	Σ/Σ	Specific Activity
Size Class II	2.0			0/2	0.0
	0.7			0/2	0.0
Size Class III					
Band 1	2.0			1/2	0.25
	0.7			0/2	0.0
Band 2	1.5	48/4	8.0		
	0.5			19/2	19.0
	0.15			6/2	18.0
	0.05			5/3	30.0
Band 3	1.0	0/4	0.0		
	2.0			7/2	1.75
	0.7			3/2	2.0
Band 4	1.5	0/4	0.0		
	2.0			0/2	0.0
	0.7			0/2	0.0
Band 5	1.5	0/4	0.0		
	2.0			0/2	0.0
	0.7			0/2	0.0

[a] See Table 1.
[b] See Table 5.

the recipient Ltk⁻ cells is $<10^{-8}$. Nevertheless, it was necessary to characterize carefully the tk activity expressed by the transformed cell clones. The tk activity of these HAT-resistant clones is at least 20 times greater than the activity detected in the Ltk⁻ parent. The biochemical and antigenic properties of this enzyme were characterized by examining the neutralization of activity by specific antisera raised against HSV-1 tk; the electrophoretic mobility of the enzymatic activity; and the selective inhibition of tk activity by agents specific for the viral enzyme. For all these parameters, the enzyme was indistinguishable from HSV-1 tk and differed from either mouse or monkey cell tk. The conclusion that the tk activity appearing in transfected Ltk⁻ cells results from the introduction and expression of viral DNA therefore appears firm.

Bam I-treated DNA, competent in transfection assays, can be resolved into about 25 bands on agarose gels. Through a series of electrophoretic fractionations in concert with transfection assays, we have identified and isolated a 3.4 kb fragment of viral DNA containing the tk gene. This fragment is capable of efficiently transfecting tk activity to Ltk⁻ cells in the absence of any additional HSV information.

The frequency of transfection from experiments 2 and 3 in Table 1 is clearly too low to allow a reliable statistical analysis of dose dependence. When transfection is performed with the purified tk fragment, the frequency of successful transfection increases and a clear dose dependence emerges. The precise nature of the dose dependence clearly requires additional experiments spanning broad ranges of gene concentrations. These experiments may provide useful information for understanding the mechanism of transfection, which is currently not understood. Nevertheless, several interesting questions emerge as a result of our data. The efficiency of transfection we observe with the purified fragment of Bam I digests of HSV-1 DNA is about 10 colonies per μg DNA per 10^6 cells. Transfer of tk activity occurs with about 2% the efficiency of lytic transfection as assayed by plaque formation using purified intact HSV-1 DNA (personal observations). It is possible that this difference results in part from the fact that lytic transfection requires only the transient expression of viral functions in recipient cells. Detection of the transfer of tk activity requires the stable expression of this enzyme through several cell doublings.

Of further interest is the observation that the isolated Bam I fragment containing the tk gene is about 20 times more effective in transfection assays than the unfractionated Bam I digest of total HSV-1 DNA. This is an unexpected result for which we have no certain explanation. It is possible that transfection with the total DNA digest results in the transfer of either lytic functions or of other viral gene products capable of regulating tk expression. Transfer of these functions need only be transient to obscure the stable expression of tk activity. Expression of these inhibitory functions may therefore be far more frequent than the stable expression of tk activity, resulting in the lower efficiency of tk transfer observed with unfractionated, Bam I-cleaved HSV DNA.

At present, we have no information on the state of the viral tk gene in our transfected cell lines. Our observations indicate that the transfected Ltk$^-$ cells continue to express tk activity for hundreds of generations under selective pressure. The cloning efficiency of these transformed cells, either in the presence or absence of selective pressure, is equivalent to the cloning efficiency of the parental Ltk$^-$ line. The cloning efficiency of the transformed lines in the presence of BdUrd, however, which selects for the tk$^-$ phenotype, was about one hundredth the efficiency observed in HAT medium. This relatively high rate of reversion to the tk$^-$ phenotype may reflect the loss of the viral information from recipient cells or may result from suppression of tk expression or an unusually high mutation rate. It is of interest that high reversion rates were similarly observed in Ltk$^-$ cells infected with ultraviolet inac-

tivated herpes virus (Davidson et al., 1973), and that tk activity in these phenotypic revertants could be reactivated with high efficiency.

A final point concerns the possible control of viral tk gene expression by the regulatory machinery of the host cell. Our data indicate that the 3.4 kb fragment, containing the tk structural gene, can be efficiently transcribed by host RNA polymerase. Transcription of this gene may proceed via initiation at a viral promoter or, if chromosomal integration has occurred, at a host promoter. The size of the viral thymidine kinase, 44,000 daltons (Honess and Watson, 1974; Summers et al., 1975), requires a DNA coding segment 1.3 kb in length. The 3.4 kb fragment is therefore likely to contain the tk structural gene as well as any regulatory elements which may exist adjacent to structural gene information. If such regulatory sequences exist, then this system may permit a direct analysis of the mechanism of control of viral gene expression in host cells and serve as a model for the regulation and expression of eucaryotic genes.

Experimental Procedures

Cell Culture
Vero cells were obtained from the American Type Culture Collection and maintained in Dulbecco's modified Eagle's medium (DME) supplemented with 10% calf serum. L1210, a mouse leukemic cell line obtained from Dr. A. Bloch, was maintained in suspension culture in RPMI 1640 supplemented with 5% fetal calf serum (FCS). Mouse Ltk$^-$ cells were obtained from Dr. P. Spear and maintained in minimal essential medium (MEM) supplemented with 5% FCS and 30 μg/ml BdUrd. Mouse A9 cells, obtained from Dr. O. J. Miller, were maintained in MEM with 5% FCS. All lines except L1210 were grown as monolayer cultures at 37°C under an humidified atmosphere of 5% CO_2.

Cloning Efficiency
For cloning efficiency experiments, cells were seeded at 10, 10^2, 10^3, 10^4 or 10^5/100 mm petri dish, at least in triplicate, directly into selective or nonselective medium. After 10–17 days, cells were fixed and stained. Only large colonies were counted, and cloning efficiency was calculated as number of colonies per plate divided by number of cells initially seeded.

Transfection and Selection of the tk$^+$ Phenotype
Transfections were performed as described by Graham and van der Eb (1973). HSV-1 DNA was mixed with salmon sperm DNA (Worthington) to a final concentration of 20 μg/ml in Hepes-buffered saline: 8.0 g/l NaCl, 0.37 g/l KCl, 0.125 g/l $NA_2HPO_4 \cdot 2H_2O$, 1.0 g/l dextrose and 5.0 g/l Hepes (pH 7.10). 2 M $CaCl_2$ were added to a final concentration of 125 mM, and the mixture was allowed to stand at room temperature for 30 min before overlaying cells. Prior to transfection, Ltk$^-$ cells were grown for three cell doublings in the absence of BdUrd. Cells were then seeded at $6 \cdot 10^5$/100 mm petri dish. 24 hr later, the cell number was 10^6 per dish. Growth media were aspirated, and cells were overlaid with 0.5 ml of the previously prepared DNA/Ca phosphate mixture. After 30 min, an additional 10 ml of MEM + 5% FCS were added. After 12 hr, media were aspirated, and cultures were fed 10 ml of MEM + 5% FCS. After another 12–24 hr, media were again aspirated, and cultures were fed HAT selection media (15 μg/ml hypoxanthine, 1 μg/ml aminopterin, 5 μg/ml thymidine, 15 μg/ml glycine in MEM + 5% FCS). Cultures were

thereafter fed every 2–3 days with HAT until the experiment was terminated.

After 2 weeks, colonies were counted in stained or unstained plates. Where HAT-resistant colonies had developed, isolated colonies of cells were picked using the sterile cylinder technique (Ham, 1972).

Assay of Thymidine Kinase Activity
Cells growing in monolayer cultures were scraped into phosphate-buffered saline. After washing with phosphate-buffered saline, the cell pellet was suspended in 5 vol of extraction buffer: 0.01 M Tris-HCl (pH 7.5), 0.01 M KCl, 1 mM $MgCl_2$, 1 mM 2-mercaptoethanol and 50 μM thymidine. The cell suspension was frozen and thawed 3 times, and the KCl concentration was adjusted to 0.15 M. After sonication, the homogenate was centrifuged at 30,000 × g for 30 min, and the supernatant was used for tk assays as previously described (Lee and Cheng, 1976). One unit of thymidine kinase is defined as the amount of enzyme which converts 1 nmole of thymidine into thymidine monophosphate (TMP) per minute.

Electrophoresis of Thymidine Kinase Activity
Polyacrylamide gel electrophoresis analyses were performed in 5% acrylamide gels as previously described (Lee and Cheng, 1976). Thymidine (50 μM), mercaptoethanol (2mM) and $MgCl_2$ (2 mM) were added to the gel polymerization mixture to stabilize the enzyme activity. At the end of the run, the gels were cut into 2.5 mm slices and assayed for activity as previously described (Lee and Cheng, 1976).

Preparation of Antisera to HSV-1-Induced Thymidine Kinase and the Enzyme Neutralization Experiments
HSV-1-induced thymidine kinase was purified by means of affinity column chromatography (Cheng and Ostrander, 1976). The purified enzyme was injected into New Zealand white rabbits, and the gammaglobulin fraction was partially purified as described (Cheng, Chadha and Hughes, 1976a). The antiserum or preimmune serum was mixed with an equal volume of an enzyme preparation, and ATP-Mg was added to 6.7 mM. The enzyme-antibody mixture was incubated for 30 min at room temperature. The mixture was centrifuged at 2000 × g for 10 min, and the supernatant was assayed for tk activities.

Effect of Pyrimidine Analogs on the Thymidine Kinase Activity of Extracts from Transformed and Normal Cells
100 μM of 5-ethyl dUrd or 5-allyl dUrd were included in the assay mixture that contains 50 μM ^{14}C-thymidine and 2 mM Mg-ATP, and the inhibition of the conversion of ^{14}C-thymidine to ^{14}C-TMP was measured (Cheng et al., 1976b). The pyrimidine analogs were a gift from Dr. M. Bobek.

Growth of Virus and Isolation of Intact Viral DNA
The F strain of HSV-1 was obtained from Dr. Bernard Roizman, and after plaque purification, grown up into stocks by passage at low moi. Virus stocks routinely had titers of 2–5 × 10^9 pfu/ml.

Vero cells grown in glass roller bottles were infected at an moi of 5 for 2 hr. At this time, the infecting fluid was removed, and cells were overlaid with 50 ml of DME with 2% calf serum. At 24 hr post-infection, the cells were shaken off the glass and harvested by centrifugation. Cells were resuspended in RSB [10 mM Tris-HCl, 10 mM NaCl, 1.5 mM $MgCl_2$ (pH 7.5)] and lysed by the addition of NP-40 to 1%. Nuclei were removed by low speed centrifugation and lysed by the addition of deoxycholate to 1%. After sonication, large debris was removed by centrifugation at 10,000 × g for 10 min, and virus from the cytoplasmic and nuclear fractions was pelleted by centrifugation for 60 min at 15,000 rpm in an SW27 rotor. The virus pellets were resuspended in virus buffer (VB) [0.1 M NaCl, 0.01 M Tris-HCl (pH 7.5)] and centrifuged through 10–50% w/v sucrose gradients made in VB for 1 hr at 23,000 rpm in an SW27 rotor. The visible band of virus was removed, and after dilution in VB, pelleted by centrifugation at 25,000 rpm for 2 hr in an SW27 rotor. The pelleted virus was resuspended in 10 mM Tris, 1 mM EDTA, and DNA was released by addition of SDS to 0.5% followed by treatment with pronase (heat-inactivated for 10 min at 80°C) at 200 μg/ml. The released DNA was centrifuged through neutral sucrose gradients (10–30% w/v) prepared in 1 M NaCl, 50 mM Tris, 10 mM EDTA (pH 7.5) containing 0.15% Sarcosyl for 12 hr at 20,000 rpm in an SW27.1 rotor. The gradients were scanned at 254 nM using an ISCO density gradient fractionator. The peak corresponding to 56S DNA (Kieff, Bachenheimer and Roizman, 1971) was collected and concentrated by ethanol precipitation. Viral DNA was monitored for contamination with host sequences by isopycnic centrifugation in CsCl and by agarose gel electrophoresis.

Restriction Endonuclease Digestion
Restriction endonucleases Bam I, Eco RI, Sal I, Hind III and Hpa I were obtained from New England Biolabs. Bgl II was a gift from Dr. R. Roberts. Reaction mixtures contained in 100λ: 50 mM NaCl, 10 mM Tris-HCl, 10 mM $MgCl_2$ and 3 units of restriction enzyme per μg of HSV-1 DNA. Reactions were for 3 hr at 37°C and were terminated by the addition of 10λ of 0.2 M EDTA. DNA was extracted with phenol/chloroform and concentrated by ethanol precipitation.

Gel Electrophoresis of DNA Fragments
DNA restriction fragments were separated by electrophoresis in 0.5% or 1.0% agarose (SeaKem) slab gels. The cooled 17 cm gel was run for 2.0 hr at 200 V in a buffer system of 0.04 M Tris-HCl, 0.004 M sodium acetate, 0.001 M Na_2EDTA (pH 7.9). Gels were stained in ethidium bromide (1 μg/ml H_2O) and photographed under 254 nm ultraviolet light. For preparative runs, DNA was electrophoresed at 350 V for 13 hr in 45 cm 1.0% agarose slab gels.

Isolation of DNA Restriction Fragments
After electrophoresis, slab gels were stained with ethidium bromide and visualized with a short-wave ultraviolet light. Gel slices containing fragments of interest were made and dissolved in 5 vol of 5 M $NaClO_4$, 0.1 M Tris-HCl (pH 8.1). The DNA/agarose solution was loaded onto an hydroxyapatite column at 60°C and washed extensively with 0.14 M phosphate (pH 6.8). The DNA was eluted in a small volume with 0.5 M phosphate and desalted on a G-50 sephadex column preequilibrated with 0.01 M Tris-HCl (pH 7.9), 0.15 M NaCl, 0.001 M EDTA. DNA extracted from a staphylococcal nuclease limit digest of calf thymus nuclei (Axel, 1975) was added as carrier to 10 μg/ml, and the DNA was concentrated by ethanol precipitation. To calculate yields throughout the procedure, Bam I-digested, ^{32}P-labeled HSV-1 DNA was added as tracer.

Acknowledgments

We thank Dr. I. B. Weinstein and Dr. Sol Spiegelman for their helpful criticism and generous support. This work was supported by contracts and grants from the National Cancer Institute, the American Cancer Society and the American Leukemia Society. Y.-c. Cheng is a scholar of the American Leukemia Society.

Received January 5, 1977; revised February 10, 1977

References

Axel, R. (1975). Biochemistry *14*, 2921–2925.

Cheng, Y. C. and Ostrander, M. (1976). J. Biol. Chem. *251*, 2605–2610.

Cheng, Y. C., Chadha, K. C. and Hughes, R. G. (1976a). Infectious Dis. Immunol., in press.

Cheng, Y. C., Domin, B., Sharma, R. A. and Bobek, M. (1976b). Antimicrobial Agents and Chemotherapy *10*, 119–122.

Cohen, S. N. and Chang, A. C. Y. (1974). In Microbiology (Washington, D.C.: American Society of Microbiology). pp. 64–73.

Davidson, R. L., Adelstein, S. J. and Oxman, M. N. (1973). Proc. Nat. Acad. Sci. USA 70, 1912–1916.

Graham, F. L. and van der Eb, A. J. (1973) Virology 52, 456–467.

Graham, F. L., Abrahams, P. J., Mulder, C., Heijneker, H. L., Warnaar, S. O., de Vries, F. A. J., Fiers, W. and van der Eb, A. J. (1975). Cold Spring Harbor Symp. Quant. Biol. 39, 637–650.

Ham, R. G. (1972). In Methods in Cell Physiology, 5, D. M. Prescott, ed. (New York: Academic Press), pp. 37–74.

Honess, R. W. and Watson, D. H. (1974). J. Gen. Virol. 22, 171–185.

Kieff, E. D., Bachenheimer, S. L. and Roizman, B. (1971). J. Virol. 8, 125–132.

Kit, S., Dubbs, D., Piekarski, L. and Hsu, T. (1963). Exp. Cell Res. 31, 297–312.

Klemperer, H. G., Haynes, G. R., Shedden, W. I. H. and Watson, D. H. (1967). Virology 31, 120–128.

Lee, L. S. and Cheng, Y. C. (1976). J. Biol. Chem. 251, 2600–2604.

Littlefield, J. (1963). Proc. Nat. Acad. Sci. USA 50, 568–573.

Macnab, J. C. M. and Timbury, M. C. (1976). Nature 261, 233–235.

Munyon, W., Kraiselburd, E., Davies, D. and Mann, J. (1971). J. Virol. 7, 813–820.

Pettersson, U., Mulder, C., Delius, H. and Sharp, P. A. (1973). Proc. Nat. Acad. Sci. USA 70, 200–204.

Summers, W. P., Wagner, M. and Summers, W. C. (1975). Proc. Nat. Acad. Sci. USA 72, 4081–4084.

Note Added in Proof

The 3.4 kb band we have isolated consists of a partial molar fragment along with a molar fragment containing the tk gene. We have subsequently purified the tk-containing fragment by isolation of a pure 8.3 kb fragment following HPA I digestion which can transfer the tk activity. Cleavage of this fragment with Bam I generates the 3.4 kb fragment containing the tk gene in pure form.

28

TRANSFORMATION OF MAMMALIAN CELLS WITH GENES FROM PROCARYOTES AND EUCARYOTES

M. Wigler, R. Sweet, G. K. Sim, B. Wold, A. Pellicer,
E. Lacy, T. Maniatis, S. Silverstein, and R. Axel

Summary

We have stably transformed mammalian cells with precisely defined procaryotic and eucaryotic genes for which no selective criteria exist. The addition of a purified viral thymidine kinase (tk) gene to mouse cells lacking this enzyme results in the appearance of stable transformants which can be selected by their ability to grow in HAT. These biochemical transformants may represent a subpopulation of competent cells which are likely to integrate other unlinked genes at frequencies higher than the general population. Co-transformation experiments were therefore performed with the viral tk gene and bacteriophage ΦX174, plasmid pBR322 or the cloned chromosomal rabbit β–globin gene sequences. Tk⁺ transformants were cloned and analyzed for co-transfer of additional DNA sequences by blot hybridization. In this manner, we have identified mouse cell lines which contain multiple copies of ΦX, pBR322 and the rabbit β–globin gene sequences. The ΦX co-transformants were studied in greatest detail. The frequency of co-transformation is high: 15 of 16 tk⁺ transformants contain the ΦX sequences. Selective pressure was required to identify co-transformants. From one to more than fifty ΦX sequences are integrated into high molecular weight nuclear DNA isolated from independent clones. Analysis of subclones demonstrates that the ΦX genotype is stable through many generations in culture. This co-transformation system should allow the introduction and stable integration of virtually any defined gene into cultured cells. Ligation to either viral vectors or selectable biochemical markers is not required.

Introduction

Specific genes can be stably introduced into cultured cells by DNA-mediated gene transfer. The rare transformant is usually detected by biochemical selection. In this manner, we have isolated cells transformed with a variety of cellular and viral genes coding for selectable biochemical markers (Wigler et al., 1977, 1978, 1979). The isolation of cells transformed with genes which do not code for selectable markers, however, is problematic since current transformation procedures are highly inefficient. This paper demonstrates the feasibility of co-transforming cells with two physically unlinked genes. Co-transformed cells can be identified and isolated when one of these genes codes for a selectable marker. We have used a viral thymidine kinase gene as a selectable marker to isolate mouse cell lines which contain the tk gene along with either bacteriophage ΦX174, plasmid pBR322 or the cloned rabbit β–globin gene sequences stably integrated into cellular DNA. The introduction of cloned eucaryotic genes into animal cells may provide a means for studying the functional consequences of DNA sequence organization.

Results

Experimental Design

The addition of the purified thymidine kinase (tk) gene from herpes simplex virus to mutant mouse cells lacking tk results in the appearance of stable transformants expressing the viral gene which can be selected by their ability to row in HAT (Maitland and McDougall, 1977; Wigler et al., 1977). To obtain co-transformants, cultures are exposed to the tk gene in the presence of a vast excess of a well defined DNA sequence for which hybridization probes are available. Tk⁺ transformants are isolated and scored for the co-transfer of additional DNA sequences by molecular hybridization.

Co-transformation of Mouse Cells with ΦX174 DNA

We initially used ΦX DNA in co-transformation experiments with the tk gene as the selectable marker. ΦX replicative form DNA was cleaved with Pst I, which recognizes a single site in the circular genome (Figure 1) (Sanger et al., 1977). 500 pg of the purified tk gene were mixed with 1–10 μg of Pst-cleaved ΦX replicative form DNA. This DNA was then added to mouse Ltk⁻ cells using the transformation conditions previously described (Wigler et al., 1979). After 2 weeks in selective medium (HAT), tk⁺ transformants were observed at a frequency of one colony per 10⁶ cells per 20 pg of purified gene. Clones were picked and grown into mass culture.

We then asked whether tk⁺ transformants also contained ΦX DNA sequences. High molecular weight DNA from the transformants was cleaved with the restriction endonuclease Eco RI, which recognizes no sites in the ΦX genome. The DNA was fractionated by agarose gel electrophoresis and transferred to nitrocellulose filters, and these filters were then annealed with nick-translated ³²P–ΦX DNA (blot hybridization) (Southern, 1975; Botchan, Topp and Sambrook, 1976; Pellicer et al., 1978). These annealing experi-

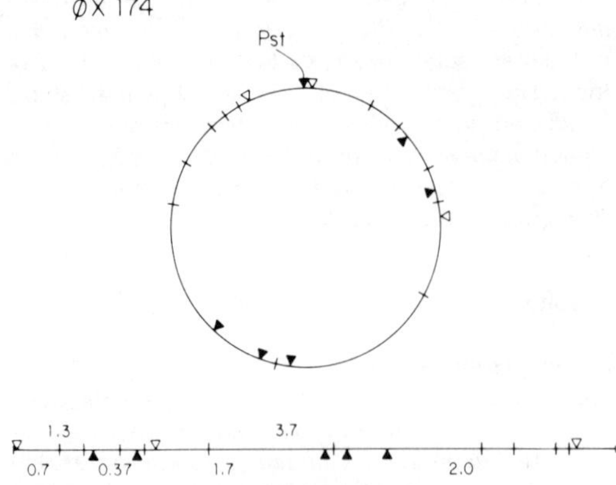

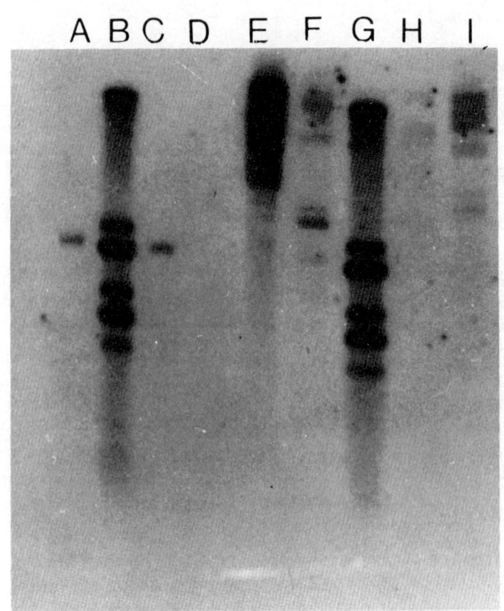

Figure 1.   Cleavage Map of the ΦX174 Genome

Cleavage sites for the restriction endonucleases Pst I, Hpa I (▽), Hpa II (▼) and Hae III (I) are shown for circular RFI and Pst I-linearized ΦX174 DNA (Sanger et al., 1977). The numbers above the line refer to the sizes of the internal Hpa I fragments in kbp, while those below the line refer to the sizes of the Hpa II fragments.

Figure 2.   Identification of ΦX Sequences in Cells Transformed with ΦX174 DNA and the HSV tk Gene

Ltk⁻ aprt⁻ cells were transformed with ΦX174 DNA and the HSV tk gene using salmon sperm DNA as carrier. Tk⁺ transformants were selected by growth in HAT, cloned and grown into mass culture in HAT. High molecular weight DNA was extracted from seven independently isolated clones; 15 μg of DNA from each were digested with Eco RI, electrophoresed through 1% agarose gels, denatured in situ and transferred to nitrocellulose filters which were then annealed with ^{32}P-ΦX174 DNA (5 × 10^8 cpm/μg) to identify co-transformants. Lanes B and G are Eco RI digests of ^{32}P-adenovirus 2 DNA; the six bands are 20.3, 4.2, 3.6, 2.6, 2.2 and 1.8 kbp. Lanes A, C, D, E, F, H and I are the seven independently isolated clones ΦX1–7, respectively. Only clone ΦX3 (lane D) lacks detectable ΦX sequences.

ments (Figure 2) demonstrate that six of the seven transformants had acquired bacteriophage sequences. Since the ΦX genome is not cut by the enzyme Eco RI, the number of bands observed reflects the minimum number of eucaryotic DNA fragments containing information homologous to ΦX. The clones contain variable amounts of ΦX sequences. Clones ΦX1 and ΦX2 (Figure 2, lanes A and C) reveal a single annealing fragment which is smaller than the ΦX genome. In these clones, therefore, only a portion of the transforming sequences persists. In lane D, we observe a tk⁺ transformant (clone ΦX3) with no detectable ΦX sequences. Clones ΦX4, 5, 6 and 7 (lanes E, F, H and I) reveal numerous high molecular weight bands which are too closely spaced to count, indicating that these clones contain multiple ΦX-specific fragments. These experiments demonstrate co-transformation of cultured mammalian cells with the viral tk gene and ΦX DNA.

Selection Is Necessary to Identify ΦX Transformants

We next asked whether transformation with ΦX DNA was restricted to the population of tk⁺ cells or whether a significant proportion of the original culture now contained ΦX sequences. Cultures were exposed to a mixture of the tk gene and ΦX DNA in a molar ratio of 1:2000 or 1:20,000. Half of the cultures were plated under selective conditions, while the other half were plated in neutral media at low density to facilitate cloning. Both selected (tk⁺) and unselected (tk⁻) colonies were picked, grown into mass culture and scored for the presence of ΦX sequences. In this series of experiments, eight of the nine tk⁺ selected colonies contained phage information (Figure 3). As

in the previous experiments, the clones contained varying amounts of ΦX DNA. In contrast, none of fifteen clones picked at random from neutral medium contained any ΦX information (data not shown). Thus the addition of a selectable marker facilitates the identification of those cells which contain ΦX DNA.

ΦX Sequences Are Integrated into Cellular DNA

Cleavage of DNA from ΦX transformants with Eco RI (Figure 2) generates a series of fragments which contain ΦX DNA sequences. These fragments may reflect multiple integration events. Alternatively, these fragments could result from tandem arrays of complete or partial ΦX sequences which are not integrated into cellular DNA. To distinguish between these possibilities, transformed cell DNA was cut with Bam HI or Eco RI, neither of which cleaves the ΦX genome. If the ΦX DNA sequences were not integrated, neither of these enzymes would cleave the ΦX fragments. Identical patterns would be generated from undigested DNA

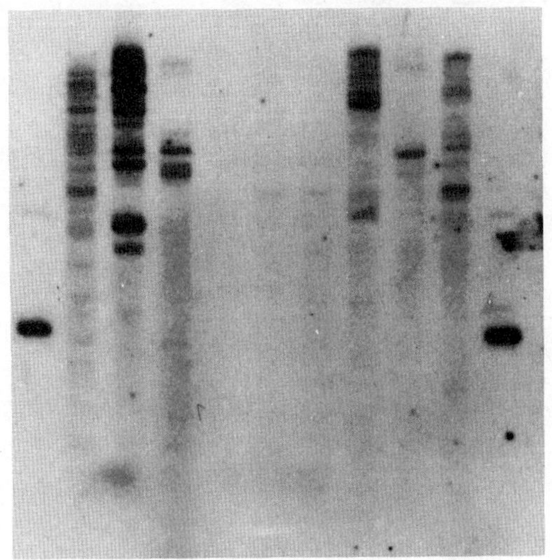

Figure 3. ΦX Sequences in tk⁺ Transformants

Cells were co-transformed as described in the legend to Figure 2, and half of the cultures were fed HAT, while the other half were replated under cloning conditions in DME. Nine colonies were selected in HAT and assayed for ΦX sequences as described (see Figure 2). Lanes A and K each contain 30 pg (2 gene equivalents) of Pst I-linearized ΦX174 DNA. Lanes B–J contain Eco RI-digested DNA from nine independently isolated tk⁺ transformants. Only one clone (lane E) does not contain ΦX sequences. None of fifteen clones isolated without selection contained ΦX sequences (blot not shown).

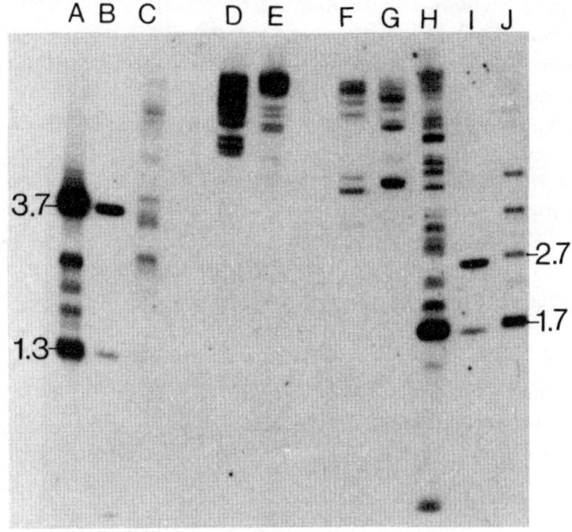

Figure 4. Extent of Sequence Representation in ΦX Co-transformants

High molecular weight DNA from co-transformant clones ΦX4 and ΦX5 was digested with either Eco RI, Bam HI, Hpa I or Hpa II and analyzed for the presence of ΦX sequences as described in the legend to Figure 2. (Lanes B and I) 50 pg (4 gene equivalents) of ΦX RFI DNA digested with Hpa I and Hpa II, respectively. (Lanes A, D, E and H) 15 μg of clone ΦX4 DNA digested with Hpa I, Eco RI, Bam HI and Hpa II, respectively, and analyzed for ΦX sequences by blot hybridization. (Lanes C, F, G and J) 15 μg of clone ΦX5 DNA digested with Hpa I, Eco RI, Bam HI or Hpa II, respectively.

and from DNA cleaved with either of these enzymes. If the sequences are integrated, then Bam HI and Eco RI should recognize different sites in the flanking cellular DNA and generate unique restriction patterns. DNA from clones ΦX4 and ΦX5 was cleaved with Bam HI or Eco RI and analyzed by Southern hybridization (Figure 4: clone 4, lanes D and E; clone 5, lanes F and G). In each instance, the annealing pattern with Eco RI fragments differed from that observed with the Bam HI fragments. Furthermore, the profile obtained with undigested DNA reveals annealing only in very high molecular weight regions with no discrete fragments observed (data not shown). Similar observations were made on clone ΦX1 (data not shown). Thus most of the ΦX sequences in these three clones are integrated into cellular DNA.

Intracellular Localization of the ΦX Sequences

The location of ΦX sequences in transformed cells was determined by subcellular fractionation. Nuclear and cytoplasmic fractions were prepared, and the ΦX DNA sequence content of each was assayed by blot hybridization. The data (not shown) indicate that 95% of the ΦX sequences are located in the nucleus. High and low molecular weight nuclear DNA was prepared by Hirt fractionation (Hirt, 1967). Hybridization with

DNA from these two fractions indicates that more than 95% of the ΦX information co-purifies with the high molecular weight DNA fraction. The small amount of hybridization observed in the supernatant fraction reveals a profile identical to that of the high molecular weight DNA, suggesting contamination of this fraction with high molecular weight DNA.

Extent of Sequence Representation of the ΦX Genome

The annealing profiles of DNA from transformed clones digested with enzymes that do not cleave the ΦX genome provide evidence that integration of ΦX sequences has occurred and allow us to estimate the number of ΦX sequences integrated. Annealing profiles of DNA from transformed clones digested with enzymes which cleave within the ΦX genome allow us to determine what proportion of the genome is present and how these sequences are arranged following integration. Cleavage of ΦX with the enzyme Hpa I generates three fragments for each integration event (see Figure 1): two "internal" fragments of 3.7 and 1.3 kb which together comprise 90% of the ΦX genome, and one "bridge" fragment of 0.5 kb which spans the Pst I cleavage site. The annealing profile observed when clone ΦX4 is digested with Hpa I is shown in Figure 4, lane A. Two intense bands are observed at 3.7 and 1.3 kb. A less intense series of bands of higher molecular weight is also observed, some of which probably represent ΦX sequences adjacent to cellular DNA. These results indicate that at least 90% of the ΦX genome is present in these cells. It is worth noting that the internal 1.3 kb Hpa I fragment is bounded by an Hpa I site only 30 bp from the Pst I cleavage site. Comparison of the intensities of the internal bands with known quantities of Hpa I-cleaved ΦX DNA suggests that this clone contains approximately 100 copies of the ΦX genome (Figure 4, lanes A and B). The annealing pattern of clone 5 DNA cleaved with Hpa I is more complex (Figure 4, lane C). If internal fragments are present, they are markedly reduced in intensity; instead, multiple bands of varying molecular weight are observed. The 0.5 kb Hpa I fragment which bridges the Pst I cleavage site is not observed for either clone ΦX 4 or clone ΦX5 (data not shown).

A similar analysis of clone ΦX4 and ΦX5 DNA was performed with the enzyme Hpa II. This enzyme cleaves the ΦX genome five times, thus generating four "internal" fragments of 1.7, 0.5, 0.5 and 0.2 kb, and a 2.6 kb "bridge" fragment which spans the Pst I cleavage site (Figure 1). The annealing patterns for Hpa II-cleaved DNA from ΦX clones 4 and 5 are shown in Figure 4 (clone ΦX4, lane H; clone ΦX5, lane J). In each clone an intense 1.7 kb band is observed, consistent with the retention of at least two internal Hpa II sites. The 0.5 kb internal fragments can also be observed, but they are not shown on this gel. Many additional fragments, mostly of higher molecular weight, are also present in each clone. These presumably reflect the multiple integration sites of ΦX DNA in the cellular genome. The 2.6 kb fragment bridging the Pst I cleavage site, however, is absent from clone ΦX4 (Figure 4, lane H). Reduced amounts of annealing fragments which co-migrate with the 2.6 kb Hpa II bridge fragment are observed in clone ΦX 5 (Figure 4, lane J). Similar observations were made in experiments with the enzyme Hae III. The annealing pattern of Hae III-digested DNA from these clones is shown in Figure 5 (clone ΦX4, lane B; clone ΦX5, lane C). In accord with our previous data, the 0.87 kb Hae III bridge fragment spanning the Pst site is absent or present in reduced amount in transformed cell DNA. Thus in general "internal" fragments of ΦX are found in these transformants, while "bridge" fragments which span the Pst I cleavage site are reduced or absent (see Discussion).

Stability of the Transformed Genotype

Our previous observations on the transfer of selectable biochemical markers indicate that the transformed phenotype remains stable for hundreds of generations if cells are maintained under selective pressure. If maintained in neutral medium, the transformed phenotype is lost at frequencies which range from <0.1 to as high as 30% per generation (Wigler et al., 1977,

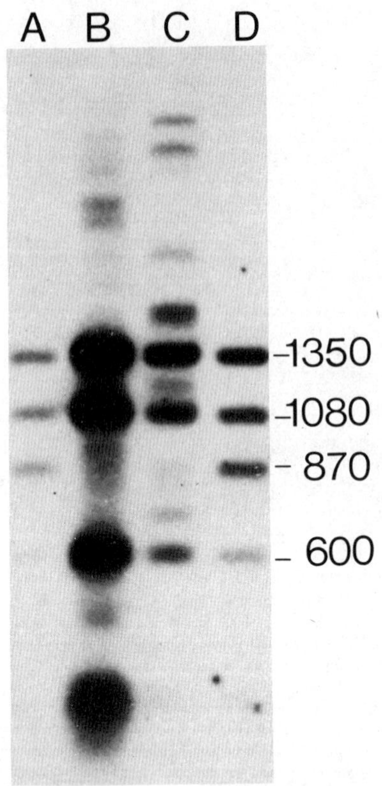

Figure 5. Annealing Pattern of DNA from Hae III-Digested ΦX Co-transformants

High molecular weight DNA from clones ΦX4 and ΦX5 was digested with Hae III, and the annealing profile was compared with that of Hae III-digested ΦXRFI DNA. Lanes A and D contain 30 and 50 pg (2 and 4 gene equivalents, respectively) of Hae III-digested ΦXRFI DNA. Lanes B and C contain 15 μg of Hae III-digested DNA from clones ΦX4 and ΦX5, respectively. The sizes of the prominent ΦX Hae III fragments in lanes A and D are 1350, 1080, 870 and 600 base pairs.

1979). The use of transformation to study the expression of foreign genes depends upon the stability of the transformed genotype. This is an important consideration with genes for which no selctive criteria are available. We assume that the presence of ΦX DNA in our transformants confers no selective advantage on the recipient cell. We therefore examined the stability of the ΦX genotype in the descendants of two clones after numerous generations in culture. Clones ΦX4 and ΦX5, both containing multiple copies of ΦX DNA, were subcloned and six independent subclones from each original clone were picked and grown into mass culture. DNA from each of these subclones was then digested with either Eco RI or Hpa I, and the annealing profiles of ΦX-containing fragments were compared with those of the original parental clone. The annealing pattern observed for four of the six ΦX4 subclones is virtually identical to that of the parent (Figure 6A). In

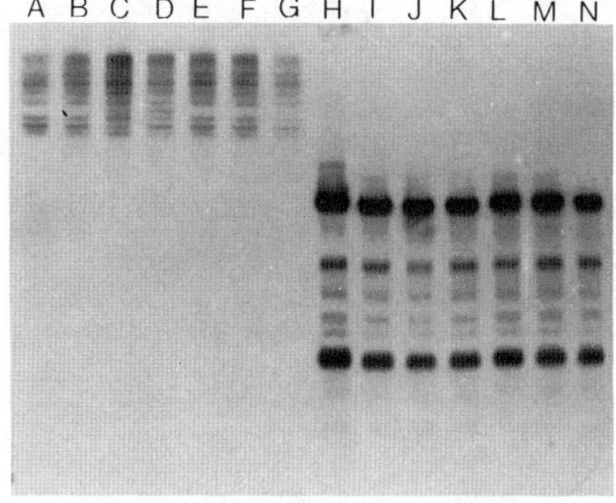

a

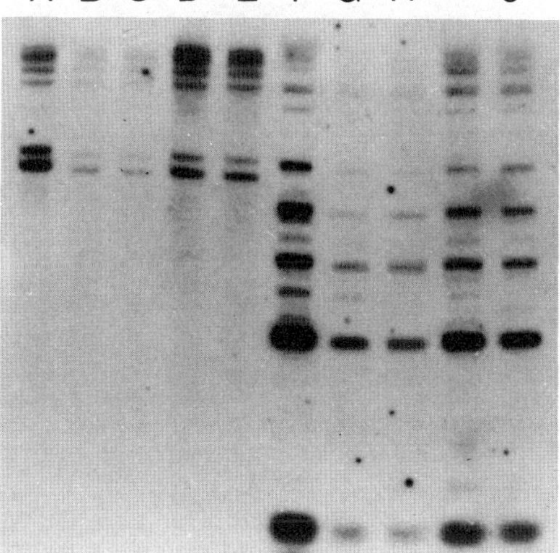

b

Figure 6. ΦX Sequences in Subclones of Co-transformants

(a) Annealing profiles of DNA from parental clone ΦX4 digested with Eco RI (lane A) and Hpa I (lane H) are compared with DNA from six independent subclones digested with either Eco RI (lanes B–G) or Hpa I (lanes I–N).

(b) High molecular weight DNA from four subclones of clone ΦX5 was isolated, cleaved with either Eco RI or Hpa II and compared with parental clone ΦX5 DNA. Lanes A and F contain clone ΦX5 DNA digested with Eco RI and Hpa II, respectively. DNA from four independently isolated subclones digested with either Eco RI (lanes B–E) or Hpa II (lanes G–J) was analyzed by blot hybridization.

two subclones, an additional Eco RI fragment appeared which is of identical molecular weight in both. This may have resulted from genotypic heterogeneity in the parental clone prior to subcloning. The patterns obtained for the subclones of ΦX5 are again virtually identical to the parental annealing profile (Figure 6B). These data indicate that ΦX DNA is maintained within the ten subclones examined for numerous generations

without significant loss or translocation of information.

Integration of pBR322 DNA into Mouse Cells

We have extended our observations on co-transformation to the EK2-approved bacterial vector, plasmid pBR322. pBR322 linearized with Bam HI was mixed with the purified viral tk gene in a molar ratio of 1000: 1. Tk$^+$ transformants were selected and scored for the presence of pBR322 sequences. The Bgl I restriction map of Bam HI linearized pBR322 DNA is shown in Figure 7. Cleavage of this DNA with Bgl I generates two internal fragments of 2.4 and 0.3 kb. The sequence content of the pBR322 transformants was determined by digestion of transformed cell DNA with Bgl I followed by annealing with ^{32}P-labeled plasmid DNA. Four of five clones screened contained pBR sequences. Two of these clones contained the 2.4 kb internal fragment (Figure 8). The 0.3 kb fragment would not be detected on these gels. From the intensity of the 2.4 kb band in comparison with controls, we conclude that multiple copies of this fragment are present in these transformants. Other bands are observed which presumably represent the segments of pBR322 attached to cellular DNA.

Transformation of Mouse Cells with the Rabbit β–Globin Gene

Transformation with purified eucaryotic genes may provide a means for studying the expression of cloned genes in a heterologous host. We have therefore performed co-transformation experiments with the rabbit β major globin gene which was isolated from a cloned library of rabbit chromosomal DNA (Maniatis et al., 1978). One β–globin clone designated RβG-1 (Lacy et al., 1978) consists of a 15 kb rabbit DNA fragment carried on the bacteriophage λ cloning vector Charon 4a. Intact DNA from this clone (RβG-1) was mixed with the viral tk DNA at a molar ratio of 100:1, and tk$^+$ transformants were isolated and examined for the presence of rabbit globin sequences. A restriction map of RβG-1 is shown in Figure 9. Cleavage of RβG-1 with the enzyme Kpn I generates 14.7 kb fragment which contains the entire rabbit β–globin gene. This fragment was purified by gel electrophoresis and nick-translated to generate a probe for subsequent annealing experiments. The β–globin genes of mouse and rabbit are partially homologous, although we do not observe annealing of the rabbit β–globin probe with Kpn-cleaved mouse DNA under our experimental conditions (Figure 10, lanes C, D and G). In contrast, cleavage of rabbit liver DNA with Kpn I generates the expected 4.7 kb globin band (Figure 10, lane B). Cleavage of transformed cell DNA with the enzyme Kpn I generates a 4.7 kb fragment containing globin-specific information in six of the eight tk$^+$ transformants examined (Figure 10). In two of the clones (Figure 10, lanes E and H), additional rabbit globin bands are observed which probably re-

PBR 322

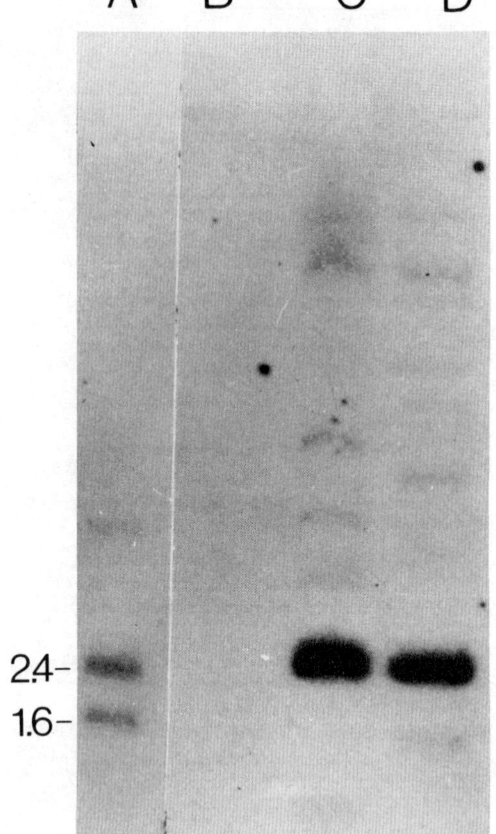

Figure 7. Cleavage Map of pBR322

The Bgl I restriction endonuclease map for Bam HI-linearized pBR322 DNA is shown. The fragment sizes are in kbp, as determined by G. Sutcliffe (personal communication).

sult from the loss of at least one of the Kpn sites during transformation. The number of rabbit globin genes integrated in these transformants is variable. In comparison with control lanes (Figure 10, lanes A and L), some clones contain a single copy of the gene (lanes I, J and K), while others contain multiple copies of this heterologous gene (lanes E, F and H). These results demonstrate that cloned eucaryotic genes can be introduced into cultured mammalian cells by co-transformation.

Transformation Competence Is Not Stably Inherited

Our data suggest the existence of a subpopulation of transformation-competent cells within the total cell population. If competence is a stably inherited trait, then cells selected for transformation should be better recipients in subsequent gene transfer experiments than their parental cells. Two results indicate that as in procaryotes, competence is not stably heritable. In the first series of experiments, a double mutant, Ltk⁻ aprt⁻ (deficient in both tk and aprt), was transformed to either the tk⁺ aprt⁻ or the tk⁻ aprt⁺ phenotype using cellular DNA as donor (Wigler et al., 1978, 1979). These clones were then transformed to the tk⁺ aprt⁺ phenotype. The frequency of the second transformation was not significantly higher than the first. In another series of experiments, clones ΦX4 and ΦX5 were used as recipients for the transfer of a mutant folate reductase gene which renders recipient cells resistant to methotrexate (mtx). The cell line A29 Mtx^RIII contains a mutation in the structural gene for dihydrofolate reductase, reducing the affinity of this enzyme for methotrexate (Flintoff, Davidson and Siminovitch, 1976). Genomic DNA from this line was used to transform clones ΦX4 and ΦX5 and Ltk⁻ cells. The frequency of transformation to mtx resistance for the ΦX clones was identical to that observed with the parental Ltk⁻ cells. We conclude that competence is not a stably heritable trait and may therefore be a transient property of cells.

Discussion

In these studies, we have stably transformed mammalian cells with precisely defined procaryotic and eucaryotic genes for which no selective criteria exist. Our chosen experimental design derives from studies of transformation in bacteria which indicate that a

Figure 8. Physical Map of pBR322 Sequences in Co-transformants

Cells were exposed to pBR322 DNA and the viral tk gene and selected in HAT. High molecular weight DNA from three independent clones was digested with Bgl I and electrophoresed on a 1% agarose gel. The DNA was denatured in situ and transferred to nitrocellulose filters which were annealed with ³²P–pBR322 DNA. (Lane A) 5 pg of pBR322 DNA digested with Bgl I; (lanes B–D) 15 μg of DNA from three independent tk⁺ transformants.

small but selectable subpopulation of cells is competent in transformation (Thomas, 1955; Hotchkiss, 1959; Tomasz and Hotchkiss, 1964; Spizizen, Reilly and Evans, 1966). If this is also true for animal cells, then biochemical transformants will represent a subpopulation of competent cells which are likely to integrate other unlinked genes at frequencies higher than the general population. Thus, to identify transformants containing genes which provide no selectable trait, cultures were co-transformed with a physically unlinked gene which provided a selectable marker. This co-transformation system should allow the introduction and stable integration of virtually any defined gene into cultured cells. Ligation to either viral vectors or selectable biochemical markers is not required.

RβG-1

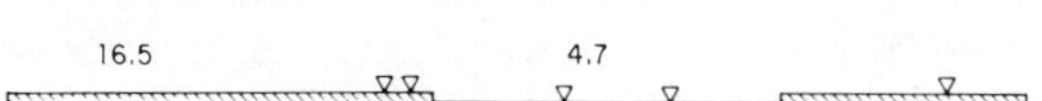

Figure 9. Physical Map of Rabbit β-Globin Phage Clone RβG-1

Cleavage sites for the restriction endonuclease Kpn I within RβG-1 are shown (Lacy et al., 1979). The numbers refer to the sizes of the fragments in kbp.

Co-transformation experiments were performed using the HSV tk gene as the selectable biochemical marker. The addition of this purified tk gene to mouse cells lacking thymidine kinase results in the appearance of stable transformants which can be selected by their ability to grow in HAT. Tk$^+$ transformants were cloned and analyzed by blot hybridization for co-transfer of additional DNA sequences. In this manner, we have constructed mouse cell lines which contain multiple copies of ΦX, pBR322 and rabbit β-globin gene sequences.

The suggestion that these observations could result from contaminating procaryotic cells in our cultures is highly improbable. At least one of the rabbit β-globin mouse transformants expresses polyadenylated rabbit β-globin RNA sequences as a discrete 9S cytoplasmic species (B. Wold et al., manuscript in preparation). The elaborate processing events required to generate 9S globin RNA correctly are unlikely to occur in procaryotes.

The ΦX co-transformants were studied in greatest detail. The frequency of co-transformation is high: 14 of 16 tk$^+$ transformants contain ΦX sequences. The ΦX sequences are integrated into high molecular weight nuclear DNA. The number of integration events varies from one to more than fifty in independent clones. The extent of the bacteriophage genome present within a given transformant is also variable; while some clones have lost up to half the genome, other clones contain over 90% of the ΦX sequences. Analysis of subclones demonstrates that the ΦX genotype is stable through many generations in culture. Similar conclusions are emerging from the characterization of the pBR322 and globin gene co-transformants.

Hybridization analysis of restriction endonuclease-cleaved transformed cell DNA allows us to make some preliminary statements on the nature of the integration intermediate. Only two ΦX clones have been examined in detail. In both clones, the donor DNA was Pst I-linearized ΦX DNA. We have attempted to distinguish between the integration of a linear or circular intermediate. If either precise circularization or the formation of linear concatamers had occurred at the Pst I cleavage site, and if integration occurred at random points along this DNA, we would expect cleavage maps of transformed cell DNA to mirror the circular ΦX map. The bridge fragment, however, is not observed or is present in reduced amounts in digests of

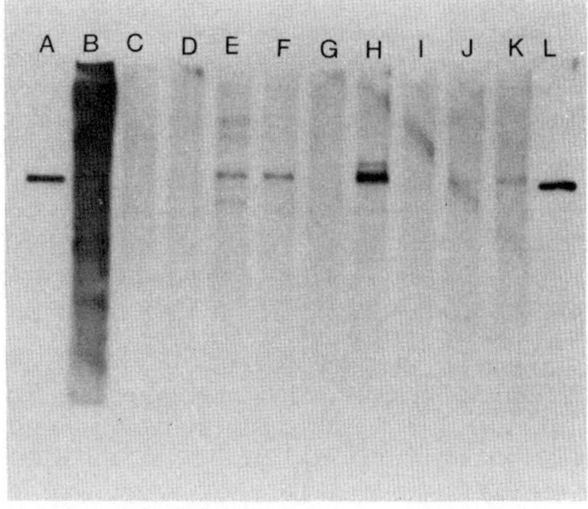

Figure 10. The Rabbit β-Globin Gene Is Present in Mouse DNA

Cells were exposed to RβG-1 DNA and the viral tk gene and selected in HAT. High molecular weight DNA from eight independent clones was digested with Kpn I and electrophoresed on a 1% agarose gel. The DNA was denatured in situ and transferred to nitrocellulose filters, which were then annealed with a ^{32}P-labeled 4.7 kbp fragment containing the rabbit β-globin gene. (Lanes A and L) 50 pg of the 4.7 kbp Kpn fragment of RβG-1; (lane B) 15 μg of rabbit liver DNA digested with Kpn; (lane C) 15 μg of Ltk$^-$ aprt$^-$ DNA; (lanes D–K) 15 μg of DNA from each of eight independently isolated tk$^+$ transformants.

transformed cell DNA with three different restriction endonucleases. The fragments observed are in accord with a model in which ΦX DNA integrates as a linear molecule. Alternatively, it is possible that intramolecular recombination of ΦX DNA occurs, resulting in circularization with deletions at the Pst termini (Lai and Nathans, 1974). Random integration of this circular molecule would generate a restriction map similar to that observed for clones ΦX4 and ΦX5. Other more complex models of events occurring before, during or after integration can also be considered. Although variable amounts of DNA may be deleted from termini during transformation, most copies of integrated ΦX sequences in clone ΦX4 retain the Hpa I site, which is only 30 bp from the Pst I cleavage site. Whatever the mode of integration, it appears that cells can be stably transformed with long stretches of donor DNA. We have observed transformants containing contiguous stretches of donor DNA 50 kb long (B. Wold et al., unpublished studies).

We have attempted to identify cells transformed with ΦX sequences in the absence of selective pressure. Cultures were exposed to ΦX and tk DNA and cells were cloned under nonselective conditions. ΦX sequences were absent from all fifteen clones picked. In contrast, 14 of 16 clones selected for the tk$^+$ phenotype contained ΦX DNA. The simplest interpretation is that a subpopulation of cells within the culture is competent in the uptake and integration of DNA. In this subpopulation of cells, two physically unlinked

221

genes can be introduced into the same cell with high frequency. At present we can only speculate on the biological basis of competence. Competent cells may be genetic variants within the culture; however, our studies indicate that the competent phenotype is not stably inherited. If we extrapolate from studies in procaryotes, the phenomenon of competence is likely to be a complex and transient property reflecting the metabolic state of the cell.

Co-transformants contain at least one copy of the tk gene and variable amounts of ΦX DNA. Although transformation was performed with ΦX and tk sequences at a molar ratio of 1000:1, the sequence ratio observed in transformants never exceeded 100:1. There may be an upper limit to the number of integration events that a cell can tolerate, beyond which lethal mutations occur. Alternatively, it is possible that the efficiency of transformation may depend upon the nature of the transforming fragment. The tk gene may therefore represent a more efficient transforming agent than phage DNA.

The usefulness of the co-transformation method will depend to a large extent on its generality. To date, we have limited experience with other cell lines. The use of tk as a selectable marker restricts host cells to tk mutants. In unpublished studies, we have demonstrated the co-transfer of plasmid pBR 322DNA into Ltk$^-$ aprt$^-$ cells using aprt$^+$ cellular DNA as donor and aprt as selectable marker. Furthermore, the use of dominant acting mutant genes which can confer drug resistance may extend the host range for co-transformation to virtually any cultured cell.

The stable transfer of ΦX DNA sequences to mammalian cells serves as a model system for the introduction of defined genes for which no selective criteria exist. We have used the tk co-transformation system to transform cells with the bacterial plasmid pBR322 and the cloned rabbit β–globin gene. Experiments which indicate that several of the pBR transformants contain an uninterrupted sequence which includes the replicative origin and the gene coding for ampicillin resistance (β–lactamase), suggest that DNA from pBR transformants may transfer ampicillin resistance to E. coli. Work in progress suggests that the rabbit β–globin gene is transcribed in at least one of the mouse transformants we have examined. Although preliminary, these studies indicate the potential value of co-transformation systems in the analysis of eucaryotic gene expression.

Experimental Procedures

Cell Culture
Ltk$^-$ aprt$^-$, a derivative of Ltk$^-$ clone D (Kit et al., 1963), was obtained from R. Hughes and maintained in Dulbecco's modified Eagle's medium (DME) containing 10% calf serum and 50 μg/ml of diaminopurine (DAP). Prior to transformation, cells were washed and grown for three generations in the absence of DAP. A Chinese hamster cell line containing an altered dihydrofolate reductase (rendering it re-

sistant to methotrexate) A29 Mtx^{RIII} (Flintoff et al., 1976) was obtained from L. Siminovitch. These cells were propagated in DME supplemented with 3X nonessential amino acids, 10% calf serum and 1 μg/ml amethopterin.

Isolation of the HSV tk Gene
Intact herpes simplex virus (HSV) DNA was isolated from CV-1-infected cells as previously described (Pellicer et al., 1978). DNA was digested to completion with Kpn I (New England Biolabs) in a buffer containing 6 mM Tris (pH 7.9), 6 mM MgCl$_2$, 6 mM 2–mercaptoethanol, 6 mM NaCl and 200 μg/ml bovine serum albumin. The restricted DNA was fractionated by electrophoresis through 0.5% agarose gels (17 $\times$ 20 $\times$ 0.5 cm) for 24 hr at 70 V, and the 5.1 kb tk-containing fragment was extracted from the gel as described by Maxam and Gilbert (1977).

Source of DNAs
ΦX174 am3 RFI DNA was purchased from Bethesda Research Laboratories. Plasmid pBR322 DNA was grown in E. coli HB 101 and purified according to the method of Clewell (1972). The cloned rabbit β major globin gene in the λ Charon 4A derivative (RβG-1) was identified and isolated as previously described (Maniatis et al., 1978).

Co-transformation of Defined DNA Sequences and the HSV tk Gene
Ltk$^-$ aprt$^-$ mouse cells were transformed with either 1–10 μg of ΦX174, 1 μg of pBR322 or 1 μg of RβG-1 DNA in the presence of 1 ng of HSV-1 tk gene and 10–20 μg of salmon sperm carrier DNA, as previously described (Wigler et al., 1979). Tk$^+$ transformants were selected in DME containing hypoxanthine, aminopterin and thymidine (HAT) and 10% calf serum. Isolated colonies were picked using cloning cylinders and grown into mass cultures.

Transformation
Methotrexate-resistant transformants of Ltk$^-$ aprt$^-$ cells were obtained following transformation with 20 μg of high molecular weight DNA from A29 Mtx^{RIII} cells and selection in DME containing 10% calf serum and 0.2 μg/ml amethopterin.

Transformation and selection for aprt$^+$ transformants were as described by Wigler et al. (1979).

Isolation of Transformed Cell DNA
Cells were harvested by scraping into PBS and centrifuging at 1000 $\times$ g for 10 min. The pellet was resuspended in 40 vol of TNE [10 mM Tris–HCl (pH 8.0), 150 mM NaCl, 10 mM EDTA], and SDS and proteinase K were added to 0.2% and 100 μg/ml, respectively. The lysate was incubated at 37°C for 5–10 hr and then extracted sequentially with buffer-saturated phenol and CHCl$_3$. High molecular weight DNA was isolated by mixing the aqueous phase with 2 vol of cold ethanol and immediately removing the precipitate that formed. The DNA was washed with 70% ethanol and dissolved in 1 mM Tris, 0.1 mM EDTA.

Nuclei and cytoplasm from clones ΦX4 and ΦX5 were prepared as described by Ringold et al. (1977). The nuclear fraction was further fractionated into high and low molecular weight DNA as described by Hirt (1967).

Filter Hybridization
DNA from transformed cells was digested with various restriction endonucleases using the conditions specified by the supplier (New England Biolabs or Bethesda Research Laboratories). Digestions were performed at an enzyme to DNA ratio of 1.5 U/μg for 2 hr at 37°C. Reactions were terminated by the addition of EDTA, and the product was electrophoresed on horizontal agarose slab gels in 36 mM Tris, 30 mM NaH$_2$PO$_4$, 1 mM EDTA (pH 7.7). DNA fragments were transferred to nitrocellulose sheets, hybridized and washed as previously described (Weinstock et al., 1978) with two modifications. Two nitrocellulose filters were used during transfer (Jeffreys and Flavell, 1977b). The lower filter was discarded, and following hybridization the filter was washed 4 times for 20 min in 2 $\times$ SSC, 25 mM

sodium phosphate, 1.5 mM Na₄P₂O₇, 0.05% SDS at 65°C and then successively in 1:1 and 1:5 dilutions of this buffer (Jeffreys and Flavell, 1977a).

Acknowledgments

We wish to thank Sharon Dana and Mary Chen for excellent technical assistance. This work was supported by grants from the NIH and the NSF.

The costs of publication of this article were defrayed in part by the payment of page charges. This article must therefore be hereby marked *"advertisement"* in accordance with 18 U.S.C. Section 1734 solely to indicate this fact.

Received January 8, 1979; revised January 29, 1979

References

Botchan, M., Topp, W. and Sambrook, J. (1976). Cell 9, 269–287.

Clewell, D. B. (1972). J. Bacteriol. 110, 667–676.

Flintoff, W. F., Davidson, S. V. and Siminovitch, L. (1976). Somatic Cell Genet. 2, 245–261.

Hirt, B. (1967). J. Mol. Biol. 26, 365–369.

Hotchkiss, R. (1959). Proc. Natl. Acad. Sci. USA 40, 49–55.

Jeffreys, A. J. and Flavell, R. A. (1977a). Cell 12, 429–439.

Jeffreys, A. J. and Flavell, R. A. (1977b). Cell 12, 1097–1108.

Kit, S., Dubbs, D., Piekarski, L. and Hsu, T. (1963). Exp. Cell Res. 31, 297–312.

Lacy, E., Lawn, R. M., Fritsch, D., Hardison, R. C., Parker, R. C. and Maniatis, T. (1979). In Cellular and Molecular Regulation of Hemoglobin Switching (New York: Grune & Stratton), in press.

Lai, C. J. and Nathans, D. (1974). Cold Spring Harbor Symp. Quant. Biol. 39, 53–60.

Maitland, N. J. and McDougall, J. K. (1977). Cell 11, 233–241.

Maniatis, T., Hardison, R. C., Lacy, E., Lauer, J., O'Connell, C., Quon, D., Sim, G. K. and Efstradiatis, A. (1978). Cell 15, 687–701.

Maxam, A. M. and Gilbert, W. (1977). Proc. Natl. Acad. Sci. USA 74, 560–564.

Pellicer, A., Wigler, M., Axel, R and Silverstein S. (1978). Cell 14, 133–141.

Ringold, G. M., Yamamoto, K. R., Shank, P. R. and Varmus, H. E. (1977). Cell 10, 19–26.

Sanger, F., Air, M., Barrell, B. G., Brown, N. L., Coulson, A. R., Fiddes, J. C., Hutchinson, C. A., Slocombe, P. M. and Smith, M. (1977). Nature 265, 687–695.

Southern, E. M. (1975). J. Mol. Biol. 98, 503–517.

Spizizen, J., Reilly, B. E. and Evans, A. H. (1966). Ann. Rev. Microbiol. 20, 371–400.

Thomas, R. (1955). Biochim. Biophys. Acta 18, 467–481.

Thomasz, A. and Hotchkiss, R. (1964). Proc. Nat. Acad. Sci. USA 51, 480–487.

Weinstock, R., Sweet, R., Weiss, M., Cedar, H. and Axel, R. (1978). Proc. Natl. Acad. Sci. USA 75, 1299–1303.

Wigler, M., Silverstein, S., Lee, L.-S., Pellicer, A., Cheng, Y.-C. and Axel, R. (1977). Cell 11, 223–232.

Wigler, M., Pellicer, A., Silverstein, S. and Axel, R. (1978). Cell 14, 725–731.

Wigler, M., Pellicer, A., Silverstein, S., Axel, R., Urlaub, G. and Chasin, L. (1979). Proc. Nat. Acad. Sci. USA, in press.

29

HIGH EFFICIENCY TRANSFORMATION BY DIRECT MICROINJECTION OF DNA INTO CULTURED MAMMALIAN CELLS

M. R. Capecchi

Summary

Direct microinjection of DNA by glass micropipettes was used to introduce the Herpes simplex virus thymidine kinase gene into cultured mammalian cells. When DNA was delivered directly into the nuclei of LMTK$^-$, a mouse cell line deficient in thymidine kinase activity, 50–100% of the cells expressed TK enzymatic activity. In contrast, no TK activity could be detected when the DNA was injected into the cytoplasm. The number of injected LMTK$^-$ cells capable of indefinite growth in a TK$^+$ selective medium (that is, transformants) depended on the nature of the plasmid DNA into which the HSV-TK gene was inserted. One cell in 500–1000 cells which received nuclear injections with pBR322/TK DNA gave rise to a viable colony when grown in HAT medium (that is, a TK$^+$ selective medium). The transformation frequency increased to one in five injected cells when specific SV40 DNA sequences were also introduced into the HSV-TK plasmid. With the microinjection procedure transformation frequency was relatively insensitive to DNA concentration and did not depend on co-injecting with a carrier DNA. Most of the transformants were stable in *nonselective* medium as soon as they could be tested.

Introduction

Specific genes can be introduced into cultured mammalian cells by chromosome-mediated gene transfer (McBride and Ozer, 1973; Willecke and Ruddle, 1975) and by purified DNA-mediated gene transfer (Bacchetti and Graham, 1977; Maitland and McDougall, 1977; Wigler et al., 1977). The uptake and expression of both the metaphase chromosomes (Spandidos and Siminovitch, 1977; Miller and Ruddle, 1978) and purified DNA (Graham and van der Eb, 1973) is enhanced by the formation of a DNA-calcium phosphate precipitate. One in 10^5–10^7 treated cells becomes transformed by either chromosome- or DNA-mediated gene transfer. The rare transformant is isolated by biochemical selection.

In the initial transformation experiments using calcium phosphate precipitation to facilitate the uptake of purified DNA, the Herpes simplex viral thymidine kinase gene (HSV-TK) was transferred into LMTK$^-$, a mouse cell line deficient in thymidine kinase (Bacchetti and Graham, 1977; Maitland and McDougall, 1977; Wigler et al., 1977). This approach has been extended to the cellular genes for thymidine kinase (Wigler et al., 1978), adenine phosphoribosyl transferase (Wigler et al., 1979a) and hypoxanthine phosphoribosyl

transferase (Graf, Urlaub and Chasin, 1979; Willecke et al., 1979).

For each of the above experiments a good selection procedure for isolating the transformants existed. It also became apparent that the transformation frequency was critically dependent on the particular cell line used as the recipient (Graf et al., 1979).

In an elegant set of experiments Wigler et al. (1979b) showed that nonselectable genes could be introduced into cultured mammalian cells by co-transformation with a unlinked but selectable gene. The nonselectable gene was mixed in a molar ratio of 1,000 to one with the selectable gene (HSV-TK) precipitated with calcium phosphate and layered onto LMTK$^-$ cells. More than 90% of the LTK$^+$ transformants contained multiple copies of the nonselectable gene.

In this study an alternative method of transferring purified genes into cultured mammalian cells is described. The DNA was directly injected into the nucleus using glass micropipettes (Diacumakos, 1973; Graessmann and Graessmann, 1976; Stacey and Allfrey, 1976). The transformation efficiency of the HSV-TK gene inserted into a number of different recombinant plasmids was compared. The transformants were characterized for the presence of HSV-TK enzymatic activity and for their stability in nonselective medium.

Results

Injection of pBR322/TK DNA

The microinjection experiments were initiated with two objectives in mind. The first was to determine the efficiency of DNA-mediated transformation obtained by microinjecting the DNA into cells and to compare this efficiency with that obtained with the more familiar calcium phosphate precipitation methods described by Bacchetti and Graham (1977), Maitland and McDougall (1977) and Wigler et al. (1977). The second objective was to attempt to discover the steps limiting the efficiency of the transformation process itself. DNA-mediated transformation of cultured mammalian cells can be divided into several steps including the DNA's entry into the cell, its transfer from cytoplasm to nucleus and its integration into the host genome. It seemed likely that microinjection of the DNA could be used to evaluate how much each step limits the frequency of transformation.

The experimental system chosen for these studies was the transfer of the Herpes simplex virus I thymidine kinase gene (HSV-TK) into a thymidine kinase deficient mouse fibroblast cell line, LMTK$^-$. The DNA injected was a purified preparation of E. coli plasmid pBR322 which carried the HSV-TK gene as an insert

(Enquist et al., 1979). Following injection of the pBR322/TK plasmid DNA into nuclei of LTK⁻ cells, thymidine kinase (TK) enzymatic activity could be detected by the incorporation of ^{3}H–thymidine into DNA followed by autoradiographic analysis (Figure 1). No TK activity was detectable by this assay in cells receiving injections of buffer alone or of pBR322 DNA not containing the HSV-TK insert.

In a number of separate experiments between 50 and 100% of the LTK⁻ cells in which a nuclear injection of pBR322/TK DNA was attempted showed TK activity. This efficiency reflects the normal rate for successful transfer of material into penetrated cells as determined either by injection of horseradish peroxidase followed by histochemical assays for peroxidase activity or by injection of radioactively labeled proteins followed by autoradiographic assays (data not shown).

In the experiment shown in Figure 1 approximately 125 molecules of plasmid DNA were injected into the nucleus of each cell. Thymidine kinase activity could be detected after injecting on the average as few as five molecules of pBR322/TK DNA per nucleus.

Nuclear vs. Cytoplasmic Injections

To assess the precision with which molecules could be delivered to selected cellular compartments, a solution of ^{125}I–IgG was injected into either the cytoplasm (Figure 2a) or the nucleus (Figure 2b). Following injection the cells were fixed with glutaraldehyde and analyzed by autoradiography. For these experiments the injections were carried out under conditions of constant flow of fluid (that is, constant pressure), the advantage of which is that by this procedure clogging the micropipette during successive injections is minimized. Under these conditions it is easy to inject fluid exclusively into the cytoplasm (Figure 2a). However, during injections intended for the nucleus the tip of the micropipette passes through the cytoplasm as it both enters and leaves the cell, depositing some fluid in the cytoplasm. Counts of exposed grains over the cytoplasm and nucleus indicate that in a nuclear injection approximately 90% of the injected material is delivered into the nucleus.

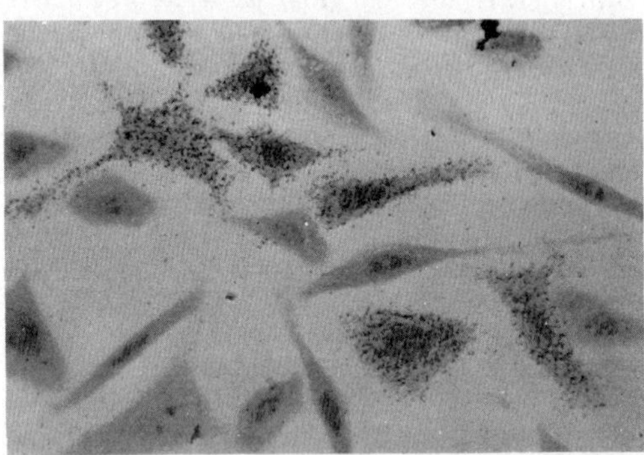

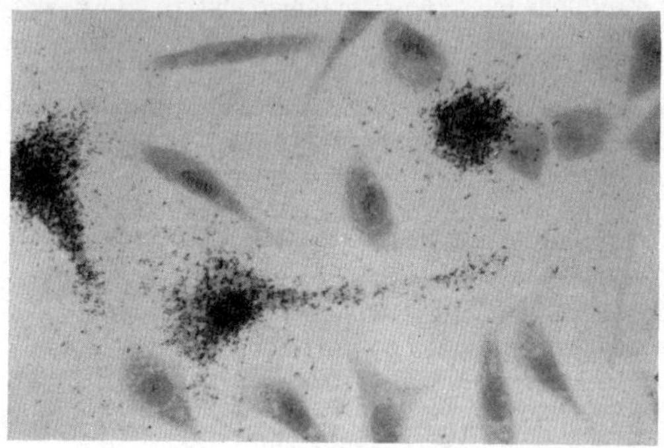

Figure 2. Autoradiographic Analysis of LTK⁻ Cells following (a) Cytoplasmic or (b) Nuclear Injection with ^{125}I–IgG

Rabbit immunoglobulins were labeled with ^{125}I to a specific activity of 10^7 dpm/μg protein using the chloramine T procedure. A solution containing 10^6 dpm/μl of ^{125}I–IgG was selectively injected into either the (a) cytoplasm or (b) nucleus of LTK⁻ cells. Following injection the cells were fixed with glutaraldehyde and processed for autoradiography as described in the legend to Figure 1. The slides were exposed in the dark for 10 days prior to being developed. The injections are done under conditions (constant pressure) where the micropipette is continuously flowing. As a result, during a nuclear injection some material is deposited into the cytoplasm as the tip of the micropipette passes through the cytoplasm on its way into and out of the nucleus. Counts of exposed silver grains over nuclear and cytoplasmic regions of the cell following nuclear injections with ^{125}I–IgG indicated that approximately 90% of the material is deposited in the nucleus. The ^{125}I–IgG was a gift from M. Rechsteiner.

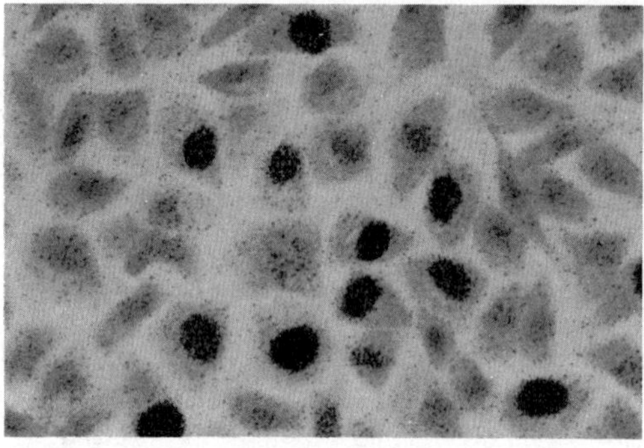

Figure 1. Autoradiographic Analysis for Thymidine Kinase Activity Present in LTK⁻ Cells following Direct Nuclear Injection with pBR322/TK DNA

Approximately 125 molecules of pBR322/TK DNA (0.1 mg/ml) were injected per nucleus of mouse LTK⁻ cells. Following injection the cells were incubated in MEM plus 10% fetal calf serum for 24 hr at 37°C in a 5% CO_2 incubator to allow for the expression of the injected HSV-thymidine kinase genes. The thymidine kinase (TK) activity present in each cell was determined by measuring the capacity of the cells to incorporate ^{3}H–thymidine into DNA. The cells were incubated for 12 hr at 37°C in MEM plus 10% fetal calf serum containing 25 μCi/ml of ^{3}H–thymidine. Following incubation in ^{3}H-thymidine medium, the cells were washed four times with PBS and fixed with 2% glutaraldehyde. After fixing, the cells were further washed two times with distilled water and two times with 30% ethanol. The cover slips containing the fixed cells were then mounted on microscope slides and dipped in NTB-2 autoradiographic emulsion. Following exposure for 72 hr, the slides were developed. In the above field thirteen cells received injections. Uninjected cells serve as an internal control.

The ability to inject pBR322/TK DNA into either the nucleus or the cytoplasm made it possible to determine whether the site of injection influences the efficiency of expression of the plasmid DNA. It will be recalled that 50–100% of the cells with pBR322/TK DNA injected into the nucleus contain detectable TK enzymatic activity. No TK activity was detected in over 1000 cells injected with pBR322/TK DNA in the cytoplasm.

Transformation of LTK⁻ Cells to LTK⁺

We next asked whether LTK⁻ cells were transformed to LTK⁺ after nuclear injection with pBR322/TK. Approximately one in 500–1000 LTK⁻ cells injected with pBR322/TK DNA gave rise to a large colony in HAT medium (Table 1). No LTK⁺ transformants were observed after LTK⁻ cells received nuclear injections of pBR322 DNA. The reversion frequency of the TK⁻ mutation in LTK⁻ cells is extremely low. We and others have never observed a spontaneous revertant of LTK⁻ cells to LTK⁺ (Wigler et al., 1978).

The LTK⁻ Transformants Contain HSV-1 Thymidine Kinase

The HSV-1 and L cell thymidine kinase activity can be distinguished by their electrophoretic separation on nondenaturing polyacrylamide gels (Figure 3). In the above experiments (Figure 3e) we show that the thymidine kinase activity in a LTK⁻ transformant, obtained after nuclear injection with pBR322/TK DNA, was indistinguishable from the HSV enzyme with respect to this property. In a like manner five other randomly chosen LTK⁻ transformants were shown to contain HSV-1 TK activity (data not shown).

Stability of the Transformants

Are the transformants following microinjection with pBR322/TK DNA stable in the sense that they can be

Table 1. Transformation Frequency with pBR322/TK DNA

DNA Injected	Concentration (mg/ml)	Number of Transformants per 10³ Cells Receiving an Injection	Number of Cells Receiving an Injection
pBR322	0.33	0	10⁴
pBR322/TK	0.33	2.2	5 × 10³
pBR322/TK	0.33	1.2	5 × 10³
pBR322/TK	0.33	2.6	5 × 10³
pBR322/TK plus salmon sperm (1:5 w/w)	1.20	0.6	5 × 10³

LTK⁻ cells were grown on a 10 × 10 mm cover slip in 35 mm petri dishes. Two hundred cells per dish received nuclear injections with the respective DNA solutions. After the injections the cells were incubated for 24 hr in nonselective medium at 37°C in a 5% CO₂ incubator and then switched to HAT medium. Two weeks later the dishes were scored for the presence of a large colony.

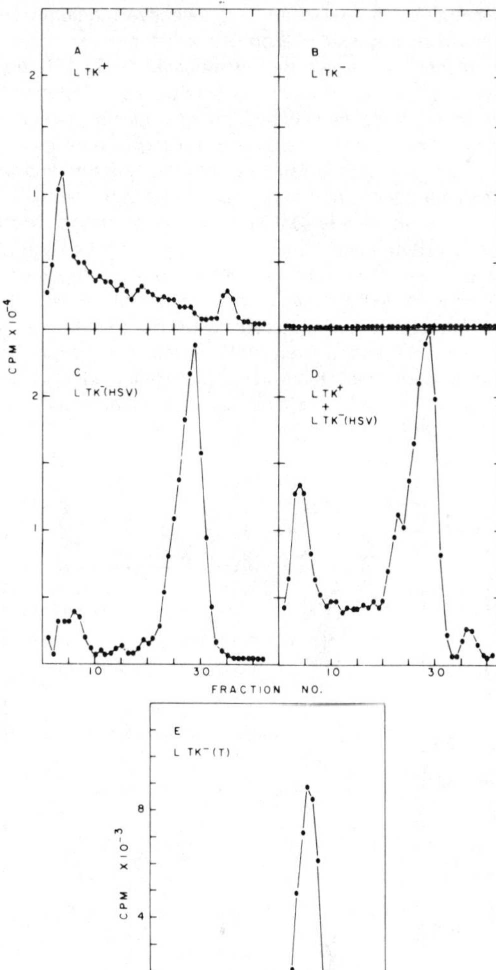

Figure 3. Electrophoretic Analysis of the Thymidine Kinase Activity in Extracts Prepared from (a) LTK⁺ Cells, (b) LTK⁻ Cells, (c) LTK⁻ Cells Infected with HSV-1, (d) a Mixture of LTK⁺ Cells and LTK⁻ Cells Infected with HSV-1 and (e) an LTK⁻ Transformant Obtained after Injecting pBR322/TK DNA into Nuclei of LTK⁻ Cells

The LTK⁺, LTK⁻ and LTK⁻ transformant cells were grown as monolayers in MEM plus 10% fetal calf serum. Some of the LTK⁻ cells were infected with Herpes simplex virus I at a multiplicity of 20 and harvested 18 hr post infection. The cells were washed twice with PBS and lysed with a buffer containing 0.5% NP40, 10% glycerol, 0.1 M KCl, 0.001 M DTT, 0.001 M MgCl₂ and 50 μM dThd. Under this lysis condition the soluble cytoplasmic proteins are released and the remainder of the cell remains adhered to the culture dish surface. A small but detectable amount of the mitochondrial TK activity is released (A and D). The extracts were centrifuged at 30,000 × g for 30 min and applied to 5% polyacrylamide disc gels. Following electrophoresis the gels were cut into 1 mm slices and assayed for thymidine kinase as described by Lee and Chen (1976). The electrophoretic mobilities of the cellular and HSV-1 TK activities relative to the mobility of the front were 0.15 and 0.6 respectively. The R꜀ of the thymidine kinase activity in the LTK⁻ transformant was 0.6, consistent with the electrophoretic mobility of the HSV enzyme.

maintained in the absence of selective medium? To answer this question each of six independent transformants was grown in nonselective medium. Every week an aliquot of each line was plated in both HAT medium (TK$^+$ selective medium) and nonselective medium. The number of colonies arising in both media was scored and the ratio was plotted as a function of the time the line had been maintained in nonselective medium (see Figure 4). Five of the transformants were observed to retain TK gene expression in the absence of selection. The sixth transformant did lose the ability to grow in HAT medium in the absence of selection at a rate of 3% per generation. Stable transformants could be selected from the unstable one simply by picking colonies which grew up in HAT medium after these cells had grown in nonselective medium for over 50 generations.

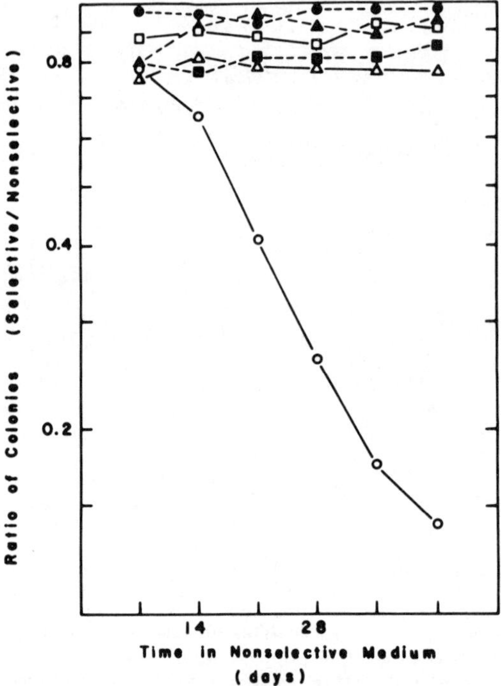

Figure 4. Stability of the TK$^+$ Phenotype of Six LTK$^-$ Transformants When Grown in Nonselective Medium

Six independent TK$^+$ transformant cell lines obtained by nuclear injection of LTK$^-$ cells with pBR322/TK DNA were grown in nonselective medium. Every week an aliquot of each line (approximately 3000 cells) was plated on six 100 mm petri dishes in both HAT medium (TK$^+$ selective medium) and nonselective medium. Ten days later the cells were fixed with methanol and stained with Giemsa. The ratio of the resultant colonies which grew up in the selective and nonselective medium was plotted as a function of the time the line had been maintained in nonselective medium (Klobutcher and Ruddle, 1979). The plating efficiency of the transformants in nonselective medium was approximately 70%.

Effect of Carrier DNA on DNA-mediated Transformation

DNA-mediated transformation of cultured mammalian cells using the calcium phosphate precipitation technique is enhanced by approximately three orders of magnitude if a carrier DNA (vertebrate DNA) is used to co-precipitate the plasmid DNA containing the desired transforming gene (Graham and van der Eb, 1973). While the role of the carrier DNA has not been fully elucidated, it appears that after the DNA is taken up by the cells as a calcium phosphate precipitate large concatemers of carrier DNA and plasmid DNA are formed. These concatemers may have the capacity to replicate independently of the host chromosome or may integrate as units into the host chromosome (Perucho, Hanahan and Wigler, manuscript submitted).

The transformation experiments in Table 1 were performed in the absence of carrier DNA. Considering the enhancement of transformation exhibited by carrier DNA with the calcium phosphate precipitation technique, it was of interest to determine whether carrier DNA would enhance transformation obtained by microinjection. As shown in Table 1 no enhancement of transformation frequency was observed when salmon sperm DNA was coinjected with pBR322/TK DNA. The same salmon sperm DNA preparation enhanced by 1000 fold pBR322/TK directed transformation of LTK$^-$ cells to LTK$^+$ using the calcium phosphate precipitation technique (P. Barry, unpublished results).

Stimulation of the Transformation Frequency by Co-injecting SV40 DNA with pBR322/TK DNA

Co-injecting pBR322/TK DNA with DNA isolated from SV40 into the nuclei of LTK$^-$ cells increases the transformation frequency to LTK$^+$ by a factor of approximately ten over that obtained by injection of pBR322/TK DNA alone (Table 2). The increased transformation frequency may result from SV40 providing, through recombination, DNA sequences which facilitate either replication or integration of the TK plasmid DNA. Alternatively, SV40 may provide products, in trans, which facilitate either replication or integration of the TK-plasmid DNA. To test these alternatives, recombinant HSV-TK plasmids containing different portions of the SV40 genome gene were prepared and their ability to transform LTK$^-$ cells to LTK$^+$ was determined (see below).

SV40/HSV TK Recombinant Plasmids

Restriction maps of the recombinant plasmids used for the experiments to be described are illustrated in Figure 5. The first step in constructing this set of plasmids was to insert the intact SV40 genome into pBR322 through their Eco RI sites. Eco RI cleaves

Table 2. Transformation Frequency by Co-injection of pBR322–TK DNA plus SV40 DNA

DNA Injected 10³	Concentration (mg/ml)	Number of Transformants per 10³ Cells Receiving an Injection	Number of Cells Receiving an Injection
pBR322/TK plus SV40 (1:1 M)	0.53	15	10³
pBR322/TK plus SV40 (1:1 M)	0.53	12	10³
SV40	0.2	0	10⁴

The experimental procedure was as described in the legend to Table 1 except that in the SV40, pBR322/TK coinjection experiments, 50 cells per dish received nuclear injections.

SV40 in the middle of the late region leaving the early region intact and functional. The orientation of the SV40 genome with respect to pBR322 was determined by cleaving the recombinant plasmid with Bam HI (Figure 6a, slot 5). The purified HSV Bam HI fragment containing the TK gene was inserted into the above pBR322/SV4O recombinant plasmid at the Bam HI sites. In the process of inserting the HSV-TK fragment a small 1 kb Bam HI fragment from pBR322/SV40 was removed. The resulting recombinant plasmid was designated pBR322/SV-O+T/TK to indicate that this plasmid contains the SV40 origin for DNA replication (Ori) as well as a functional SV40 early region (that is, capable of synthesizing T antigens). The orientation of the HSV TK gene was determined from a number of restriction enzyme digests shown in Figure 6a.

12 hr after injecting 100 molecules of pBR322/SV-O+T/TK DNA per cell into the nuclei of LTK⁻ cells one can detect SV40 T antigen by indirect immunofluorescence assays (Figure 7). The level of T antigen per cell was comparable to that observed in SV80 cells, a human cell line which was transformed by SV40 and overproduces T antigen (Todaro, Green and Swift, 1966; Henderson and Livingston, 1974).

To generate a recombinant plasmid which still retained the SV40 origin of DNA replication but lacked sequences coding for large and small T antigen, pBR322/SV-O+T/TK DNA was partially hydrolyzed with Hind III followed by re-ligation. This DNA was used to transform E. coli. Ampicillin resistant colonies were screened on agarose gels for the presence of a recombinant plasmid missing a 1.7 kb fragment encoding for most of the early region of SV40 as well as the splicing junctions for large and small T mRNA. The resulting recombinant plasmid still retains the SV40 origin for DNA replication and was designated pBR322/SV-O/TK.

No material which cross reacts with T antigen-antiserum is observed in LTK⁻ cells which were microinjected with pBR322/SV-O/TK DNA (data not shown).

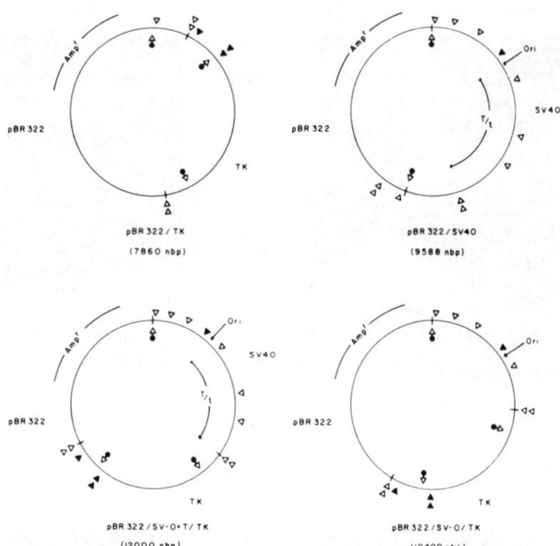

Figure 5. Restriction Maps of the Recombinant Plasmids pBR322/TK, pBR322/SV40, pBR322/SV-O+T/TK and pBR322/SV-O/TK

pBR322/TK (obtained from G. F. Vande Woude; see Enquist et al., 1979) contains the HSV-1 thymidine kinase gene as part of a 3.5 kb HSV-1 Bam HI fragment. The entire SV40 genome was ligated into pBR322 at the single Eco RI site to form pBR322/SV40. The early region of SV40 remains intact and functional in this recombinant plasmid. pBR322/SV-O+T/TK was obtained by inserting the purified 3.5 kb HSV-1 Bam HI fragment into the pBR322/SV40 recombinant plasmid. In the process of inserting the HSV-TK fragment into pBR322/SV40, a 1 kb fragment is removed which contains 0.75 kb of the coding region for SV40-VP1 and a 0.25 kb fragment of the pBR322 Tet' gene. pBR322/SV-O+T/TK contains sequences for the SV40 origin of DNA replication (Ori), the coding sequences for large and small SV40 T antigen and the HSV-TK gene. pBR322/SV-O/TK was obtained from pBR322/SV-O+T/TK by partial hydrolysis with Hind III followed by religation. In the process two SV40 Hind III fragments (1169 and 526 nbp) were removed. These later fragments contain most of the coding sequences for SV40 large and small T antigen as well as the junctions for splicing their respective mRNAs. The symbols used to represent some of the characteristic endonuclease restriction sites of the above recombinant plasmids are: ( ) Eco RI; ( ) Bam HI; (△) Hind III; (▲) Kpn I; ( ) Bgl II.

Transformation Frequency of LTK⁻ Cells to LTK⁺ following Injection with Either pBR322/SV-O+T/TK or pBR322/SV-O/TK DNA

Approximately one LTK⁻ cell in five which received nuclear injection of either pBR322/SV-O+T/TK or pBR322/SV-O/TK DNA gave rise to a cell line capable of indefinite growth in HAT medium (Table 3). For these experiments five cells per 35 mm plate were injected with one of these recombinant plasmid DNAs. The location of the five separate injected cells in each plate was recorded and the plates were monitored for colony growth in HAT medium. In some plates two or three colonies were observed growing up at the sites of the injected cells. For the data presented in Table 3 a plate was recorded as positive if it contained one or more large colonies (greater than 1000 cells per colony). This method of recording obviates problems associated with secondary colonies arising from pri-

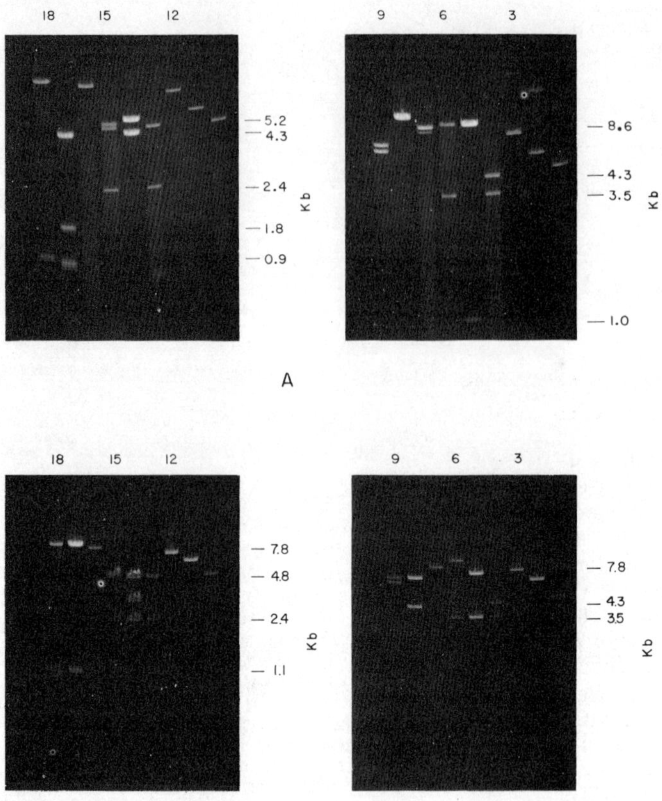

Figure 6. Restriction Fragment Analysis of the Recombinant Plasmids (a) pBR322/TK, pBR322/SV40 and pBR322/SV-O+T/TK and (b) pBR322/TK, pBR322/SV-O/TK and pBR322/SV-O+T/TK

The plasmid DNAs (0.5–1.0 μ) were digested with 2–3 units of restriction enzyme using the conditions described by Bethesda Research Lab and New England Biolabs. The DNA restriction fragments were resolved by slab gel electrophoresis in 0.8% agarose. The gels were then stained with ethidium bromide and visualized on an ultraviolet box. In (a) the uncut plasmid DNAs pBR322/TK, pBR322/SV40 and pBR322/SV-O+T/TK were applied to gel slots 1–3 and 10–12 respectively. Slots 4–6 contain the Bam HI digests of pBR322/TK, pBR322/SV40 and pBR322/SV-O+T/TK. Slots 7–9 contain Kpn I and Bgl II double digests of pBR322/TK, pBR322/SV40 and pBR322/SV-O+T/TK. Slots 13–15 contain Eco RI digests and slots 16–18 contain Hind III digests of pBR322/TK, pBR322/SV40 and pBR322/SV-O+T/TK, respectively. In (b) the uncut plasmid DNAs pBR322/TK, pBR322/SV-O/TK and pBR322/SV-O+T/TK were applied to gel slots 1–3 and 10–12, respectively. Slots 4–6 contain Bam HI digests of pBR322/TK, pBR322/SV-O/TK and pBR322/SV-O+T/TK. Slots 7–9 contain Kpn I and Bgl II double digests of pBR322/TK, pBR322/SV-O/TK and pBR322/SV-O+K. Slots 13–15 contain Eco RI digests and slots 16–18 contain Hind III digests of pBR322/TK, pBR322/SV-O/TK and pBR322/SV-O+T/TK, respectively.

mary colonies but probably underestimates the actual efficiency of transformation.

Surprisingly, the same efficiency of transformation was observed with either pBR322/SV-O+T/TK or pBR322/SV-O/TK DNA. The presence of SV40 T antigens did not influence the efficiency of transforming LTK$^-$ cells to LTK$^+$.

The efficiency of transformation was also relatively insensitive to the number of molecules of pBR322/SV-O+T/TK or pBR322/SV-O/TK DNA injected into each LTK$^-$ cell (that is, over the range from 10 to 400 molecules per cell) (Table 3).

Stability of the LTK$^-$ Transformant Resulting from Nuclear Injections with pBR322/SV-O+T/TK or pBR322/SV-O/TK DNA

The high efficiency of transformation of LTK$^-$ cells by the recombinant SV40 plasmids could result from SV40 donating to the plasmids an origin of DNA replication functional in mouse cells, thereby allowing them to be replicated efficiently without integration into the host genome. If this were the case one might anticipate that LTK$^+$ transformants arising from nuclear injections with pBR322/SV-O+T/TK or pBR-322/SV-O/TK DNA would be unstable when grown

in a nonselective medium. As illustrated in Figure 8 this does not appear to be the case. The stability in nonselective medium of three independent LTK$^+$ transformants arising after nuclear injections of pBR322/SV-O+T/TK and of three arising after injection of pBR322/SV-O/TK DNA was tested. All six clones were found to be stable.

The above result does not exclude efficient plasmid replication since the frequency of LTK$^-$ segregants could be very low if the cellular plasmid population were large. A segregation frequency of 0.1% per generation would have been detected. Direct evidence for the integration of the pBR322/SV-O+T/TK and pBR322/SV-O/TK plasmids into host DNA sequences by blotting analysis will be presented in a separate communication.

Extracts from the above cell lines were shown to contain TK enzymatic activities, whose electrophoretic mobility was indistinguishable from the HSV-1 enzyme (data not shown).

Discussion

When pBR322/TK DNA was injected into nuclei of mouse LTK$^-$ cells, 50–100% of the cells express

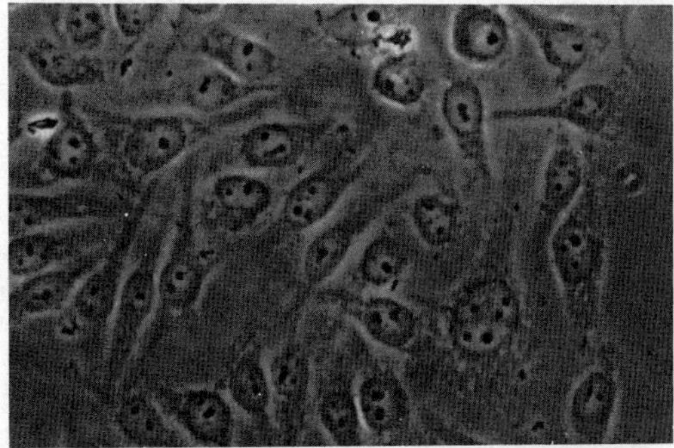

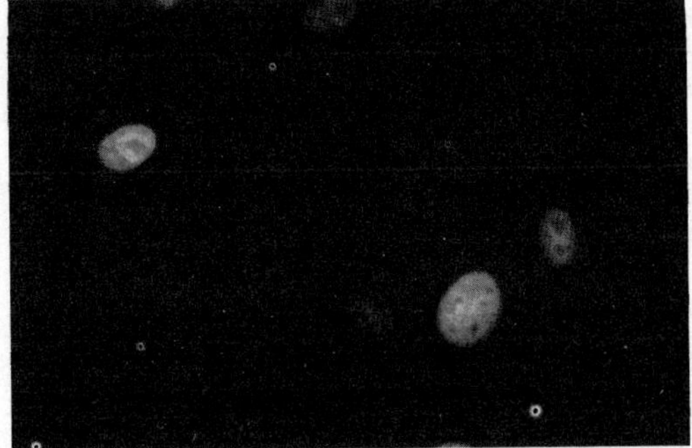

Figure 7. Immunofluorescent Assay for the Presence of SV40 T Antigen in Cells Injected with pBR322/SV-O+T/TK (a) Cells Photographed by Phase-Contrast Microscopy and (b) Cells Photographed by Fluorescent Microscopy

LTK⁻ cells were injected with pBR322/SV-O+T/TK DNA (0.15 mg/ml) and incubated for 12 hr at 37°C in a 5% CO₂ incubator. The cells were then washed with PBS plus 1% calf serum and fixed at O°C with acetone-ethanol (2:1). The fixed cells were then incubated with hamster anti-T serum for 30 min at 25°C, washed with PBS plus 1% calf serum and incubated with FITC-rabbit anti hamster IgG for an additional 30 min. After washing the cells with PBS plus calf serum, the coverslip containing the cells was fixed to a microscope slide upside down with 90% glycerol. For each set of experiments cover slips containing SV80 cells (human cells which were transformed with SV40 and produce high levels of T antigen) and uninjected LTK⁻ cells were processed in parallel. Seven cells in the above field were injected with pBR322/SV-O+T/TK DNA. The amount of fluorescence observed in the injected cells was comparable to that observed in parallel reactions with SV80 cells.

Table 3. Transformation Frequency Obtained by Injection of pBR322-SV-0+T/TK and pBR322-SV-0/TK DNA

DNA Injected	Concentration (mg/ml)	Molecules per Cell	Number of Plates Containing One or More Colonies[a]
pBR322/SV-0+T/TK	0.5	400	17/20
pBR322/SV-0+T/TK	0.125	100	19/20
pBR322/SV-0+T/TK	0.125	100	16/20
pBR322/SV-0+T/TK	0.0125	10	18/20
pBR322/SV-0+T/TK	0.0125	10	15/20
pBR322/SV-0/TK	0.4	400	13/20
pBR322/SV-0/TK	0.1	100	16/20
pBR322/SV-0/TK	0.1	100	16/20
pBR322/SV-0/TK	0.01	10	17/20
pBR322/SV-0/TK	0.01	10	18/20

[a] Five cells per plate received nuclear injections with the respective recombinant plasmid DNA. The cells were then incubated for 24 hr in nonselective medium and then switched to HAT medium. A plate was scored as positive if after two weeks of incubation in HAT medium it contained one or more large colonies (that is greater than 1000 cells per colony).

thymidine kinase enzymatic activity. Of over one thousand cells injected with pBR322/TK DNA into the cytoplasm none contained detectable TK activity. Increasing the number of plasmid DNA molecules injected into the cytoplasm of each cell or increasing the time period between the injection and the assay for TK activity did not alter the results. These experiments indicate that the plasmid DNA cannot be expressed in the cytoplasm and that it does not readily gain access to the nucleus. It is not known whether the nuclear membrane itself acts as a physical barrier or whether the plasmid DNA is rapidly degraded when injected into the cytoplasm.

In light of the results described above it is interesting to speculate that the formation of the DNA-calcium precipitate may enhance transformation by protecting the DNA as it traverses the cytoplasm. After uptake by phagocytosis a small fraction of the DNA may be shuttled into the nucleus as a precipitate where it is dissolved, releasing active genes. In such a model "competence" could reflect the cell's capacity to transport the DNA from the cytoplasm to the nucleus rather than its capacity for DNA uptake.

Although 50–100% of the cells injected with pBR322/TK DNA express TK activity, only one cell in 500–1000 injected cells can be propagated indefinitely in HAT medium. The above results can be interpreted in a number of ways. For example, it might be that only in a small proportion of the injected cells is the plasmid DNA properly integrated into the host chromosome (that is, integrated in a site that allows the HSV-TK gene to be transcribed). Alternatively, it might be that only a few of the cells that received injections provide the proper environment for independent replication of the plasmid DNA prior to integration. Such replication could result from modification of the plasmid DNA or the presence of unique factors required for replication. The above explanations are not mutually exclusive.

Most of the LTK⁺ transformants following injection with pBR322/TK DNA were found to be stable even when grown in nonselective medium. The stability studies were initiated as soon after microinjection as possible (that is, when approximately 10^7 cells of each cell line were available). These results differ from those obtained by the calcium phosphate precipitation

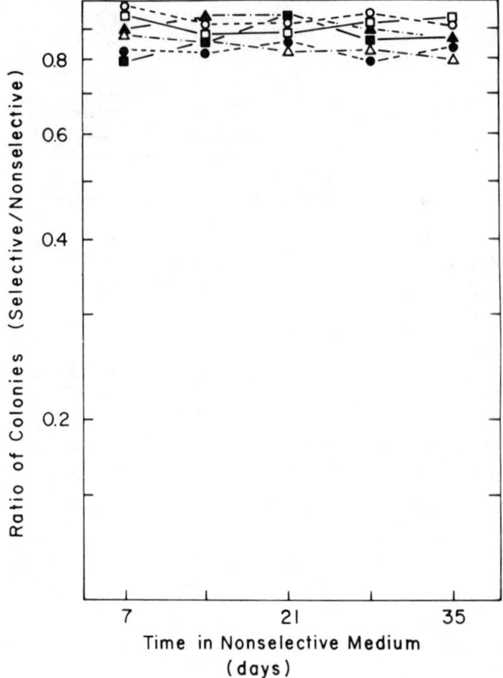

Figure 8. Stability of the TK$^+$ Phenotype of Six LTK$^-$ Transformants When Grown in Nonselective Medium

Three of the transformants were obtained by nuclear injection with pBR322/SV-O+T/TK DNA and three by nuclear injection with pBR322/SV-O-TK DNA.

The experimental protocol was as described in the legend for Figure 4. Open symbols designate transformants obtained by nuclear injections with pBR322/SV-O+T/TK DNA, closed symbols designate transformants obtained by injection with pBR322/SV-O/TK DNA.

method where stability of the transformed phenotype, in the absence of selection, is often not observed (Wigler et al., 1977; Graf et al., 1979; Wigler et al.,1979a; Willecke et al., 1979). The difference in the results obtained by the two methods could be explained if an early event in the calcium phosphate precipitation method were the incorporation of the transforming gene into the carrier DNA. The carrier-transforming gene complex may be capable of independent replication and therefore susceptible to loss due to unequal segregation during cellular division, or if incorporated into the host chromosome as a large package, the unit itself may be sensitive to loss via a recombinational mechanism. Since carrier DNA is not used with the microinjection procedure, these options are not available.

Using the microinjection procedure the number of gene copies injected into each cell can be controlled. In the range of 1–100 molecules, the number of pBR322/SV-O/TK incorporated into the transformant host DNA sequences is roughly proportional to the number of molecules injected into each cell (unpublished results). The transformation frequency to LTK$^+$ is relatively insensitive to the number of gene copies

injected into each cell. The system therefore rapidly saturates and high transformation frequencies can be obtained by injecting only a few molecules per cell (that is, less than five).

Approximately 20% of the LTK$^-$ cells which received injections with either pBR322/SV-O+T/TK or pBR322/SV-O/TK DNA were transformed to LTK$^+$. The presence or absence of a functional SV40 early region did not alter the transformation frequency. The increased transformation frequency observed with these recombinant plasmid DNAs does not appear to result from independent replication. Loss of the plasmid is not observed when the transformants are grown in nonselective medium. One could argue that so many copies of the self replicating recombinant plasmid DNAs are present in each transformant that loss due to segregation could not be detected. Southern blotting analysis of these transformants indicates, however, that in most of the transformants the plasmid DNA is integrated into the host DNA sequences. Free plasmid DNA is not observed (unpublished results).

How do SV40 DNA sequences near the origin enhance the transformation frequency of the TK recombinant plasmids? A working hypothesis is that these DNA sequences facilitate the integration of the plasmid DNA into the host chromosome either because they contain palindromes and/or because they contain regions of homology to middle-repetitive host sequences. Jelinek and his colleagues (Jelinek et al., 1980) have described ubiquitous, interspersed, repeated DNA sequences in mammalian genomes which have homology to the SV40 origin of DNA replication. The SV40 sequences are not unique in their ability to stimulate the transformation efficiency of TK recombinant plasmids. Comparable transforming efficiencies have been obtained by inserting the 3′ R 5′ terminal repeats from the Schmidt-Ruppin-A strain of the avian sarcoma virus into the pBR322/TK plasmid (our unpublished results, in collaboration with P. Luciw, S. M. Bishop and H. E. Varmus). The 3′ R 5′ terminal repeat sequences of RNA tumor viruses do not have homology to the SV40 origin of DNA replication. However, they do have an intriguing DNA sequence similarity to bacterial IS sequences and are postulated to be used for integrating the provirus into the host chromosome (Hughes et al., 1978).

In this study I have emphasized the use of the microinjection technology to obtain and characterize DNA-mediated transformants. For many experiments it will be advantageous to look simply for gene expression rather than to wait for a transformant to grow up. Injection of DNA into the nuclei of cultured mammalian cells provides a convenient and sensitive bioassay for gene expression. The equivalent of 1% of a mammalian genome can be injected into a single cell. When coupled with immunofluorescent, radioimmune and autoradiographic techniques, detecting the expression of a single gene copy is possible.

In summary, microinjection of DNA using glass micropipettes provides an alternative procedure for introducing genes into cultured mammalian cells. When appropriate recombinant plasmids are used, transformation frequencies approaching unity are obtained. With these high transformation frequencies the application of the technique to embryological problems becomes feasible.

Experimental Procedures

The methods used for culturing the cells, preparing cell extract and autoradiography have been described in detail elsewhere (Sharp, Capecchi and Capecchi, 1973; Wahl, Hughes and Capecchi, 1975; Capecchi et al., 1977).

Microinjection

The recipient cells were grown on small glass slides (10 × 10 mm). The solution of macromolecules was injected into the cells via glass micropipettes having tip diameters ranging from 0.1 to 0.5 μ. The pipettes were prepared from glass capillaries (Omega Dot Tubing, 1.2 mm OD, W. P. Instruments) on a Model P77 Brown-Flaming micropipette puller (Sutter Instruments). Injections were carried out under direct visual control on a fixed stage of an inverted phase contrast microscope (Leitz Diavert, 400X). Movement of the micropipettes was controlled with micromanipulators (Narishige MO-15, modified) which have a hydrolic microdrive along the pipette axis. The fluid containing the macromolecules is forced into the cells under constant pressure supplied by a Hamilton threaded plunger syringe (Model 87000). The amount of fluid injected into each cell was controlled with moderate precision (within a factor of two) by visually monitoring changes in the cellular refractive index as the fluid enters the cell and regulating the time that the micropipette remained in the cell. The average volume (10–20 femtoliters) injected into each cell was determined by injecting ^{3}H–dTTP (10 $\mu c/\mu l$) into five thousand cells, washing the cells with PBS and then measuring the radioactive content by liquid scintillation counting. For L cells this volume corresponds to 1–2% of the cell volume. With practice one can attempt injections into 500–1000 cells per hour with a successful transfer of material being assured in 50–100% of them. Similar procedures for injecting macromolecules into cultured mammalian cells have been described by Graessmann and Graessmann (1976), by Diacumakos (1973) and by Stacey and Allfrey (1976).

Preparation of Recombinant Plasmids

Restriction enzymes and T4 DNA ligase were obtained from BRL and New England Biolabs and used under conditions recommended by the vendors. Plasmid DNAs were isolated from cultures of E. coli HB101, grown to saturation in NZYPD medium (per liter: 5 g NaCl, 2 g MgCl$_2$, 10 g NZ amine A, 1 g caseamino acids and 5 g yeast extract). The cells were lysed with 0.03 M NaOH, 0.003 M EDTA and 0.5% SDS. The chromosomal DNA was precipitated with 1 M NaCl and 3% PEG. Following the addition of NaCl and PEG the extracts were incubated for 2 hr at 4°C and centrifuged at 20,000 × g for 30 min. The supernatants containing the plasmid DNA were extracted twice with chloroform-phenol. The RNA was removed by gel filtration on a Biorad A50 column.

SV40 DNA was isolated from lytically infected TC7 cells and purified by the procedure of Trilling and Axelrod (1970). The HSV-1 Bam HI fragment which contains the HSV-TK gene was purified from pBR322/TK DNA by digestion to completion with Bam HI and fractionation on a 0.9% agarose gel.

SV40 was inserted into pBR322 by digesting both to completion with Eco RI. The pBR322 DNA was then treated with alkaline phosphatase [(pH 9) 68°C for 30 min] to prevent self ligation. After ligation of pBR322 and SV40 DNA with T4 ligase it was used to transform E. coli strain HB101. The resulting ampicillin resistant colonies were screened for recombinant plasmids on agarose gels. pBR322/SV-O+T/TK was prepared as described above by inserting the purified HSV-Bam HI fragment into pBR322/SV40. Following ligation the DNA was used to transform E. coli HB101. 90% of the ampicillin-resistant colonies contained the desired recombinant plasmid. pBR322/SV-O/TK was prepared from pBR322/SV-O+T/TK DNA by partial hydrolysis with Hind III followed by ligation with T4 ligase.

Acknowledgments

I would like to thank L. Fraser and R. Myers for expert technical assistance and E. Sparks and J. Harshman for help in the initial phases of these experiments. I also wish to express my gratitude to L. Okun for many thoughtful and encouraging discussions concerning the setting up of the microinjection apparatus. Technical advice concerning cloning procedures from P. Luciw is gratefully acknowledged. The DNA-mediated transformation studies using the calcium phosphate precipitation method were carried out by P. Barry.

The costs of publication of this article were defrayed in part by the payment of page charges. This article must therefore be hereby marked "*advertisement*" in accordance with 18 U.S.C. Section 1734 solely to indicate this fact.

Received August 11, 1980; revised September 5, 1980

References

Bacchetti, S. and Graham, F. L. (1977). Proc. Nat. Acad. Sci. USA *74*, 1590–1594.

Capecchi, M. R., Vonder Haar, R. A., Capecchi, N. E. and Sveda, M. M. (1977). Cell *12*, 371–381.

Diacumakos, E. G. (1973). Methods in Cell Biology, *7*, D. M. Prescott, ed. (New York: Academic Press), pp. 287–311.

Enquist, L. W., Vande Woude, G. F., Wagner, M., Smiley, J. R. and Summers, W. C. (1979). Gene *7*, 335–342.

Graessmann, M. and Graessmann, A. (1976). Proc. Nat. Acad. Sci. USA *73*, 366–370.

Graf, L. H., Urlaub, G. and Chasin, L. (1979). Som. Cell Genet *5*, 1031–1044.

Graham, F. L. and van der Eb, A. J. (1973). Virology *52*, 456–467.

Henderson, I. C. and Livingston, D. M. (1974). Cell *3*, 65–70.

Hughes, S. H., Shank P. R., Spector, D. H., Kung, M. J., Bishop, S. M. and Varmus, H. E. (1978). Cell *15*, 1397–1410.

Jelinek, W. R., Toomey, T. P., Leinwand, L., Duncan, C. H., Biro, P.A., Choudary, P. V., Weissman, S. M., Rubin, C., Houck, C. M., Deininger, P. L. and Schmid, C. W. (1980). Proc. Nat. Acad. Sci. USA *77*, 1398–1402.

Klobutcher, L. A. and Ruddle, F. H. (1979). Nature *280*, 657–660.

Lee, L. and Cheng, Y. (1976). J. Biol. Chem. *251*, 2600–2604.

McBride, O. W. and Ozer, H. C. (1973). Proc. Nat. Acad. Sci. USA *70*, 1258–1262.

Maitland, N. J. and McDougall, J. K. (1977). Cell *11*, 233–241.

Miller, C. L. and Ruddle, F. H. (1978). Proc. Nat. Acad. Sci. USA *75*, 3346–3350.

Preston, C. M. (1977). J. Virol. *23*, 455–460.

Sharp, J. D., Capecchi N. E. and Capecchi, M. R. (1973). Proc. Nat. Acad. Sci. USA *70*, 3145–3149.

Southern, E. M. (1975). J. Mol. Biol. *98*, 503–517.

Spandidos, D. A. and Siminovitch, L. (1977). Proc. Nat. Acad. Sci. USA *74*, 2943–2947.

Stacey, D. W. and Allfrey, V. G. (1976). Cell *9*, 725–732.

Todaro, G. J., Green, H. and Swift, M. C. (1966). Science *153*, 1252–1254.

Trilling, D. M. and Axelrod, D. (1970). Science *168*, 268–271.

Wahl, G. M., Hughes, S. H. and Capecchi, M. R. (1975). J. Cell Physiol. *85*, 307–320.

Wigler, M., Silverstein, S., Lee, L., Pellicer, A., Cheng, Y. and Axel, R. (1977). Cell *11*, 223–232.

Wigler, M., Pellicer, A., Silverstein, S. and Axel, R. (1978). Cell *14*, 725–731.

Wigler, M., Pellicer, A., Silverstein, S., Axel, R., Urlaub, G. and Chasin, L. (1979a). Proc. Nat. Acad. Sci. USA 76, 1373–1376.

Wigler, M., Sweet, R., Sim, G. K., Wold, B., Pellicer, A., Lacy, E., Maniatis, T., Silverstein, S. and Axel, R. (1979b). Cell 16, 777–785.

Willecke, K. and Ruddle, F. H. (1975). Proc. Nat. Acad. Sci. USA 72, 1792–1796.

Willecke, K., Klomfass, M., Mierau, R. and Dohmer, J. (1979). Mol. Gen. Genet. 170, 179–185.

Editor's Comments
on Papers 30 and 31

30 McBRIDE and OZER
Transfer of Genetic Information by Purified Metaphase Chromosomes

31 DEWEY et al.
Mosaic Mice with Teratocarcinoma-Derived Mutant Cells Deficient in Hypoxanthine Phosphoribosyltransferase

OTHER GENETIC TRANSFER SYSTEMS

The studies discussed in the previous section focused on the use of DNA for transferring genes. However, even before reliable DNA-mediated gene transfer systems were developed for mammalian cells, McBride and Ozer in 1973 (Paper 30) demonstrated that metaphase chromosomes can be used to transfer genetic information. Metaphase chromosomes were purified from HGPRT positive Chinese hamster cells, and the intact chromosomes were incubated with HGPRT deficient mouse cells. HGPRT positive cells were selected in HAT medium, and the presumptive transformants were analyzed enzymatically and karyologically. The HGPRT activity in the selected cells was shown to correspond to the donor (hamster) type, but no intact Chinese hamster chromosomes were observed in the transformants. Thus, these results provided a new system for transferring limited amounts of genetic information from one cell to another.

In general, only a very small chromosomal fragment is transferred by means of metaphase chromosomes. However, the co-transfer of a nonselected gene closely linked to the selected marker has been observed in some cases. For example, in experiments using purified human metaphase chromosomes, the human gene for galactokinase was co-transferred with the gene for thymidine kinase 25% of the time (Willecke et al., 1976). The human genes for galactokinase and thymidine kinase are known to be closely linked on chromosome 17.

A very different type of system from those just described has been developed for transferring the genetic information of cultured somatic cells into embryos. This system involves the use of terato-carcinoma (TC) cells injected into blastocysts. TC cells, although highly tumorigenic, can participate in the development of chimeric mice (Brinster, 1974). In the chimeric mice, TC cells contribute to the formation of many normal tissues and can even produce normal germ cells (Mintz and Illmensee, 1975). The use of TC cells as a vector for transferring specific genetic characteristics into animals was demonstrated by Mintz and co-workers (Paper 31). TC cells growing in culture were mutagenized and selected for HGPRT deficiency. The mutant cells were microinjected into blastocysts and chimeric mice developed. Since the TC cells and recipient embryos were from different inbred strains of mice, it was possible to show that mutant TC cells participated in the development of many normal tissues. Furthermore, the HGPRT deficiency introduced into the TC cells in culture was evident in the chimeric mice. These studies, combining the techniques of somatic cell genetics and developmental biology, demonstrated that specific genetic alterations introduced into TC cells in culture could be transferred into animals. Although not yet accomplished, such experiments offer the possibility of germ line transmission of genetic alterations that were predetermined in cell culture.

In addition to mutagenesis, other techniques of somatic cell genetics have been used to alter the genetic makeup of TC cells in culture. In one study, the HSV-TK gene and the human β globin gene were introduced into TC cells by DNA-mediated transformation (Pellicer et al., 1980). In another study, mouse TC cells were fused with rat hepatoma cells (Illmensee and Croce, 1979). Hybrid cells were injected into mouse blastocytes, and a small percentage of the mice that developed appeared to be chimeric. These few mice showed evidence of hybrid cell contributions to organs of endodermal origin, as indicated by the expression of rat enzymes. Thus, the techniques of somatic cell genetics apparently can be used to break down the species barrier during embryonic development. More recently, it was shown that genes of foreign species can be introduced into animals by direct microinjection of DNA into the pronuclei of fertilized eggs. The injected genes could be detected in the adult animals that developed, and could even be transmitted through the germ line (Gordon and Ruddle, 1981).

REFERENCES

Brinster, R. L., 1974, The Effect of Cells Transferred into the Mouse Blastocyst on Subsequent Development, *J. Exp. Med.* **140:**1049–1056.

Gordon, J. W., and F. H. Ruddle, 1981, Integration and Stable Germ Line Transmission of Genes Injected into Mouse Pronuclei, *Science* **214:**1244–1246.

Illmensee, K., and C. M. Croce, 1979, Xenogeneic Gene Expression in Chimeric Mice Derived from Rat-Mouse Hybrid Cells, *Natl. Acad. Sci. (USA) Proc.* **76:**879–883.

Mintz, B., and K. Illmensee, 1975, Normal Genetically Mosaic Mice Produced from Malignant Teratocarcinoma Cells, *Natl. Acad. Sci. (USA) Proc.* **72:**3585–3589.

Pellicer, A., E. F. Wagner, A. E. Kareh, M. J. Dewey, A. J. Reuser, S. Silverstein, R. Axel, and B. Mintz, 1980, Introduction of a Viral Thymidine Kinase Gene and the Human β-Globin Gene into Developmentally Multipotential Mouse Teratocarcinoma Cells, *Natl. Acad. Sci. (USA) Proc.* **77:**2098–2102.

Willecke, K., R. Lange, A. Krüger, and T. Reber, 1976, Cotransfer of Two Linked Human Genes into Cultured Mouse Cells, *Natl. Acad. Sci. (USA) Proc.* **73:**1274–1278.

30

Reprinted from *Natl. Acad. Sci. (USA) Proc.* **70:**1258–1262 (1973)

Transfer of Genetic Information by Purified Metaphase Chromosomes

(Chinese hamster and mouse fibroblasts/HeLa cells/hypoxanthine phosphoribosyl transferase)

O. WESLEY McBRIDE AND HARVEY L. OZER*

Laboratory of Biochemistry, National Cancer Institute, National Institutes of Health, Bethesda, Maryland 20014

Communicated by Alton Meister, February 15, 1973

ABSTRACT Transfer of genetic information from isolated mammalian chromosomes to recipient cells has been demonstrated. Metaphase chromosomes isolated from Chinese hamster fibroblasts were incubated with mouse A_9 cells containing a mutation at the hypoxanthine–guanine phosphoribosyl transferase (*hprt*) locus. Cells were plated in a selective medium, resulting in death of all unaltered parental A_9 cells. However, colonies of cells containing hypoxanthine phosphoribosyl transferase (EC 2.4.2.8) appeared with a variable frequency of about 10^{-6} to 10^{-7}. The enzyme from these cells was indistinguishable from that from Chinese hamster cells, as shown by DEAE-cellulose chromatography and gel electrophoresis, and differed clearly from the mouse enzyme. The colonies, thus, did not result from reversion of A_9 parental cells to wild type, but appeared to represent progeny of individual cells that had ingested chromosomes, replicated, and expressed the *hprt* gene. These colonies differed from each other in stability of expression of the transferred gene.

A means for genetic mapping of mammalian chromosomes is provided by a combination of the technique of cell fusion and karyotypic analysis of resultant hybrid clones by the recently developed quinacrine (3) and Giemsa banding procedures (4). Techniques for the direct transfer of genetic information from subcellular particles to cells could provide a complementary method for genetic mapping. This would eliminate the necessity of awaiting segregation of chromosomes, thereby reducing the possibility of chromosomal rearrangements. Mammalian metaphase chromosomes appear eminently suitable for this purpose since a meaningful biological fractionation of genes is present in chromosomes, and numerous methods (5–7) have been described for isolation of these particles. Chromosomal DNA might be somewhat better protected from degradation during cellular uptake than free DNA due to its compact structure and its association with proteins and RNA. The introduction of intact chromosomes into cells could circumvent problems of integration of DNA into the host genome; subsequent replication and expression of chromosomal genes should be analogous to the steps following cell fusion.

Evidence exists that isolated metaphase chromosomes can penetrate into mammalian cells *in vitro* (8–16), but most of the chromosomal DNA is subsequently degraded (8–12). Previous information suggesting that mammalian chromosomes can be replicated after uptake is extremely sparse (14–16), and no evidence has been provided for expression of this new genetic information by the host cell.

Abbreviations: HPRT, hypoxanthine phosphoribosyl transferase (EC 2.4.2.8); *hprt*, gene directing synthesis of HPRT; HAT, hypoxanthine–amethopterin–thymidine, selective growth medium of Littlefield (1); MEM, Eagle's minimal essential medium (2).

*Present address: Worcester Foundation for Experimental Biology, Shrewsbury, Mass. 01545.

This paper presents the first evidence that both replication and expression of chromosomal genes can occur after the uptake of mammalian metaphase chromosomes. Moreover, permanent transfer of this new genetic information results, although the frequency of this gene transfer is low.

MATERIALS AND METHODS

Cell Cultures. Cells used were: (*1*) wild-type Chinese hamster fibroblasts (V-79), recently cloned; (*2*) mouse fibroblasts (L_{929}); (*3*) HeLa cells; and (*4*) mouse L-cell lines A_9 and B_{82}, deficient in hypoxanthine phosphoribosyl transferase (HPRT; EC 2.4.2.8) and thymidine kinase (EC 2.7.1.21), respectively (17). Cells were maintained in monolayer cultures at 37° in a gas-flow (7% CO_2–air), humidified incubator, in Eagle's minimal essential medium (MEM) containing twice the usual concentration of amino acids and vitamins. Cells were also grown in suspension culture, in Eagle's medium without calcium, or in Ham's F-10 medium (18). All media were supplemented with 10% fetal-calf serum, 4 mM glutamine, penicillin (50 μg/ml), and streptomycin (50 μg/ml).

Isolation and Purification of Metaphase Chromosomes. Chromosomes were isolated under sterile conditions from [^{3}H]-dT-labeled cells (0.2 mCi/liter) as described (19), by slight modifications of either the procedure of Mendelsohn *et al.* (5) at pH 3 or the method of Maio and Schildkraut (6) at pH 7. Chromosomes were subsequently separated from intact cells and most debris by ultracentrifugation through a layer of 80% sucrose (w/v). Nuclei were then removed by unit-gravity sedimentation at pH 7, essentially as reported for fractionation of nuclei (20). Final chromosome preparations contained about one nucleus per thousand cell equivalents of chromosomes (i.e., one nucleus per 25,000 chromatid pairs). For each preparation, the molecular weight of dissociated, single-stranded, chromosomal DNA was determined by velocity sedimentation in alkaline sucrose density gradients (21), with ^{14}C-labeled simian virus 40 (SV40) DNA I and II (22) as markers. The molecular weight of chromosomal DNA decreases progressively on storage of the chromosomes, even at 5°, although chromosome morphology remains good. Thus, chromosomes were used in gene-transfer experiments immediately after isolation.

Incubation of A_9 Cells with Chromosomes. Purified metaphase chromosomes (about 1 cell equivalent per recipient cell) from Chinese hamster cells were dispersed with A_9 mouse fibroblasts (6 $\times$ 10^6/ml) in complete Eagle's MEM spinner medium, containing 12 μg/ml of poly-L-ornithine (molecular weight 70,000; Mann Research) in a sterile, siliconized, glass culture tube. The tube was equilibrated with 5% CO_2–air and incubated for 2 hr at 37° while rolling in a nearly horizontal position at 10 rpm. Aliquots of 5 $\times$ 10^5 cells were transferred to 100-mm plastic dishes (Falcon) containing 10 ml of complete MEM. After 3 days of incubation, the medium was replaced with HAT medium, and the plates were refed with this selective medium at 3- to 4-day intervals for 6 weeks. Colonies that appeared during this interval were cloned in metal cylinders, removed by treatment with trypsin, and recultured in HAT medium.

Enzyme Assays. Cells were washed with 0.15 M NaCl, suspended in 0.01 M Tris·HCl (pH 7.4) (6 × 10^7 cells per ml), and lysed by freezing and thawing. The assay for HPRT activity was basically that described by Harris and Cook (23), involving conversion of [8-^{14}C]hypoxanthine substrate to [^{14}C]IMP product, which was collected on DEAE-cellulose disks (Whatman DE-81). The reaction product of the hamster and wild-type mouse HPRT, as well as that of extracts of the experimental clones, was confirmed to be [^{14}C]IMP by thin-layer chromatography on cellulose; both solvent systems B and C of Ciardi and Anderson (24) were used. Purity (>97–99%) of the [^{14}C]hypoxanthine substrate was ascertained in the same manner.

DEAE-Cellulose Chromatography. Micro-granular DEAE-cellulose (Whatman DE-52) was washed (25), equilibrated with starting buffer [0.01 M Tris·HCl (pH 8.7)], and packed in a 5 × 140 mm column. Enzyme extract (about 5 mg of protein) was applied to the column at 5° and sequentially eluted at a constant flow-rate (8 ml/hr) with 4 ml of starting buffer, a 60-ml linear (0–225 mM) NaCl gradient followed by 3 ml of 0.4 M NaCl (both containing starting buffer), and finally with 6 ml of 2 M NaCl–0.1 M Tris·HCl (pH 8.8). 1-ml Fractions were assayed immediately for HPRT activity and later for protein concentration, conductivity, and pH.

Gel Electrophoresis. Gels were prepared by a modification of the procedure of Bakay and Nyhan (26). An 8% poly-

TABLE 1. *HPRT-positive colonies after incubation of A_9 cells with chromosomes*

Experiment	Total no. of cells[a] ($\times 10^{-6}$)	Positive plates/ total plates[b]	DNA[c] mol. wt. ($\times 10^{-6}$)
1	6	5/12	30
2	25	2/50[d]	3
3	25	2/49[e]	30
4	50	0/100[f]	20
5	6	1/12	25
6	10	0/20[f]	1.5–30[g]
7	9	1/18	30
8A	10	1/20[h,i]	30–130
8B	10	3/20[h]	30–130

Chromosomes used in experiments 2, 4, and 8 were isolated at pH 7; otherwise chromosomes were isolated at pH 3. [a] Total number of A_9 cells incubated and subsequently plated. [b] Number of plates with one or more colonies per total number of plates inoculated. [c] Molecular weight of single-stranded DNA in the chromosome preparations (see *Methods*). [d] Colonies not confirmed by cloning and growth in HAT medium. [e] HeLa chromosomes incubated with A_9 recipient cells. [f] Incubation medium contained 2 mM $CaCl_2$ (monolayer medium). [g] Very heterogeneous molecular weight. [h] Ratio of cell equivalents of chromosomes to recipient cells was 10:1 in experiment 8B and 1:1 in experiment 8A; the ratio was about 1:1 in the other experiments. [i] This colony was a revertant.

acrylamide gel (70 × 100 × 2 mm) was formed between two glass plates, and a 5% stacking gel was added. A cellulose acetate strip containing sample slots was pushed into the upper gel. Protein extracts (about 15 μg of protein) were mixed with bovine-serum albumin (25 μg per application) and sucrose (10% w/v) and layered under upper-tray buffer in the slots. Bromphenol blue (0.05 ml saturated solution per liter) was included in the upper tray as a tracking dye. Electrophoresis was conducted at 300 V until albumin migrated to the bottom of the running gel. The gel was reacted with substrate (30 min at 37°), and the [^{14}C]IMP product was precipitated with 0.1 M $LaCl_3$–0.1 M Tris·HCl (pH 7.0) (26). Autoradiography was performed after repeated washing and dehydration of the gel (27).

Other Assays. Protein concentration was determined by the procedure of Lowry *et al.* (28) with bovine-serum albumin as the standard, and most assays were kindly performed by Mr. Miles Otey with a Technicon Autoanalyzer. Conductivity and pH measurements were performed at room temperature (24°). Particle concentrations were determined with an electronic counter (Celloscope).

Karyotypes. Cells were exposed to colcemid (0.2 μg/ml) for 3 hr, swollen in 1% Na citrate (20 min at 37°) or 75 mM KCl (20 min at 25°), and fixed with methanol–acetic acid (3:1). The fixed cells were applied to a cold, moist slide and spread by flaming before staining with crystal violet.

RESULTS

Isolation of HPRT-Positive Colonies after Incubation of A_9 Cells with Chromosomes. Isolated Chinese hamster chromosomes were incubated in suspension (see *Methods*) with mouse A_9 cells before plating the cells and the subsequent addition of selective medium (Table 1). Colonies appeared at a relatively low frequency of 10^{-6} in experiment 1 and about 10^{-7} in the combination of all experiments. A positive result was also obtained (experiment 3) with chromosomes isolated from HeLa cells. Since migration of cells may occur, resulting in satellite colonies, only one colony was scored for any plate, irrespective of the actual number of colonies observed, and each colony that was further analyzed was cloned from a separate plate. Local overgrowth of unaltered cells may also occur early before the addition of HAT, and slowly regress or persist for long intervals. Therefore colonies were cloned and cultured in the same selective medium and colonies lost in the cloning process were considered false positives. In two experiments designed to detect gene transfer from hamster chromosomes to A_9 cells, we used inactivated Sendai virus to mediate the transfer, but were unsuccessful.

Reversion of A_9 Cells. A control incubation performed in experiment 1 (Table 1), by use of identical procedures without chromosomes, resulted in no colonies. Larger numbers of cells (1.37 × 10^9) have also been plated in selective HAT medium, and only two revertant colonies were found. Thus, the A_9 cells have a very low rate of reversion that appears to be lower than the gene-transfer frequency, even considering that under the experimental conditions, cells could have doubled about three times before the HAT selective medium was added.

Detection of the Product of Gene Transfer In Vitro. The clones obtained in experiment 1 of Table 1 were propagated

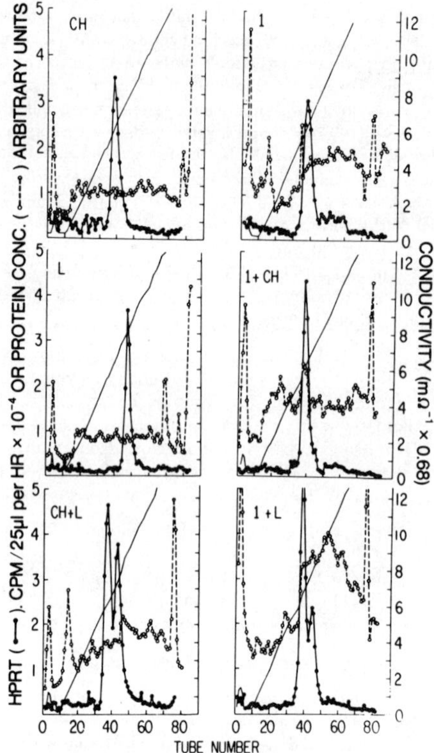

FIG. 1.   Chromatography of crude enzyme extracts isolated rom Chinese hamster (*CH*), mouse (*L*), Clone 1 cells, and artificial mixtures of these solutions. Extracts (3–15 mg of protein) were applied to columns of DEAE-cellulose and eluted with a gradient of NaCl in 0.01 M Tris·HCl (pH 8.7). Fractions of 1 ml were collected and assayed for HPRT activity (●——●), protein (O‑ ‑ ‑O), and conductivity (——). Enzyme activity is plotted at 0.5 the normal scale for L, 1 +CH, and 1 + L. Each unit of protein concentration (*left ordinate*) represents 100 μg/ ml (CH, L, 1 +CH) or 70 μg/ml (CH + L, 1, 1 + L).

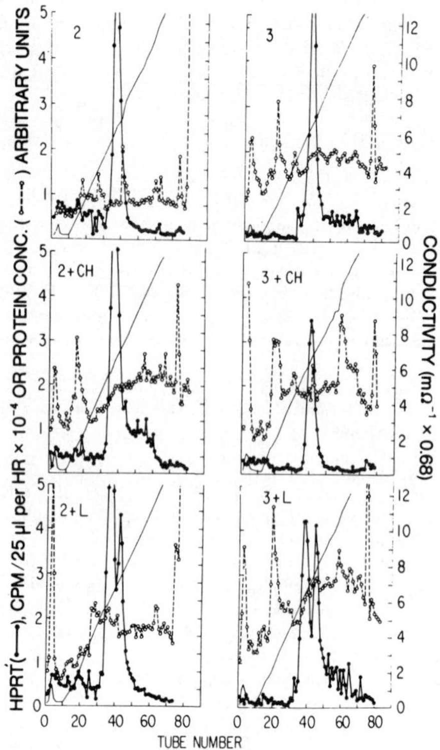

FIG. 2.   DEAE-cellulose chromatography at 5° of HPRT extracts isolated from clones 2 and 3, and artificial mixtures of these solutions with Chinese hamster (*CH*) or mouse (*L*) extracts. Enzyme activity is plotted at 0.5 the usual scale for 3 +CH. Each unit of protein concentration (*left ordinate*) represents 100 μg/ml (2, 2 + L, 3 +CH) or 70 μg/ml (2 +CH, 3, 3 + L). See Fig. 1 and *Methods* for further details.

in suspension culture in HAT medium, and the high-speed (100,000 × *g*) supernatant fluids of freeze–thaw lysates were examined directly for HPRT activity (Table 2). The specific activities were similar to those obtained from extracts of Chinese hamster fibroblasts or wild-type mouse L₉₂₉ cells, or the closely related, B₈₂ thymidine-kinase-mutant L cells.

DEAE-Cellulose Chromatography of HPRT Extracts. Chromatography demonstrated a single peak of HPRT activity for both the mouse and hamster parental species (Fig. 1). However, the hamster HPRT is adequately resolved from mouse (L) enzyme when compared directly by elution position or conductivity at the point of emergence, or by mixture of the extracts before chromatography (*lower left*, Fig. 1). Clones from experiment 1 of Table 1 were similarly analyzed. Chromatography of an extract of clone 1 revealed a single peak of HPRT activity occurring in the position appropriate for hamster enzyme. The mixture of clone 1 extract with hamster-HPRT again resulted in the single peak of activity,

whereas the mixture with mouse enzyme disclosed two peaks of HPRT at the appropriate positions for each species.

Similar results are presented in Fig. 2 with two other clones (from experiment 1) alone or as artificial mixtures with hamster or mouse HPRT. Approximately equal quantities (enzyme activity) of each of the two components were used in all mixtures. The elution positions and half-widths as measured by elution volumes or conductivities of eluate for all of these clones, individually or mixed with hamster extract, are virtually identical with that obtained with the hamster enzyme alone. It therefore appears unlikely that the patterns result from three revertant clones that all fortuitously show chromatographic behavior very similar to that of the hamster enzyme. Furthermore, HPRT in extracts of thymidine-kinase-mutant L-cells (B₈₂) and the single A₉ revertant exhibit chromatographic behavior (not shown) identical with that of wild-type mouse HPRT. A fourth clone from experiment 1 (Table 1) was lost before chromatography, but it appeared to contain hamster enzyme, as determined by electrophoresis on cylindrical acrylamide gels. The fifth clone was lost before further study.

Chromatographic analyses (not illustrated) of the clone

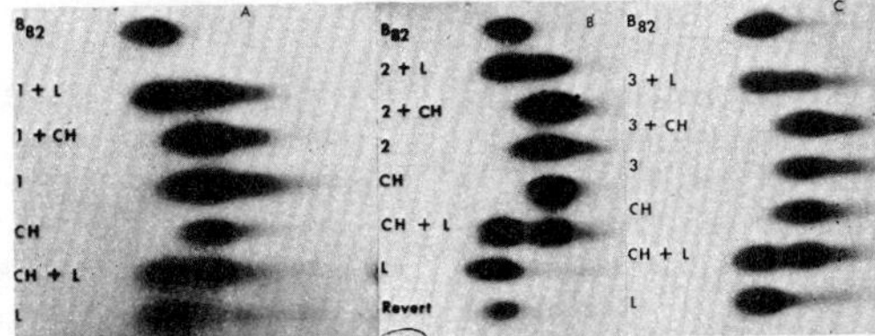

Fig. 3. Gel electrophoresis of HPRT extracts on vertical slabs of polyacrylamide at 5°. The individual extracts and artificial mixtures are identified by the same symbols as Figs. 1 and 2, and the A₉ revertant is also shown (*B*).

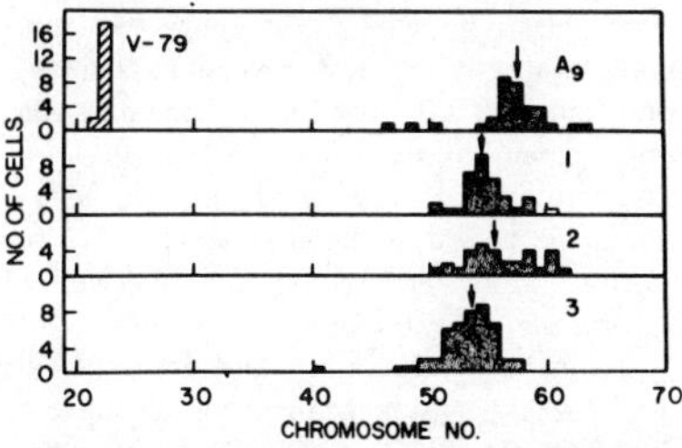

Fig. 4. Histogram of chromosomes in parental cell lines (V-79 and A₉) and in clonal lines (1, 2, and 3) from experiment 1 of Table 1. *Arrows* indicate the median number of chromosomes in each line.

from experiment 5 of Table 1 and the three clones from experiment 8*B* of Table 1 also demonstrated the hamster-enzyme profile, while the clone from experiment 8*A* exhibited a profile that is identical with that of mouse HPRT, and therefore represents a revertant.

Acrylamide Gel Electrophoresis of HPRT. Mouse HPRT has a greater electrophoretic mobility than the hamster enzyme, from which it is adequately separated (Fig. 3). HPRT in extracts of clones 1, 2, and 3 from experiment 1 of Table 1 were all identical in mobility with hamster enzyme, when run alone or when mixed with hamster extract, whereas two bands resulted when artificial mixtures of these extracts with mouse HPRT were subjected to electrophoresis. Furthermore, the HPRT produced by the A₉ revertant (Fig. 3*B*) is not electrophoretically distinguishable from the wild-type mouse enzyme. Evidence (not shown) that the radioactive spots reflect the location of HPRT activity is provided by the fact that the spots were markedly attenuated when the gel was reacted with substrate at 5° rather than 37°, as well as by the fact that no radioactivity could be detected when 5-phosphoribosylpyrophosphate was omitted from the reaction mixture. No radioactivity was observed under any condition when an extract of A₉ cells was subjected to electrophoresis. Gel electrophoresis (not shown) also demonstrated that the HPRT products of the clone from experiment 5 and the three clones from experiment 8*B* of Table 1 were indistinguishable from the Chinese hamster enzyme, whereas the product of the clone from experiment 8*A* had the same electrophoretic mobility as the mouse HPRT.

Karyotypes. Histograms of the numbers of total chromosomes (Fig. 4) and biarmed chromosomes (not shown) in the clones from experiment 1 of Table 1 were closely similar to that of the parental A₉ cells. Karyotypes of all experimental cell lines clearly differ from the Chinese hamster karyotype, which exhibits a narrow mode of 23 chromosomes,

indicating that none of these lines could have arisen by contamination of the cultures with hamster cells.

Stability of Genotype after Chromosome Transfer. The clones of experiment 1 of Table 1 were grown in selective HAT medium for several generations. After a shift to nonselective MEM spinner medium, the growth of each line was continued in suspension cultures for 2 months. Aliquots were removed at intervals for determination of plating efficiencies in MEM, HAT, and 20 µM 8-azaguanine, and the plating efficiencies in HAT relative to those in MEM are shown in Fig. 5. Clones 1 and 2 exhibited no detectable change in plating efficiency in the selective HAT medium during this entire interval, whereas there was a very rapid accumulation of HPRT-deficient cells when clone 3 was cultured in nonselective medium. The curve for clone 3 suggests that about 10–20% of the cells lose the *hprt* gene at each division. Similar reversion behavior has been reported by Schwarz *et al.* (29) for cells containing *hprt* on a chromosome fragment. The instability exhibited by clone 3 would be highly unlikely if it had arisen by reversion (back-mutation) of A₉ cells rather than by gene transfer, unless the parental A₉ cells had very marked selective growth advantage in MEM relative to the revertant.

DISCUSSION

The evidence for transfer of genetic information from ingested metaphase chromosomes to recipient cells and expression of this information by recipient mammalian cells can be summarized as: (**1**) A relatively high frequency of appearance

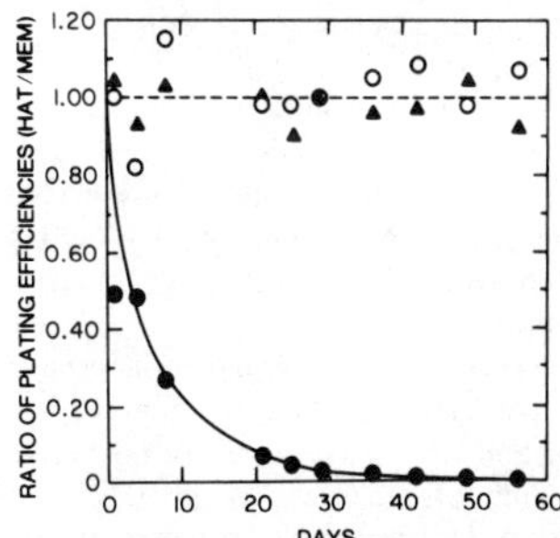

Fig. 5. Stability of the *hprt*⁺ genotype in colonies after removing selective pressure. The plating efficiency in selective HAT medium compared to that in nonselective medium is plotted as a function of the time interval after the cells were removed from HAT medium. The cell lines depicted are clones 1 (O), 2 (△), and 3 (●———●) from experiment 1 of Table 1.

Table 2. *HPRT activity*

Cell type	Specific activity*	Cell type	Specific activity*
CH-V-79	196, 226, 354, 322	Clone 2	274, 254
L₉₂₉	134, 143, 151, 193	Clone 3	149
B₈₂	43, 142, 105	Clone 4	155
A₉	<0.01, <0.02	A₉ revertant	141
Clone 1	65, 160		

* nmol of IMP/hr per mg of protein.

241

of colonies in selective medium when chromosomes are present, compared to the very low reversion frequency under similar conditions. (2) The rapid loss of the *hprt* gene by one of the clones (clone 3 of experiment 1), which is unexpected if the colony arose by reversion of parental cells. (3) The physical characterization of the enzyme (HPRT) product as indistinguishable from the chromosomal species and clearly different from the parental species, as shown by DEAE-cellulose column chromatography and acrylamide gel electrophoresis. This last point is the most convincing one.

Other possible explanations that have been considered, but appear extremely unlikely or completely inconsistent with the results, include the following:

(*1*) Reversion is inconsistent with any of the three points mentioned above and is especially refuted by the physical characterization of the gene product. Some revertants could occur involving mutation at a locus for a charged residue, resulting in a gene product that differed from the parental type. However, the possibility that the product of all revertants could be completely indistinguishable from that of the donor chromosome species by two methods of characterization seems remote. Furthermore, the single authentic revertant that was analyzed produced an HPRT that was not distinguishable from that of the parental (mouse) species. Schwarz *et al.* (29) also reported that an A_9 revertant produced HPRT that was electrophoretically identical to wild-type mouse enzyme.

(*2*) The possibility that the cultures were contaminated with a few wild-type (L_{929}) mouse cells is excluded by the characterization of the enzyme product and by the absence of similar colonies in control cultures.

(*3*) Contamination of the cultures with a few Chinese hamster cells or incomplete removal of these cells during the process of chromosome isolation is excluded by the fact that the karyotypes of the resultant clones were similar to that of the mouse species and totally different from that of the hamster species. Furthermore, no viable intact cells were detected under the conditions of chromosome isolation.

(*4*) It is unlikely that spontaneous fusion of any intact cells (surviving chromosome isolation and purification procedures) or nuclei with A_9 mouse cells is responsible for the observed results in view of the low number of intact cells and nuclei in the preparations. Furthermore, experiments performed under conditions more favorable for cell fusion, involving the use of inactivated Sendai virus, failed to result in colony formation. Also, no evidence for persistence of large numbers of hamster chromosomes was found in karyograms.

(*5*) Transformation of the cells by naked DNA or nucleoprotein cannot be excluded but it is considered unlikely. Most DNA or nucleoprotein would have been removed during the chromosome isolation by several centrifugations at 1000 $\times g$ for 30 min.

(*6*) Some nonspecific effect of added chromosomes is excluded by the physical characterization of the enzyme product as hamster type and the absence of gene transfer when chromosomes containing low molecular weight DNA were used (19). Any stimulation by degraded chromosomal products is unlikely since they would be rapidly diluted out during growth.

(*7*) Contamination of cultures with viruses or mycoplasma coding for an active HPRT is excluded by inability to culture mycoplasma from the clones, failure to observe HPRT-positive colonies in control cultures, and physical characterization of the enzyme product. However, the possibility that a transducing virus was present cannot be excluded.

There are several possible explanations for the low frequency of gene transfer observed (see ref. 19). The ability to demonstrate any gene transfer in the present experiments results from the use of a selective system using recipient cells with an extremely low reversion frequency and chromosomes isolated from a different species, thereby permitting positive identification of the species of origin of the gene product.

Mammalian chromosome uptake *in vitro*, particularly combined with the use of fractionated chromosomes, could provide a powerful tool for genetic mapping. However, the utility of this procedure would be increased by the development of methods for a greater efficiency of transfer and expression of the genetic information. The possible application of this technique to "gene modification" is open to considerably greater skepticism (30).

We thank Mrs. Susan Bridges for expert technical assistance.

1. Littlefield, J. W. (1964) *Science* **145**, 709–710.
2. Eagle, H. (1959) *Science* **130**, 432–437.
3. Caspersson, T., Zech, L. & Johansson, C. (1970) *Exp. Cell Res.* **60**, 315–319.
4. Drets, M. E. & Shaw, M. W. (1971) *Proc. Nat. Acad. Sci. USA* **68**, 2073–2077.
5. Mendelsohn, J., Moore, D. E. & Salzman, N. P. (1968) *J. Mol. Biol.* **32**, 101–112.
6. Maio, J. J. & Schildkraut, C. L. (1969) *J. Mol. Biol.* **40**, 203–216.
7. Hearst, J. E. & Botchan, M. (1970) *Annu. Rev. Biochem.* **39**, 151–182.
8. Chorazy, M., Bendich, A., Borenfreund, E., Ittensohn, O. L. & Hutchison, D. J. (1963) *J. Cell Biol.* **19**, 71–77.
9. Whang-Peng, J., Tjio, J. H. & Cason, J. C. (1967) *Proc. Soc. Exp. Biol. Med.* **125**, 260–263.
10. Ittensohn, O. L. & Hutchison, D. J. (1969) *Exp. Cell Res.* **55**, 149–154.
11. Kato, H., Sekiya, K. & Yosida, T. H. (1971) *Exp. Cell Res.* **65**, 454–462.
12. Burkholder, G. D. & Mukherjee, B. B. (1970) *Exp. Cell Res.* **61**, 413–422.
13. Ebina, T., Kamo, I., Takahashi, K., Homma, M. & Ishida, N. (1970) *Exp. Cell Res.* **62**, 384–388.
14. Yosida, T. H. & Sekiguchi, T. (1968) *Mol. Gen. Genet.* **103**, 253.
15. Sekiguchi, T., Sekiguchi, F., Satake, S. & Yosida, T. H. (1969) *Symp. Cell Biol.* **20**, 223.
16. Sekiguchi, T., Sekiguchi, F., Satake, S. & Yosida, T. H. (1969) *Jap. J. Med. Sci. Biol.* **22**, 72–73.
17. Littlefield, J. W. (1966) *Exp. Cell Res.* **41**, 190–196.
18. Ham, R. G. (1963) *Exp. Cell Res.* **29**, 515–526.
19. McBride, O. W. & Ozer, H. L. (1973) in *Possible Episomes in Eukaryotes Le Petit Colloquia on Biology and Medicine* (North-Holland, Amsterdam), Vol. 4, in press.
20. McBride, O. W. & Peterson, E. A. (1970) *J. Cell Biol.* **47**, 132–139.
21. Abelson, J. & Thomas, C. A., Jr. (1966) *J. Mol. Biol.* **18**, 262–291.
22. Sebring, E. D., Kelly, T. J., Jr., Thoren, M. M. & Salzman, N. P. (1971) *J. Virol.* **8**, 478–490.
23. Harris, H. & Cook, P. R. (1969) *J. Cell Sci.* **5**, 121–133.
24. Ciardi, J. E. & Anderson, E. P. (1968) *Anal. Biochem.* **22**, 398–408.
25. Himmelhoch, S. R. (1971) in *Methods in Enzymology*, ed. Jakoby, W. B. (Academic Press, New York), Vol. 22, pp. 273–286.
26. Bakay, B. & Nyhan, W. L. (1971) *Biochem. Genet.* **5**, 81–90.
27. Fairbanks, G., Jr., Levinthal, C. & Reeder, R. H. (1965) *Biochem. Biophys. Res. Commun.* **20**, 393–399.
28. Lowry, O. H., Rosebrough, N. J., Farr, A. L. & Randall, R. J. (1951) *J. Biol. Chem.* **193**, 265–275.
29. Schwartz, A. G., Cook, P. R. & Harris, H. (1971) *Nature New Biol.* **230**, 5–8.
30. Fox, M. S. & Littlefield, J. W. (1971) *Science* **173**, 195.

Reprinted from *Natl. Acad. Sci. (USA) Proc.* **74:**5564–5568 (1977)

Mosaic mice with teratocarcinoma-derived mutant cells deficient in hypoxanthine phosphoribosyltransferase

(Lesch–Nyhan disease/cell selection/trisomy/allophenic mice/blastocyst injection)

MICHAEL J. DEWEY*, DAVID W. MARTIN, JR.[†], GAIL R. MARTIN[‡], AND BEATRICE MINTZ*[§]

* Institute for Cancer Research, Fox Chase Cancer Center, Philadelphia, Pennsylvania 19111; and [†] Departments of Medicine and Biochemistry, and [‡] Department of Anatomy and Cancer Research Institute, University of California, School of Medicine, San Francisco, California 94143

Contributed by Beatrice Mintz, August 23, 1977

ABSTRACT   Mutagenized stem cells of a cultured mouse teratocarcinoma cell line were selected for resistance to the purine base analog 6-thioguanine. Cells of a resistant clone were completely deficient in activity of the enzyme hypoxanthine phosphoribosyltransferase (HPRT, IMP:pyrophosphate phosphoribosyltransferase, EC 2.4.2.8), the same X-linked lesion as occurs in human Lesch–Nyhan disease. After microinjection into blastocysts of another genetic strain, the previously malignant cells successfully participated in normal embryogenesis and tumor-free, viable mosaic mice were obtained. Cells of tumor lineage were identified by strain markers in virtually all tissues of some individuals. Mature function of those cells was evident from their tissue-specific products (e.g., melanins, liver proteins). These mutagenized teratocarcinoma cells are therefore developmentally totipotent. Retention of the severe HPRT deficiency in the differentiated state was documented in extracts of mosaic tissues by depressed specific activity of the enzyme, and also by presence of unlabeled clones in autoradiographs of explanted cells incubated in [³H]hypoxanthine. Some mosaic individuals had mutant-strain cells in only one or a few tissues. Such animals may provide unique opportunities to identify the tissue sources of particular aspects of the complex disease syndrome. The tissue distribution of HPRT-deficient cells suggests that selection against them is particularly strong in blood of the mosaic mice, as is already known to be the case in human heterozygotes. This phenotypic parallelism supports the expectation that afflicted F_1 male mice that might be obtained from mutant germ cells can serve as a model of the human disease.

The embryonal stem cells in transplantable mouse teratocarcinomas may continue to proliferate and also to differentiate into a disorganized, limited assortment of tissues (1, 2). When stem cells from an *in vivo* transplanted tumor (OTT 6050) were microinjected in small numbers (1–5 cells) into early embryos at the blastocyst stage, they underwent normal, stable differentiation (3–5). Healthy mosaic mice with both tumor-derived and embryo-derived functional somatic tissues were obtained. Functional germ cells were also sometimes formed from this chromosomally X/Y tumor, in males, and gave rise to normal F_1 progeny. The conclusion was reached that the stem cells of that tumor line had retained the normal genetic complement and were developmentally totipotent. Whether they behaved in a normal or a malignant fashion was a matter of change in gene expression rather than gene structure, and the choice was determined by local environmental influences.

The above tests for totipotency of "wild-type" teratocarcinoma cells were undertaken in part to establish the feasibility of converting *mutant* teratocarcinoma cells into mice bearing specific mutations (6, 7). Gametes or fertilized eggs, from which development ordinarily occurs, are not self-propagating in culture and therefore cannot yield clones in which a desired mutation may be selected following exposure to a mutagenic agent. Because teratocarcinoma cells are in fact developmentally totipotent, their adaptability to culture makes them promising vehicles for the deliberate introduction of specific mutant genes into mice.

One of the possible uses of such an experimental system would be the construction of animal models of human genetic diseases (6, 7). An example is the Lesch–Nyhan syndrome (8), due to a severe deficiency of hypoxanthine phosphoribosyltransferase (HPRT, IMP:pyrophosphate phosphoribosyltransferase, EC 2.4.2.8) (9), which functions in the salvage pathway of purine biosynthesis. The defective (recessive) form of the X-linked gene leads, in affected males, to excess formation of uric acid and to severe neurological symptoms, including self-mutilation, spasticity, and mental retardation (10). The HPRT gene is apparently also X-linked in the mouse (11), but no mutation has yet been detected in the laboratory species.

We report here that cells from an HPRT-deficient *in vitro* mouse teratocarcinoma stem-cell line have proven to be normalizable and developmentally totipotent when injected into genetically marked blastocysts. In the mosaic mice that have resulted, mature functional cells of the tumor strain have retained the HPRT deficiency and are found in essentially all tissues of some animals. In other individuals, only one or a few tissues have cells derived from the mutant cell lineage; these cases may help to identify the tissue sources of particular aspects of the disease.

MATERIALS AND METHODS

Teratocarcinoma Cell Line. The wild-type starting material was the PSA 1 teratocarcinoma cell line (refs. 12 and 13; unpublished data) clonally derived from primary cultures of the OTT 5568 mouse teratocarcinoma; that tumor was produced (14) almost a decade ago by the method of grafting a blastocyst-stage embryo of the *agouti black* 129/Sv Sl^J C P inbred strain (to be referred to as 129) under the testis capsule of an adult (Fig. 1). PSA 1 cells were maintained on a mitomycin C-treated feeder layer (12) of the mouse embryo fibroblast STO cell line (from Alan Bernstein of the Ontario Cancer Institute). STO cells are 6-thioguanine-resistant. To obtain mutant cells, PSA 1 cultures were exposed for 2 hr to medium containing 20 μM N-methyl-N'-nitro-N-nitrosoguanidine, and then seeded at 10^7 cells per 10-cm² confluent STO feeder layer. After growth for 7 days in the absence of a selective agent, to allow for expression of mutant phenotypes, HPRT-deficient mutants

---

Abbreviations: HPRT, hypoxanthine phosphoribosyltransferase; APRT, adenine phosphoribosyltransferase; GPI, glucosephosphate isomerase; MUP, major urinary protein complex.
[§] To whom reprint requests should be addressed.

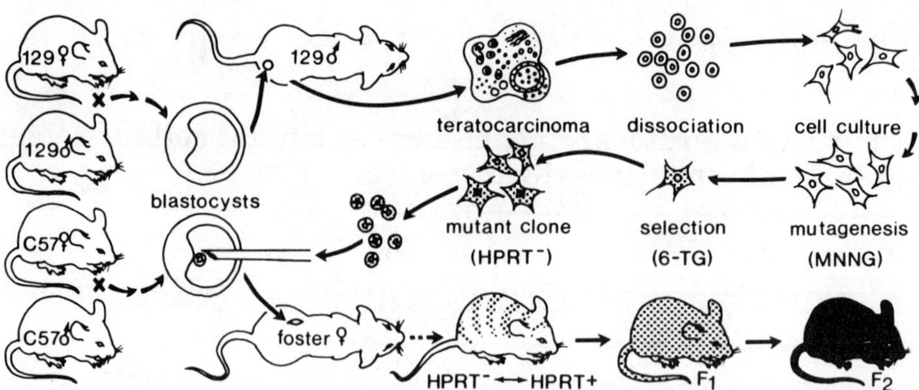

FIG. 1. The plan of the experiment, spanning almost a decade, is diagrammed, starting at the upper left. A blastocyst from a mating of 129-strain mice was grafted under the testis capsule of a syngeneic host (14). The graft formed a malignant teratocarcinoma. After dissociation, tumor cells were explanted and the stem cells were established as an *in vitro* culture line (ref. 12, and unpublished data). Following exposure to the mutagen N-methyl-N'-nitro-N-nitrosoguanidine (MNNG), HPRT⁻ cells (stippled) were selected for their resistance to 6-thioguanine (6-TG). Cells from a resistant *clone* were then microinjected into the cavity of genetically marked blastocysts (e.g., of the C57BL/6 strain). The injected embryos were transferred to the uterus of a pseudopregnant foster mother (previously mated to a sterile vasectomized male) and live mice were born. Some were mosaics comprising HPRT⁻ cells derived from the mutant teratocarcinoma lineage, along with blastocyst-derived cells, in their coats (striped) and/or internal somatic tissues. Tissue-specific effects of the deficiency are analyzable in the mosaics. If mutant cells are of X/O (as in this case) or X/X sex chromosome type and contribute to the germ line of females, affected "Lesch–Nyhan" males would be obtained in the F₁ generation; if the mutant cells are X/Y, in mosaic (non-Lesch–Nyhan) males, affected males would occur in the F₂. (Based upon figure 17, ref. 6.)

were selected in medium containing 0.1 mM 6-thioguanine. One of the resistant clones, NG 2, was used for blastocyst injections.

Blastocyst Injections. The microinjection procedure was as previously described (4). Three to five NG 2 cells were introduced into the cavity of blastocysts that had numerous genetic markers distinguishing them from the 129 tumor strain. Blastocysts were from ICR sublines of the following inbred strains: C57BL/6 (C57), WH, CBA with the T6/T6 homozygous translocation, and F₁ (CBA-T6/T6 × C57) hybrids; in some cases, color-marked *brown* (*b/b*) or *beige* (*bg^J/bg^J*) strains coisogenic with C57 were used. Injected blastocysts were surgically transferred to uteri of pseudopregnant ICR randombred females, for development to term.

Analyses of Tissue Genotypes. All animals were screened at 7–10 days for coat mosaicism and at approximately 4 weeks for blood and liver mosaicism. Blood-cell lysates were analyzed for strain-specific electrophoretic variants of glucosephosphate isomerase (GPI; D-glucose-6-phosphate ketol-isomerase, EC 5.3.1.9), as described elsewhere (15). Liver genotypes were revealed by electrophoretic patterns (16) of proteins normally excreted in the urine (major urinary protein complex, MUP) but synthesized in the liver (17) and thus indicative of hepatic parenchyma genotypes (18). At autopsy, blood and soft tissues were frozen in dry ice, stored at −70°, and tested for GPI isozymes.

HPRT Specific Acitvity. Tissue homogenates of some experimentally derived mice were thawed, immediately sieved on columns of Sephadex G-25, and assayed as coded specimens for HPRT and also for adenine phosphoribosyltransferase (APRT, EC 2.4.2.7) activity, as previously described (19, 20), except that 4 mM thymidine triphosphate was present in all assays. The assays were linear with both time and protein concentrations; the latter were estimated (21) with bovine serum albumin as a standard.

Autoradiography. Skin and subcutaneous connective tissue from two mice with coat mosaicism were explanted. The fibroblastic cell cultures were incubated in [³H]hypoxanthine at 2.5 μCi/ml (0.35 μM) for 14 hr, followed by standard methanol/acetic acid fixation and processing. Slides were coated with Kodak NTB-2 emulsion, exposed in the dark for 7 days at 4°, developed, and stained with May–Grünwald–Giemsa.

RESULTS

Extracts prepared from thioguanine-resistant cultured NG 2 teratocarcinoma cells showed no detectable HPRT activity. Tests aimed at discriminating between presence of enzymatically inactive HPRT protein and actual loss of HPRT—hence possible deletion of the locus—are inconclusive to date.

Forty-four mice were obtained from injected blastocysts. Included were 21 females, 22 males, and 1 (living) male with possible intersexuality. Twelve animals were autopsied between the ages of 3 days and 6 weeks. Of the 32 mice still alive, the oldest are 9 months of age; all appear healthy and have no visible tumors.

Among the 12 animals autopsied, as many as nine (cases 1–9) had teratocarcinoma-derived cells in one or more of their tissues, as evidenced by 129-strain-specific markers (Table 1). Among the 32 living animals, in which only the coat, blood, and liver have been tested, one (case 10) has liver mosaicism; and the abnormal male (not listed) is likely to be a sex chromosome mosaic (22, 23). Some of the remaining 30 mice may prove to have 129-strain cells in some internal tissues.

In the most striking case (case 1, Table 1), 18 tissues (including hair follicles as well as melanocytes in the coat) were genotypically analyzed and, with the notable exception of blood, all contained 129-strain cells. Examples are seen in Fig. 2 *left*, with the almost-ubiquitous GPI isozyme marker. In another animal with extensive mosaicism (case 2, Table 1), 15 tissues were analyzed and 13 of them—again excluding blood, and also salivary glands—had some cells of the tumor strain.

Evidence for normal specialized tissue functions of 129-strain cells was found in hair follicles, with the phaeomelanin (*agouti* locus) discriminant (Table 1, cases 1 and 2); in melanocytes, with the 129 *non-beige* eumelanin marker (case 1); and in

Table 1.   Percent tissue contributions* derived from HPRT-deficient teratocarcinoma cells (129 strain) after injection into blastocysts (C57BL/6 strain)

Case, sex	Autopsy age	Coat	Blood cells	Liver	Spleen	Thymus	Heart	Lungs	Kidneys	Gut	Pancreas	Salivary glands	Muscle	Brain	Gonads	Repr. tract[†]
1 ♀[‡]	5 weeks	5	0	15	10	30	55	5	50	40	5	10	25	10	25	20
2 ♂	2 weeks	5	0	30	10	10	50	10	20	40	5	0	40	15	15	25
3 ♀	3 days		0	5	15	0	60	10	40	20	0	0	20	0	20	20
4 ♂	6 weeks	0	5	10	5	0	0	0	0	0	0	0	0	10	0	15
5 ♂	5 days			0	0	0	20	0	5	0		0	0	0	0	0
6 ♂	3 weeks	0	0	0	5	0	0	0	0	0	0	0	0	5	0	0
7 ♀	3 weeks	0	0	0	0	0	0	0	0	0	0	0	0	15	0	0
8 ♀	3 weeks	0	0	0	0	0	0	0	0	0	0	0	0	15	0	0
9 ♂	3 weeks	0	0	0	0	0	0	0	0	0	0	0	0	10	0	0
10 ♂	Alive	0	0	5[§]												

* Strain-specific markers in the coat were at the *A* (*agouti*) locus, expressed in hair-follicle dermis, in all; in addition, gene substitutions at single pigmentary loci supplied melanocyte markers in cases 3, 5, 6, 7, 9, 10 (*b/b* blastocysts) and 1, 4 (*bg^J^/bg^J^* blastocysts). Other tissue analyses were based on GPI isozymic strain differences.
[†] Accessory reproductive glands and ducts.
[‡] Additional GPI analyses revealed a 15% 129-strain tissue contribution in adrenals and 10% in skin.
[§] Liver genotypes in *living* animals were determined by the MUP marker in urine samples.

hepatocytes, with the MUP strain variant (case 10; Fig. 2 *right*). In addition, a normal developmental feature unique to myogenesis was seen in muscle: an intermediate or heteropolymeric GPI band (cases 1–3, Table 1; Fig. 2 *left*) representing heterokaryon formation by fusion of uninucleated myoblasts (24).

Progeny tests of living animals, to detect any germ-line transmission from the tumor strain of origin, are still incomplete. Karyotype analyses show that the NG 2 cell line has a modal class (72%) of 40 chromosomes; the sole apparent anomalies are trisomy of chromosome 6 and presence of only one sex chromosome, i.e., a single X (C. Cronmiller and B. Mintz, unpublished). If X/O cells from the NG 2 line are in fact able to contribute germ cells *in vivo*, these should become female rather than male gametes: The Y chromosome is needed for maleness, the X/O mouse being a fertile female (25); and functional reversal of germ cells of one sex chromosome type to the opposite sex phenotype does not occur or is exceedingly rare in allophenic mice (22).

Among the known mosaics already autopsied, three (cases 1–3, Table 1) have some tumor-strain cells in their gonads. Two are phenotypic females with some 129-strain ovarian cells (Fig. 2) and one is a male with some testicular cells of that strain. The gonadal 129-type cells may have been germinal and/or somatic and, if germinal, might have become functional eggs in the females.

In order to learn whether the 129-strain cells in differentiated tissues of mosaic animals had continued to be HPRT-deficient *in vivo*, two kinds of assays were conducted. The first measured HPRT specific activity of tissue homogenates after their 129-type tissue contribution had been ascertained from GPI isozyme tests. HPRT activity was expressed in relation to APRT activity, because the latter is unaffected in thioguanine-resistant cells of non-erythroid tissues of Lesch–Nyhan patients (10). All samples were encoded before HPRT and APRT assays. In the experimental animal designated case 1 (Table 2), there is a clear correlation between the presence of an appreciable proportion (50–55%) of putatively HPRT⁻ (129-strain GPI type) cells in heart and kidney and a marked depression of the HPRT/APRT ratios in those tissues, as compared with HPRT⁺ control tissues of the blastocyst and tumor strains. But when only minor amounts (5–10%) of 129 cells occurred, as in the brain and pancreas of case 1, no significant lowering of the ratio was seen. In case 2, results for three of the five tissues (heart, pancreas, and salivary gland) substantiated these trends.

Independent confirmation was obtained by autoradiographic

Table 2.   HPRT specific activity* in tissues of mice partially derived from HPRT-deficient teratocarcinoma cells

| Tissue | HPRT/APRT | | | % 129-type GPI[†] |
	C57 control	129 control	Exp.[‡]	
Heart[§]	3.08	1.90	0.50	55
Kidney[§]	0.79	1.01	0.48	50
Brain	1.58	1.79	1.81	10
Pancreas	0.56	0.58	0.48	5
Heart[§]	0.54	0.52	0.26	50
Liver	0.57	1.25	0.51	30
Spleen[§]	0.73	0.78	0.42	10
Pancreas	0.22	0.28	0.25	5
Salivary gland	0.36	0.19	0.29	0

* The standard deviations of the specific activities of the indicated (§) samples were less than 20% ($n = 8$ determinations). Tissue extracts from the two experimental animals, and their respective controls, were stored frozen for two different periods of time prior to assay; this accounts for the different ratios for heart and pancreas in the two experiments.
[†] The teratocarcinoma strain contribution; data from Table 1.
[‡] The experimental animal in the upper section is case 1 (Table 1); the one in the lower section is case 2.

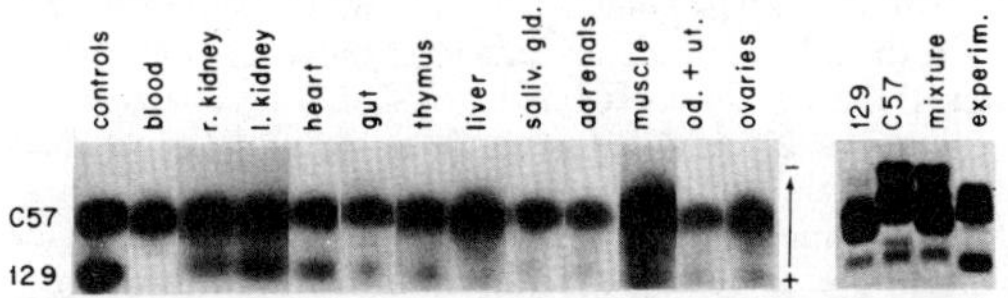

FIG. 2.   (*Left*) Glucosephosphate isomerase allelic strain variants in starch gel electrophoresis of tissue extracts from female mosaic *mouse no. 1*. The HPRT-deficient teratocarcinoma cell strain (129) is absent from its blood but present in all other tissues (including five additional tissues listed in Table 1 but not included here). Three bands in muscle denote normal myoblast fusion leading to heterokaryons. Note the 129-strain cells, which are chromosomally X/O female, in the ovaries. (od. + ut., oviduct plus uterus.) (*Right*) Acrylamide gel electrophoretic patterns of allelic strain variants of the major urinary protein complex in experimental *mouse no. 10*, and controls. A 129-strain contribution in the mosaic indicates that normally functioning parenchyma cells of that strain are present in its liver, where the protein is produced.

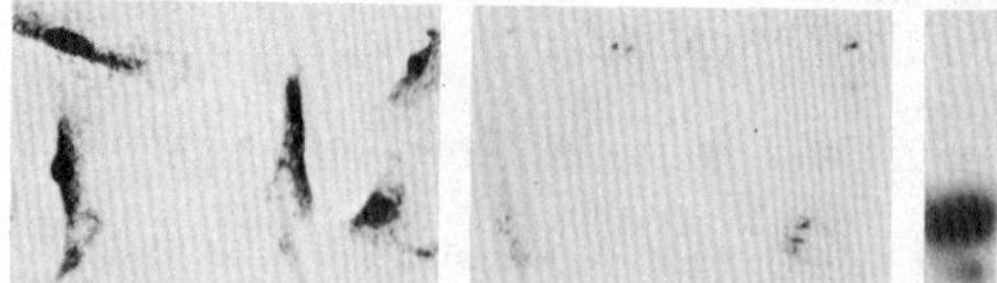

FIG. 3.   In a culture of fibroblastic cells grown from subcutaneous connective tissue of *mouse no. 2*, a part of a normal colony (*Left*) and of a markedly HPRT-deficient colony (*Center*) are seen in autoradiographs after [³H]hypoxanthine labeling. (× 126.) The GPI starch gel at the *Right* confirms the presence of a majority of the fast-moving (C57, normal) and a minority of the slow-moving (129, HPRT⁻) cell strains in a lysate of the same culture.

visualization of HPRT⁻ colonies in cultures of subcutaneous connective tissue taken from mouse no. 2 at autopsy. The cells were plated at low density, cultured for 1 week, and incubated 14 hr in [³H]hypoxanthine. The autoradiographs (Fig. 3 *left* and *center*) clearly showed two kinds of colonies: Some had many silver grains over the cells, as in wild-type (HPRT⁺) control cultures; others were virtually devoid of grains, as in the HPRT⁻ STO cultures. The low frequency of the unlabeled colonies from the mosaic animal conforms to the minor representation of the slow-migrating 129-type of GPI in a cell extract of the same culture (Fig. 3 *right*).

DISCUSSION

When mouse teratocarcinoma cells with severe HPRT deficiency—comparable to that in human Lesch–Nyhan disease (8, 9)—are introduced into wild-type blastocysts (Fig. 1), they cease to be malignant and proceed to contribute to normal embryogenesis. In the viable mice thus obtained, normally functioning differentiated cells of the tumor strain have been found, by means of strain-specific markers, in virtually all tissues of some individuals. The stem cells of this *in vitro* line, known as NG 2, are therefore developmentally totipotent.

Maturation *in vivo* of cells from any teratocarcinoma line initially characterized by a biochemical lesion does not in itself ensure that the lesion is still present many cell generations later in the differentiated population. Assays of HPRT specific activity were therefore carried out and did in fact document retention of the deficiency (Table 2): When appreciable populations of NG 2-derived cells were present, a marked depression of HPRT activity was found. In addition, autoradiographs of [³H]hypoxanthine-incubated cultures from connective tissue of mosaics showed some unlabeled (HPRT⁻) and some labeled (HPRT⁺) colonies (Fig. 3). The HPRT change in NG 2 thus appears to constitute a bona fide mutation. Germ-line transmission would afford final proof.

The present experiment initiates the experimental application of a scheme previously proposed (6, 7) for analyzing mammalian differentiation and disease. The objective is to select *in vitro* for specific mutations in developmentally totipotent cells, so that the effects of a biochemically defined change may then be followed *in vivo* at all levels of biological organization. With this system, mouse models of human genetic diseases might be produced (6, 7). Lesch–Nyhan disease was chosen as the first candidate partly because of the ease with which the relevant X-linked enzymatic lesion (9) is selectable in culture. While the actual model, an F₁ "Lesch–Nyhan" male mouse, is not yet on hand, the mosaics themselves (both females and males) have certain unique and useful features not available in "models" whose genetic lesion is in all their cells. As seen in Table 1, a given mosaic animal may have HPRT⁻ cells in only a single tissue, such as brain (cases 7–9), or in only a few tissues,

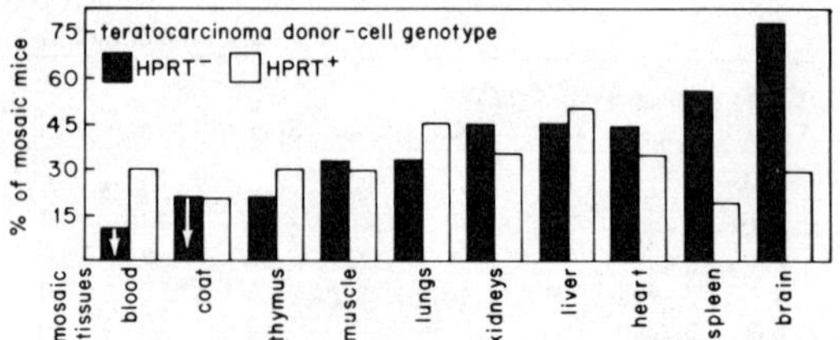

FIG. 4.   Percent *of mosaic mice* with mosaicism of each of the tissues indicated. Data are based on 9 mosaics with HPRT⁻ cells (derived from the OTT 5568 teratocarcinoma; this series) and 20 mosaics with HPRT⁺ cells (derived from the OTT 6050 teratocarcinoma; refs. 4 and 5; further unpublished data). Only mosaic animals with virtually all tissues genotypically analyzed are included. The definitive frequency of blood and of coat mosaicism in the final population of HPRT⁻ mosaics is certain to be *less* than shown, as designated by downward arrows (see *text*). Results suggest specific selection against HPRT⁻ cells in blood.

such as kidneys and heart (case 5). The metabolic and clinical attributes of such individuals could help to identify a tissue in which the complex syndrome of the disease may originate. No obvious behavioral defects have been seen in the mosaics. This might signify that the brain, despite its very high normal levels of HPRT (10), is surprisingly little dependent on the salvage pathway of purine biosynthesis; or that the brain regions where these HPRT⁻ cells (a small minority, in any case) are located are not crucial to the behavioral manifestations in affected individuals; or that systemically distributed products from other tissues are responsible for secondary effects on the nervous system. The deleterious accumulation of excessive purine degradation products is of course a major characteristic of the disease. The mosaic mice may also help to define the chief tissue sources of that problem.

Some preliminary indications of tissue-specific phenotypic effects are already detectable through cell selection in the mosaic animals. The first of these, in the blood, is of interest because it parallels selective trends in human HPRT⁺/HPRT⁻ heterozygotes. In those females, there is cellular phenotypic, but not genotypic, mosaicism owing to expression of only one X-linked allele per cell. Another difference as compared to HPRT⁺ ↔ HPRT⁻ mice with cellular genotypic mosaicism is that the two phenotypic heterozygous populations of cells are probably approximately equal early in development, whereas sporadic incorporation of teratocarcinoma-derived cells may result in their scattered representation in tissues of allophenic mice (Table 1). The human deficiency may be severe or moderate; though heterozygous carriers of the severe type have a normal and a deficient class of fibroblasts (26), they have normal erythrocytes (27). Only carriers of the moderate deficiency have individual erythrocytes of the two separate phenotypes (28). Thus, in human erythrocytes, HPRT deficiency is critical at the cellular level and causes a greater competitive disadvantage as it becomes more severe. Carriers of the gross deficiency do, however, have some HPRT⁻ lymphocytes (29).

These two trends appear to be mirrored in our experimental mice. In Fig. 4, the frequency with which each tissue has any teratocarcinoma-strain cells is compared in the mosaic mice of two groups: the present one, in which the cells of donor origin are HPRT⁻ (9 mosaics; Table 1), and an earlier group, in which they are HPRT⁺ (20 mosaics; refs. 4 and 5, and further unpublished data). The data from the HPRT⁻ experiment are arranged (left to right) from the least frequently mosaic tissue, blood (11%), to the most frequently mosaic, brain (78%). However, the *final* frequency in this experiment is certain to diminish still further in blood (and coat): When all living mice have been autopsied and analyzed, no new blood mosaics will have been identified, but the size of the mosaic population will have increased by at least one (case 10). As seen in Fig. 4, teratocarcinoma-derived cells are less often found in blood if the cells are HPRT$^-$ than if they are HPRT$^+$. On the other hand, in the spleen, which is composed largely of lymphocytes, HPRT$^-$ cells are frequently included (56% of mosaic mice). Though other factors (e.g., origin of HPRT$^-$ and HPRT$^+$ cells from separate tumors, or trisomy-6, or hidden mutations in NG 2) may play a role in cell selection, the parallelism to the selective trends in human heterozygotes is an encouraging indication that the "synthesis" of mice with HPRT$^-$ cells, whether in some or all of their tissues, will be a useful way to examine the basis for the human disease, and possibly to attempt its cure.

While the brain is most frequently mosaic, it has one of the lowest levels of HPRT$^-$ cells (Table 1, Fig. 4). Therefore, selection in the brain may involve two steps: first, relative ease of entrance of HPRT$^-$ cells into the precursor pool; second, relative discrimination against those cells as brain development progresses.

Of the two karyotype changes in NG 2 cells (C. Cronmiller and B. Mintz, unpublished), the occurrence of a single X chromosome and no Y is not necessarily disadvantageous, because X/O mice are viable fertile females (25). However, trisomy-6, like all autosomal trisomies of mice, is usually lethal prenatally. It leads (in non-mosaic embryos) to retardation and hypoplasia, and death between 12 and 14 days of gestation (30). The full functional maturation of cells of trisomic lineage in the animals described here implies that this trisomy is not a *cell* lethal.

Animals whose tissues comprise NG 2-derived cells now range up to 9 months of age and have remained free of tumors. This reinforces our earlier observations following injections of small numbers of cells from another tumor line (OTT 6050), in which very few animals developed teratomas (refs. 3–5, and further unpublished data). Those tumors apparently arose only from teratoma cells that failed to become integrated into the embryo: When single cells were injected into blastocysts, mutually exclusive results were obtained, leading either to participation in normal differentiation or to retention of teratomatous growth. In the former cases, mosaic normal tissues remained normal, even if transplanted to new hosts. Thus, induction of normal differentiation of teratocarcinoma stem cells in blastocysts stably terminated their malignancy.

The teratocarcinoma cell line used in the present experiment was *clonally* derived; it exhibited typical malignant behavior in subcutaneous grafts. The stem cells introduced into blastocysts therefore *all* presumably possessed malignant potential yet were successfully normalized.

We thank Mrs. Claire Cronmiller and Mrs. Mariann Durante for technical assistance. This work was supported by the following: to B.M., U.S. Public Health Service Grants HD-01646, CA-06927, and RR-05539, and an appropriation from the Commonwealth of Pennsylvania; to D.W.M., Jr., U.S. Public Health Service Grant GM-19527; to G.R.M., American Cancer Society (California Division) Grants VC-246 and 854. D.W.M., Jr. is an Investigator, Howard Hughes Medical Institute.

1. Stevens, L. C. (1967) *Adv. Morphog,* **6,** 1–31.
2. Pierce, G. B. (1967) in *Current Topics in Developmental Biology,* eds. Moscona A. A. & Monroy, A. (Academic Press, New York), Vol. 2, pp. 223–246.
3. Mintz, B., Illmensee, K. & Gearhart, J. D. (1975) in *Symposium on Teratomas and Differentiation,* eds. Sherman, M. I. & Solter, D. (Academic Press, New York), pp. 59–82.
4. Mintz, B. & Illmensee, K. (1975) *Proc. Natl. Acad. Sci. USA* **72,** 3585–3589.
5. Illmensee, K. & Mintz, B. (1976) *Proc. Natl. Acad. Sci. USA* **73,** 549–553.
6. Mintz, B. (1976) *Harvey Lect.* **71,** in press.
7. Mintz, B. (1977) in *Cell Differentiation and Neoplasia, 30th Symposium, M. D. Anderson Hospital and Tumor Institute* (Williams and Wilkins, Baltimore, MD), in press.
8. Lesch, M. & Nyhan, W. L. (1964) *Am. J. Med.* **36,** 561–570.
9. Seegmiller, J. E., Rosenbloom, F. M. & Kelly, W. N. (1967) *Science* **155,** 1682–1684.
10. Seegmiller, J. E. (1976) *Adv. Hum. Genet.* **6,** 75–163.
11. Chapman, V. M. & Shows, T. B. (1976) *Nature* **259,** 665–667.
12. Martin, G. & Evans, M. (1976) *Proc. Natl. Acad. Sci. USA* **72,** 1441–1445.
13. Martin, G. & Evans, M. (1975) in *Symposium on Teratomas and Differentiation,* eds. Sherman, M. I. & Solter, D. (Academic Press, New York), pp. 59–82.
14. Stevens, L. C. (1970) *Dev. Biol.* **21,** 364–382.
15. Gearhart, J. D. & Mintz, B. (1972) *Dev. Biol.* **29,** 27–37.
16. Wilcox, F. H. (1975) *Biochem. Genet.* **13,** 243–245.
17. Finlayson, J. S., Asofsky, R., Potter, M. & Runner, C. C. (1965) *Science* **149,** 981–982.
18. Condamine, H., Custer, R. P. & Mintz, B. (1971) *Proc. Natl. Acad. Sci. USA* **68,** 2032–2036.
19. Graf, L. H., McRoberts, J. A., Harrison, T. M. & Martin, D. W., Jr. (1976) *J. Cell. Physiol.* **88,** 331–342.
20. Green, C. D. & Martin, D. W., Jr. (1973) *Proc. Natl. Acad. Sci. USA* **70,** 3698–3702.
21. Lowry, O. H., Rosebrough, N. J., Farr, A. L. & Randall, R. J. (1951) *J. Biol. Chem.* **193,** 265–275.
22. Mintz, B. (1969) in *Birth Defects: Original Article Series 5* (National Foundation, New York), pp. 11–22.
23. Mintz, B. (1974) *Annu. Rev. Genet.* **8,** 411–470.
24. Mintz, B. & Baker, W. W. (1967) *Proc. Natl. Acad. Sci. USA* **58,** 592–598.
25. Welshons, W. J. & Russell, L. B. (1959) *Proc. Natl. Acad. Sci. USA* **45,** 560–566.
26. Migeon, B. R., Der Kaloustian, V. M., Nyhan, W. L., Young, W. J. & Childs, B. (1968) *Science* **160,** 425–427.
27. Nyhan, W. L., Bakay, B., Connor, J. D., Marks, J. F. & Keele, D. K. (1970) *Proc. Natl. Acad. Sci. USA* **65,** 214–218.
28. Johnson, L. A., Gordon, R. B. & Emmerson, B. T. (1976) *Nature* **264,** 172–174.
29. Albertini, R. J. & DeMars, R. (1974) *Biochem. Genet.* **11,** 397–411.
30. Gropp, A. (1975) *Clin. Genet.* **8,** 389.

Part III

MUTANT MAMMALIAN CELLS

Editor's Comments
on Papers 32, 33, and 34

32   FISCHER
Increased Levels of Folic Acid Reductase as a Mechanism of Resistance to Amethopterin in Leukemic Cells

33   SCHIMKE et al.
Gene Amplification and Drug Resistance in Cultured Murine Cells

34   HAKALA
Prevention of Toxicity of Amethopterin for Sarcoma-180 Cells in Tissue Culture

RESISTANCE TO FOLIC ACID ANTAGONISTS

Studies on mutant mammalian cells had their origins over three decades ago in attempts to understand the mechanisms by which tumor cells become resistant to anti-tumor drugs. It was not until several years later that there developed a specific interest in using mammalian cell mutants for genetic analysis. Nevertheless, much of our basic knowledge of mammalian cell mutants involves drugs that were investigated initially because of their anti-tumor activity. Over the years, the study of mutant cells expanded into increasingly diverse areas. Because of this diversity, a focus was needed for the papers on mammalian cell mutants to be presented in this volume. Therefore, in choosing the papers, emphasis was placed on mutant cell studies relevant to the other areas covered in this volume—cell hybridization and gene transfer. For a detailed discussion of early studies on mutant mammalian cells, the reader may refer to the book by Harris (1964).

Some of the earliest studies on drug resistant cells involved the folic acid antagonist, amethopterin (methotrexate), which inhibits dihydrofolate reductase (DHFR) activity. Mammalian cells resistant to amethopterin were observed over thirty years ago in leukemic tumors treated with the drug, and cultured cell lines resistant to amethopterin were selected several years later (Fischer, 1959; Aronow, 1959). Subsequent studies showed that there are at least three different mechanisms by which cells can develop resistance to amethopterin. The development of amethopterin resistance as the result of increased DHFR activity was demonstrated by Fischer (Paper 32) and Hakala et al. (1961). In the studies by Fischer, cultured mouse leukemic cells were selected stepwise for increasing resistance to amethopterin. Enzyme assays revealed that the drug resistant cells selected in the first and second steps had increased levels of DHFR activity and that the increase in DHFR activity correlated well with the increased levels of amethopterin resistance.

Almost two decades after these results were obtained, a genetic mechanism for increased DHFR activity in drug resistant cells was provided by Schimke and co-workers (Paper 33). These studies involved a mouse sarcoma line highly resistant to amethopterin. The resistant cells showed more than a 200-fold increase in DHFR activity, and this increased activity was due to an increased rate of synthesis of DHFR. Through the use of a DHFR-specific molecular probe and nucleic acid hybridization techniques, it was shown that there was more than a 200-fold increase in the level of DHFR-specific messenger RNA in the drug resistant cells and, most significantly, a corresponding increase in the number of copies of the DHFR gene. These results thus demonstrated that gene amplification is one mechanism by which mammalian cells can develop resistance to drugs.

Early studies of amethopterin resistant leukemic cells provided evidence for a second mechanism of drug resistance, involving decreased uptake of the drug by resistant cells (Fischer, 1962). A third mechanism of resistance involved an alteration in the structure of the DHFR molecule, so that the binding of amethopterin to the enzyme was decreased in the resistant cells (Albrecht et al., 1972).

For the development of somatic cell genetics, one of the most important results from the early work on folic acid antagonists came not from the study of amethopterin resistant cells but from the analysis by Hakala (Paper 34) of compounds that prevented amethopterin toxicity. Mouse sarcoma cells were cultured in the presence of amethopterin plus other compounds, and the ability of these compounds to reverse the growth-inhibitory effects of amethopterin was determined. It was found that the cells were able to grow in the presence of toxic concentrations of amethopterin only if both hypoxanthine and thymidine were added to the medium. These three compounds (the inhibitor plus the purine and pyrimidine sources) represent precisely the components of the medium now referred to as HAT medium (see Paper 25). As described in other parts of this book, selection with HAT medium has played a major role in studies on all aspects of somatic cell genetics. In addition to defining the components of HAT medium, the studies by Hakala demonstrated how the addition of a metabolic inhibitor could force mammalian cells to depend on exogenous nutrients that they normally did not require. This is the basis of selection not only in the HAT system but also in other selective systems subsequently developed for use with cultured cells.

REFERENCES

Albrecht, A. M., J. L. Biedler, and D. J. Hutchinson, 1972, Two Different Species of Dihydrofolate Reductase in Mammalian Cells Differentially Resistant to Amethopterin and Methasquin, *Cancer Res.* **32:**1539–1546.

Aronow, L., 1959, Studies on Drug Resistance in Mammalian Cells I. Amethopterin Resistance in Mouse Fibroblasts, *J. Pharm. Exp. Ther.* **127:**116–121.

Fischer, G. A., 1959, Nutritional and Amethopterin-Resistant Characteristics of Leukemic Clones, *Cancer Res.* **19:**372–376.

Fischer, G. A., 1962, Defective Transport of Amethopterin (Methotrexate) as a Mechanism of Resistance to the Antimetabolite in L5178Y Leukemic Cells, *Biochem. Pharm.* **11:**1233–1237.

Hakala, M. T., S. F. Zakrzewski, and C. A. Nichol, 1961, Relation of Folic Acid Reductase to Amethopterin Resistance in Cultured Mammalian Cells, *J. Biol. Chem.* **236:**952–958.

Harris, M. 1964, *Cell Culture and Somatic Variation,* Holt, Rinehart, and Winston, New York, 547p.

INCREASED LEVELS OF FOLIC ACID REDUCTASE AS A MECHANISM OF RESISTANCE TO AMETHOPTERIN IN LEUKEMIC CELLS

G. A. Fischer

STUDIES concerning amethopterin-resistance in murine leukemic cells (L5178Y) have been described.[1] Using a medium developed to permit the clonal reproduction of single isolated leukemic cells,[2] it has been established that independently derived "first-step" mutants (presumed to be the result of single independent mutational events), had separate "low" levels of resistance. The concentration of amethopterin required for 50% inhibition of growth of 20 first-step mutant clones ranged from 1·4 to 4·3 × 10^{-8} M, or 1·5 to 4·8 times that concentration (0·9 × 10^{-8} M) required for comparable inhibition of the parent strain. This finding suggests that, with respect to this enzymic activity, a multiplicity of mutable sites or alternate stable forms of a single mutable site may be present in these cells. With one such first-step mutant clone, selected in an amethopterin-containing medium which completely prevented the reproduction of sensitive cells, approximately twice the concentration of amethopterin ("2-fold mutant") was needed for 50% inhibition of growth, as was the case with the parent strain. From this first-step clone, a "second-step" clone ("16-fold mutant") was isolated. These first and second-step clones had a 2- and 17-fold increase, respectively, in their levels of folic acid reductase activity (Table 1).[6] Increased folic acid reductase activity has been reported in amethopterin-resistant cells from (a) sublines of mouse leukemia L5178 selected in culture[1, 6] (b) various leukemic lines selected in mice,[7] (c) cultured sarcoma-180 cells,[8] and (d) human leukemic individuals treated with the drug.[9]

TABLE 1. FOLIC ACID REDUCTASE LEVELS OF AMETHOPTERIN-SENSITIVE AND AMETHOPTERIN-RESISTANT CELLS

Line tested	Relative level of resistance*	Relative activity of folic acid reductase†
Sensitive	1·0	1·0
"First-step" clone	2·3	2·0
"Second-step" clone	16·0	17·0

A mixture containing 0·02 M phosphate buffer (pH 6·0), 1·8 × 10^{-2} M isocitric acid, 1·8 × 10^{-2} M magnesium chloride, 2 × 10^{-4} M triphosphopyridine nucleotide and a preparation of isocitric dehydrogenase[3] was incubated for 5 min at 37 °C. Of this mixture, 0·15 ml was added to 0·30 ml of a preparation of folic acid reductase (this enzyme was prepared by the addition of 4 volumes of water to packed cells from culture, dialysis for 18 hours against 1,000 volumes of 0·002 M phosphate buffer (pH 6·0), centrifugation at 100,000 g for 30 min, recovery of the supernatant fraction, and adjustment of the phosphate concentration to 0·02 M (pH 6·0)); the, 0·05 ml of 0·0012 M folic acid was added. After 20 min the reaction was terminated by the addition of 0·2 ml of 4 N hydrochloric acid, 0·8 ml of acetone and 0·5 ml of water. The diazotizable product was determined by the method of Bratton and Marshall.[4]

* Concentration required for 50% inhibition of sensitive line = 0·9 × 10^{-8} M.

† Activity of sensitive line = 2·6 units of folic acid reductase activity per mg of protein.[5] One unit of folic acid reductase activity = 1 × 10^{-3} μmoles of product (measured as diazotizable amine) per 30 min.

"Pseudo-irreversible"[10] complete inhibition of one unit of folic reductase activity (see Table 1) required the same amount of amethopterin (2·9 $\times$ 10^{-12} moles) whether the enzyme was prepared from drug-sensitive or drug-resistant cells. Since amethopterin is bound to the catalytic site of folic acid reductase,[11, 12, 13] and apparently to no other cell component, it is suggested that an increased number of such catalytic sites, with identical turnover number and "affinity" for amethopterin, are present in the resistant cells.

Resistant cells, grown in the presence of tritium-labeled amethopterin (103 μc/mole) bound the antagonist in a manner which suggested that the entry of amethopterin into the cell was less rapid than that of the synthesis of folic acid reductase (Table 2). From such "bound" enzyme, a minimum of 80% of the ^{3}H-amethopterin was liberated by dialysis against 100 volumes of 0·2 M phosphate buffer (pH 6·0) and enzyme activity was regenerated. Although the amount of bound radioactivity (sufficient for the binding of 8 units of enzyme activity per 10^7 cells) which was release by dialysis was not in close agreement with the units of enzyme recovered after dialysis (15 units per 10^7 cells), it is conceivable that, under the conditions of the experiment, some tritium may have been lost from the amethopterin. In any case, the increased levels of enzyme activity found in the resistant lines correlated directly with the degree of resistance, a circumstance which appears to reflect an increased amount of enzyme synthesis in the resistant cells. It is suggested that such synthesis provides excess enzyme which "binds" nearly all of the amethopterin which is capable of entering the leukemic cells under the conditions described. The enzyme which remains "unbound" under these circumstances is sufficient to reduce folic acid to coenzyme forms essential for cell reproduction.

TABLE 2. BINDING OF FOLIC ACID REDUCTASE DURING THE REPRODUCTION IN CULTURE OF A "16-FOLD" AMETHOPTERIN-RESISTANT MUTANT

Molar concentration ($\times$ 10^{-8}) of ^{3}H-amethopterin in growth medium	cpm/10^7 cells		Units of enzyme/10^7 cells	
	cpm	Indicated units bound	Before dialysis	After dialysis
14	290	11		
7·0	190	8	1·6	17
3·5	160	6		
1·8	80	3		
0·9	20	0·8		
0·0	0	0	21	20

For each level of ^{3}H-amethopterin (103 μc/μmole), 2·5 $\times$ 10^7 cells were collected after seven generations in culture. The enzyme was prepared from washed cells and the radioactivity was measured in a liquid scintillation counter. The enzyme activity, expressed in units (see Table 1), was measured by determination of the product formed after incubation for 10 min at varying concentrations of the enzyme. The inhibited enzyme was regenerated by equilibrium-dialysis against 100 volumes of 0·2 M phosphate buffer (pH 6·0) for 18 hr.

In the absence of amethopterin the resistant lines have retained their levels of resistance quantitatively for a period of 6 months. Furthermore, no gross chromosomal abnormalities have been detected in the resistant lines, when these were compared with the near-diploid clone of origin. Comparative nutritional studies of various aspects of the metabolism of folic acid have disclosed no other differences between amethopterin-sensitive and -resistant cells. Thus, the requirement for very high levels of folic acid, which characterizes the drug-sensitive cells,[1, 14] remains elevated in the amethopterin-resistant cells. Also, in the presence of very high levels of amethopterin (10^{-6} M), the same levels of thymidine, hypoxanthine and serine are required to support the growth of either the sensitive or the resistant cells. These findings suggest that the increase in folic acid reductase activity may reflect a single mutational event, associated with a single biochemical alteration in the resistant cell (possibly affecting the extent to which a single enzyme is formed).

REFERENCES

1. G. A. Fischer, *Cancer Res.* **19**, 372 (1959).
2. E. E. Haley, G. A. Fischer and A. D. Welch, *Cancer Res.* In press.
3. A. L. Grafflin and S. Ochoa, *Biochim. Biophys. Acta* **4**, 205 (1950).
4. A. C. Bratton and E. K. Marshall, *J. Biol. Chem.* **128**, 537 (1939).
5. L. C. Mokrasch and R. W. McGilvery, *J. Biol. Chem.* **221**, 909 (1956).
6. G. A. Fischer, *Proc. Am. Assoc. Cancer Res.* **3**, 111 (1960).
7. D. K. Misra, S. R. Humphreys, M. Freidkin, A. Goldin and E. J. Crawford, *Nature* **189**, 39 (1961).
8. M. T. Hakala, S. F. Zakrzewski and C. A. Nichol, *Proc. Am. Assoc. Cancer Res.* **3**, 115 (1960).
9. J. R. Bertino, F. M. Huennekens and B. W. Gabrio, *Clin. Res.* **9**, 103 (1961).
10. W. W. Ackerman and V. R. Potter, *Proc. Soc. Exp. Biol. Med.* **72**, 1 (1949).
11. S. F. Zakrzewski and C. A. Nichol, *J. Biol. Chem.* **235**, 2984 (1960).
12. J. Peters and D. M. Greenberg, *Biochim. Biophys. Acta* **32**, 273 (1959).
13. W. C. Werkheiser, *Proc. Am. Assoc. Cancer Res.* **3**, 72 (1959).
14. G. A. Fischer and A. D. Welch, *Science* **126**, 1018 (1957).

Reprinted from *Science* **202**:1051–1055 (1978)

GENE AMPLIFICATION AND DRUG RESISTANCE IN CULTURED MURINE CELLS

R. T. Schimke, R. J. Kaufman,
F. W. Alt, and R. F. Kellems

Summary. Resistance of mouse cells to the folate analog, methotrexate, results from selection of increasingly resistant cells on progressive increases of methotrexate in the culture medium. High-level resistance is associated with high rates of synthesis of dihydrofolate reductase and correspondingly high numbers of reductase genes. In some variants high resistance and gene copy number are stable in the absence of selection pressure, whereas in others they are unstable. Analogies are made to antibiotic and insecticide resistance wherein selection of organisms with increased capacity to counteract the drug effect results in emergence of resistance. Gene amplification may underlie many such resistance phenomena.

In this article we review studies in our laboratory concerning the molecular mechanism for the development of resistance of cultured mouse cells to the 4-amino analog of folic acid methotrexate (MTX). We show that this resistance results from a selection of cells with higher contents of a specific enzyme, dihydrofolate reductase (DHFR) and corresponding increases in the number of copies of the gene coding for this enzyme, that is, gene amplification. In some cells the amplified genes are stable in the absence of selection pressure (MTX), whereas in others the genes are unstable. The properties of this resistance are analogous to many cases of the emergence of drug resistance in bacteria and insects (*1*). Thus, gene amplification may be also involved in these instances.

Methotrexate Resistance in Cultured Cells

Methotrexate, a 4-amino analog of folic acid, is commonly used in the treatment of malignancy (*2*). It kills cells by specifically inhibiting dihydrofolate reductase (DHFR), the enzyme that catalyzes the reduction of dihydrofolate to tetrahydrofolate (*3*). Tetrahydrofolate is required for single carbon transfer (such as $-CH_3$ or $-CH_2^-$) reactions for glycine, purine, and thymidylate synthesis. Thus inhibition of DHFR prevents de novo synthesis of key precursors of proteins and nucleic acids.

Continued administration of MTX to patients often results in the emergence of drug-resistant tumors; these tumors are generally associated with an increased content of DHFR (*2*). Methotrexate resistance can also be obtained in cultured murine and hamster cells. Three mechanisms for this resistance have been described: (i) An increase in DHFR occurs (*4–7*); highly resistant variants are obtained only by a stepwise selection with progressive increments in MTX in the medium. Treatment of cells with common mutagens does not increase the frequency of emergency of resistance as a result of elevation in DHFR. Resistance accompanying high DHFR levels results from the fact that at any concentration of MTX in the medium, if there is more enzyme, some enzyme molecules will be in an uninhibited state and hence will permit cell growth. (ii) Alterations in the structure of DHFR such that the high affinity for MTX is lost; hence enzyme is no longer effectively inhibited (*8, 9*). (iii) Alterations in transport of MTX into cells such that the intracellular concentration of MTX is minimal; hence DHFR is not inhibited (*10*). Resistance by mechanisms (ii) and (iii) rarely occurs spontaneously but can be enhanced by mutagenesis (*11*). Cultured cells can also become resistant to MTX by combinations of the above mechanisms.

Molecular Mechanism for Increased Dihydrofolate Reductase

We have been studying the molecular mechanism responsible for the increased content of DHFR in resistant cells derived from the murine sarcoma cell line S-180 (*6*). Resistant cells (AT-3000 and clones derived from that cell line) are resistant to 3000 times the MTX concentration that kills sensitive cells (S3) from which the resistant line was derived. The AT-3000 cells contain approximately 200 times as much DHFR as the S3 cells, as judged by enzyme assays and stoichiometric binding of MTX to DHFR (*6*) as well as by immunologic criteria in which a highly specific antibody is used (*12*).

Experiments on the labeling of cellular protein with radioactive amino acids and isolating DHFR by specific antibody precipitation indicate that the 200-fold increase in enzyme can be accounted for entirely by an increased rate of its synthesis (*12*). A similar conclusion has been made by Hanggi and Littlefield for a MTX-resistant cell line derived from baby hamster kidney (BHK) cells (*13*). The only protein that is detectably different in extracts of sensitive and of resistant cells is a peak corresponding to a molecular weight of 21,000, the size of the DHFR molecule (Fig. 1). In the resistant cells, DHFR constitutes approximately 3 to 4 percent of the total cell protein. Thus, by this admittedly imprecise criterion, the alteration in resistant cells appears to be limited to the synthesis of a single protein.

In order to quantify the cell content of DHFR specific messenger RNA (mRNA) and the number of DHFR genes, we have used nucleic acid hybridization techniques. The molecular probe used is DNA complementary (cDNA) to dihydrofolate reductase mRNA (*14*). Figure 2 shows the results of kinetic hybridizations of the cDNA with (i) an excess of total mRNA to quantitate the relative content of DHFR mRNA and (ii) an excess of cellular DNA to determine the relative number of DHFR genes in sensitive and resistant cells. The degree of acceleration of hybridization of cDNA measures the relative differences in the mRNA contents and the number of gene copies. There is an approximately 200-fold acceleration in the rate of hybridiza-

Dr. Schimke is a professor of biology, Department of Biological Sciences, Stanford University, Stanford, California 94305. Mr. Kaufman is a graduate student in the Department of Pharmacology, Stanford University. Dr. Kellems is an assistant professor of biochemistry, Baylor Medical School, Houston, Texas. Dr. Alt is a postdoctoral fellow in the Department of Biology, Massachusetts Institute of Technology, Cambridge, Massachusetts.

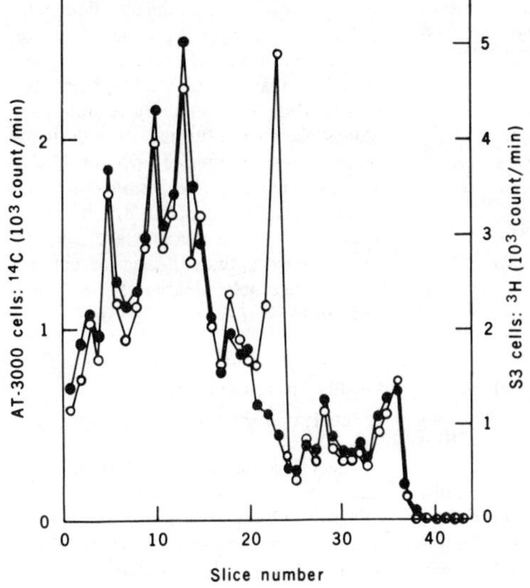

Fig. 1. Comparison of pulse-labeled soluble proteins from sensitive and resistant S-180 cells. Growing cultures of S3 and AT-3000 cells were labeled with [³H]leucine (●---●) or [¹⁴C]leucine (○---○), respectively. Soluble protein extracts were subsequently mixed and subjected to electrophoresis on sodium docecyl sulfate polyacrylamide gels (*12*). [Courtesy of the *Journal of Biological Chemistry*]

tion of cDNA with RNA from resistant cells as compared to the rate for sensitive cells. Thus, the difference in enzyme content and rate of its synthesis results from a comparable increase in the amount of DHFR mRNA. There is a comparable acceleration in the rate of hybridization of cDNA with DNA, indicating that there is a comparable amplification of the number of DHFR genes.

Table 1 summarizes data for two murine cell lines, indicating the correspondence between enzyme levels, mRNA content, and gene copy number. More recently we have found in other MTX-resistant cell lines with increased DHFR levels, including murine L 5178Y, hamster BHK, and hamster CHO, an increased gene copy number commensurate with the increased enzyme levels (*15*).

Stability and Instability of Cell

Resistance to Methotrexate

The AT-3000 line, as well as a number of cloned sublines, were found originally to retain high enzyme levels only when maintained continually in the presence of MTX (*12*). Instability in the absence of selection pressure has been reported with a number (*4, 6, 7*) but not all (*7, 11*) resistant cell lines. High enzyme levels and rates of enzyme synthesis decline rapidly when a clone (R1) of the AT-3000 cells that contain 100 times as much DHFR as S3 cells is grown in the absence of MTX. After approximately 20 cell doublings, the resulting population

of cells contains only 50 percent of the initial enzyme activity (Fig. 3). The same R1 cells that were cultured in 50 μM MTX continuously for 2 years were ex-

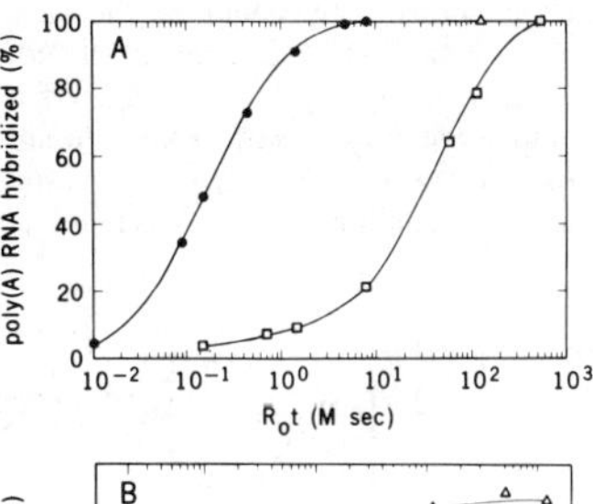

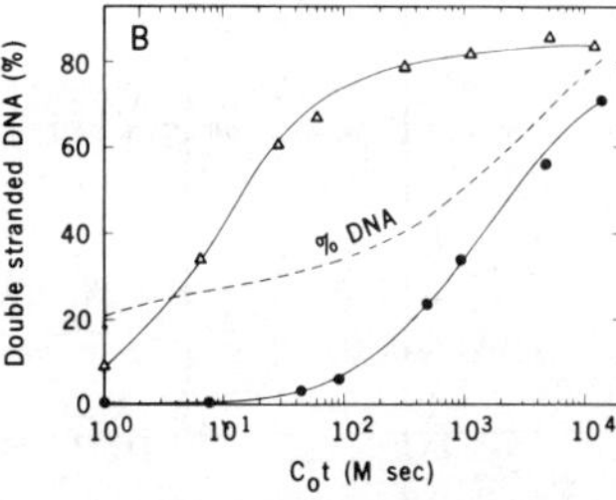

Fig. 2. Comparisons of the relative amount of dihydrofolate reductase RNA and gene number. (A) Hybridizations of tritiated DNA complementary to DHFR mRNA were undertaken with poly(A) RNA from MTX-sensitive (□) cells and MTX-resistant (●) cells. (B) Same as (A) except that DNA was used instead of poly(A) RNA. The dotted line indicates the percent of double-stranded DNA determined from absorbancy; ○, sensitive cells; △, resistant cells. The difference in the rate of hybridization is an indication of the relative differences between the two cell populations (12). [Courtesy of the *Journal of Biological Chemistry*]

amined for the rate of reversion of resistance in the absence of MTX (Fig. 3). After 2 years of continuous culture in MTX, the cells reverted to lower-level resistance more slowly and became stably resistant, with enzyme levels approximately 50 times those of sensitive cells. This difference appears to result from changes associated with growth of the cells continually in MTX over a 2-year period, since R1 cells that had been frozen away at the time of the original experiment showed the same rapid reversion to low-level resistance observed originally. Thus, a cell population that was originally unstable with respect to MTX resistance has subsequently become stabilized.

The question arises as to whether the revertant cell population at the point where enzyme levels are 50 percent of the initial levels consists of 50 percent of cells synthesizing DHFR at the high rate and 50 percent at the rate of sensitive cells, or whether it consists of a population that synthesizes DHFR at an intermediate rate. We have studied this question in two ways. We have determined (Fig. 4) the resistance of the cells to MTX (*7*). When R1 cells were grown in the absence of MTX for more than 400 cell doublings (designated R1A-400), they contained ten times as much enzyme as S3 cells, and they were correspondingly more resistant than the S3 cells to MTX. The R1 clone of AT-3000 was highly resistant, and a 50:50 proportion of R1 and R1A-400 cells showed the expected mixed sensitivity. The R1 cell population (circles of Fig. 3) that has reverted to a point where the synthesis of DHFR was only at 50 percent that of the R1 cells grown in MTX displays a killing curve of cells with approximately 50 percent of the DHFR content of the R1 cells. However, the MTX killing curve of the partially reverted R1 cells is not that of a unique cell population in which each cell contains a uniform content of DHFR. This is in keeping with the heterogeneity in enzyme content and cell (see below). We have also used a method that allows for quantifying DHFR in individual cells (see below) and have come to the same conclusion—namely, that, in the partially reverted population, cells contain approximately 50 percent of the enzyme level of the R1 cells.

Quantifying Dihydrofolate Reductase

in Individual Cells

Since the cell populations that we study are not likely to be uniform with respect to enzyme content (and gene copy number), we have devised a method for ascertaining the amount of DHFR in individual cells (*16*). This method is based on the ability of a fluorescein derivative of MTX to bind specifically, and with high affinity, to DHFR. The quantitative binding allows for the use of the Fluorescent-Activated Cell Sorter both to quantify the enzyme per cell and to separate cell populations on the basis of differing enzyme content. The distribution of cells is plotted as a function of fluorescence per cell (Fig. 5). We have found that fluorescence is a linear function of the specific enzyme activity and the rate of enzyme synthesis (*16*). In addition we have found that fluorescence is linearly related to gene copy number (*15*). The S3 cells (Fig. 5C) constitute a uniform population with virtually no fluorescence, whereas the R1A-400 cells constitute a separate and nonoverlapping population with higher enzyme content. Figure 5A shows the fluorescence distribution of the R1 cells grown for 2 years continuously in MTX (see Fig. 4) and shows a skewed distribution at the high fluorescence end (note change in scale). The highly fluorescent cells have been sorted under sterile conditions to obtain cells with high fluorescence (see arrow) and analyzed immediately for their fluorescence distribution (dotted lines). When these cells are grown for ten cell doublings in the absence of MTX, the enzyme levels are not noticeably changed. However, after 7 weeks of growth in the

Table 1. Dihydrofolate reductase, mRNA, and gene copy number in various murine cell lines. The cell lines are described in the text (*13*). The values are relative.

Cell line	Specific enzyme activity	mRNA sequences	Gene copies
S-180			
S3	1	1	1
AT-3000	250	250	250
L1210			
S	1		1
R	35		35

absence of MTX, enzyme levels are reduced by approximately 40 percent and the fluorescence distribution (Fig. 5B) is correspondingly altered as two discrete cell populations begin to appear; these populations are stable with respect to enzyme levels and MTX resistance in that they show approximately 20 and 50 times as much enzyme as the S3 cells.

By means of these techniques, we have been able to demonstrate a heterogeneity in DHFR levels in cells in which the amplified genes were unstable. When these cells were grown without selection pressure, several different cell populations emerged with differing enzyme content per cell. Inasmuch as the cells sorted for high enzyme content did not overlap with the distribution of cells that eventually stabilized with respect to enzyme content, it is unlikely that our original sorting contained a few cells that had

less enzyme and became the dominant cells upon subsequent growth. We have also cloned a number of cells from the sorted population with very high enzyme content (Fig. 5). Each of these individual clones likewise reverted progressively when grown in the absence of MTX, strengthening our conclusion that individual cells lose capacity for high enzyme synthesis. Reversion with respect to enzyme content is also associated with comparable decreases in resistance to MTX (*15*).

Possible Mechanisms of Gene Amplification and Loss

The following is an interpretation of our results in the context of current thoughts on how gene duplication-amplification events normally occur in organisms. The generation of increasingly resistant cells by stepwise selection with increasing concentrations of MTX suggests that the process resulting in a high degree of gene amplification occurs by steps, beginning with an initial duplication, then selection of those cells with the duplicated DNA sequence, and finally further amplification and further selection.

Gene duplications have been demonstrated in various organisms, including *Escherichia coli*, bacteriophage lambda, *Salmonella typhimurium*, and *Drosophila melangaster* (*17, 18*). A high selection

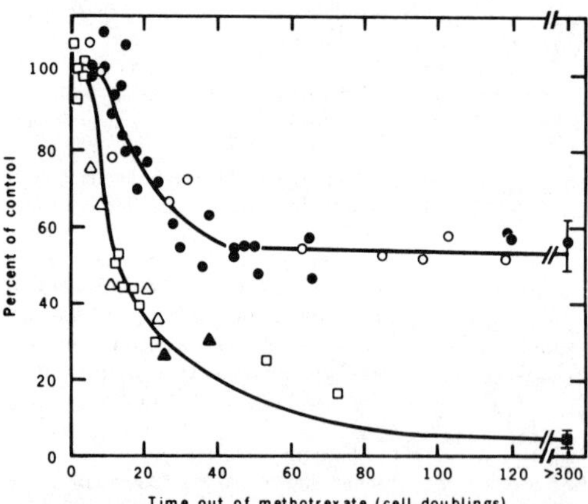

Fig. 3. Loss of dihydrofolate reductase activities in S-180 cells grown in the absence of methotrexate. R1 cells (a clone of AT-3000) were grown in the absence of MTX for various cell doublings. Enzyme activities or rates of DHFR synthesis were determined (*15*). Squares are cells studied 2 years ago; ■, enzyme levels; □, rates of synthesis. Triangles are those same cells studied in the last 6 months but stored frozen in the interim; ▲, enzyme levels; △, rates of synthesis. Circles are R1 cells grown continuously in MTX for more than 2 years prior to removal from MTX; ●, enzyme levels; ○, rates of synthesis. [Courtesy of the *Cold Spring Harbor Symposium on Quantitative Biology*]

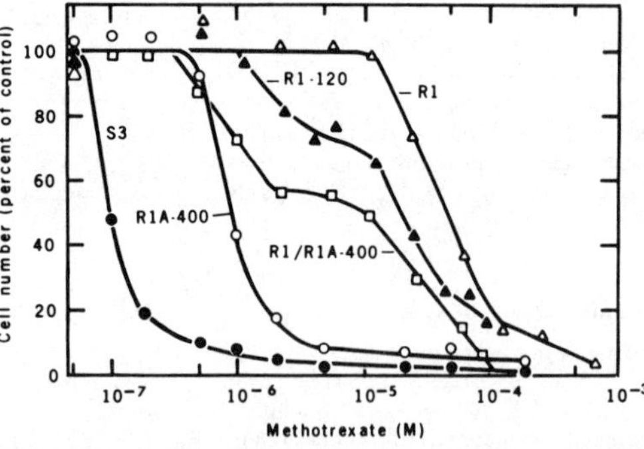

Fig. 4. Methotrexate sensitivity of sensitive, resistant, and revertant cell lines. Methotrexate sensitivity was determined by a growth assay in which 10⁵ cells were inoculated into media containing different concentrations of methotrexate. When the exponentially growing cultures reached late log phase, the cells were washed, scraped, and counted (Coulter counter). The number of each concentration of methotrexate was calculated from duplicate flasks all of which varied by less than 10 percent. The results were then expressed as a percent of control, which was determined from the growth rate of that line in the absence of methotrexate. The cloned sensitive cell line S3 had a doubling time of 16 hours, whereas all other lines had doubling times of 23 hours. △—△, the resistant R1 line; ●—●, the sensitive line; ○—○ and ▲—▲, low-level revertants and high-level revertants; □—□, a 50:50 mixture of low-level revertants and R1 resistant cells. [Courtesy of Cold Spring Harbor Press]

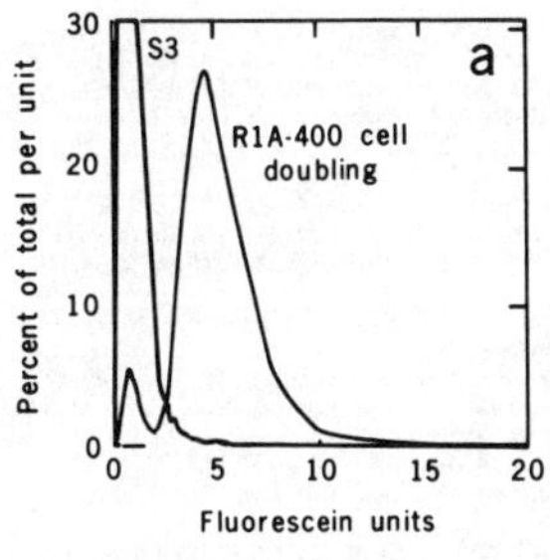

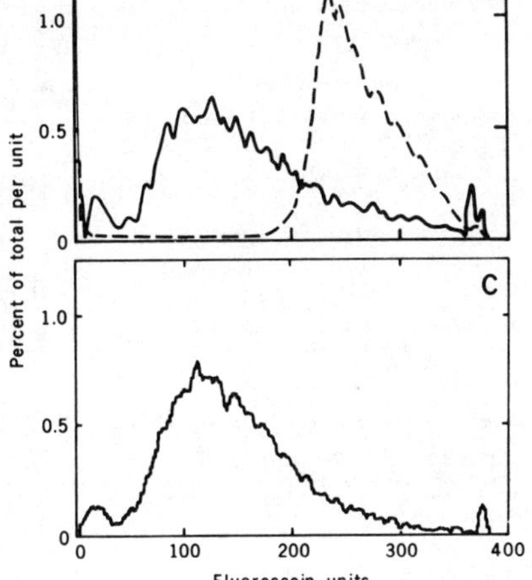

Fig. 5. Dihydrofolate reductase in individual methotrexate sensitive and resistant cell populations. (A) Distribution of sensitive (S3) and cells derived from AT-3000 (see Fig. 3) cells grown over 400 cell doublings in the absence of MTX. (B) R1 cells similar to those which revert slowly as indicated in Fig. 3 (circles). The arrow indicates the cutoff point at which cells with high fluorescence were sorted. The dotted line indicates the fluorescence distribution as analyzed immediately after the sort. (C) The sorted cells when grown 120 cell doublings in the absence of MTX.

pressure was exerted to demonstrate their existence. Anderson *et al.* (*18*) have estimated that in *S. typhimurium* duplications of genes in the histidine operon occur with a frequency as high as 0.1 percent of the cell population. Interestingly, such duplicated genes can be highly unstable (*18*) and therefore would not be detected by standard genetic analyses.

We propose that, in the cultured cells that we have been studying, a random duplication of genes occurs as an infrequent event. When grown in low concentrations of MTX, those occasional cells with the duplicated gene will have a growth advantage. An initial DHFR gene duplication could occur by a number of different mechanisms, including unequal crossing over (*19*) or uptake of a DNA segment from killed cells (*20*) which is integrated into a chromosome adjacent to the resident gene (*21*) or disproportionate replication (*22*).

Once a gene is initially duplicated tandemly, further amplification can occur by unequal crossing over, uptake of DNA from lysed cells, or generation of extrachromosomal sequences from rolling circle replication (*23*). Various combinations of these mechanisms could result in different genomic organizations for the amplified genes. Cytological studies of Biedler (*24*) indicate specific abnormalities which were described as "expanded homogeneously staining chromosomal regions" in single chromosomes in MTX-resistant hamster lung cell lines, which are not present in the sensitive parental cells. We have recently shown that a similar homogenously staining region on a single chromosome of a MTX-resistant Chinese hamster ovary cell line contains the DHFR genes by in situ hybridization with mouse DHFR cDNA (*25*). Thus at least in the Chinese hamster ovary cell line, the genes, which are stably amplified, are located to a chromosome.

The loss of genes from cells can occur by the reversal of the processes producing amplification. Unequal crossing over would result in one daughter cell with more genes and one with less. Excision of chromosomal genes with subsequent loss would result in all cells having decreased gene numbers. Thus, a population of cells generated from any given cell with amplified genes would vary with respect to gene number, thereby resulting in a heterogeneous cell population, as is indicated in Fig. 5. When a population of cells is placed under more stringent selection, that is, increased MTX concentration, those cells with more highly amplified genes will be selected. We have found that within a population of cells unstably amplified for DHFR genes, cells with a lower gene copy number have a slightly shorter generation time (*15*) than those with high gene copy numbers. Hence, when cells are grown without MTX, the population of cells ultimately emerging will appear to have lost genes because of differential growth rates.

Why is it, then, that resistant S-180 cells grown in the presence of MTX con-tinuously for 2 years have acquired the property of stability with respect to high enzyme content and resistance (and presumably high gene number)? We propose that fixation of stable resistance can be explained on the basis of selection of that population of cells that has maximal growth potential at a fixed MTX concentration. If cells are both amplifying and losing genes continuously, cells that lose too many genes (and lose high DHFR levels) will be killed (Fig. 4) and those cells with highly amplified DHFR genes (relative to that required for growth at the MTX concentration) will be at a growth disadvantage. Consequently, for maximal growth of the cell population, those occasional cells that have a number of genes appropriate for survival at the MTX concentration and that do not undergo loss or amplification will become dominant within the population after a number of generations.

At present we have no definitive information concerning the molecular mechanism (or mechanisms) for fixation of the DHFR genes. Possible mechanisms include (i) translocation of genes from a tandem array such that they cannot undergo recombination events resulting in gene loss; (ii) loss (deletion) of a protein or enzyme involved in recombination processes (*19*); (iii) integration of unstable, extrachromosomal genes into a chromosomal site (or sites); and (iv) loss of sequences that flank the DHFR genes and are involved in recombination processes. Such proposed sequences would be analogous to "insertion sequences" found in bacterial genomes (*26*).

Is the amplification of genes in cultured mammalian cells unique to DHFR, or do comparable phenomena occur with other genes? A number of investigators have described the selection of cultured cells resistant to normally lethal concentrations of 25-hydroxycholesterol, deoxynucleosides, and *N*-(phosphoacetyl)-L-aspartate (PALA), which have been found to contain from 2- to 100-fold increases in activities of hydroxymethylglutaryl CoA reductase (E.C. 4.1.3.4) (*27*), ribonucleotide reductase (E.C. 1.17.4.1) (*28*), and aspartate transcarbamylase (E.C. 2.1.3.2) (*29*), respectively.

Stark and his colleagues have recently found that PALA resistance is associated with an amplification of aspartate transcarbamylase genes in their cultured cells (*30*). Whether other instances of resistance in cultured mammalian cells results from gene amplification awaits further study.

The extent to which amplification of genes underlies other instances of drug resistance is still unknown. However, the properties of resistance of cultured cells to MTX, including (i) a stepwise selection of progressively resistant cells; (ii) an increase in a specific protein present at low levels in sensitive cells, which, when present in larger amounts, results in resistance; and (iii) stable or unstable resistance in the absence of selection pressure, have analogies both in antibiotic (*31*) and insecticide resistance (*32*). Recently Normark *et al.* (*33*) have, in fact, shown that penicillin resistance in *E. coli* K$_{12}$ obtained by stepwise selection results in chromosomal amplification of the gene for β-lactamase (penicillinase).

Our studies with MTX resistance provide further rationale for the principles of drug therapy (whether for bacteria, malignancies, or insects), including the use of multiple drugs, each in sufficient amounts to effect killing separately; treatment for only as long as necessary and with drugs not retained in the environment; and use of a second set of multiple drugs if resistance develops (*1, 34*). On the basis of the concept of gene amplification as a mechanism of drug resistance, the drugs used should not be counteracted by amplification of a single DNA sequence. Our results suggest that the prolonged administration of a single drug in ever increasing concentrations, which is retained in the environment, is precisely that form of administration

most likely to result in amplification of genes in a stable state, thereby imparting stable resistance.

References and Notes

1. A. Goldstein, L. Aronow, S. M. Kalman, *Principles of Drug Action* (Wiley, New York, 1974), chap. 8.
2. J. R. Bertino, D. M. Donohue, B. Simmons, B. W. Gabrio, R. Silber, F. M. Huennekens, *J. Clin. Invest.* **42**, 466 (1963).
3. W. C. Werkheiser, *J. Biol. Chem.* **236**, 888 (1961).
4. G. A. Fischer, *Biochem. Pharmacol.* **7**, 75 (1961).
5. M. Freidkin, E. J. Crawford, S. R. Humphreys, H. Goldin, *Cancer Res.* **22**, 600 (1962); J. P. Perkins, B. L. Hillcoat, J. R. Bertino, *Mol. Pharmacol.* **5**, 213 (1969).
6. M. T. Hakala, S. F. Zakrezewski, C. H. Nichol, *J. Biol. Chem.* **236**, 952 (1961).
7. J. W. Littlefield, *Proc. Natl. Acad. Sci. U.S.A.* **62**, 88 (1969).
8. A. Albrecht, J. L. Biedler, D. J. Hutchison, *Cancer Res.* **32**, 1539 (1972).
9. W. E. Flintoff, S. V. Davidson, L. Siminovitch, *Somatic Cell Genet.* **2**, 245 (1976).
10. G. A. Fischer, *Biochem. Pharmacol.* **11**, 1233 (1962); F. M. Sirotnak, S. Kurita, D. J. Hutchinson, *Cancer Res.* **28**, 75 (1968).
11. W. F. Flintoff, S. M. Spindler, L. Siminovitch, *In Vitro* **12**, 749 (1976).
12. F. W. Alt, R. E. Kellems, R. T. Schimke, *J. Biol. Chem.* **251**, 3063 (1976).
13. U. J. Hanggi and J. W. Littlefield, *ibid.*, p. 3075.
14. F. W. Alt, R. E. Kellems, J. R. Bertino, R. T. Schimke, *ibid.* **253**, 1357 (1978).
15. R. J. Kaufman and R. T. Schimke, in preparation.
16. R. J. Kaufman, J. R. Bertino, R. T. Schimke, *J. Biol. Chem.* **253**, 5852 (1978).
17. R. P. Anderson and J. R. Roth, *Annu. Rev. Microbiol.* **31**, 473 (1977); W. R. Folk and P. Berg, *J. Mol. Biol.* **58**, 595 (1971); S. W. Emmons, V. MacCoshan, R. L. Baldwin, *ibid.* **91**, 133 (1975); A. H. Sturtevant, *Genetics* **10**, 117 (1925); K. D. Tartoff, *Annu. Rev. Genet.* **9**, 370 (1975).
18. R. P. Anderson, C. G. Miller, J. R. Roth, *J. Mol. Biol.* **105**, 201 (1976).
19. G. P. Smith, *Cold Spring Harbor Symp. Quant. Biol.* **38**, 507 (1973); S. Wolff, *Annu. Rev. Genet.* **11**, 183 (1977).
20. W. O. McBride and H. L. Ozer, *Proc. Natl. Acad. Sci. U.S.A.* **70**, 1258 (1973); K. Willecke and F. H. Ruddle, *ibid.* **72**, 1792 (1975); D. A. Spandidos and L. Siminovitch, *Cell*, in press.
21. A. Hinnen, J. B. Hicks, G. R. Fink, *Proc. Natl. Acad. Sci. U.S.A.*, in press.
22. K. D. Tartoff, *Annu. Rev. Genet.* **9**, 370 (1975).
23. D. Hourcade, D. Dressler, J. Wolfson, *Cold Spring Harbor Symp. Quant. Biol.* **38**, 537 (1973).
24. J. L. Biedler and B. A. Spengler, *Science* **191**, 185 (1976).
25. J. H. Nunberg, R. J. Kaufman, R. T. Schimke, G. Urlaub, L. A. Chasin, *Proc. Natl. Acad. Sci. U.S.A.*, in press.
26. S. N. Cohen, *Nature (London)* **263**, 731 (1976).
27. M. Sinensky, *Biochem. Biophys. Res. Commun.* **78**, 863 (1977).
28. M. Meuth and H. Green, *Cell* **3**, 367 (1974).
29. T. D. Kemp, E. A. Swyryd, M. Bruist, G. R. Stark, *ibid.* **9**, 541 (1976).
30. G. N. Wahl, P. A. Padgett, G. R. Stark, in preparation.
31. M. Demerec, *J. Bacteriol.* **56**, 63 (1948); V. Bryson and S. Szybalski, *Adv. Genet.* **7**, 1 (1955); D. Perlman and R. H. Rownd, *J. Bacteriol.* **123**, 1013 (1975); M. Demerec, *Proc. Natl. Acad. Sci. U.S.A.* **31**, 16 (1945); K. D. Eriksson, H. G. Bowman, J. A. T. Jansson, S. Thoren, *J. Bacteriol.* **90**, 54 (1965); A. S. Breeze, P. Sims, K. A. Stacey, *Genet. Res.* **25**, 207 (1975).
32. W. D. McEnroe and A. Lakocy, *J. Econ. Entomol.* **62**, 283 (1969); M. W. Service, *J. Trop. Med. Hyg.* **67**, 190 (1964); G. P. Georghian, *Annu. Rev. Entomol. Syst.* **3**, 133 (1972); L. C. Terriere, *Annu. Rev. Entomol.* **13**, 75 (1968); M. Tsukamotom, *Residue Rev.* **25**, 289 (1969); L. Gil, *Entomol. Exp. Appl.* **11**, 15 (1968); J. Sternbury, *J. Agr. Food. Chem.* **2**, 1125 (1954); A. Grigoio and F. J. Oppenworth, *Genetica* **37**, 159 (1966); V. Thomas, *J. Econ. Entomol.* **59**, 779 (1966); J. F. Crow, *ibid.* **47**, 393 (1966); J. C. King, *ibid.*, p. 387; C. R. Walker and L. C. Terriere, *Entomol. Exp. Appl.* **13**, 260 (1970).
33. S. Normark, T. Edlund, T. Grundstrom, S. Bergstrom, H. Wolf-Watz, *J. Bacteriol.* **132**, 912 (1977).
34. J. R. Bertino and R. E. Skeel, *Pharmacologic Basis of Cancer Chemotherapy* (Williams & Wilkens, Baltimore, 1974), p. 681.
35. Supported by grants from the American Cancer Society (NP 148) and the Cancer Institute (CA-16318).

Reprinted from *Science* **126**:255 (1957)

PREVENTION OF TOXICITY OF AMETHOPTERIN FOR SARCOMA-180 CELLS IN TISSUE CULTURE

M. T. Hakala

The present paper is a preliminary report on the finding that sarcoma-180 (S-180) cells grow normally when the function of folic acid is prevented by amethopterin (Methotrexate), if the medium is supplemented with products of the biosynthetic reactions dependent on folic acid cofactors. It is known that certain microorganisms which require folic acid for growth—for example, *Streptococcus faecalis* 8043—can grow in the absence of this vitamin in a medium containing thymine and adenine or guanine (*1*). That folic acid is one of the nutritional requirements for the growth of mammalian cells in tissue culture was shown by Eagle (*2*). The inhibition of the growth of sarcoma-180 cells in such a medium by amethopterin has been demonstrated (*3*).

The techniques of Eagle (*2, 4*) were used. The medium contained 5 percent thoroughly dialyzed horse serum. The cells were grown in the experimental media for 7 days. The growth of the cells in the presence of amethopterin in a medium containing hypoxanthine, thymidine, and glycine is shown in Table 1. Disintegration of the cells occurred if hypoxanthine or thymidine was omitted from the mixture. In the absence of glycine, some growth was observed, indicating either a small amount derived from the dialyzed horse serum or some formation of glycine, possibly from the exogenous L-threonine (*5*). When glycine was added to this medium, the growth was comparable to that of the control. The complete mixture fully supported the growth of sarcoma-180 cells even in the presence of amethopterin at concentrations 10,000 times that ordinarily required for complete inhibition (Table 1). In body fluids, the concentrations of such compounds could be critical to the effectiveness of amethopterin on neoplastic cells *in vivo*. Similar compounds in the crude medium (chicken plasma clot) might also explain the failure of amethopterin to inhibit sarcoma-180 cells, as reported by Biesele (*6*).

When the function of folic acid was prevented by amethopterin, it was found that sarcoma-180 cells were able to utilize adenine, adenosine, deoxyadenosine, hypoxanthine, and inosine equally well; guanosine supported slower growth of sarcoma-180 cells under these conditions; that is, there was only a threefold increase in 7 days, whereas xanthine and xanthosine were inactive. These results indicate that some "adenine" was derived from guanosine, but the extent to which "guanine" or "adenine," or both, were derived from xanthine and xanthosine was insignificant.

In the presence of amethopterin, thymidine, and glycine, the cells disintegrate if purines are not supplied in the medium (Table 1). Under such conditions it is unlikely that purine synthesis *de novo* occurs. Thus, the single purine in the medium must serve as the sole source, not only of adenine and guanine of nucleic acids, but also of all the coenzymes containing purines as structural constituents. Accordingly, this technique appears to be useful for the study of the pathways of purine metabolism. The present work demonstrates for the first time that mammalian cells (sarcoma-180) are fully capable of using exogenous purines for growth and multiplication.

Further work demonstrated that thymidine could be replaced by thymidylic acid for the growth of sarcoma-180 cells in the presence of amethopterin, glycine, and a purine, but thymine and thymine-riboside had no activity in this respect. A mixture of thymine and deoxyadenosine did not replace a mixture of thymidine and adenine, indicating that this type of transdeoxyribosidation did not occur.

It is seen that the presence of amethopterin creates new requirements for the growth of the tumor cells *in vitro*. When these requirements are met by preformed purines, thymidine, and glycine, inhibition of growth might still be achieved if the utilization of even one of these compounds were prevented. Logical combinations for chemotherapy are thus suggested. The new requirements created by amethopterin probably differ in different species and tissues. It is already known that differences exist in the abilities of various tissues and species to utilize preformed purines *in vivo* (*7*). In addition, rabbit fibroblasts, unlike

Table 1. Growth of sarcoma-180 cells in tissue culture in the presence of amethopterin. Folic acid (pteroylglutamic acid) was present at a concentration of $2 \times 10^{-7} M$.

Varied supplements in the medium				Degree of cellular multiplication†
Amethopterin* (M)	Hypoxanthine $3 \times 10^{-5} M$	Thymidine $3 \times 10^{-5} M$	Glycine $1 \times 10^{-4} M$	
0				5.7‡
3×10^{-8}				0.70
3×10^{-7}				0.34
3×10^{-7}	+	+	+	5.7§
3×10^{-4}	+	+	+	6.1
3×10^{-7}		+	+	0.60
3×10^{-7}	+		+	0.43
3×10^{-7}	+	+		1.8

* 4-Amino-10-methyl-pteroylglutamic acid.
† Referred to inoculum as 1; determined by the method of Oyama and Eagle (*11*); correlation of the protein determinations with cell counts is discussed by Oyama and Eagle (*11*).
‡ Control.
§ The culture has been carried for 3 weeks under these conditions and is being maintained.

sarcoma-180 cells, require exogenous L-serine for growth in tissue culture (*8*).

Thymine or thymidine increased the rate of the development of amethopterin resistance in *Streptococcus faecalis* 8043 (*9*). The effect of similar factors on the development of amethopterin resistance in mammalian cells is under study (*10*).

References and Notes

1. E. E. Snell and H. K. Mitchell, *Proc. Natl. Acad. Sci. U.S.* 27, 1 (1941).
2. H. Eagle, *Science* 122, 501 (1955).
3. ——— and G. E. Foley, *Am. J. Med.* 21, 739 (1956).
4. H. Eagle, *J. Biol. Chem.* 214, 839 (1955); H. Eagle *et al.*, *Science* 123, 845 (1956).
5. H. L. Meltzer and D. B. Sprinson, *J. Biol. Chem.* 197, 461 (1952).
6. J. J. Biesele, *Ann. N.Y. Acad. Sci.* 58, 1129 (1954).
7. C. Heidelberger, *Ann. Rev. Biochem.* 25, 589 (1956); L. L. Bennett and H. E. Skipper, *Arch. Biochem. and Biophys.* 54, 566 (1955).
8. R. F. Haff and H. E. Swim, *Federation Proc.* 15, 591 (1956).
9. M. T. Hakala, *Suomen Kemistilehti* 28, 30 (1955).
10. The study reported here was supported in part by the Dorothy H. and Lewis Rosenstiel Foundation.
11. V. I. Oyama and H. Eagle, *Proc. Soc. Exptl. Biol. Med.* 91, 305 (1956).

Editor's Comments
on Papers 35, 36, and 37

35   SZYBALSKI and SMITH
*Genetics of Human Cell Lines I. 8-Azaguanine Resistance, a
Selective "Single-Step" Marker*

36   LIEBERMAN and OVE
Enzyme Studies with Mutant Mammalian Cells

37   CHU and MALLING
*Mammalian Cell Genetics, II. Chemical Induction of Specific
Locus Mutations in Chinese Hamster Cells In Vitro*

RESISTANCE TO PURINE ANALOGS

As was the case with resistance to folic acid antagonists, mammalian cells resistant to the toxic guanine analog 8-azaguanine (AZG) were observed over thirty years ago in leukemic tumors treated with the drug. Several years later, cultured cell lines resistant to AZG were selected by Szybalski and Smith (Paper 35) and by Lieberman and Ove (1959). In the studies by Szybalski and Smith, human cells were cultured in the presence of AZG and resistant cells were selected. AZG resistance appeared to develop as a "single step" event, and statistical methods were employed to measure the mutation rate for AZG resistance. As described in other sections of this volume, AZG resistant cells have played an important role in somatic cell genetics, not only by providing a selectable marker for cell hybridization and gene transfer studies, but also by serving as a focus for studies on the mechanisms of genetic variation in cultured cells.

The biochemical mechanism of AZG resistance in mammalian cells was first elucidated with drug resistant tumors (Brockman et al., 1959) and subsequently confirmed with AZG resistant cells selected in culture. It was shown that AZG resistant cells lack the ability to convert AZG, as well as guanine and hypoxanthine, to their ribonucleotides. This reaction is catalyzed by the enzyme inosinic acid pyrophorphorylase, which is now known as hypoxanthine-guanine phosphoribosyltransferase (HGPRT). More recently, it was found that cells derived from patients with the Lesch-Nyhan syndrome also are deficient in HGPRT activity (Seegmiller et al., 1967). Because of their enzyme deficiency, cells of Lesch-Nyhan patients are inherently resistant to AZG. Thus, through the use of cells cultured from patients, mutant cells with a specific biochemical lesion are immediately available, without selection for drug resistance. In addition to the Lesch-Nyhan syndrome, several other diseases can provide mutant cells with well-defined lesions for genetic studies in vitro.

Studies discussed in the preceding section of this book showed that cells cultured in the presence of amethopterin require both hypoxanthine and thymidine in the medium in order to survive (see Paper 34). Lieberman and Ove (Paper 36) took advantage of this observation in studies on the growth of drug resistant cells in the presence of aminopterin, a very close analog of amethopterin. The cells in this study were selected for resistance to 6-mercaptopurine (MP), which selects cells with the same enzyme deficiency as does AZG. The MP resistant cells were unable to phosphorylate hypoxanthine and lacked HGPRT activity. Most significantly, these HGPRT deficient cells were unable to grow in the presence of aminopterin even when the medium was supplemented with hypoxanthine and thymidine (i.e., HAT medium). In contrast, the parental cells expressing HGPRT activity were able to grow in HAT medium. Thus, these studies demonstrated that HAT medium can be used to select against HGPRT deficient cells, because of the conditional hypoxanthine dependence caused by aminopterin and the inability of HGPRT deficient cells to incorporate hypoxanthine from the medium.

Another step in the development of HAT medium as an important tool for somatic cell genetics was the demonstration by Szybalski and co-workers (see Paper 25) that HAT medium can be used to select HGPRT positive cells from a population of HGPRT deficient cells. Selection for HGPRT activity, however, represents only one side of the HAT system. In addition to hypoxanthine, cells must incorporate thymidine from the medium in order to survive in HAT medium. The utilization of exogenous thymidine by cells requires the conversion of thymidine to its nucleotide, a reaction catalyzed by the enzyme thymidine kinase (TK).

Studies by Kit and co-workers (1963) showed that cells deficient in TK activity and unable to incorporate thymidine can be selected with the thymidine analog 5-bromodeoxyuridine. Thus, selection for TK deficient cells is comparable to selection for HGPRT deficient cells. As described earlier in this volume, Littlefield (see Paper 3) demonstrated that TK deficient cells, like HGPRT deficient cells, are unable to survive in HAT medium, and that HAT medium can be used to select for cells with TK activity as well as HGPRT activity. The selection with HAT medium for cells expressing TK and HGPRT activity has played a predominant role in the development of somatic cell genetics.

By the mid-1960s, a large number of mammalian cell mutants had been isolated and their enzymatic alterations elucidated. However, chemical mutagenesis, which had proven very useful in mutational studies with microorganisms, had not yet been demonstrated in mammalian systems. In early attempts to demonstrate chemical mutagenesis in mammalian cells, a large number of known bacterial mutagens were tested for their ability to increase the frequency of AZG resistant cells (Szybalski et al., 1964). However, no definite or consistent mutagenic effects were observed.

In 1968 chemical mutagenesis with mammalian cells was demonstrated by Chu and Malling (Paper 37) and by Kao and Puck (1968). In the studies by Chu and Malling, AZG resistance was used as the marker to monitor the induction of mutants. A significant increase in the frequency of drug resistant cells was observed when cultures were treated with various chemical mutagens. Factors such as cell density and expression time (the time between mutagenesis and selection) were found to be critical for the induction and isolation of mutants. In the studies by Kao and Puck (1968), which also provided evidence for chemical mutagenesis, a then newly developed method for the isolation of auxotrophic mutants (see Paper 40) was utilized. Chemical mutagenesis of cultured mammalian cells not only has facilitated the isolation of mutant cell lines, but has also permitted studies on the mechanisms of mutagenesis in mammalian cells.

REFERENCES

Brockman, R. W., L. L. Bennett, M. S. Simpson, A. R. Wilson, J. R. Thompson, and H. E. Skipper, 1959, A Mechanism of Resistance to 8-Azaguanine II. Studies with Experimental Neoplasms, *Cancer Res.* **19:**856–859.

Kao, F.-T., and T. T. Puck, 1968, Genetics of Somatic Mammalian Cells, VII. Induction and Isolation of Nutritional Mutants in Chinese Hamster Cells, *Natl. Acad. Sci. (USA) Proc.* **60:**1275–1281.

Kit, S., D. R. Dubbs, L. J. Piekarski, and T. C. Hsu, 1963, Deletion of Thymidine Kinase Activity from L Cells Resistant to Bromodeoxyuridine, *Exp. Cell Res.* **31:**297–312.

Lieberman, I., and P. Ove, 1959, Isolation and Study of Mutants from Mammalian Cells in Culture, *Natl. Acad. Sci. (USA) Proc.* **45:**867–872.

Seegmiller, J. E., F. M. Rosenbloom, and W. N. Kelley, 1967, Enzyme Defect Associated with a Sex-Linked Human Disorder and Excessive Purine Synthesis, *Science* **155:**1682–1684.

Szybalski, W., G. Ragni, and N. K. Cohn, 1964, Mutagenic Response of Human Somatic Cell Lines, in *Cytogenetics of Cells in Culture*, R. J. Harris, ed., International Society on Cell Biology Symposium, vol. 3, Academic Press, New York, pp. 209–221.

35

Reprinted from *Soc. Exp. Biol. Med. Proc.* **101**:662–666 (1959)

Genetics of Human Cell Lines I. 8-Azaguanine Resistance, a Selective "Single-Step" Marker.* (25053)

WACLAW SZYBALSKI AND M. JOAN SMITH (Introduced by Vernon Bryson)

Inst. of Microbiology, Rutgers State University, New Brunswick, N. J.

From the methodological point of view, the mammalian cell grown *in vitro* can be regarded as a unicellular microorganism. Thus, methods developed for quantitative work with microbes can be adapted to the genetic study of human cells, provided suitable selective markers are available. Plating and colony-counting technic, so useful for assay of viable microbial cells, was introduced into tissue culture methodology by Puck *et al.*(1). Our purpose was to find a suitable mutational system for mammalian cells in which, under selective conditions, mutant cells survive and form well developed colonies while parental population is completely inhibited or eliminated. Mutation from 8-azaguanine (AG) sensitivity to resistance satisfied the foregoing criteria(2). Moreover, the property of 8-azaguanine resistance appears to be an excellent genetic marker, since it is not associated with modifications in morphological or cultural characteristics of cells either in presence or absence of the selective agent.

Materials and methods. Strain *Detroit-98* (D98), derived from human sternal marrow by Berman and Stulberg(3), was kindly supplied by Dr. H. Moser of Cold Spring Harbor

Labs. A single-cell-derived clone D98S was used throughout these studies. *Media.* The basic medium was essentially that described by Eagle(4), containing 10% complete horse serum. Sterilization was effected by filtration through Selas No. 2 filter. *Cultivation and plating procedures.* The methods were based on those developed by Puck *et al.*(1) and adapted for strain D98 by Moser and Tomizawa(5). Cells were grown on glass surface of 60-mm Petri dishes at 37°C in 5-10% CO_2 atmosphere. The inoculum was prepared from a 3- to 6-day-old culture of vigorously growing cells, washed serum-free with balanced salt solution before detachment and dispersion in the same solution containing 0.25% pancreatin (Nutritional Biochemical Corp., Cleveland) (5-minute incubation at 37°C followed by gentle manual shaking). The action of pancreatin was terminated by addition of serum-containing medium, after which the cell suspension was filtered aseptically through fine cheesecloth to remove any cell clumps. The cell density of this essentially monodispersed suspension was assayed in a hemacytometer. For colony counts, cells appropriately diluted in 5 ml of serum-containing medium were plated and incubated 6 to 12 days, with a complete medium change every 2 to 3 days,

* This investigation was supported by Research Grant No. 3492 of U. S. Public Health Service, Nat. Inst. of Health.

Growth was assayed either by counting and measuring colonies or by protein determination. *Protein determination* is based in part on protein staining method of Durrum(6) and its modifications for electrophoretic(7) and cytochemical(8) purposes. Adaptation of these methods for quantitative protein determination in mammalian cell cultures, suggested by Dr. A. W. Kozinski of this laboratory, permits determination of total protein without disturbing colonies on the glass. Thus a colony count may be performed on the same plate after completion of protein determination. The assay is based on spectrophotometric determination of bromphenol blue selectively bound by the protein component of $HgCl_2$-fixed cells and subsequently eluted with alkaline acetone. The procedure consists of: (1) thorough washing of glass-attached cells with balanced salt solution to remove all serum proteins; (2) 15-minute staining at 37°C with aqueous solution containing 0.1% bromphenol blue and 5% $HgCl_2$; (3) three consecutive washes with 0.5% acetic acid each lasting 6 minutes; (4) quantitative elution with 3-ml portions of aqueous acetone (20 ml acetone 0.2 ml 10 N NaOH, 5 ml water, freshly prepared); (5) adjusting the collected extracts to 10-ml volume and measuring optical density at 595 mμ wave length against a blank prepared similarly from cell-free plate preincubated with serum-containing medium. The actual amount of cell protein/plate is ascertained from predetermined calibration curves. The same procedure was also used for cell suspensions, with centrifugation as an added step. *Colony count.* After fixation in Bouin's fluid or after protein determination, colonies were stained with Giemsa solution. The dry dishes were then inserted into a projector (Bausch & Lomb, "Tri-Simplex" micro-projector), and well-focused images of colonies were scored by an electronic counter (built by Philip D. Mintz of this laboratory).

Results. When above procedures are followed, strain D98S exhibits a consistent plating efficiency of greater than 0.6, *i.e.*, over 60% of microscopically assayed cells give rise to well defined colonies (Fig. 1), which can easily be counted, providing their number

does not exceed 5000/plate. Above this figure, colonies coalesce, and total protein synthesis after 14 days' incubation becomes less than proportional to the number of cells in the inoculum (Fig. 2).

A series of plates was prepared, each seeded with approximately 5000 cells, to which AG was added in graded concentrations. Fig. 3 represents the survival curve 1 of the wild-type strain D98S. It is apparent that in presence up to 0.1 μg AG/ml survival and colony-forming ability of each cell is virtually unaffected. There is a sharp drop in plating efficiency between 0.1 and 1 μg AG/ml, followed by a plateau extending to approximately 10 μg AG/ml, at which level 1 to 2% of cells survive and form colonies. These colonies were isolated and found to be stable, AG-resistant clones (D98/AG) as reflected by survival curve 2. Their growth, as measured by protein synthesis, was not affected by AG in concentrations up to 12 μg/ml (curve 4), while protein synthesis of the wild-type strain was increasingly suppressed above 0.1 μg/ml.

Thus the wild-type population contains approximately 1 to 2% mutant cells, exhibiting to 100-fold increase in AG resistance without impairment of plating efficiency or growth rate either in the presence or in the absence of AG (Fig. 2 and 3). Fig. 4 shows appearance of colonies on 2 plates seeded with equal inoculum of AG-sensitive strain D98S, incubated in absence (A) and in presence (B) of AG (6 μg/ml).

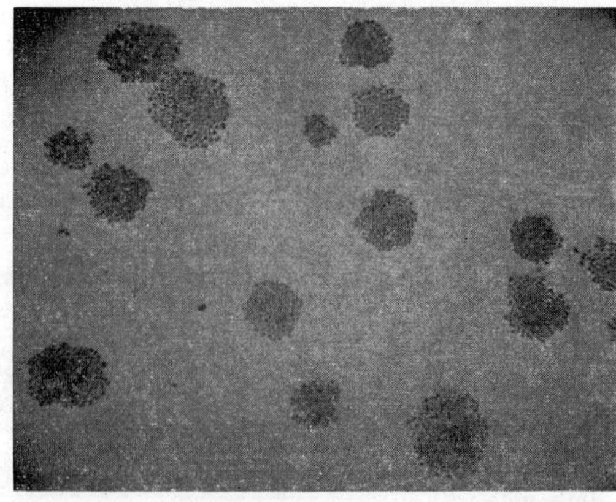

FIG. 1. 6-day-old colonies of strain D98S grown on glass, Bouin fixed, Giemsa stained.

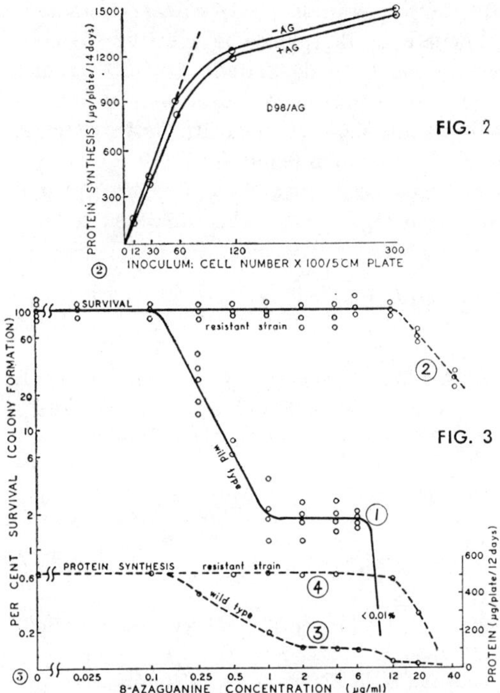

FIG. 2. Relationship between inoculum size (colony-forming cells) and total growth determined as amount of protein synthesized (14 days) by AG-resistant strain D98/AG grown in presence (+AG) or absence (−AG) of 8 μg AG/ml medium.

FIG. 3. 12-day growth determined as amount of protein synthesized and survival (colony formation) of AG-sensitive (wild type—1, 3) and AG-resistant (2, 4) isolates derived from strain D98S grown at increasing concentrations of AG. Inoculum was 5000 cells/plate.

Resistance to AG appears to be a highly stable property. Subculture for over 150 generations, either in absence (curve 3) or in presence (curve 2) of AG (6 μg/ml) did not alter level of resistance (Fig. 5). By subculturing resistant lines at high concentrations of AG, strains with slightly higher resistance could be isolated (Fig. 5, curve 4).

Determination of mutation rates was based on Luria-Delbrück variance test(9), modified for mammalian cells grown on glass. Each colony grown in absence of AG and firmly attached to glass may be considered as analogous to one "test tube culture"(9). On addition of the drug (AG), only colonies containing at least one resistant cell would give rise to secondary resistant colonies, while others would slough off the glass and be lost during medium changes. Mutation rate based on

preliminary experiments, employing the P_0 method(9) corresponds to 5 x 10^{-4}/cell/generation.

Properties of AG-resistant cells. AG-sensitive cells were found to exhibit the same inhibitory threshold for AG and 8-azaguanosine (10^{-7} M). Neither guanine (1 to 50 μg/ml) nor adenine nor thymine antagonize the inhibitory action of 1 to 50 μg/AG/ml. AG-resistant lines are approximately 100 times more resistant to AG, but only 2-3 times more resistant to 8-azaguanosine.[†] They were also found to be more resistant to 5-fluorouracil and to ultraviolet (UV) light (Fig. 6 and Fig. 7, curve 3). It was therefore of interest to determine whether UV would increase mutation frequency from AG sensitivity to resistance or would select for the more UV-resistant, AG-resistant cells(10). The wild-type population was irradiated with increasing doses of UV and plated in absence (curve 1) or in the presence (curves 2 and 2a) of 6 μg AG/ml. It is apparent that AG-resistant cells not exposed to AG prior to irradiation, exhibit a "single-hit" survival curve (curves 2 and 2a), while UV survivals of parental strain D98S (curve 1) and of the established resistant line D98/AG (curve 3) are characterized by "multi-hit" curves. Curve 4 represents UV survival of the AG-resistant line at slightly inhibitory concentrations of AG (12 μg/ml). The shape of survival curve appears unchanged. It may therefore be concluded that the newly emerging AG-resistant cells, pre-existing in the wild-type population, contain only one AG-resistant "complement" (chromosome?), while the number of these "complements" increases during establishment of the resistant sublines. Further genetic and cytological studies may explain this phenomenon and in-

---

† Strain D98/AG contains approximately 0.01% mutants (D98/AGR) resistant in a single (genetic) step to over a 100-fold higher concentration of 8-azaguanosine (AGR), with concomitant further increase in AG resistance (over 10-fold). Direct mutations from AG sensitivity (D98S) to AGR resistance (D98/AGR) were not detected in populations as large as 10^7 cells. Thus AG and AGR resistances behave as two sequential mutational steps.

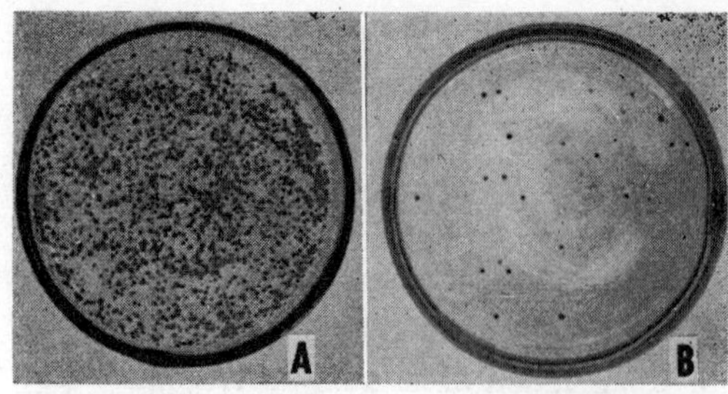

FIG. 4. Colony formation on 2 plates seeded with identical inocula (4000 cells) of AG-sensitive cells (strain D98S) in absence (A) or presence (B) of 6 μg AG/ml medium.

dicate whether processes analogous to non-disjunction or mitotic crossing-over could be involved.

Summary. A cloned wild-type population (strain D98S) derived from human bone marrow cells, which is inhibited by 8-azaguanine (AG) in concentrations in excess of 0.1 μg/ml, contains approximately 1 to 2% cells resistant to AG in 100-fold higher concentrations. A single-step mutational process (in broad meaning of the word) appears to be involved. The AG-resistant lines are stable upon prolonged subculture either in presence or absence of drug. Mutation rate from sensitivity to resistance is of the order of 5 x 10^{-4}/cell/generation. AG resistance is a useful genetic marker, since at selective AG concentrations (a wide plateau at 1-10 μg/ml) colony formation by resistant cells is not impaired and sensitive cells are destroyed. A method for protein determination is described, based on measurement of eluted bromphenol blue from $HgCl_2$-fixed, glass-attached cells.

1. Puck, T. T., Marcus, P. I., Cieciura, S. J., *J. Exp. Med.*, 1956, v103, 273.

2. Szybalski, W., *Microb. Genetics Bull.*, 1958, No. 16, 30.

3. Berman, L., Stulberg, C. S., PROC. SOC. EXP. BIOL. AND MED., 1956, v93, 730.

4. Eagle, H., *J. Exp. Med.*, 1955, v102, 37.

5. Moser, H., Tomizawa, K., *Ann. Rep. Biol. Lab. Cold Spring Harbor*, 1957-1958, v68, 33.

6. Durrum, E. L., *J. Am. Chem. Soc.*, 1950, v72, 2943.

7. Kunkel, H. G., Tiselius, A., *J. Gen. Physiol.*, 1951, v35, 89.

8. Mazia, D., Brewer, P. A., Alfert, M., *Biol. Bull.*, 1953, v104, 57.

9. Luria, S. E., Delbrück, M., *Genetics*, 1943, v28, 491.

10. Szybalski, W., Smith, M. J., *Fed. Proc.*, 1959, v18, 336.

Received May 22, 1959.   P.S.E.B.M., 1959, v101.

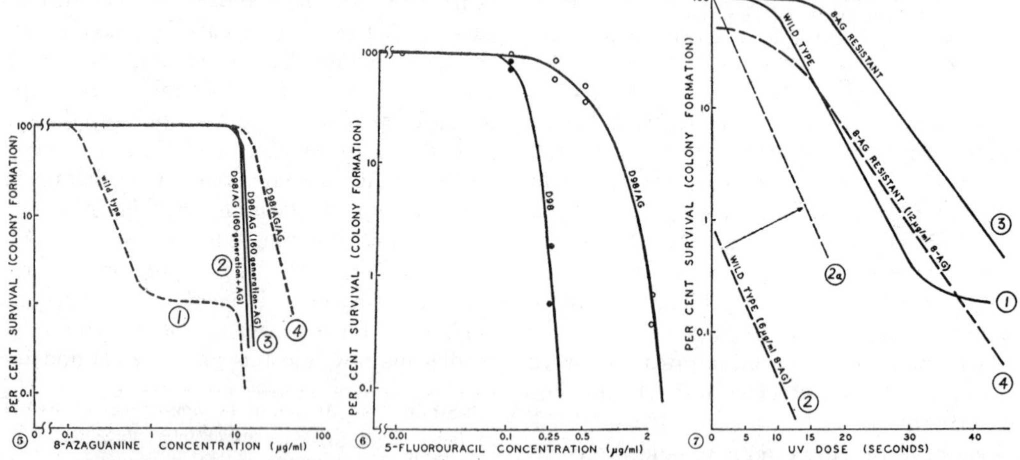

FIG. 5. Survival of AG-sensitive (1) and of AG-resistant mutant lines at increasing concentrations of AG. Curves 2 and 3 represent survival of AG-resistant strains after subculture for 160 generations in presence (2) or absence (3) of 5 μg AG/medium. Survival of more resistant line D98/AG/AG, obtained by serial subculture of D98/AG at 20 to 40 μg AG/ml, is depicted by curve 4.

FIG. 6. Survival of AG-sensitive (D98) and AG-resistant (D98/AG) strains at increasing concentrations of 5-fluorouracil.

FIG. 7. Ultraviolet survival of AG-sensitive (1, 2) and AG-resistant (3, 4) cell lines grown in absence (1, 3) or presence (2, 4) of AG. Curve 2a is parallel to curve 2. UV intensity 120 μW/cm².

ENZYME STUDIES WITH MUTANT MAMMALIAN CELLS*

I. Lieberman and P. Ove

The availability of cultured mutant[1] mammalian cells (1–3) manipulable by procedures similar to those used with other microorganisms has provided a tool which may facilitate studies of various aspects of mammalian genetics. One question of interest which can be readily explored with them concerns the nature of genetic changes at the biochemical level. To gain more insight into this problem, some of the enzyme activities involved in purine metabolism were examined in susceptible parent AMK2-2[2] cells and in mutants selected from them resistant to 2,6-diaminopurine or 6-mercaptopurine.

Investigations with a streptococcus (6, 7) and with neoplastic cells grown in mice (8) have indicated that resistance to 8-azaguanine (and simultaneously to 6-mercaptopurine) is associated with a decreased ability to form the nucleotide, 8-azaguanylic acid (or 6-mercaptopurine ribonucleotide). With the mammalian cells this was correlated with what may be about a 75% depression in the activity which catalyzes the synthesis of IMP from hypoxanthine and ribosylpyrophosphate α-5-phosphate (IMP pyrophosphorylase (9–11)). Although enzyme studies have not been reported, similarly, with DP,[3] the resistance of bacterial mutants seems to derive from their inability to synthesize the 5′-nucleotide (12, 13).

Experiments with cell-free preparations of mutant AMK cells have indicated a similar mechanism of resistance for them. Only the parent possesses both AMP and IMP pyrophosphorylase; cells resistant to DP lack the AMP enzyme, whereas those resistant to MP lack the IMP one. The results obtained with growing cultures have been consistent with the loss in the mutants of one or the other enzyme.

The purpose of this report is to present the evidence for the nature of the enzyme changes which are associated with mutation to resistance to DP and MP in a permanently cultivatable mammalian cell.

EXPERIMENTAL

Materials—Cells were grown on glass surfaces in a medium similar to the one of Healy *et al.* (14) and supplemented with

---

* This investigation was supported by a research grant from the National Institutes of Health, United States Public Health Service.

[1] The term "mutant" is used to describe a cell with a new trait stable for many generations in the absence of the specific selective agent. Until recombinational analysis can be made, it is recognized that no rigorous proof of mutation is possible.

[2] Parent AMK2-2 cells were kindly supplied by Dr. L. Siminovitch and Dr. R. C. Parker. They were originally considered to be "altered" monkey kidney cells, but subsequent studies (4) have suggested their derivation from Earle's mouse fibroblast strain L (5).

[3] The abbreviations used are: DP, 2,6-diaminopurine; MP, 6-mercaptopurine.

15% bovine serum. To prepare extracts, when maximal growth was approached, the cells were harvested by scraping with a rubber spatula and washed 3 times by centrifugation with an ice-cold solution containing NaCl (0.14 M), KCl (0.03 M), CaCl$_2$ (0.001 M), MgCl$_2$ (0.001 M), and sodium phosphate (0.01 M, pH 7.4). The packed washed cells, suspended in water, were disrupted by sonic vibration as previously described (15).

Unlabeled purines and their derivatives were obtained from the Sigma Chemical Company and from the Nutritional Biochemicals Corporation. 8-C[14]-Labeled adenine and hypoxanthine were products of the Volk Radiochemical Company and 8-C[14]-adenosine was prepared by Schwarz Laboratories. The purine analogues, DP and MP, were obtained from the Nutritional Biochemicals Corporation, and 4-aminopteroylglutamic acid (Aminopterin) was a gift from Dr. S. M. Hardy, Lederle Laboratories. PP-ribose-P was prepared as previously described (16).

Determinations—Cell counts were made with a Levy counting chamber. Radioactivity measurements were made with a gas flow counter after drying the samples in stainless steel dishes, or if they contained high concentrations of acid, on round cover glasses. Protein was estimated by the procedure of Lowry *et al.* (17).

Assay Procedures—To measure the pyrophosphorylase activities, C[14]-labeled purine substrates were used, and the nucleotide products were estimated after isolation by ion-exchange chromatography.

AMP Pyrophosphorylase—The reaction mixtures (0.5 ml) contained 0.05 ml of 0.25 M tris(hydroxymethyl)aminomethane buffer (pH 8.0), 0.02 ml of 0.1 M MgCl$_2$, 0.05 ml of 0.001 M PP-ribose-P, 0.05 ml of 1.6×10^{-3} M 8-C[14]-adenine (2.2×10^5 c.p.m. per μmole) and the enzyme preparation. After 45 minutes at 37°, the mixtures were heated in a boiling water bath for 1 minute, cooled, and 2 ml of 2 M HCl were added. Insoluble material was discarded by centrifugation and the supernatant solutions were put on columns of Dowex 50, H$^+$ form, 2 cm in height and 1 cm in diameter, at room temperature. AMP was eluted with 2 M HCl in a total effluent volume of 15 ml and radioactivity measurements were made on 1-ml aliquots dried on cover glasses. A unit of enzyme was defined as the amount yielding 1 μmole of AMP per hour and specific activity was defined as units per mg of protein.

Under the conditions of the assay, no AMP synthesis occurred in the absence of added PRPP, and nucleotide formation was proportional to the amount of enzyme preparation over the range tested. Thus, with 0, 0.005, 0.01, and 0.02 ml of extract 0, 1245, 1950, and 4270 c.p.m. were found in the eluates, respectively.

Evidence consistent with the identity of the radioactive product with AMP was obtained by comparing its elution pattern

269

TABLE I

Effect of purine analogues on parent and mutant cells

The cells were harvested by scraping with a rubber spatula and were suspended in growth medium to yield 12,500 per ml. Two-milliliter aliquots of the suspensions were delivered to tubes which had previously received the indicated amounts of the purine analogues. The tubes were incubated (37°) in an almost horizontal position and after 5 days, growth was estimated by counting the cells.

Analogue		Culture		
DP	MP	Parent	DP[r]*	MP[r]†
μmole per ml		+Δ *cells (thousands)*		
0		360	308	308
0.012		95	283	31
0.024		65	323	17
0.048		0	267	0
0.600		0	181	0
	0.006	25	29	371
	0.012	0	0	295
	0.024	0	0	343
	0.600	0	0	347

* DP-resistant cells.
† MP-resistant cells.

TABLE II

AMP and IMP pyrophosphorylases in parent and mutant extracts

The cell-free extracts were prepared and assayed as described under "Experimental."

Purine substrate	Extract			Specific activity
	Parent	DP[r]*	MP[r]†	
	mg protein tested			*units/mg protein*
Adenine	0.11			0.19
		0.53		<0.001
			0.17	0.23
	0.11	0.53		0.14‡
Hypoxanthine	0.032			0.48
		0.032		0.39
			0.87	<0.001
	0.032		0.87	0.36‡

* DP-resistant cells.
† MP-resistant cells.
‡ Only the protein contributed by the parent extract was used to calculate specific activity.

from Dowex 1 with that of the authentic compound. Further, treatment with adenylate deaminase yielded a compound with the chromatographic properties (Dowex 1) of IMP.

IMP Pyrophosphorylase—The test mixtures were the same as for the AMP enzyme assay except that the purine substrate was 0.02 ml of 2×10^{-3} M 8-C^{14}-hypoxanthine (1.1×10^6 c.p.m. per μmole). After 45 minutes at 37°, the reaction was stopped by heating in a boiling water bath for 1 minute. Two milliliters of water were then added, insoluble material was discarded by centrifugation, and the supernatant solutions were put on columns of Dowex 1, chloride form, 2% cross-linked, 2 cm in height and 1 cm in diameter, at room temperature. Hypoxanthine was eluted with 35 ml of 0.01 M HCl, IMP with 5 ml of the 1 M acid.

Radioactivity measurements were made as described above and the enzyme unit and specific activity were similarly defined.

No IMP was formed in the absence of added PP-ribose-P and nucleotide synthesis was proportional to the amount of enzyme preparation over the range tested. Thus, with 0, 0.003, 0.01, and 0.03 ml of extract 59, 1096, 3245, and 8770 c.p.m. were found in the 1 M HCl eluates, respectively. Experimental values were corrected for the small, consistent blank. The chromatographic properties of the radioactive product on Dowex 1 were identical with those of authentic IMP.

RESULTS

Isolation and Properties of Mutants—Cells resistant to DP were selected simply by plating 3×10^5 or more parent cells in growth medium containing the analogue (0.15 to 0.3 μmole per ml). The cultures were fed with the same medium after 4 and again after 8 days. After 12 days, colonies of resistant cells were transferred to medium free from DP, and their descendants were cloned and maintained in the absence of the analogue. Cells resistant to MP, on the other hand, were not so readily obtainable. Mutants were selected as previously described (2) by repeated exposure to 8-azaguanine (1 μg per ml) and regrowth in its absence. The mutants thus isolated were resistant to both the selective agent and MP. Cross-resistance with these two purine analogues has been observed previously (see, for example, Brockman *et al.* (8)).

The ease of isolation of the DP-resistant cells derives from the relatively high frequency of the mutation. With the use of the method of Luria and Delbrück (18) and the procedures previously described (3), the rate was calculated to be approximately 2.5×10^{-6} mutation per cell per generation. Resistance to MP appears to occur with insufficient frequency to be measurable with the methods now available.

The behavior of the parent and mutant clonal isolates in the presence of various concentrations of the purine analogues is shown in Table I. As shown in the table, resistance to one of the analogues was not accompanied by resistance to the other.

Estimation of AMP and IMP Pyrophosphorylases—Estimation of the pyrophosphorylases in cell-free extracts of the parent and mutant cells revealed that the parent cells possess both enzymes, the DP-resistant cells have only the IMP pyrophosphorylase, and the MP-resistant cells have only the AMP enzyme (Table II). The table also shows that mixtures of the parent extract with the deficient mutant ones were active, indicating that stable inhibitors are not responsible for the inactivity of the mutant preparations. The apparent depressed activities of the parent extract in the mixtures probably resulted from the phosphatases contributed by the mutant extracts.

Incorporation of C^{14}-Labeled Purines—The inactivity of their cell-free preparations does not, of course, prove that the mutants are enzymatically deficient. To obtain additional information, the ability of growing cells to take up C^{14}-adenine and hypoxanthine was studied (Table III). As shown in the table, consistent with the enzyme data, although parent cells became labeled with both purines, only hypoxanthine labeled the DP-resistant cells and only adenine the MP-resistant ones.

Effect of Adenine, Hypoxanthine, and Derivatives in Presence of Aminopterin—Further evidence for the absence of one or the other pyrophosphorylase activity in the mutants was obtained with Aminopterin in an assay procedure similar to the one described by Hakala and Taylor (19). Although the AMK cells

TABLE III

Incorporation of purines by parent and mutant cells

To suspensions of the various cells (12,500 cells per ml) in growth medium was added per ml 0.04 μmole of 8-C^{14}-adenine (2.2 $\times$ 10^5 c.p.m. per μmole) or 8-C^{14}-hypoxanthine (1.1 $\times$ 10^6 c.p.m. per μmole) and 2-ml aliquots were delivered to culture tubes. When 4 to 6 $\times$ 10^5 cells were present per culture, the attached cells were washed three times with 2-ml aliquots of growth medium and twice with serum-free medium. The cells were then detached into 1 ml of serum-free medium, and aliquots were used to estimate growth and incorporated radioactivity.

Purine	Culture		
	Parent	DP^r*	MP^r†
	c.p.m. per 10^5 cells‡		
Adenine..................	1460	25	2120
Hypoxanthine............	1390	1480	0

* DP-resistant cells.

† MP-resistant cells.

‡ When the washed, detached cells were treated with perchloric acid (final concentration 1%), more than 85% of the radioactivity was recovered in the insoluble residue. It would appear, therefore, that most of the radioactivity was present in nucleic acid.

TABLE IV

Effective purines and derivatives in presence of Aminopterin

To suspensions of the various cells (12,500 per ml) in growth medium (containing glycine) were added per ml 0.015 ml of thymidine (5 $\times$ 10^{-4} M) and 0.005 ml of aminopterin (2 $\times$ 10^{-5} M). Two-milliliter aliquots of the suspensions were distributed to culture tubes containing 0.01 ml of the indicated test solutions (0.01 M) and after 5½ days (37°), growth was estimated by counting cells.

Additions	Culture		
	Parent	DP^r*	MP^r†
	cell number (thousands)		
None...........................	5	11	11
Adenine........................	340	13	486
Hypoxanthine..................	385	420	18
Adenosine.....................	376	480	5
Inosine.......................	334	410	8
Adenine + hypoxanthine.......	391	432	502
No Aminopterin...............	412	463	483

* DP-resistant cells.

† MP-resistant cells.

do not ordinarily need the preformed compounds, absolute growth requirements for glycine, thymidine, and a purine can be readily demonstrated in the presence of the folic acid analogue, Aminopterin. Under these conditions, both adenine and hypoxanthine served as a source of purines for the parent cells whereas only hypoxanthine supported the growth of the DP-resistant mutant and only adenine was effective with the MP-resistant one (Table IV). The same results were obtained with higher levels (0.3 μmole per ml) of the purines.

As shown in the table, for the MP-resistant cells, neither adenosine nor inosine could replace adenine.[4] Although these

[4] AMP, IMP, and adenine deoxyriboside and deoxyribotide were likewise active only with the parent and DP-resistant cells.

observations suggested an additional deficiency in the MP-resistant cells (*e.g.* lack of a permease), the inactivity of the ribosides appears to result instead from their rapid conversion to hypoxanthine, a compound which can serve as a purine source for all but the MP-resistant cells. Thus, for example, when 3 $\times$ 10^6 MP-resistant cells were incubated (37°) with adenosine (0.4 μmole, 54,000 c.p.m.) for 1 hour, chromatography with Dowex 50, H$^+$ form, revealed no radioactivity in the adenosine area and 22,930 c.p.m. in the hypoxanthine area. Most of the remaining radioactivity was found in the inosine area (20,500 c.p.m.). After an additional 15 hours of incubation, more than 98% of the radioactivity was recoverable in the hypoxanthine area.

Adenosine deaminase and purine nucleoside phosphorylase could account for the formation of the major products. It would require only initial hydrolysis by a phosphatase to explain the inactivity of the 5'-nucleotides for the MP-resistant cells. Regardless of the enzymes involved, however, these observations suggest that caution should be used in interpreting the results obtained with purine derivatives fed to intact cultured mammalian cells.

DISCUSSION

The relationship between resistance to a purine analogue and a decrease in a nucleotide pyrophosphorylase activity was recognized by Brockman *et al.* (8), who worked with 8-azaguanine and experimental neoplasms in mice. The pyrophosphorylases convert the relatively innocuous free bases of the analogues to their nucleotide derivatives (8, 11, 13, 20), and clearly, it is the latter compounds which are highly toxic (21, 22).

The ability of the mutant AMK cells to grow in the presence of levels of DP or MP lethal to parent cells appears to depend upon the same mechanism. The DP-resistant cells lacking AMP pyrophosphorylase cannot convert adenine and presumably its analogue to their respective 5'-nucleotides, whereas the MP mutant is inactive with hypoxanthine and its analogues.[5] Interestingly, this suggests that pathways of purine nucleotide synthesis involving nucleoside phosphorylase and a phosphokinase do not play a significant role in the AMK cells.

Such a pathway, however, might explain the high residual activity of the azaguanine-resistant neoplastic cells (8) which appeared to synthesize about 25% as much IMP + inosine as the susceptible ones. Other possibilities, on the other hand, include heterogeneous cell populations and partially blocked mutants.

It is noteworthy that the lack of a nucleotide pyrophosphorylase in the mutant cells had no effect upon their growth in a medium containing no detectable purines or derivatives; the doubling time for the parent and mutant cultures was about 22 hours. Mutants resistant to both purine analogues, however, grew at less than half this rate but this may have been related to an additional heritable change.

SUMMARY

Mutants capable of growth in the presence of levels of 2,6-diaminopurine or 6-mercaptopurine lethal for the parent cells have been isolated from cultures of mammalian cells, AMK2-2. The mutant resistant to diaminopurine has no detectable adenosine 5'-phosphate pyrophosphorylase and is incapable of in-

[5] With 8-C^{14}-azaguanine, in 30 hours intact MP-resistant cells incorporated less than 5% as much radioactivity as the parent AMK cells.

corporating adenine, but it is indistinguishable from parent cells with regard to hypoxanthine.   Conversely, the mutant resistant to mercaptopurine has no detectable inosine 5′-phosphate pyrophosphorylase and can utilize adenine but not hypoxanthine as a source of purine.

Acknowledgment—The authors gratefully acknowledge the technical assistance of Miss Marcia Heasley.

REFERENCES

1. VOGT, M., *J. Cellular Comp. Physiol.*, **52**, Suppl. 1, 271 (1958).
2. LIEBERMAN, I., AND OVE, P., *Proc. Natl. Acad. Sci. U. S.*, **45**, 867 (1959).
3. LIEBERMAN, I., AND OVE, P., *Proc. Natl. Acad. Sci. U. S.*, **45**, 872 (1959).
4. ROTHFELS, K. H., AXELRAD, A. A., SIMINOVITCH, L., McCULLOCH, E. C., AND PARKER, R. C., IN R. W. BEGG, *Canadian cancer conference, Vol. 3*, Academic Press, Inc., New York, 1959.
5. SANFORD, K. K., EARLE, W. R., AND LIKELY, G. D., *J. Natl. Cancer Inst.*, **9**, 229 (1948).
6. BROCKMAN, R. W., SPARKS, M. C., AND SIMPSON, M. S., *Biochim. et Biophys. Acta*, **26**, 671 (1957).
7. BROCKMAN, R. W., SPARKS, C., HUTCHISON, D. J., AND SKIPPER, H. E., *Cancer, Research*, **19**, 177 (1959).
8. BROCKMAN, R. W., BENNETT, L. L., JR., SIMPSON, M. S., WILSON, A. R., THOMSON, J. R., AND SKIPPER, H. E., *Cancer Research*, **19**, 856 (1959).
9. KORNBERG, A., LIEBERMAN, I., AND SIMMS, E. S., *J. Biol. Chem.*, **215**, 417 (1955).
10. KORN, E. D., REMY, C. N., WASILEJKO, H. C., AND BUCHANAN, J. M., *J. Biol. Chem.*, **217**, 875 (1955).
11. LUKENS, L. N., AND HERRINGTON, K. A., *Biochim. et Biophys. Acta*, **24**, 432 (1957).
12. ELION, G. B., SINGER, S., AND HITCHINGS, G. H., *J. Biol. Chem.*, **202**, 647 (1953).
13. REMY, C. N., AND SMITH, M. S., *J. Biol. Chem.*, **228**, 325 (1957).
14. HEALY, G. M., FISHER, D. C., AND PARKER, R. C., *Proc. Soc. Exptl. Biol. Med.*, **89**, 71 (1955).
15. LIEBERMAN, I., *J. Biol. Chem.*, **225**, 883 (1957).
16. KORNBERG, A., LIEBERMAN, I., AND SIMMS, E. S., *J. Biol. Chem.*, **215**, 389 (1955).
17. LOWRY, O. H., ROSEBROUGH, N. J., FARR, A. L., AND RANDALL, R. J., *J. Biol. Chem.*, **193**, 265 (1951).
18. LURIA, S. E., AND DELBRÜCK, M., *Genetics*, **28**, 491 (1943).
19. HAKALA, M. T., AND TAYLOR, E., *J. Biol. Chem.*, **234**, 126 (1959).
20. WAY, J. L., AND PARKS, R. E., JR., *J. Biol. Chem.*, **231**, 467 (1958).
21. SALSER, J. S., AND BALIS, M. E., *Federation Proc.*, **18**, 315 (1959).
22. HAKALA, M. T., AND NICHOL, C. A., *J. Biol. Chem.*, **234**, 3224 (1959).

37

Reprinted from *Natl. Acad. Sci. (USA) Proc.* **61:**1306-1312 (1968)

MAMMALIAN CELL GENETICS, II. CHEMICAL INDUCTION OF SPECIFIC LOCUS MUTATIONS IN CHINESE HAMSTER CELLS IN VITRO*

By E. H. Y. Chu and H. V. Malling

BIOLOGY DIVISION, OAK RIDGE NATIONAL LABORATORY, OAK RIDGE, TENNESSEE

Chemical induction of point mutations in mammalian tissue culture cells will make possible a study in these cells of genetic events at the molecular level similar to previous studies with microorganisms. Most reports on chemical induction of mutations in mammals describe studies of chromosome aberrations,[1] dominant lethals,[2, 3] variegated-type position effect,[4] and, just recently, a qualitative method to isolate nutritional mutants in tissue culture cells after treatment with chemical mutagens.[5]

The present report presents quantitative methods for measuring mutation frequency in chemically treated Chinese hamster cells in culture. Spontaneous point mutations that resulted in L-glutamine auxotrophy (gln^-) or 8-azaguanine resistance (azg^r) were observed earlier in cultures of Chinese hamster cells.[6] We have shown that the reverse-mutation frequency of gln^- and both the forward- and reverse-mutation frequencies of azg increased significantly over the spontaneous frequencies after treatment with ethyl methanesulfonate (EMS), methyl methanesulfonate (MMS), or N-methyl-N'-nitro-N-nitrosoguanidine (MNNG). Factors such as inoculum size, expression time, and level of selective agent were found to have a profound influence on the mutation frequency. Some preliminary results of this study have been published elsewhere.[7]

Materials and Methods.—The origin, culture procedures, and properties of the Chinese hamster cell line V79-122D1 have been described previously.[6] The basic medium used in cell maintenance and plating was Eagle's minimal essential medium. After mutagenic treatment of the cells, selective agents were added to the cultures at various times. For recovering reversions from gln^- to gln^+, medium lacking glutamine but containing glutamic acid was used. For recovering the forward mutations from azaguanine sensitivity (azg^s) to resistance (azg^r), 5–30 μg/ml of the drug were added to the medium. For recovering reversions from azg^r to azg^s, the selective "THAG" medium contained thymidine ($10^{-7}\ M$), hypoxanthine ($10^{-5}\ M$), aminopterin ($3.2 \times 10^{-6}\ M$), and glycine ($10^{-4}\ M$). Both the survival and the mutation frequencies were determined on the basis of the frequencies of colonies formed.

Any spontaneous mutations that might have accumulated were removed by subjecting the cell population to appropriate selection media before the induction experiments. To remove the spontaneous gln^+ revertants, the same "thymineless death" method originally used for selecting gln^- mutants was used.[8, 6] To remove the spontaneous azg^r mutants, the azaguanine-sensitive, wild-type population was maintained for 6–7 days in THAG medium before being returned to the standard medium.

The general experimental design used in mutagen treatment and in assaying mutants is depicted in Figure 1. The three chemical mutagens initially used were chosen because of their potent mutagenicity and their use in parallel experiments with *Neurospora*[9, 10] and mice[8] in this laboratory. EMS and MMS were obtained from Eastman Organic Chemicals, Rochester, New York; MNNG was from Koch-Light Laboratories, Colnbrook, Great Britain. EMS and MNNG were used directly, and 0.2 M of MMS in absolute methanol was prepared. Appropriate further dilutions were then made in

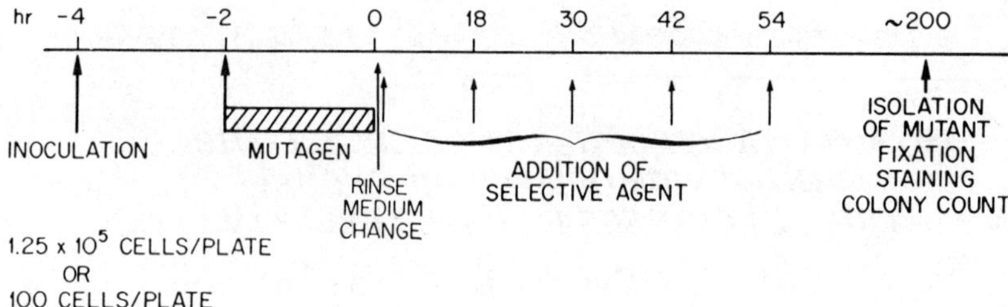

Fig. 1.—Experimental scheme for chemical induction of mutations in Chinese hamster cell cultures.

Hanks' balanced saline (HBSS) and applied to cells within an hour of preparation. All operations were conducted under strict light control at wavelengths longer than 700 mμ.

Plastic Petri dishes (60 or 100 mm, from Falcon Plastics, Los Angeles, California) were used for all plating experiments. Cells became attached to the surface of the plate usually within an hour after inoculation. After being exposed to a mutagen, the cells were rinsed twice with HBSS and transferred to the growth medium.

Results and Discussion.—Table 1 summarizes the results of inactivation experiments on V79 cells by the three mutagens. The two-hour treatment was chosen because of its facility of operation and the short half life of certain chemicals in dilute solution. Increased time of treatment at a given concentration reduced cell survival.

The spontaneous mutation frequencies of gln^- to gln^+ and azg^s to azg^r varied from experiment to experiment and increased with time between purification (for removal of pre-existing spontaneous mutations) and the mutation experiment.

To date, evidence has been obtained in the hamster cells for a significant increase in forward-mutation frequency from azg^s to azg^r and in reverse-mutation frequencies from gln^- to gln^+ and from azg^r to azg^s with EMS, MMS, and MNNG. It has been found that the recovery of the mutants after chemical treatment depends on the conditions of the experiment. Therefore, for quantitative studies it is necessary to define these conditions in order to maximize this recovery. The

TABLE 1. *Cytotoxicity of three chemical mutagens to Chinese hamster cells in culture.*

Mutagen treatment*		No. of cells plated	No. of Colonies per Plate		Per Cent Survival	
			Expt. A	Expt. B	Expt. A	Expt. B
None		100	86	117	100	100
		200		190		100
EMS	$10^{-3}\,M$	100	85	142	99	121
	$5 \times 10^{-3}\,M$	100	75	105	87	89
		200		195		103
	$10^{-2}\,M$	100	65	47	47	40
		200		103		54
MMS	$1 \times 10^{-3}\,M$	100	7	14	6	14
	$2 \times 10^{-3}\,M$	100	0		0	
MNNG	$5 \times 10^{-6}\,M$	100		43		37
	$10^{-5}\,M$	100	58		67	
	$5 \times 10^{-5}\,M$	100	1	0	1	0
	$10^{-4}\,M$	100	0		0	

* Two hours at 37°C, followed by two rinses with Hanks' balanced saline and medium change.

following factors have been shown to influence the mutation frequency after treatment of these cells with chemical mutagens.

Inoculum size: Harris[11] has reported that the spontaneous mutation frequency to puromycin resistance in a line of pig kidney cells increases with the population density in culture. He attributed this phenomenon to the "feeder-layer effect," which *enhanced* the growth and recovery of the newly emerged mutants. Since mutagens at the concentrations used in the present study appreciably reduced the cell survival (e.g., 30–40 per cent killed), thus creating a disparity in population density between the treated and untreated cells, the relation between the size of the cell inoculum and the induced mutation frequency was investigated. The results of one experiment are shown in Table 2. Since

TABLE 2. *The relationship between inoculum size and the frequency of mutation (from azaguanine sensitivity to resistance) in Chinese hamster cells in culture.*

No. of cells per plate	Mutagen treatment	Mutation frequency per 10^5 survivors	Induced: spontaneous ratio
2.5×10^5	None	2.6	4.3
	EMS	11.3	
5×10^5	None	1.2	3.6
	EMS	4.2	
10×10^5	None	0.3	1.3
	EMS	0.4	

Expt. 333, V79, EMS: $10^{-2} M$, 2 hr; survival = 69%. Azaguanine: 9 μg/ml. Expression time: 18 hr.

a maximum plating efficiency of 100 per cent has been achieved for this line of hamster cells,[6] crowded conditions in culture apparently did not provide the feeder-layer effect; instead, crowding *impaired* the recovery of mutations. This result was confirmed and extended to lower cell numbers in another experiment. It was then decided that an inoculum size at 1.25×10^5 cells per 100-mm plate would be suitable, since further reduction in cell inoculum would necessarily increase the number of plates per experimental series and thus reduce efficiency.

Mutation expression time: The next problem was to decide when the selective agent should be added. The "time of mutation expression" refers here to the interval between quenching the mutagen (rinsing with HBSS and medium change) and adding the selective agent.

At various times after removal of mutagen, the number of cells per colony was determined by direct microscopic observation of living cultures or by counting cells in fixed preparations. Fifty colonies per time period, in either treated or untreated plates, were counted. The results in three experiments were in full agreement; data from one such experiment on the number of cells per colony at various times are shown in Figure 2. Only the results in the untreated controls are illustrated, since they did not differ significantly from those in the mutagen-treated series.

The average time for cell doubling in this cell line is shown to be approximately 12 hours, which agrees with previous determinations arrived at through auto-radiographic techniques under our laboratory conditions (Chu, unpublished data), as well as with the results of Sinclair.[12] At 18 hours after mutagen re-

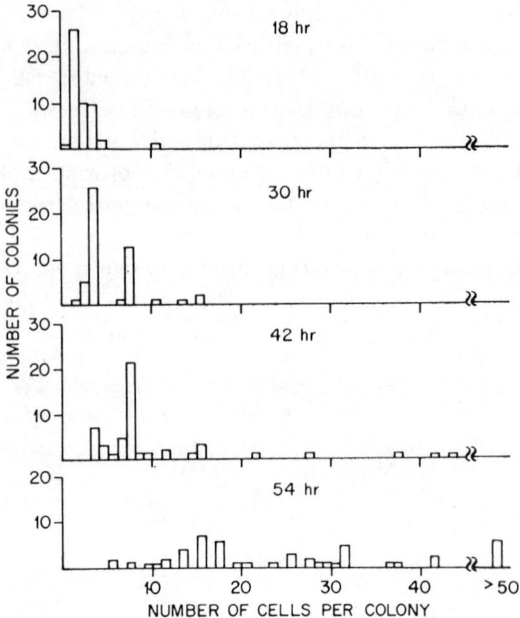

FIG. 2.—Number distribution of colonies, each consisting of varying numbers of hamster cells after different periods of growth at 37°C. All were untreated except sham operations of a rinsing with balanced saline and medium change at time zero (expt. 326).

moval, most cells divided once, and a small proportion of them divided twice. This initial mitotic delay observed in both treated and untreated cultures could have been the result of handling.

In three experiments, when the selective agent, azaguanine in this case, was added at the zero time immediately after mutagen quenching, we found no increase in mutation frequency in the treated series, in comparison to the control. However, when the addition of the selective agent was delayed, thus permitting cells to grow and divide, the mutation frequency seemed to increase with time up to 42 hours. The relationship between the time of mutation expression and mutation frequency observed in one of the two repeat experiments is shown in Table 3 and graphically depicted in Figure 3. It is clear that, under experimental conditions, a maximum number of mutations could be recovered 42 hours after the removal of the mutagen—a time interval permitting approximately three cell divisions to occur. At 54 hours, a decline in the frequencies of both

TABLE 3. *The relationship between time of mutation expression and mutation frequency in Chinese hamster cells in culture.*

Expression time (hr)	Mutagen treatment	Plating efficiency	Per cent survival	Mutation frequency per 10^5 survivors
18	None	143	100	36.6
	EMS	106	74	81.2
30	EMS	117	82	130.1
42	None	132	92	17.4
	EMS	112	78	170.8
54	None	138	97	15.3
	EMS	110	77	120.9

Expt. 345, V79*, 1.25 × 10^6 cells/plate. EMS: 10^{-2} M, 2 hr. Azaguanine: 30 μg/ml.

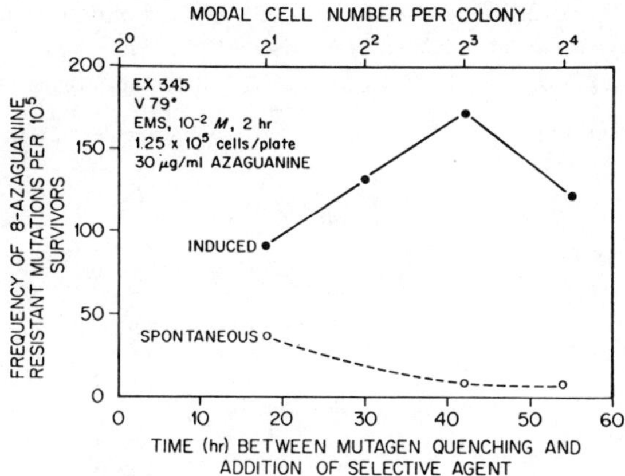

FIG. 3.—Time of mutation expression in relation to mutation frequency in cultured hamster cells.

spontaneous and induced mutations occurred. This was probably due to the crowding effect, as discussed earlier.

Biochemical evidence indicates that alkylating agents such as MMS and EMS react with the four bases of DNA in the following decreasing order: guanine (G) > adenine (A) > cytosine (C).[13] The predominant mutagenic effect of alkylation is probably due to base-pair transition from G:C to A:T.[14] The pairing-error theory[14] requires that, before the induced mutant phenotype can appear in the population, at least two rounds of DNA replication occur after alkylation. Although we have no chemical data on alkylation of DNA in these cells, our data on time of mutation expression seem to be consistent with this requirement.

Another possible explanation for the delayed expression of the mutant phenotype is the presence in the cytoplasm of the mutated cell or its immediate progeny of sufficient quantities of the wild-type gene messenger or products. Several cell divisions might be required before such a messenger or products became diluted or nonfunctional.

Concentration of selective agent: Table 4 summarizes the result of an experiment in which the influence of the concentration of azaguanine on the frequency of forward mutation from azg^s to azg^r was studied. We used a newly isolated clonal line, V79-4, which showed 100 per cent plating efficiency and which was azaguanine-sensitive. The optimal experimental conditions for recovery of mutations were provided to obtain a maximum frequency of mutations. It can be seen in this table that any one of the three chemical mutagens led to a highly significant increase in mutation (azg^s to azg^r) frequency, in comparison to the untreated control. Since the survival in this experiment was 52.5 per cent in EMS-treated and 41.1 per cent in MNNG-treated cells, these increases in the mutation frequency of the treated populations over the untreated controls could not have been due to selective killing of the wild-type parental cells, but rather to experimental induction at this locus. Although the cell survival in MMS-treated series in this particular experiment was low, the highly significant in-

TABLE 4. *Frequency of spontaneous and chemically induced mutations in Chinese hamster cells resistant to various levels of 8-azaguanine.*

Mutagen treatment	Concentration of 8-azaguanine (μg/ml)	Mutation Frequency per 10^5 Survivors Definitive colonies	All colonies
EMS 10^{-2} M, 2 hr ($S = 52.5\%_C$)	5	150.4	198.7
	10	89.6	114.4
	30	27.3	34.4
MMS 10^{-3} M, 2 hr ($S = 5.8\%$)	5	157.2	657.9
	10	24.8	53.8
	30	0.8	13.8
MNNG 10^{-5} M, 2 hr ($S = 41.1\%$)	5	181.4	256.4
	10	90.9	132.9
	30	36.6	58.4
None ($S = 100\%$)	5	2.6	14.1
	10	2.2	7.0
	30	1.9	1.9

Expt. 382, V79-4, 1.25 × 10^5 cells/plate. Expression time: 42 hr. S = survival.

crease in the frequency of the induced mutation clearly renders improbable the possibility of selective killing.

At the roughly comparable levels of cell survival and mutation yield and on an equimolar basis, MNNG is the most efficient one of the three mutagens used in mutation induction. MMS is more cytotoxic but also more mutagenic than EMS.

This experiment also shows that, at different concentrations of the selective agent 8-azaguanine, significantly different numbers of mutations might be recovered. Figure 4 illustrates the number and morphology of colonies in the EMS-treated and untreated cultures. It can be seen that the colony size varies, thus reflecting the rate of cell multiplication during the same period of incubation. The number of large, definitive colonies and that of all colonies are listed in Table 4. Some of the EMS-induced mutants from a similar experiment were isolated and further tested. Only two of the nine isolates found resistant to 5 μg/ml of azaguanine were likewise resistant to 30 μg/ml. A preliminary test on C^{14}-hypoxanthine incorporation[6] into each of these nine resistant mutants indicated that the incorporated radioactivity per cell was variable. This last finding suggests that independent forward-mutational events had occurred, each presumably representing a different isoallele with a different level of enzymatic activity controlled by the *azg* locus. Further work on the possible induction of "leaky" mutations by chemicals and on mutagenic specificity is in progress.

Summary.—Quantitative measurements of the frequencies of gene mutations induced in mammalian somatic cells in culture have not been previously done. We have isolated from populations of Chinese hamster cells mutant clones which are either auxotrophic for L-glutamine (*gln*$^-$) or resistant to 8-azaguanine (*azg*^{r}). The present experiments clearly demonstrate that ethyl methanesulfonate (EMS), methyl methanesulfonate (MMS), and N-methyl-N'-nitro-N-nitrosoguanidine (MNNG) significantly increase the mutation frequency at both loci. Furthermore, we have found that the following factors affect the mutation yield:

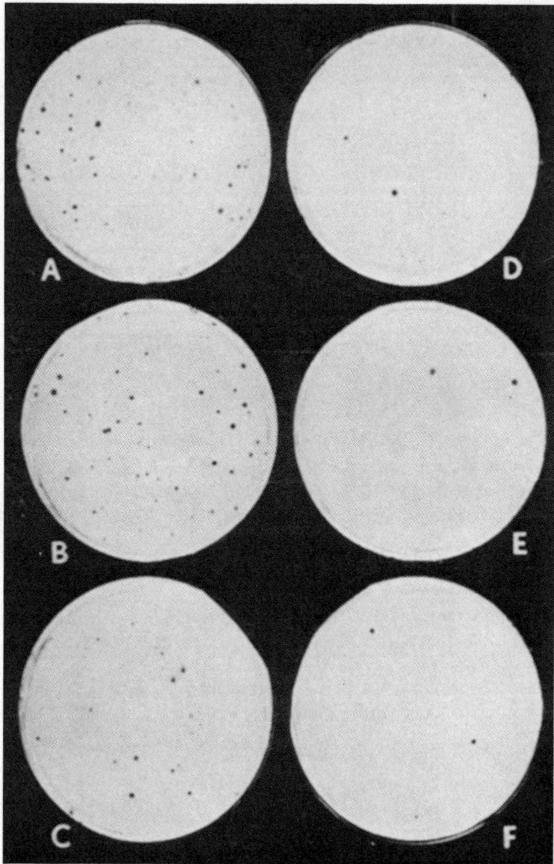

Fig. 4.—Surviving mutant colonies of hamster cells, either spontaneous or induced, in the presence of various concentrations of 8-azaguanine. To each plate, 1.25×10^5 drug-sensitive V79-4 cells were inoculated. All plates were incubated for 8 days. (*A–C*) Ethyl methanesulfonate-treated. (*D–F*) Untreated controls. Concentrations of 8-azaguanine (μg/ml) are as follows: 5 in *A* and *D*, 10 in *B* and *E*, 30 in *C* and *F* (expt. 382).

(1) inoculum size, (2) time period between mutagen quenching and the addition of a selective agent (mutation expression time), and (3) concentration of the selective agent.

* This research was sponsored jointly by the National Cancer Institute and by the U.S. Atomic Energy Commission under contract with Union Carbide Corporation. The skillful technical assistance of Miss Patricia Brimer is gratefully acknowledged.

[1] Ostertag, W., *Akad. Wiss. Lit. (Mainz), Abhandl. Math. Nat. Kl.* (1966), p. 1.

[2] Bateman, A. J., *Genet. Res.*, 1, 381 (1960).

[3] Ehling, U. H., R. B. Cumming, and H. V. Malling, *Mutation Res.*, 5, 417 (1968).

[4] Cattanach, B. M., *Z. Vererbungslehre*, 92, 165 (1961).

[5] Kao, F. -T., and T. T. Puck, *these* PROCEEDINGS, 60, 1275 (1968).

[6] Chu, E. H. Y., P. Brimer, K. B. Jacobson, and E. V. Merriam, submitted for publication.

[7] Chu, E. H. Y., and H. V. Malling, in *Proceedings of the Twelfth International Congress of Genetics* (Tokyo: Science Council of Japan, 1968), vol. 1, p. 102.

[8] De Mars, R., and J. L. Hopper, *J. Exptl. Med.*, 111, 559 (1960).

[9] Malling, H. V., and F. J. de Serres, *Mutation Res.*, 6, 181 (1968).

[10] Malling, H. V., and F. J. de Serres, *Genetics*, in press.

[11] Harris, M., *J. Natl. Cancer Inst.*, 38, 185 (1967).

[12] Sinclair, W. K., *Radiation Res.*, 21, 584 (1964).

[13] Lawley, P. D., and P. Brookes, *Exptl. Cell Res.*, Suppl., 9, 512 (1963).

[14] Freese, E., *Molecular Genetics*, ed. J. H. Taylor (New York and London: Academic Press, 1963), part I, p. 207.

Editor's Comments
on Papers 38 Through 42

38 PUCK and FISHER
Genetics of Somatic Mammalian Cells   I. Demonstration of the Existence of Mutants with Different Growth Requirements in a Human Cancer Cell Strain (HeLa)

39 DE MARS and HOOPER
A Method of Selecting for Auxotrophic Mutants of HeLa Cells

40 PUCK and KAO
Genetics of Somatic Mammalian Cells, V. Treatment with 5-Bromodeoxyuridine and Visible Light for Isolation of Nutritionally Deficient Mutants

41 NAHA
Temperature Sensitive Conditional Mutants of Monkey Kidney Cells

42 THOMPSON et al.
Isolation of Temperature-Sensitive Mutants of L-Cells

CONDITIONAL LETHAL MUTANTS

Genetic studies with microorganisms profited greatly from the ability to isolate conditional lethal mutants, for example, nutritional mutants and temperature sensitive mutants. Selection for conditional lethal mutants provides large numbers of mutants of different types, and their analysis has been useful in many biochemical and genetic studies. Techniques for isolating nutritional mutants of mammalian cells were introduced over two decades ago, at approximately the same time as the first drug resistant cells were being selected in culture, but techniques for isolating temperature sensitive mutants were developed only much later.

The analysis of mutant cells depends on the ability to isolate genetically pure lines derived from single cells. Studies on mammalian cell mutants were greatly facilitated by a major advance in cell culture methodology in the mid-1950s, namely, the establishment by Puck and co-workers (1956) of techniques for single cell plating (cloning) of cultured mammalian cells. These new techniques were applied immediately by Puck and Fisher (Paper 38) to isolate cellular variants with different nutritional requirements. The human cervical carcinoma line HeLa was cloned and the growth requirements of two clones were analyzed. The two clones were found to have very different serum requirements, and the different requirements appeared to be stable, heritable characteristics. These studies demonstrated that nutritional variants exist in cultured cell populations and that they can be isolated as stable clones. The variants in this study were detected by screening randomly isolated clones, a procedure that obviously would be unfeasible for rare variants. Soon after these studies, it was demonstrated by Haff and Swim (1957) that rare variants with altered nutritional requirements could be isolated by selection with medium deficient in various components. Because of the nature of the media in this study, the specific deficiencies in the selective medium that the variant cells were able to overcome were not defined.

The first system for selecting auxotrophic mutants of mammalian cells was described by DeMars and Hooper in 1960 (Paper 39). The selective system was based on the results of studies with bacteria. It had been shown that thymine-requiring bacteria die in medium lacking thymine but containing all other necessary growth factors. However, when another essential nutrient was deleted from the thymine-deficient medium, the bacteria remained viable. The viability of the bacteria under such conditions provided the basis for the experiments with mammalian cells. The folic acid antagonist, aminopterin, was used to create a thymidine deficiency in the cells (see Paper 34). By analogy with the bacterial system, cells in the presence of aminopterin should remain viable in thymidine-free medium if another essential nutrient is missing. Thymidine-starved cells not requiring the other missing nutrient should be killed whereas cells requiring the other missing nutrient should remain viable. Thus, it should be possible to specifically select a minority cell type that requires the other missing nutrient.

In these studies, two lines of HeLa cells were used. One could grow when glutamine in the medium was replaced with glutamic acid, while the other line had an absolute requirement for glutamine. When the glutamine-requiring cells were incubated in medium containing aminopterin and glutamic acid, they remained viable, whereas the cells able to grow on glutamic acid died in that medium. Reconstruction experiments proved that this method could be used to select glutamine auxotrophs from an excess of the other cell type in a mixed population. These results demonstrated the feasibility of selecting auxotrophic mutants of mammalian cells. Furthermore, this approach, based on the establishment of conditions in which wild type cells are killed while mutant cells become dormant but remain viable, has been used to develop other selective systems for auxotrophic mutants as well as selective systems for temperature sensitive mutants.

Most of the auxotrophic mutants studied to date have been isolated with the *bromodeoxyuridine-light* technique developed by Puck and Kao (Paper 40). This technique, representing a modification of the approach just described, combined the use of a deficient medium in which auxotrophs could not grow plus an agent (5-bromodeoxyuridine, BUDR) to kill the wild type cells that continued to divide. These experiments involved two Chinese hamster cell lines, one of which required proline for growth (although it had not been selected for proline auxotrophy). When the wild type cells were cultured in proline-deficient medium that contained BUDR, the cells grew, incorporated BUDR into DNA, and were killed by subsequent irradiation with visible light. (It was known that bacteriophage with BUDR in their DNA become photosensitive and can be killed by light.) In contrast, when the proline auxotrophs were cultured in proline-deficient medium containing BUDR, the cells could not grow, did not incorporate BUDR into DNA, and thus were not killed when irradiated with visible light. Reconstruction experiments demonstrated that nutritionally deficient cells could be enriched by several thousand-fold in a single cycle of selection with this method. A wide variety of auxotrophic mutants have since been selected with this technique.

Temperature sensitive (TS) mutants represent another useful type of conditional lethal mutant. The selection of TS mutants offers the opportunity of isolating especially large numbers of mutants, because selection is not applied for any specific alteration (other than the inability to grow at elevated temperatures). TS mutants of mammalian cells were first isolated by Naha in 1969 (Paper 41), using the same principles as just discussed for selecting auxotrophic mutants. Mouse cells were cultured at 39°C (the "nonpermissive" temperature for TS mutants) in the presence of BUDR, and the cells were irradiated with visible light (see Paper 40). This protocol was designed to kill wild type cells able to grow and incorporate BUDR into DNA at 39°C, whereas TS mutants would not grow at that temperature, would not incorporate BUDR into DNA, and should remain viable.

Following irradiation to kill the cells that had incorporated BUDR at 39°C, the cells were shifted to 35°C, surviving clones were isolated, and these clones were tested for the ability to grow at 39°C. This procedure led to the isolation of several TS mutants that were unable to grow when shifted to elevated temperatures. Similar protocols, differing primarily in the agents utilized to kill wild type cells at the nonpermissive temperature, have been used to isolate many different mammalian cell TS mutants.

The first characterization of mammalian TS mutants was provided by Siminovitch and co-workers (Paper 42). In these studies, radioactive thymidine and cytosine arabinoside were used to kill wild type cells at the nonpermissive temperature. Several TS mutants were isolated, and one was studied in detail. In this TS mutant, the rates of RNA and protein synthesis appeared to be relatively unaffected by a shift to the nonpermissive temperature. In contrast, DNA synthesis in the mutant dropped very rapidly at the nonpermissive temperature, decreasing approximately 1000-fold within twenty-four hours after an increase in temperature. Thus, this TS mutant appeared to have an alteration in some function necessary for DNA synthesis. Since these early studies, TS mutants have been used to study a wide variety of processes involved in the growth of mammalian cells in culture.

REFERENCES

Haff, R. F., and H. E. Swim, 1957, Isolation of a Nutritional Variant from a Culture of Rabbit Fibroblasts, *Science* **125:**1294.
Puck, T. T., P. I. Marcus, and S. J. Cieciura, 1956, Clonal Growth of Mammalian Cells In Vitro. Growth Characteristics of Colonies from Single HeLa Cells With and Without a "Feeder" Layer, *J. Exp. Med.* **103:**273–284.

38

Reprinted from *J. Exp. Med.* **104**:427–434, Plate 31 (1956)

GENETICS OF SOMATIC MAMMALIAN CELLS

I. Demonstration of the Existence of Mutants with Different Growth Requirements in a Human Cancer Cell Strain (HeLa)*

By T. T. PUCK, Ph.D., and H. W. FISHER

(*From the Department of Biophysics, Florence R. Sabin Laboratories, University of Colorado Medical Center, Denver*)

PLATE 31

(Received for publication, May 28, 1956)

In previous communications (1, 2) methods have been described for growth of single mammalian cells into colonies of any desired size. These methods are simple and rapid, and have been applied to cells originating from both cancerous and normal tissues (2, 3). Plating efficiencies (*i.e.* the proportion of the single cells plated which yield macroscopic colonies) close to 100 per cent have been readily obtained with a variety of human cells. These techniques make possible study of the mammalian somatic cell as a microorganism and promise to permit bridging of at least some of the gaps between genetic operations hitherto confined to microorganisms on the one hand, and multicellular forms on the other.

Study of such genetic processes requires methods for examination of the hereditary behavior of individual cells, and the isolation of mutants with deviant metabolic characteristics. Such operations are extremely difficult with techniques involving only many celled populations, because of uncertainties regarding the genetic purity of the starting population, and the degree to which a given characteristic may depend on the interaction of two or more cells. Micromanipulative techniques with individual cells become laboriously prohibitive for studies requiring screening of huge numbers of cells in order to isolate rare mutants. When large numbers of single cells can be grown with high efficiency into macroscopic colonies in a single plating operation, the isolation and identification of mutants becomes greatly simplified and analogous to the corresponding procedure in bacteriology (4).

The present paper describes application of the single cell plating procedure to the isolation of mutants in the population constituting the HeLa strain—a human, cervical carcinoma first isolated by Gey and his associates (5) and maintained in tissue culture for a period of several years. The characterization of two stable mutants with different growth requirements is described. In addition, further observations on the operation of a "feeder" system of x-rayed

* Contribution No. 45. This study has been aided by a grant from The Commonwealth Fund, and from The National Foundation for Infantile Paralysis, Inc.

cells on the growth of single cells are presented, and their implications discussed.

Definitions.—The terms *clone* and *plating efficiency* are used as previously defined (2). *Mutation*—any change in heredity not ascribable to sexual segregation and recombination (6). (See Discussion.) *Parental population*—a cell strain arising directly from material obtained by a macroscopic biopsy, with or without subsequent cultivation in tissue culture, but which has not been developed as a clone from a single cell, *in vitro*. *Feeder layer*—a layer of cells x-rayed with at least 2000 r, which exhibits no multiplication itself, but supports growth of an inoculum of single cells deposited over it (2, 7). *Average generation time*—an index of the rate of multiplication of single cells, obtained by dividing the incubation time in hours by the number of generations represented by the average number of cells per colony. This calculation ignores the effect of lag periods. But since these are almost always less than 24 hours in our experience (2) and the incubation periods employed are never less than 9 days, we use this approximate figure instead of the actual generation time which requires construction of a complete growth curve, except in experiments demanding maximal preciseness.

Methods and Materials

The methods employed are described in detail in previous publications (2). In all the experiments described here, the medium employed for cell growth contained 3 components: our standard solution of salts, glucose, amino acids, vitamins, and growth factors (2), which was always present in a fixed concentration of 40 per cent; mammalian serum—human, animal, or a mixture of both—present in variable concentration which is always specified; and Hanks's saline, added in quantity to make the total 100 per cent. Since, for all the present experiments, the serum component is the only significant variable, the medium employed in each case will be identified by listing its serum composition. Human serum was obtained from healthy, young, adult donors. Animal serum was obtained from fasted, healthy horses or pigs.

The use of a "feeder" system involves plating the single cells which are to be grown into colonies on a Petri dish already seeded with a layer of x-irradiated cells, which themselves do not multiply. These non-reproducing cells can change the environmental medium so as to permit colony formation by subsequently added viable cells which otherwise might not reproduce (1, 2). Approximately 10^5 HeLa cells are pipetted into a 60 mm. Petri dish in 4.0 cc. of the growth medium to be tested. After a period of incubation which can vary from 2 to 18 hours, the dish is irradiated with 4000 r, a dose sufficient to insure complete suppression of sustained multiplication in the feeder layer (7). The inoculum of cells whose reproductive capacity is under test is then added and the plates are incubated. The medium may require replacement after 6 to 7 days.

EXPERIMENTAL RESULTS

It was earlier shown (2) that when cells of the parental HeLa population are singly plated on top of a feeder layer, the plating efficiency approximates 100 per cent, yielding colonies which are dense and fairly uniform in appearance. In the absence of a feeder system, however, colonial morphologies are more heterogeneous and, although the plating efficiency still approximates 100 per cent, an appreciable fraction of the colonies which develop have quite small cell numbers. It was also demonstrated that when one particular clone had been picked and subcultured, its single cells regularly produced large, densely staining colonies of much greater uniformity than those of the parental population. (*Cf.* Figs. 3 and 4 of Plate 10 in reference (2)). These observations suggested that the parental population consists of individuals with different, hereditable growth characteristics.

From plates containing colonies grown from single cells of the parental population, four different clones were selected and isolated for subculture by means of the steel cylinder technique described earlier (2). This report compares the properties of two of these clonal stocks, respectively designated S1 and S3, which were isolated from the parental Hela cell population.

Growth in Different Serum Concentrations without a Feeder System.—

Both S1 and S3 cell strains have the same typical HeLa cell morphology, and both readily display the ability to grow either in tightly packed, columnar form, or as loose, migratory, stretched cells, depending on whether the serum component is equine, bovine, or porcine, on the one hand, or human on the other (2).

Studies of the plating efficiency and average growth rate of the two cell types in various serum concentrations were carried out. Such experiments revealed that S1 has a much lower plating efficiency than S3 in any concentration of human serum employed. At concentrations of human serum between 5 and 10 per cent, the difference in behavior of the two clonal strains becomes maximal; *i.e.*, the plating efficiency of S3 is 100 per cent, while that of S1 is practically zero. These relationships are illustrated in Text-Fig. 1.

Even those colonies of S1 which do form in the presence of 20 to 30 per cent human serum contain many fewer cells than S3 grown under identical conditions. Thus, the average number of cells per colony formed after 9 days' incubation of S1 in 30 per cent human serum was 96, and in 20 per cent serum the value was only 60 cells per colony. In contrast, S3 colonies averaged 1300 cells in 20 or 30 per cent serum. Even in 5 per cent serum, in which S1 does not grow at all, S3 colonies contained an average of 300 cells after 9 days' growth, a value equivalent to an average generation time of 26 hours, and only slightly longer than the optimal reproductive time for S3. Figs. 1 a to 1 d illustrate typical platings of each cell type in two different serum concentrations.

When an animal serum like equine or porcine is used instead of human serum the general pattern of events is the same, except that the growth of S1 tends to be somewhat better in high concentrations of animal serum than it is in human. Thus, in 20 per cent horse serum S1 may occasionally exhibit a plating efficiency as high as 60 per cent. However, the number of cells per colony still tends to be low. Moreover, as the animal serum concentration is diminished, the difference in behavior of S3 and S1 becomes just as great as with human serum. In general, concentrations in the neighborhood of 2 to 6 per cent of horse or porcine serum usually act like human serum in producing 100 and 0 per cent plating efficiencies of S3 and S1, respectively. In many experiments, the serum component employed was a mixture of 2 parts human to 1 part horse serum, which regularly gives clean separation of the two strains.

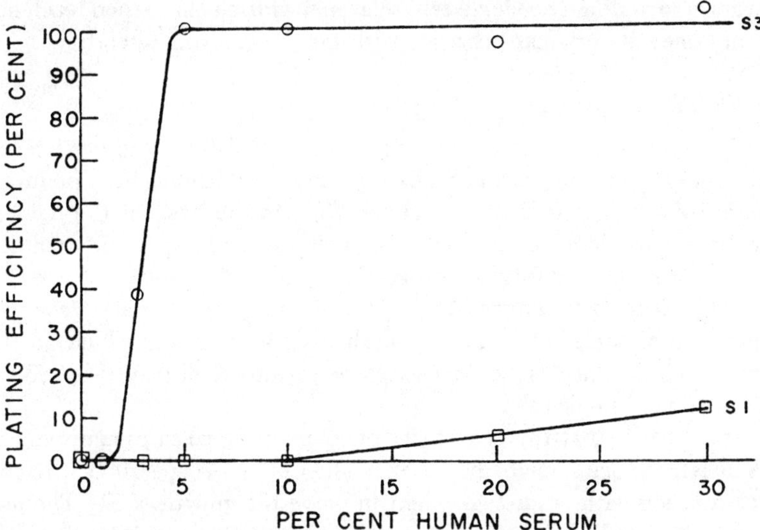

TEXT-FIG. 1. Typical experiment demonstrating the change of plating efficiency of the S1 and S3 clones with change in concentration of human serum in a medium constant in the composition of other components. The range of serum concentration at which maximum difference in behavior of the 2 strains occurs varies somewhat with different sera, but usually encompasses the interval of 2 to 10 per cent. Horse or porcine sera behave similarly, though with a slightly narrower concentration range for optimal differentiation. A mixture of 2 parts human to 1 part horse serum can also be employed to differentiate the two strains, with excellent results.

Stability of the Genetic Trait.—

Experiments were performed to test whether the differing growth response to reduced serum concentration of these two clonal strains is a trait sufficiently stable to serve as a useful marker for genetic studies. Both cell strains have been grown as routine in bottles, using massive inocula for transfer during the 5 months after their first isolation from single cells. During this time no change in any characteristic of either population had been observed, and the differential action of low serum concentrations had been established. As a more rigorous test of the stability of the genetic trait under study further single cell isolations were performed. Two new clonal isolations were made from S1 and one from S3. These were subcultured, regrown, and tested, and found in every case to duplicate the behavior of the parental clonal stock from which they had originated. When a single cell is grown up in this way to a population of 10^7, approximately 23 generations have elapsed, so that the test of genetic stability is equivalent to a large number of serial transfers of large inocula in bottles. Each of these cell strains has been in continuous cultivation for 13 months in this laboratory and has passed through a total of approximately 100 successive generations, including two single cell isolations. During this period, each strain has maintained its original behavior with no recognizable alterations.

Effect of a Feeder System.—

The foregoing experiments pose a problem in explaining the persistence of the S1 genotype in the parental HeLa population, which has been maintained for years by cultivation in human serum (8). On the basis of the data here presented, one would have expected S3 to overgrow and displace S1 completely in a period of perhaps 100 to 200 generations. Since this has not occurred, and since the difference between the S1 and S3 traits is apparently stable genetically, the operation of some compensating mechanism is suggested which, in tissue culture procedures, neutralizes the selective advantage displayed by S3 when grown as isolated single cells.

It seemed likely that the persistence of S1 in tissue cultures maintained by means of large inocula might be an expression of cell cooperation whereby a close association with other cells might improve the growth of S1. The use of feeder systems readily lends itself to exploration of this possibility by test of the growth of S1 in a deficient medium supplemented with a feeder layer of metabolizing, but non-reproducing, irradiated cells. Experiments were carried out in which S1 was plated with a feeder layer of x-rayed S3 cells, in a medium whose serum composition produces no growth of S1 whatever in the absence of feeders. The presence of the feeder cells raised the plating efficiency of S1 from 0 to almost 100 per cent. In addition, the growth rate achieved was comparable to the optimal exhibited by S3. Thus in a typical experiment in which the total serum concentration was 10 per cent, the average colony size after 10 days' incubation was 500, a value equivalent to an average generation time of 27 hours, and very close to that exhibited by S3 in the same medium. It follows that the presence of the feeder system produced optimal growth of S1 in a medium which yielded no growth in its absence, and so eliminated the selective advantage exhibited by S3 in the absence of feeders. Figs. 2 a and 2 b present photographs demonstrating the growth obtained under these conditions.

DISCUSSION

The foregoing experiments demonstrate that the parental HeLa population, while morphologically uniform, contains a mixture of genetically stable cells differing in their growth requirements. Of the two types studied, only one can grow with 100 per cent plating efficiency in the absence of a feeder system, in any of the media so far employed. These studies demonstrate that the genetic purity of a cell strain cannot be taken for granted on the basis either of its morphologic purity, or of having grown for a period of time, however long, in tissue culture by procedures involving massive inocula.

The availability of alternative growth conditions which result in plating efficiencies for the deficient strain in the neighborhood of 100 or 0 per cent, respectively, makes possible quantitative determination of each cell type in their mixtures. Experiments using this technique are in progress to determine whether high energy irradiation can effect mutation of these two types of growth response.

The present experiments give no information on the mode of origin of the differences between S1 and S3. All the cells from a given mammalian individual arose initially from the same fertilized egg, and so, at least in a formal sense, clonal cell lines of divergent genetic constitution isolated from a mammalian individual directly or indirectly, may be called mutants. Such a designation contains no implications concerning the mechanism by which such changes have arisen, their frequency, degree of reversibility, or randomness. The notation is convenient and in accord with the operational definition current in microbiological genetics. Once such a mutation has been demonstrated, the question must next be attacked as to whether the observed change involves nuclear or other genetic determinants; whether viruses, transforming principles, or related agents may be operative; or whether some complex mechanism which chooses for expression one of a variety of possible latent hereditary characters, as has been demonstrated in *Paramecium* (9), may be involved.

The presence in a cell population of diverse mutants, one or more of which may find a selective advantage as a result of a changed metabolic situation, and so outgrow other cells, has been invoked to explain changes in cell characteristics related to invasiveness (10). If such mutant cells can be readily characterized, methods like those described here should prove valuable in cellular analysis of the population dynamics of such systems. Single cell plating should permit identification of situations in which large numbers of cells are suddenly changed in their behavior, as opposed to gradual change due to shift in the proportion of mutants present in an ininitially heterogenous population. It would be of interest to study by these methods situations demonstrating gradual progression to the carcinomatous state (11).

The cooperative action between S1 and S3 mutants demonstrated by the present experiments with a feeder system raises many possibilities for exploring cell-cell interaction between similar, as well as different, cell types. These results establish that a feeder system can overcome a genetically controlled deficiency for growth in a particular medium. Further studies with this technique are under way, testing the abilities of different feeder systems to influence various cell operations. This approach may permit study of a variety of phenomena in addition to multiplication, wherein one cell supplies another with the conditions necessary for particular functions. Experiments are also planned to identify the serum constituents which are capable of differentiating S1 from S3 and which become unnecessary in the presence of a feeder system.

SUMMARY

The parental HeLa cell population, a morphologically uniform, human cancer cell strain, grown for several years in tissue culture by procedures always involving massive inocula, has been shown to contain different mutant cell types.

Two clonal lines have been isolated and established as reliable stock cultures. Both strains exhibit 100 per cent plating efficiency in high or low serum concentrations in the presence of a feeder system. In the absence of a feeder system and in low serum concentrations, the two strains are quantitatively differentiable: S3 still exhibits 100 per cent plating efficiency, while that of S1 lies in the neighborhood of zero.

These differences have remained stable throughout 100 successive generations of growth of each strain including 2 single cell isolations.

Application of these techniques to studies in the genetics of mammalian somatic cells and to specific cell-cell interactions has been indicated.

BIBLIOGRAPHY

1. Puck, T. T. and Marcus, P. I., *Proc. Nat. Acad. Sc.*, 1955, **41**, 432.
2. Puck, T. T., Marcus, P. I., and Cieciura, S. J., *J. Exp. Med.*, 1956, **103**, 273.
3. Marcus, P. I., Cieciura, S. J., and Puck, T. T., J. Exp. Med., in press.
4. Lederburg, J., Microbial Genetics, Madison, University of Wisconsin Press, 1951.
5. Gey, G. O., Coffman, W. D., and Kubicek, M. T., *Cancer Research*, 1952, **12**, 264.
6. Darlington, C. D. and Mather, K., The Elements of Genetics, New York, The Macmillan Co., 1950, 405.
7. Puck, T. T., and Marcus, P. I., *J. Exp. Med.*, 1956, **103**, 653.
8. Scherer, W. F., Syverton, J. T., and Gey, G. O., *J. Exp. Med.*, 1953, **97**, 695.
9. Sonneborn, T. M., *Harvey Lecture*, 1950, **44**, 145.
10. Klein, E., *Exp. Cell Research* 1955, **8**, 188.
11. Rous, P. and Beard, J. W., *J. Exp. Med.*, 1935, **62**, 523.

EXPLANATION OF PLATE 31

FIGS. 1 a to 1 b. Differential growth of S3 and S1 in different serum concentrations. 200 cells were inoculated on each plate as follows:—

Fig. 1 a. S3 in 15.0 per cent serum; Fig. 1 b. S1 in 15.0 per cent serum; Fig. 1 c. S3 in 2.5 per cent serum; Fig. 1 d. S1 in 2.5 per cent serum.

The lower serum concentration produced clearest differentiation between the two strains. The serum used in these experiments consisted of a mixture of 2 parts human to 1 part horse. All photographs are actual size.

FIGS. 2 a and 2 b. Demonstration that a feeder system raises the plating efficiency and growth rate of S1 to values approximating that of S3. Each plate received an inoculum of 100 S1 cells, and was incubated for 10 days.

Fig. 2 a. 3 per cent serum, no feeders. No growth has occurred in 10 days. (Actual size.) Fig. 2 b. 3 per cent serum plus a feeder system. The colonies are clearly seen as the densely staining areas, which stand out against the speckled grey background of giant feeder cells (7). The plating efficiency on such plates averaged 84 per cent. Note that the colony size is similar to that of the S3 colonies shown in *c* of Fig. 1, which were grown in approximately the same serum concentration for almost the same period of time (9 days as compared with 10). (Actual size.)

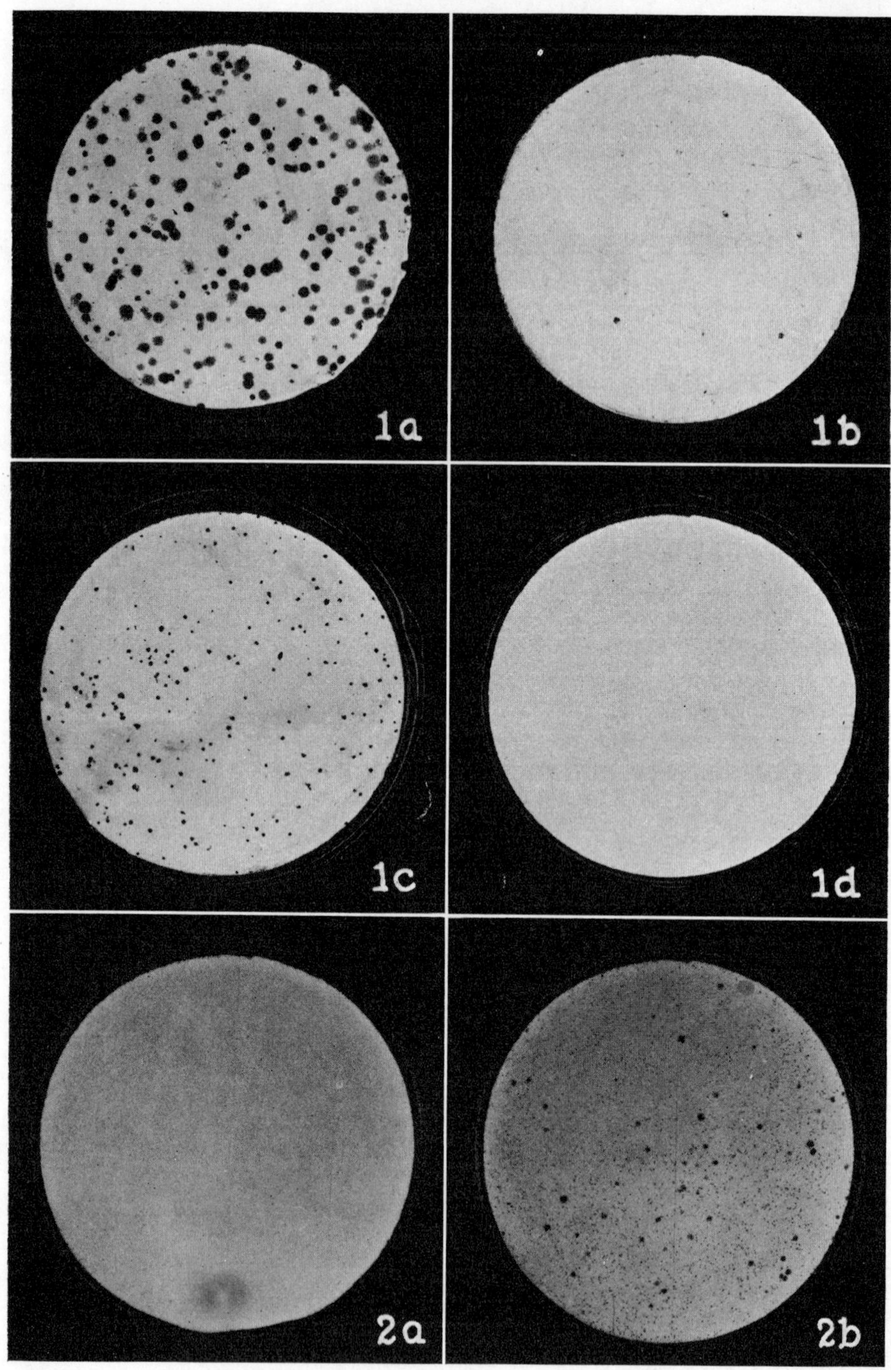

A METHOD OF SELECTING FOR AUXOTROPHIC MUTANTS OF HeLa CELLS

By ROBERT DE MARS*, Ph.D. AND J. LEONARD HOOPER

(*From the Laboratory of Cell Biology, National Institute of Allergy and Infectious Diseases, National Institutes of Health, Bethesda, Maryland*)

PLATE 47

(Received for publication, October 9, 1959)

The experiments described here suggest a method of selecting for auxotrophic mutants[1] in cultures of human or other types of cells. They are based mainly on three findings made by others:

(*a*) A variety of cultivated mammalian cells are unable to proliferate in growth media containing aminopterin or amethopterin (1).

(*b*) The growth-inhibitory effects of amethopterin can be completely reversed with a mixture of adenine, thymidine, and glycine (2, 3).

(*c*) Bacteria that are unable to synthesize thymine, either because of a mutation or because of treatment with sulfanilamide, lose viability when in a thymine-free medium that otherwise contains all factors necessary for growth. In contrast, if the thymine-free medium also lacks one other factor ordinarily necessary for growth, the bacteria remain viable (4). This preferential survival of thymine-starved cells that are also deprived of an additional growth factor should permit their selection from large populations of thymine-starved cells that do not ordinarily require the additional growth factor.

The general plan of this work was to create a thymine deficiency in HeLa cells (5) with the use of aminopterin and to compare the survival of cells in a growth-supporting medium with that of cells in a medium deficient in a single essential nutrient.

Methods and Materials

Media.—Stock cultures were maintained with Eagle's medium (7) supplemented with 5 per cent undialyzed human serum. All experimental media contained, instead, 5 per cent human serum that was dialyzed 24 hours against cold running tap water and 4 hours against cold running distilled water. Except when indicated, media contained 2×10^{-3} M L-glutamine. When glutamine was replaced with 2×10^{-2} M monosodium *L*-glutamate (Experiments 5 and 6) the concentration of NaCl was reduced by the same amount. In Experiment 4 certain media contained 10^{-4} M L-citrulline in place of arginine.

---

* Present Address: Department of Medical Genetics, University of Wisconsin, Madison.

[1] An auxotrophic mutant is considered here to be a genetically altered cell having at least one more nutritional requirement than the cell type from which it was derived.

Aminopterin (AP), obtained from Lederle Laboratories, was sterilized by filtration and stored frozen.

Cell Strains.—HeLa I-11: a clonal population that can initiate and maintain growth indefinitely when transferred from glutamine-containing medium to medium in which glutamic acid replaces glutamine.

HeLa S3-1: a clonal population isolated at random from HeLa S3 (6) by Dr. Royce Lockart. It is unable to initiate growth when transferred to medium in which glutamic acid replaces glutamine. Experiments 1 to 5 were done with this strain alone.

Growth Experiments.—Cells were grown adherent to glass. Inocula for experimental cultures were prepared by removing cells from the glass of stock cultures with a versene solution (8). Replicate aliquots (about 5×10^4 cells each) of the resultant cell suspension were inoculated into T15 culture flasks. After 2 days, permitting about a twofold increase in cell protein, experimental media were added. Fresh experimental media were provided on days 2, 4, and 5, and then every 2nd day when experiments lasted longer than 6 days. When cell populations were less than about 5×10^5 per T15 the cultures received enough CO_2 at each feeding to adjust the pH to about 7.4. No CO_2 was added at higher cell densities. At intervals, duplicate flasks of each experimental series were removed for growth determinations.

Experiments 2 to 5 had the same basic format. The growth of cells in various experimental media was determined during the first portion of the experiment. All experimental media were then removed from the remaining cultures and were replaced with Eagle's medium containing 5 per cent dialyzed human serum. Determinations of the subsequent growth of the cultures gave some measure of the ability of the cells to resume growth after sojourn in the various experimental media employed during the first portion of the experiment.

Measurement of Growth.—Growth was usually expressed in terms of the protein content of the cultures. Protein determinations were made according to Oyama and Eagle (9). Each value cited is the average of duplicate determinations of duplicate flasks. Duplicate flasks agreed within ±10 per cent of the average.

In some cases the numbers of cells in the cultures were determined by removing the cells from the glass with versene and counting them with a hemocytometer. At least 400 cells from each flask were counted and duplicate flasks usually agreed within ±20 per cent of their average count.

RESULTS

Experiment 1. Growth of HeLa Cells as a Function of AP Concentration.—The object of this experiment was to determine the minimum concentration at which AP would inhibit cell growth in Eagle's medium. Text-fig. 1 shows that at AP concentrations of 2×10^{-9} M and lower (not shown) there was no detectable inhibition of growth. At AP concentrations of 4×10^{-9} M and higher there was an initial doubling in the protein content of the cultures followed by a slow decline. The curves for concentrations of 8×10^{-9} M and higher (not shown) are indistinguishable from each other. The minimum inhibitory concentration found here (4×10^{-9} M) agrees well with that previously reported (1).

Experiment 2. The Ability of Cells to Resume Growth after Exposure to AP.—This experiment was an attempt to determine if the growth-inhibitory effect of AP, described in Experiment 1, is a transient one or if it leads to a permanent loss of growth ability. Text-fig. 2 shows that after 133 hours of exposure to AP in Eagle's medium, HeLa cells failed to show any measurable ability to grow

294

during a 190 hour recovery period in AP-free medium. A 40 hour exposure to AP failed to affect the ability of the cells to grow after its removal while an 87 hour exposure prevented a measurable increase in the protein content of the cultures for approximately 100 hours after the removal of AP.

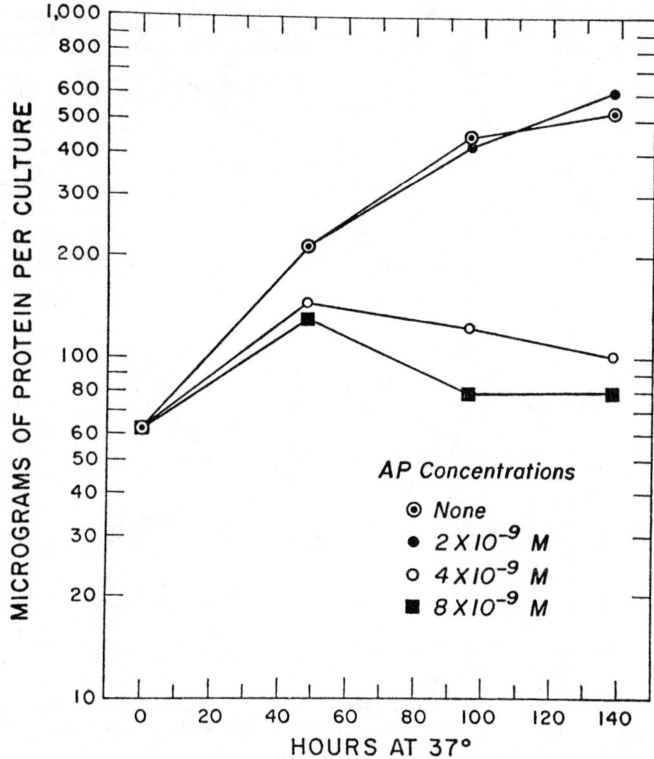

TEXT-FIG. 1. (Experiment 1) Growth of HeLa cells as a function of the concentration of AP. Replicate cultures of HeLa S3-1 were grown in Eagle's medium containing AP at the indicated concentrations. Growth was determined by protein determinations on duplicate flasks.

In this experiment the ability of the cells to recover from AP treatment was crudely expressed in terms of protein determinations, which give little information about the number of cells that were actually growing. The experiments cited below show that this measure is a usable rough index of the number of viable cells in a culture.

Observation of cultures treated with AP for 5 or more days reveals a few patches of proliferating cells. If these originate from single cells they represent 10^{-5} to 10^{-6} of the original population. Many of these survivors are mutants that are resistant to AP. With the exception of such cells, this experiment shows that AP does not merely inhibit the growth of HeLa cells, but eventually destroys their ability to grow.

Experiment 3. Reversal of AP-Action and Evidence for Thymineless Death.— Hakala (2, 3) showed that a mixture of thymidine, adenine, and glycine could

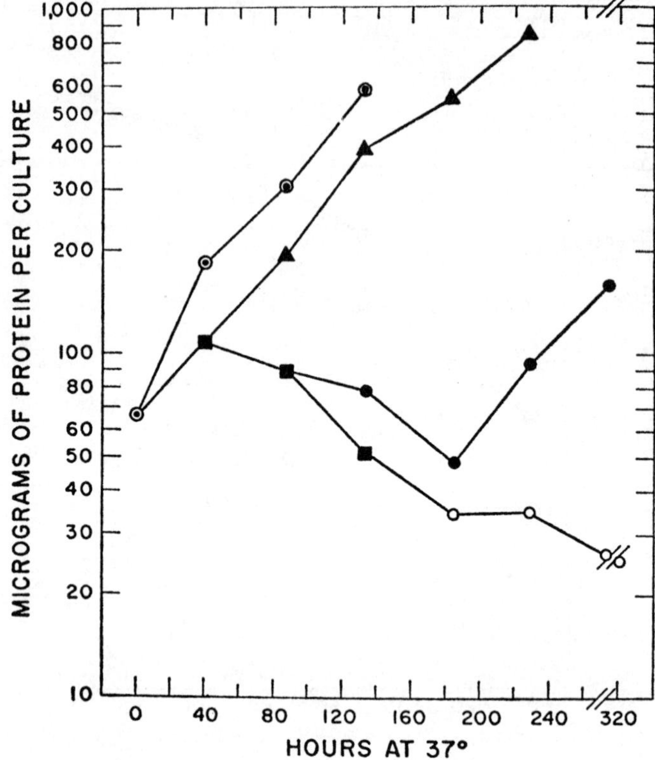

TEXT-FIG. 2. (Experiment 2) The ability of cells to resume growth after exposure to AP. HeLa S3-1 was cultivated in an AP-free medium (◉) or in medium containing 10^{-8} M AP (■). After 40 hours (▲), 87 hours (●), and 133 hours (○) of exposure to AP, sets of cultures were given AP-free Eagle's medium in order to permit the growth of surviving cells. Growth is expressed in terms of the protein content of the cultures.

overcome the growth-inhibiting effect of amethopterin. Text-fig. 3 (curves A and B) shows that these compounds also reverse the effect of AP on HeLa cells. Moreover, a comparison of curves E and F shows that it is the thymine deficiency created by AP that leads to its characteristic effects; *i.e.*, an initial doubling in the protein content of the cultures followed by a slow decline and inability to resume growth after the removal of AP. In contrast, HeLa cells in a medium containing AP, adenine, and thymidine, but no glycine, (curve C) grew almost as well as untreated control cultures. When adenine alone was omitted in the presence of AP (curve D) there was an initial doubling in protein followed by a slight decline. Removal of AP after 136 hours allowed growth to

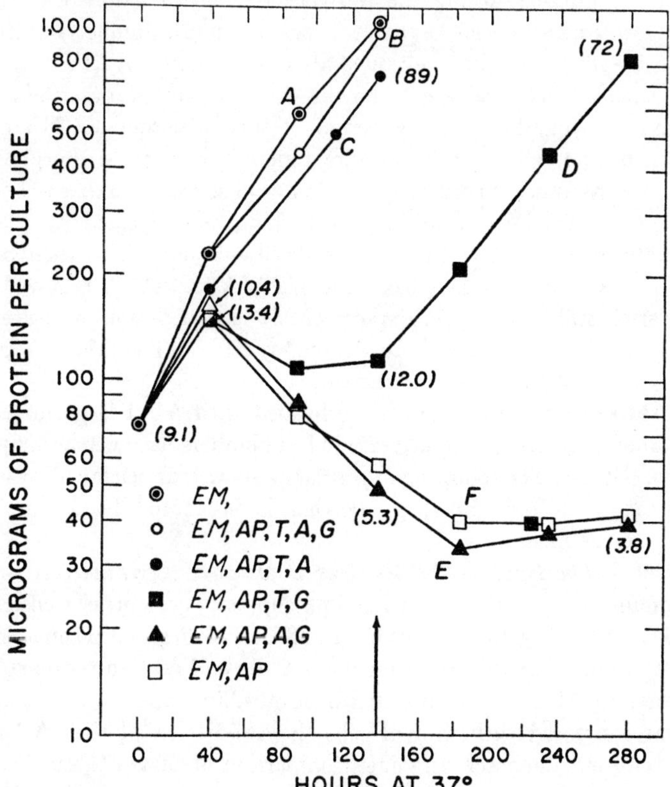

Text-Fig. 3. (Experiment 3) The reversal of AP action and evidence for thymineless death. HeLa S3-1 was cultivated in the media indicated in the figure. *EM*, Eagle's medium. *AP*, aminopterin, 10^{-8} M. *T*, thymidine, 3×10^{-5} M. *A*, adenine, 3×10^{-5} M. *G*, glycine, 10^{-} M. Starting at 134 hours (arrow) all remaining cultures were fed AP-free Eagle's medium in order to determine the residual growth ability of cells treated with the experimental media used during the first 134 hours of the experiment.

Growth is expressed in terms of the protein content of the cultures or in terms of cell number. The numbers in parentheses represent the number of cells ($\times 10^{-4}$) per culture at the indicated points.

resume without a lag, indicating that most of the cells had retained the ability to grow.

In this experiment the protein determinations were supplemented with cell counts at selected times. The omission of glycine (curve C) permitted a tenfold increase in cell number that corresponds to the increased protein content of the cultures. When adenine was omitted (curve D) the cell number increased

50 per cent during the first 48 hours and then remained almost constant until AP was removed. Most of these cells were viable as indicated by the sixfold increase in cell number during the recovery period. Omission of thymidine (curve E) permitted an initial 15 per cent increase in cell number and the subsequent decline in cell number continued after the removal of AP.

The persistence of countable cells on the glass explains in part the residual protein found during the recovery period of such experiments. Microscopic examination of the cells reveals a characteristic sequence of morphological changes in the thymine-starved cells that is not observed in untreated control cells (Fig. 2 *a*). They swell initially (Fig. 2 *b*), as suggested by the noncommensurate increases in cell protein and cell number. They then become distorted (Fig. 2 c) and detach from the glass. After AP is removed some of the cells swell still further (see Experiment 5) so that protein measurements become the resultant of loss due to detaching cells and continuing protein synthesis by some of the cells that remain.

In every experiment that we have performed with AP a large increase in protein without a corresponding increase in cell numbers has been accompanied by extensive cell loss. The following experiments show that cells tend to survive AP treatment when this disparate increase is prevented by a nutritional insufficiency of the medium.

Experiment 4. The Effect of AP on Cells Deprived of Arginine.—HeLa cells require arginine for growth (10). Citrulline supports growth as well as does arginine (11), indicating that the cells can effect the enzymatic conversion of citrulline to arginine. In this experiment the ability of cells to resume growth after exposure to AP in a medium in which citrulline was the sole source of arginine was compared with that of cells that were treated with AP in the absence of both citrulline and arginine. Comparison of curves C and E (Text-fig. 4) shows that cells completely deprived of arginine, whether AP was present (E) or absent (C), behaved similarly; they grew very little, failed to swell, and resumed growth when AP was removed and arginine provided. This indicates that cells that are unable to grow because of an amino acid deficiency are not subject to the lethal effects of AP. In contrast, cells treated with AP in a medium containing citrulline as arginine source (curve D) underwent typical swelling and distortion, and failed to resume growth when AP was removed. Presumably, mutants unable to use citrulline as an arginine source would tend to survive sojourn in medium containing citrulline and AP.

Experiment 5. The Effect of AP on Cells Deprived of Glutamine.—The glutamine requirement of HeLa I-11 cells can be satisfied with glutamic acid (12). These cells die in a medium containing glutamic acid and AP but survive AP treatment when both glutamine and glutamic acid are absent. The behavior of HeLa S3-1 is in contrast to this, as described in Text-fig. 5. Curve B shows that this strain does not initiate growth in a glutamic acid-containing medium that supports good growth of HeLa I-11. In such a glutamine-free medium, HeLa

S3-1 cells behaved similarly whether AP was present (curve C) or absent (curve B). The cultures declined in both protein content and cell number, but resumed growth when AP was removed and glutamine provided; *i.e.*, they behaved like cultures of HeLa I-11 that are deprived of both glutamine and

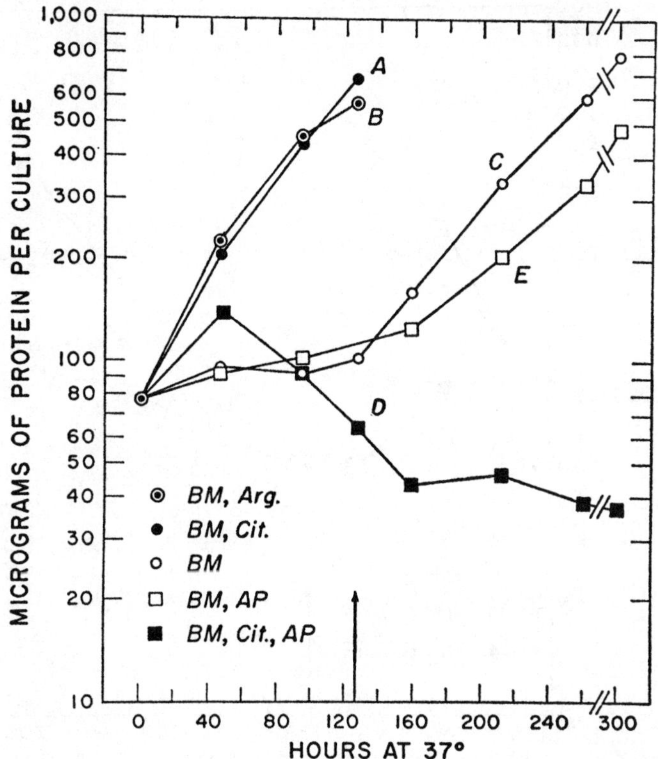

TEXT-FIG. 4. (Experiment 4) The effect of AP on cells deprived of arginine. HeLa S3-1 was cultivated in the media indicated in the figure. *BM*, Eagle's medium minus arginine. *Arg.*, arginine, 10^{-4} M. *Cit.*, citrulline, 10^{-4} M. *AP*, aminopterin, 10^{-8} M. Starting at 127 hours (arrow) all remaining cultures were fed complete Eagle's medium without supplements in order to assess the residual growth ability of the cells. Growth is expressed in terms of the protein content of the cultures.

glutamic acid. Curve A shows that HeLa S3-1 grew at about the same rate as HeLa I-11 in medium containing glutamine and, when this medium contained AP (curve D), the cells showed the usual effects of AP. In this case, the cells were exposed to AP for only 120 hours, permitting more cells than usual to persist. These did not increase in number but did synthesize protein, the average protein content of the cells at the end of the experiment being about tenfold greater than at the beginning.

This result permitted us to test the usefulness of thymine starvation in selecting for infrequent auxotrophs (HeLa S3-1 type) in a large population of non-auxotrophic (HeLa I-11 type) cells.

Experiment 6. Selection for Glutamine-Requiring Cells.—Petri dishes (60 mm. diameter) were inoculated with a mixture of 5×10^4 HeLa I-11 and 2.5

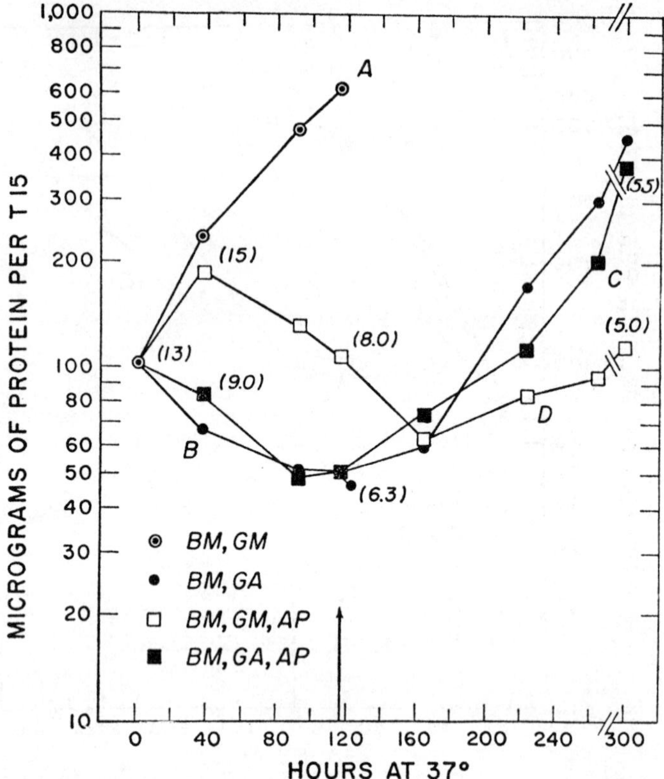

TEXT-FIG. 5. (Experiment 5) The effect of AP on cells deprived of glutamine. HeLa S3-1 was cultivated in the media indicated in the figure. *BM*, Eagle's medium minus glutamine. *GM*, glutamine, 2×10^{-3} M. *GA*, monosodium glutamate, 2×10^{-2} M. *AP*, aminopterin, 10^{-8} M. Starting at 116 hours (arrow) all remaining cultures were given complete Eagle's medium without supplements in order to assess the residual growth ability of the cells. Growth is expressed in terms of the protein content of the cultures or in terms of cell number. The numbers in parentheses represent the number of cells ($\times 10^{-4}$) per culture at the indicated points.

$\times 10^2$ HeLa S3-1 cells per dish. Control dishes contained only HeLa I-11. These were grown for 2 days in a medium containing glutamine to permit the formation of small clones of HeLa S3-1. On the 3rd day this medium was replaced with one containing glutamic acid in place of glutamine. One day was allowed for the HeLa I-11 cells to start growing in this medium and then AP (2×10^{-8} M) was added. At this time the cultures contained 2.5×10^5 cells and about 2.5×10^2 small clones of HeLa S3-1 per dish. The medium was replaced every day for 6 days and on the 7th day AP was removed and glutamine was provided. Surviving cells were allowed to proliferate in this medium for 12 days. At this time unseeded control dishes appeared as in Fig. 1 *A* while plates seeded with HeLa S3-1 appeared as in Fig. 1 *B*. The photographs in Fig. 1 are of plates in a second experiment of this kind in which about 500 HeLa S3-1 cells per dish were used. Every seeded plate had many large clumps of cells visible to the naked eye. Twenty-two clumps were picked and propagated into large populations (about 10^6 cells) in medium containing glutamine. Small clumps of viable cells were detected with the microscope in unseeded control plates. Fifteen of these were picked and propagated in glutamine. The 37 populations were then tested for their ability to grow in medium free of glutamine but containing glutamic acid. Nineteen of the cultures derived from seeded plates were unable to grow, one was able to grow, and two were unscorable. Fourteen of the populations derived from the unseeded control plates were able to grow and one could not. The nature of this single glutamine-requiring population (HeLa I-11a) was studied further.

When poliovirus Type I is plated on cell monolayers it forms 5 times as many plaques (average diameter 5 mm.) on HeLa I-11 as it does on HeLa S3-1 (average plaque diameter 2 mm.) (J. Darnell, personal communication). HeLa I-11a responded as did HeLa I-11 when tested in this way, indicating that it probably represented a new isolation of a glutamine-requiring strain rather than a chance contamination of the control plate with HeLa S3-1. Each of 58 clones picked at random derived from HeLa I-11 have been tested and shown to be able to initiate growth in glutamic acid medium.

DISCUSSION

Two situations have been used here to study the effects of aminopterin on the growth of HeLa cells. The first situation (Experiments 3 to 5) involves a comparison of the behavior of a single cell strain (HeLa S3-1) in aminopterin-containing media that either contain all the nutritional factors necessary for growth of the cells or, in contrast, are deficient in a single essential nutritional factor. In complete growth medium, aminopterin evokes a characteristic set of changes: cessation of cellular multiplication, doubling of the average protein content of the cells, cellular distortion and detachment of most of the cells from the glass of the culture vessel. Few of the cells that remain can resume proliferation when aminopterin is removed. Cells of the same strain in a deficient medium, while failing to increase in number, also fail to increase their protein content, and tend to remain attached to the glass. Most of these cells can multiply when aminopterin is removed and the missing growth factor is supplied.

The second situation (Experiments 5 and 6) compares two different cell strains in the same medium. HeLa I-11 can grow when transferred from glutamine-containing medium to one in which glutamic acid replaces glutamine and it is subject to the lethal action of aminopterin in such medium. In this same medium, HeLa S3-1 is unable to initiate growth and tends to survive aminopterin treatment. When an artificially mixed population, containing about one per thousand HeLa S3-1, is treated with aminopterin in a glutamine-free medium containing glutamic acid, almost all the HeLa I-11 cells are killed. Removal of the aminopterin and provision of glutamine then permit the HeLa S3-1 cells to resume proliferation so that they can be recovered as discrete colonies with high efficiency. This, and the isolation of a new glutamine-requiring strain from HeLa I-11, demonstrates the practicality of using aminopterin in selecting for auxotrophic mutants of HeLa cells.

Our experiments show that aminopterin creates growth requirements for thymine and adenine. They also show that it is mainly the thymine deficiency that leads to cell death in a complete growth medium. In this respect, and in the disparate increase in protein relative to cell number, the effects of aminopterin on HeLa cells resemble those of thymine deficiency in bacteria (4). In contrast, the aminopterin-induced purine deficiency permits a high survival of HeLa cells under our experimental conditions.

The creation by aminopterin of requirements for thymine and a purine accords, on the one hand, with the participation of reduced folic acid coenzymes in their synthesis, and, on the other, with the reported inhibition of folic acid reduction by aminopterin (13, 14). It is unexpected, therefore, to find that aminopterin-treated cultures can increase tenfold in both protein content and cell number in the absence of added glycine or serine. The cellular reserves of these amino acids could account for no more than a 30 per cent increase in protein (15) and it is possible that they are derived in appreciable amounts from sources not involving folic acid coenzymes. These ambiguities make it desirable to use an agent more specific than aminopterin for inducing thymine deficiency. Preliminary experiments by ourselves, Dr. L. A. Herzenberg, and by Dr. N. P. Salzman (personal communication) indicate that 5-fluorouracil deoxyriboside will be suitable with the HeLa cells as it is with other animal cells (16).

Experiments 4 and 5 show that aminopterin-treated cells that are also deprived of an essential nutrient tend to survive the period of thymine starvation. The survival of these cells depends in part on the care taken in depleting cellular reserves of the nutrient to be omitted. Appreciable unbalanced growth and cell loss occur when the cells are not washed with and incubated in deficient medium prior to the addition of aminopterin. Survival also depends on the ability of the cells to withstand starvation for substances, other than thymine, that are necessary for growth. This varies with the particular nutrient omitted, glutamine starvation (Experiment 5) usually evoking more rapid cell loss than starvation for arginine (Experiment 4) or adenine (Experiment 3). The period of exposure to aminopterin, which will, in part, determine the efficiency of selection for auxotrophs, must be adjusted to this limitation, but Experiment 2 shows that most non-deficient cells are lost after a 4 day treatment. Repeated cycles of short term exposure to aminopterin in minimal medium followed by growth of survivors in complete medium without aminopterin should permit a progressive increase in the frequency of auxotrophs to the point at which the testing of individual clones becomes feasible. Experiment 6 shows that a single 6 day cycle of glutamine starvation and aminopterin treatment permits efficient recovery of glutamine-requiring cells. It also shows that in this case not enough cross-feeding occurs in the presence of aminopterin to seriously impair the recovery of these cells.

The morphological difference between unswollen, auxotrophic and swollen, non-auxotrophic cells during thymine starvation might permit early detection of small clones of auxotrophs by systematic microscopic examination of mass cultures. This difference is readily apparent after about 48 hours of thymine starvation when, on the one hand, the deleterious effects of starvation for nutrients other than thymine are minimized and, on the other, there is a high survival of non-auxotrophs if thymine starvation is terminated. The recovery of viable non-auxotrophs in association with auxotrophs might be important in detecting homozygous auxotrophic and prototrophic segregants arising together from heterozygous cells.

Selection for auxotrophs with aminopterin is confused by the occurrence of aminopterin-resistant mutants. However, resistance occurs at several different levels in these cells and most of the resistant mutants selected for at 10^{-8} M aminopterin are sensitive at 5×10^{-8} M. Unbalanced growth and cell loss follow the same course at both concentrations so that the background of resistant cells can be largely eliminated by selection at the higher concentration.

We have been unable to determine cell viability directly by clone formation. The combined effects of starvation, aminopterin treatment, and the procedures for removing the cells from the glass reduce the cloning efficiency almost to zero. Some of these difficulties might be avoided by the use of cell suspensions. Despite this difficulty in our work, in every experiment in which there was a large increase in the protein content of the cultures after the removal of aminopterin there was a corresponding increase in cell number. This, and the efficient recovery of the glutamine-requiring cells in experiment 6, lend validity to the use of thymine starvation in selecting for auxotrophs.

The basis of the glutamine requirement of HeLa S3-1 is not known. It takes up glutamic acid from the medium at almost exactly the same rate as does HeLa I-11. HeLa I-11, grown with glutamine has a low but detectable level of glutamine synthase (17), which rises about fifteenfold during growth in glutamic acid (18 and unpublished data). The behavior of this glutamine-synthesizing enzyme in HeLa S3-1 is now being studied. Reversions to glutamine independence occur in this strain with a frequency of about 10^{-7} in newly recloned populations.

SUMMARY

1. Aminopterin prevents the multiplication of HeLa cells in Eagle's medium. The joint addition of adenine and thymidine removes the inhibition.

2. The aminopterin-induced thymine deficiency specifically results in a cessation of cell division and a doubling in the average protein content of the cells. Continued starvation for thymine results in the inability of cells to proliferate after aminopterin is removed. After 6 days only 10^{-6} to 10^{-5} of the original population proliferates.

3. The omission of a single essential amino acid, such as arginine or glutamine, from the medium during deprivation of thymine prevents a net increase in protein and results in about a 10^5-fold greater cell survival after 6 days, when aminopterin is removed and the missing amino acid is supplied.

4. Glutamine-requiring, auxotrophic mutants of HeLa exhibit a high survival after exposure to aminopterin in a glutamine-free medium. Glutamine-independent HeLa cells show a much lower survival in the same medium. Low frequencies of the auxotrophs can be specifically and efficiently selected for in artificial mixtures of the two cell types by treatment with the glutamine-free medium containing aminopterin.

BIBLIOGRAPHY

1. Eagle, H., and Foley, G. E., The cytotoxic action of carcinolytic agents in tissue culture, *Am. J. Med.*, 1956, **21**, 739.
2. Hakala, M. T., Prevention of toxicity of amethopterin for Sarcoma–180 cells in tissue culture, *Science*, 1957, **126**, 255.
3. Hakala, M. T., Tissue culture studies on mechanisms of action of some purine and thymine analogues, *Fed. Proc.*, 1958, **17**, 236.
4. Cohen, S. S., and Barner, H. D., Studies on unbalanced growth in *Escherichia coli*, *Proc. Nat. Acad. Sc.*, 1954, **40**, 885.
5. Scherer, W. F., Syverton, J. T., and Gey, G. O., Studies on the propagation of poliomyelitis viruses, *J. Exp. Med.*, 1953, **97**, 695.
6. Puck, T. T., and Fisher, H. W., Genetics of somatic mammalian cells. I. Demonstration of the existence of mutants with different growth requirements in a human cancer cell strain (HeLa), *J. Exp. Med.*, 1956, **104**, 427.
7. Eagle, H., Oyama, V. I., Levy, M., and Freeman, A. E., Myo-inositol as an essential growth factor for normal and malignant human cells in tissue culture, *J. Biol. Chem.*, 1957, **226**, 191.
8. Mandel, B., Studies on the interaction of poliomyelitis virus, antibody and host cells in a tissue culture system, *Virology*, 1958, **6**, 424.
9. Oyama, V. I., and Eagle, H., Measurement of cell growth in tissue culture with a phenol reagent (Folin-Ciocalteau), *Proc. Soc. Exp. Biol. and Med.*, 1956, **91**, 305.

10. Eagle, H., The specific amino acid requirements of a human carcinoma cell (strain HeLa) in tissue culture, *J. Exp. Med.*, 1955, **102**, 37.
11. Eagle, H., Amino acid requirements of mammalian cell cultures, *Science*, 1959, in press.
12. Eagle, H., Oyama, V. I., Levy, M., Horton, C. L., and Fleischman, R., The growth response of mammalian cells in tissue culture to L-glutamine and L-glutamic acid, *J. Biol. Chem.*, 1956, **218**, 607.
13. Futterman, S., Enzymatic reduction of folic acid and dihydrofolic acid to tetrahydrofolic acid. *J. Biol. Chem.*, 1957, **228**, 1031.
14. Osborn, M. J., Freeman, M., and Huennekens, F. M., Inhibition of dihydrofolic reductase by aminopterin and amethopterin, *Proc. Soc. Exp. Biol. and Med.*, 1958, **97**, 429.
15. Piez, K. A., and Eagle, H., The free amino acid pool of cultured human cells, *J. Biol. Chem.*, 1958, **231**, 533.
16. Bosch, L., Harbers, E., and Heidelberger, C., Studies on fluorinated pyrimidines. V. Effects on nucleic acid metabolism *in vitro*. *Cancer Research*, 1958, **18**, 334.
17. Elliott, W. H., Isolation of glutamine synthetase and glutamotransferase from green peas, *J. Biol. Chem.*, 1953, **201**, 661.
18. De Mars, R. I., The inhibition by glutamine of glutamyl transferase formation in cultures of human cells, *Biochim. et Biophysica Acta*, 1958, **27**, 435.

EXPLANATION OF PLATE 47

FIG. 1. Selection for glutamine-requiring cells. See text, Experiment 6 for explanation. (*a*) control plates containing only HeLa I-11. (*b*) similar plates seeded with about one per thousand of HeLa S3-1. Stained with methylene blue.

FIG. 2. The microscopic appearance of HeLa cells in Eagle's medium (*a*) or in Eagle's medium containing 10^{-8} M AP after 44 hours (*b*) and 135 hours (*c*). See text, Experiment 3 for explanation. $\times$ 140.

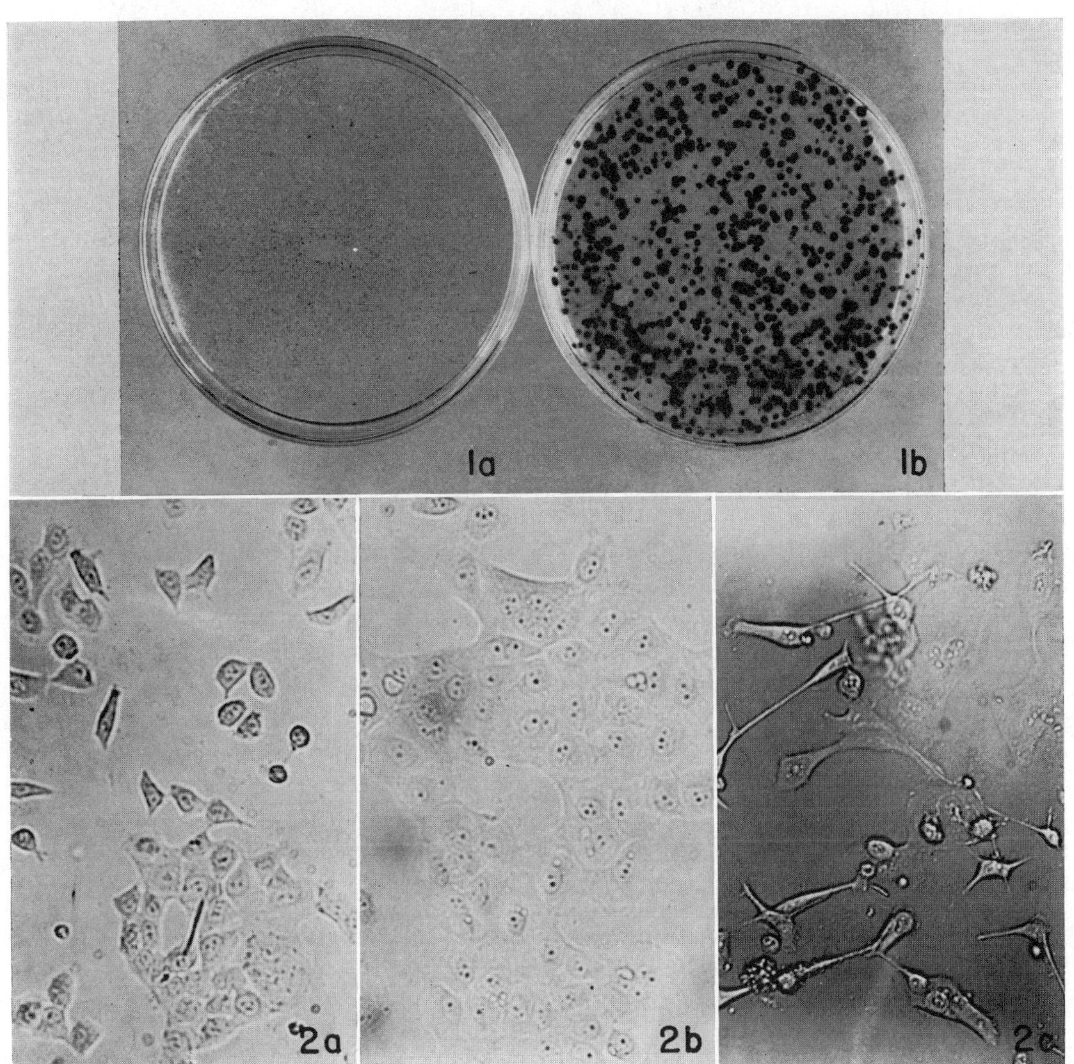

40

Reprinted from *Natl. Acad. Sci. (USA) Proc.* **58**:1227–1234 (1967)

GENETICS OF SOMATIC MAMMALIAN CELLS, V. TREATMENT WITH 5-BROMODEOXYURIDINE AND VISIBLE LIGHT FOR ISOLATION OF NUTRITIONALLY DEFICIENT MUTANTS*

By Theodore T. Puck† and Fa-ten Kao

The preceding paper of this series described genetic studies on a Chinese hamster cell which exhibited a specific growth requirement for proline. The nutritional deficiency was shown to be a mutation causing lack of the enzyme converting glutamic acid to the γ-semialdehyde. The character was demonstrated to be stable, exhibiting a reversion frequency of 10^{-6} per generation as expected of a single gene mutation.[1] The present study was undertaken in order to develop methods for the detection and isolation of additional nutritionally deficient mutants for the purpose of location of genes on chromosomes, measurement of linkage, and study of gene regulatory phenomena. The principle employed, which has been widely used in microbial genetics, involves placing a mixed cell population in a limited medium in which nutritionally deficient forms cannot grow, and applying an agent which will kill only dividing cells.[2]

Two agents were studied in these experiments. The first was H^3-thymidine, which can readily be incorporated in lethal amounts into the DNA of mammalian cells.[3] DNA synthesis has previously been shown to be dependent on protein synthesis.[4] Experiments demonstrated that differential incorporation of H^3-thymidine into proline-sufficient and -deficient cells could readily be achieved in proline-free media, with a consequent enrichment of the deficient mutant. However, the large amounts of radioactivity required for the execution of such experiments on a routine basis made this method objectionable. Consequently, another approach was adopted, in which cells were incubated with 5-bromodeoxyuridine (BUdR) for a period sufficient to ensure its incorporation into DNA.[5] Thereafter, the cells were exposed to light in the hope that those cells which had incorporated BUdR into their DNA would become light-sensitive in the visible or near-visible (sometimes referred to as "visible") region and be killed as occurs in bacteriophage.[6] Hence, if a mixed cell population is so treated in a minimal medium, one may expect more BUdR incorporation and subsequent killing among the nutritionally sufficient cell population than among the deficient forms, which should be concentrated among the survivors. This paper describes experiments which quantitate the response of mammalian cells to BUdR followed by "visible" light and which also demonstrate that proline-deficient cells can be enriched to a degree which appears useful in genetic experiments.

Materials and Methods.—Two Chinese hamster cell cultures were employed, the ovary cell (CHO Pro⁻), which exhibits no growth whatever in the absence of added proline, and the lung cell (CHL Pro⁺), whose growth is completely independent of the presence of added proline.[1] Cells with intermediate proline responses will be studied later.

The basal medium was F10, or more commonly F12,[7] with proline supplied or omitted as dictated by the needs of specific experiments. Unless otherwise indicated, the media were supplemented with 2 mg/ml of the purified, macromolecular fraction of fetal calf serum as previously described.[1] During the period of BUdR incorporation into DNA, thymidine was uniformly omitted from the F12, to maximize uptake of the brominated analogue. The cells were cultivated in glass bottles and trypsinized before use, as previously described.[8, 9] The generation time of the cells under these conditions is approximately 12 hr.

Illumination with "visible" light was accomplished by exposing cells attached in a monolayer to plastic Petri dishes (Falcon) containing medium to a depth of approximately 2 mm, to light from two fluorescent lamps (40-watt, Westinghouse "Cool White" tubes, 120-cm long), which were mounted in a fixture with an approximately parabolic reflector, so that the lamps were 10 cm apart, and 10 cm above the dishes. The lights were kept on for 30 min before the beginning of the cell exposure in order to permit stabilization of the lamp operation. Control plates shielded from the light were placed near the illuminated ones; they demonstrated the absence of any extraneous effect due to the sojourn in the room atmosphere for the periods indicated.

A typical experimental procedure is as follows: A culture bottle of actively growing cells is given a preliminary period of incubation in proline-free medium to ensure depletion of that amino acid in the proline-dependent cells. The medium is then removed and replaced with new, prewarmed medium from which both proline and thymidine have been omitted, and to which BUdR has been added in the requisite concentration. Incubation is then resumed for either 12 or 24 hr in order to achieve completely single-strand, or partly single- and partly double-strand labeling as desired. The cells are then trypsinized and appropriate inocula deposited into Petri dishes in the complete medium. Cell inocula are adjusted to permit the counting of sufficient colonies on each plate so that reproducibility of replicate plating is about 5%. The plates are incubated for 3 hr, and then illuminated with fluorescent light for measured periods. Thereafter the plates are replaced in the incubator and maintained for 7 days, after which colonies are scored.

Experimental Results.—(1) *The responses of the deficient mutant to proline starvation, BUdR, and "visible" light, alone and in combination:* The survival curve of CHO Pro⁻ cells incubated in proline-free growth medium was described earlier[1] and shown to have an initial shoulder followed by an exponential drop with a half life of about 31 hours. Life cycle experiments[10] were carried out which demonstrated that DNA synthesis in CHO Pro⁻ cells transferred from complete to proline-free medium falls, and within 18 hours reaches a level that is undetectably low by H³-thymidine radioautography. Therefore the preliminary incubation in proline-free medium to achieve a proline-starved condition in the deficient cells was set at 24 hours. Since the subsequent BUdR incorporation was maintained for either 12 or 24 hours, the total period of proline starvation of the Pro⁻ cells was either 36 or 48 hours.

TABLE 1

Experiment demonstrating the effects of proline starvation, BUdR incorporation, and "visible" light exposure, alone and in combination, on the viability of CHO Pro⁻ cells. The plates inoculated with "starved" cells received a preliminary incubation in proline-free medium, containing BUdR at the indicated levels. The proline-fed cells had a full complement of proline[1] during these incubation periods. Light was applied where indicated for 60 min after the BUdR incorporation period. Thereafter all plates were restored to complete medium without BUdR and scored after 7 days of incubation. The average plating efficiencies shown have been calculated relative to controls which contained complete growth medium, and were never subjected to proline starvation, BUdR, or light exposure, and whose absolute plating efficiencies varied between 74% and 90% in individual experiments.

| BUdR concentrations (M) | Average Plating Efficiency (%) | | | |
| | Proline-Fed Cells | | Proline-Starved Cells | |
	Without light	With light	Without light	With light
0	100.0	100.0	51.0	51.0
10^{-6}	89.4	0	41.5	34.0
10^{-5}	72.4	0	45.6	35.4

Conditions were then determined which permit cells to survive exposure to BUdR followed by "visible" light while starved for proline, but which would kill cells in which proline was supplied (either exogenously or endogenously) in an amount sufficient for maximal growth: (*a*) Proline-starved cells, unexposed to BUdR, showed the expected survival of about 50 per cent after a 48-hour period of proline starvation. Their survival remains reasonably constant regardless of

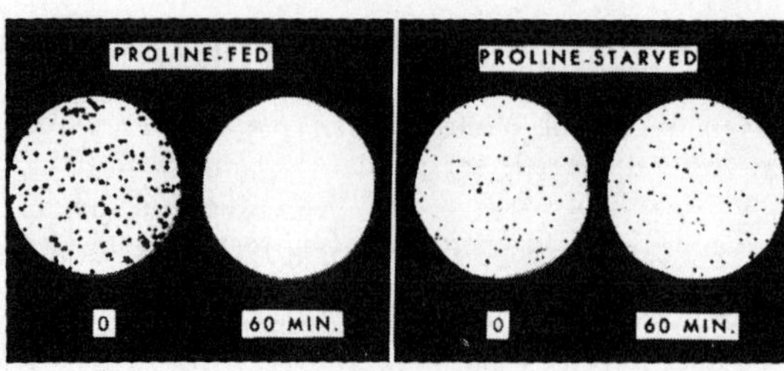

Fig. 1.—Typical appearance of plates from an experiment like that described in Table 1. All plates received an equal inoculum of cells which had been exposed to 10^{-6} M BUdR for equal periods of time. The numbers underneath plates indicate the illumination times. In the proline-starved case, the starvation process lowers the plating efficiency to about 50%, but there is no effect of BUdR or light exposure. In the presence of proline which permits normal growth, BUdR alone does not affect the plating efficiency, but BUdR plus light kills the cells.

whether or not they are exposed to BUdR in any concentration up to 10^{-5} M, and is reduced only slightly, if at all, by subsequent illumination (Table 1). (b) In contrast, under conditions of proline sufficiency, i.e., using either CHL Pro$^+$ cells or CHO Pro$^-$ cells supplied with exogenous proline, BUdR-cells are formed which are killed on subsequent illumination with "visible" light (Table 1). In the absence of light, exposure of the growing cells to any concentration of BUdR up to and including 10^{-5} M, for either 12 or 24 hours, resulted in a plating efficiency of at least 70 per cent of that of untreated controls. However, if the cells were illuminated following the BUdR treatment, extensive killing resulted. A typical set of plates is shown in Figure 1.

(2) *Survival curves of BUdR-treated cells exposed to visible light:* Detailed survival curves were constructed for illuminated CHO Pro$^-$ BUdR cells, grown in complete medium in which thymidine was replaced by either 10^{-6} M or 10^{-5} M BUdR for periods of either 12 hours or 24 hours, respectively (Fig. 2). The simple exponential character of the curves is noteworthy.[11]

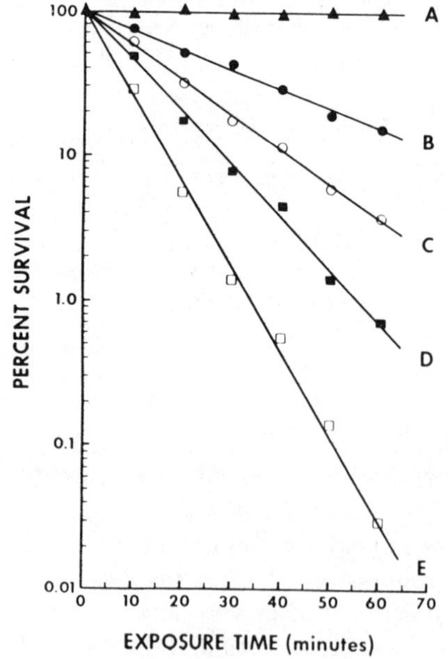

Fig. 2.—Survival curves for CHO Pro$^-$ cells which were supplied with proline, allowed to incorporate BUdR, and then exposed to "visible" light for the periods indicated. The per cent survivals are expressed relative to control plates which had the same exposure to BUdR but no illumination.
(A) No BUdR.
(B) 10^{-6} M BUdR for 12 hr.
(C) 10^{-6} M BUdR for 24 hr.
(D) 10^{-5} M BUdR for 12 hr.
(E) 10^{-5} M BUdR for 24 hr.
The values for D°, the exposure time needed to reduce the cell survivor to the fraction $1/e$, are as follows:
(B) 32.2 ± 2.2 min, (C) 17.9 ± 1.0 min; (D) 12.1 ± 0.8 min; and (E) 7.4 ± 0.4 min.

(3) *Chromosomal examination of the irradiated BUdR-cells:*   An abundance of experimental evidence exists demonstrating that mammalian cell killing by X irradiation is due to chromosomal damage.[11c, 12]   Chu[13] showed that both ultraviolet and near-visible light can produce chromosomal aberrations in BUdR-cells of the Chinese hamster.   Experiments were carried out in the present study to determine whether the frequency of such aberrations at various light doses is sufficiently high to account for the degree of cell killing here observed in the survival curves.   CHO cells treated with 10^{-6} and 10^{-5} M BUdR, respectively, were illuminated for various periods, reincubated for nine hours to permit lag periods[10, 14] to run their course, and treated with 0.10 μg/ml vinblastine sulfate (Velban, Eli Lilly Co.) for the last three hours of the incubation period to collect mitoses which were then analyzed for chromosomal aberrations.

BUdR cells which had grown for 12 hours in the presence of 10^{-6} M BUdR, but not illuminated, exhibited a frequency of chromosomal aberrations not appreciably different from untreated cells: about 5 per cent of such cells exhibited single chromatid breaks, and no chromosomal interchanges characteristic of multi-hit events were evident.   On exposure to "visible" light, however, large numbers of both single- and multi-hit aberrations were obtained.   The frequency of both types of aberrations increased steadily with light dose, and the numbers of abnormalities observed were consistent with the thesis that these underlie the killing action.   For example, in preliminary studies, a 15-minute illumination which reduced the number

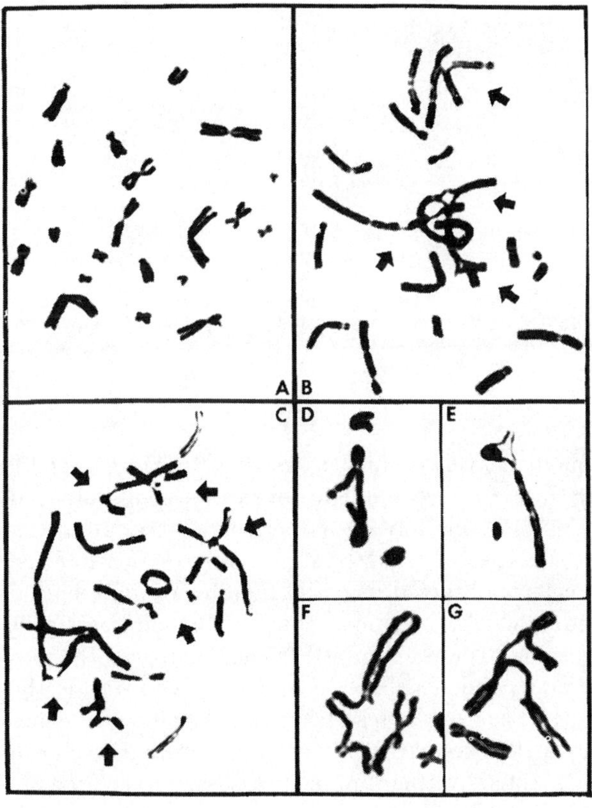

Fig. 3.—Typical metaphases obtained from cells exposed to BUdR in the concentrations indicated for 12 hr, and then illuminated for the periods shown.
(*A*) Untreated controls.
(*B*) 10^{-6} M BUdR, 60 min illumination, 15% survival.
(*C*) 10^{-5} M BUdR, 30 min illumination, 8% survival.
(*D*), (*E*), (*F*), and (*G*) show typical selected aberrant chromosomes from metaphases obtained from cell populations treated with BUdR and "visible" light to yield cell survivals varying from 0.7% to 28%.
The arrows in (*B*) and (*C*) indicate some of the chromosomal exchanges produced.

of survivors to 60 per cent increased the average number of single chromatid breaks per cell in the mitotic population examined to a value of about 0.55, and produced an average frequency of chromosomal rearrangements of approximately 0.40.   Figure 3 presents typical metaphases demonstrating the bizarre, multi-hit chromosomal malformations which occur with high frequency after illumination of BUdR-cells with near-visible light.   Detailed analysis of the dose dependence of these chromosomal anomalies will be presented later.

(4) *Auxotroph recovery from mixed populations:*   Tests were carried out to determine whether the conditions defined by the preceding experiments would permit deficient cells to be detected in and isolated from large populations of nutritionally sufficient forms.   Artificial mixtures were prepared containing varying numbers of cells from our proline-sufficient (CHL Pro$^+$) and proline-deficient (CHO Pro$^-$) cultures.   Mixtures of the two cell types were plated in proline-free medium, and incubated for 24 hours.   At the end of this period, the medium was changed to F12 from which both proline and thymidine were absent, and 10^{-5} M BUdR was added.   The results shown in Table 2 demonstrate: (*a*) The recovery

TABLE 2

Typical experiment indicating colony recovery from artificial mixtures of proline-sufficient and -deficient cells incubated with proline-free medium for 24 hr, then treated with BUdR without proline for another 24 hr, and finally exposed to "visible" light, as indicated.   Thereafter the medium was changed to complete F12 supplemented with 10% fetal calf serum, and incubation was continued for 7 days, after which the colonies were scored.   Plates with 10^3 or more CHL Pro$^+$ cells which are not treated with both BUdR and visible light yield confluent growth under these conditions.   The absolute plating efficiency of unstarved CHO Pro$^-$ cells was about 85%.

Number of Cells Plated		Exposure to 10^{-5} M BUdR	Illumination time (min)	Average no. of colonies per plate
CHL Pro$^+$	CHO Pro$^-$			
0	20	+	60	9
0	100	+	60	42
0	200	+	60	92
0	200	0	0	111
0	200	0	60	102
0	200	0	90	107
0	200	+	0	88
0	200	+	90	83
10^3	0	+	60	3
10^3	20	+	60	10
10^3	100	+	60	52
10^3	200	+	60	98
10^3	0	+	90	0
10^3	200	+	90	94
10^4	0	+	90	2
10^4	200	+	90	86

of CHO Pro$^-$ cells plated alone quantitatively follows expectation, and is essentially constant regardless of whether BUdR, near-visible light, or both are administered under the prescribed conditions.   Approximately 40–50 per cent of the initial cell inoculum is recovered as colonies, as expected from the starvation experience alone. (*b*) The best killing of the Pro$^+$ cells is obtained using 90 minutes of illumination after the BUdR treatment, under which conditions the fraction of these cells surviving is 2×10^{-4}.   (*c*) The colony counts obtained from mixtures of the two cell types, but with Pro$^+$ cells always in huge excess, quantitatively yields the numbers of colonies expected on the basis of the results with each cell type alone.

Colonies were picked from such plates seeded with various numbers of each cell type, grown into new cultures, and tested for their proline response.   In every case, the frequency of the proline-deficient mutant agreed well with the expected value calculated from the results in Table 2.

Discussion.—A method for auxotrophic mutant selection in HeLa cells was previously proposed in 1960[15] which utilized the fact that aminopterin is more toxic for growing than for nongrowing cells, and which succeeded in isolating glutamine-requiring cells from a mixed population which initially had a 200-fold larger inoculum of the glutamine-independent variety. This method involved repeated treatment with aminopterin during a seven-day period. The combined effects of starvation, aminopterin treatment, and the cell manipulative procedures employed studies of the single auxotrophic cells were not possible. The present method is more rapid, involves much less manipulation, and permits survival of, and quantitative clone formation from, single cells of the auxotroph. The data indicate that with the BUdR procedure proline-deficient cells can be recovered from large admixtures of sufficient cells with an efficiency of about 0.40, while the sufficient cells themselves have a recovery efficiency of about 2×10^{-4}. The method appears sufficiently simple to permit its repeated application to a cell population, so that even greater differential recovery of auxotrophs may be achievable.

Earlier we proposed that radiation death of mammalian cells is due in large part to anomalous action of repair processes on primary chromosome breaks resulting in production of abnormal chromosome configurations.[11c, 20, 21] These structures might block normal progression of cells along the life cycle by preventing the particular order of protein syntheses, DNA replication, chromosomal condensation, the progression of the mitotic phases, or other processes required for continuing cell reproduction. The radiation survival curves presented here appear.to be simple exponentials for cells grown in different concentrations of BUdR; for growth periods designed to produce single- or multiple-stranded substitution in the DNA; and for radiation doses leading to survivals as low as 0.04 per cent. Evidence from other laboratories indicates that BUdR incorporation into DNA inhibits certain chromosomal repair mechanisms in mammalian cells[17a] as in bacteria.[17b, c] In BUdR-mamalian cells where such inhibited repair mechanisms presumably exist, the hit numbers of the X-ray and UV survival curves often decrease and may approach unity as was found here.[16a, 17a, 18] Therefore we tentatively infer that the present survival curves also involve at least partially similar inhibition of certain aspects of chromosomal repair processes.

Figure 3 illustrates that even though certain chromosomal repair mechanisms are depressed in the BUdR cells, extensive rejoining of broken ends has occurred so that highly abnormal chromosomal configurations have resulted. The chromosomal structure of mammalian cells is complex, and several different enzymes must operate presumably in an orderly sequence before normal repair is completed. Lethal configurations might result from a distortion of the normal sequence of these repair mechanisms. Such an interpretation is supported by the recent demonstration of Phillips and Tolmach[22] that temporary, postirradiation inhibition of protein synthesis increases the survival of X-irradiated mammalian cells while similar inhibition of DNA synthesis does the reverse, a finding which suggests the need for a particular order of chromosonal repair action for viability.

The one-hit curves exhibited here suggest that a single radiation-induced event can produce a lethal process, and that cells in each of the major divisions of the life cycle[19] are similar in their average radiosensitivities. It is not yet possible to decide whether single chromosome breaks, improperly repaired, are sufficient for lethality or whether multiple lesions are necessary. In the latter case the one-hit curve may mean that the newly produced breaks interact with other preexisting chromosomal defects. The use of BUdR cells and near-visible light provides a useful system for studying these and other problems, because the energy absorption from the radiation is largely confined to the specific sites in the DNA already substituted by a BUdR moiety.

Summary.—(1) Study of the kinetics of the killing of Chinese hamster cells by near-visible light following BUdR incorporation in DNA is described. Killing occurs only if the cells have taken up the BUdR and are also illuminated. The survival curves obtained are one-hit for all modes of BUdR incorporation studied, in contrast to the multiple-hit curves which are obtained when these cells are irradiated with X rays or ultraviolet light, without previous BUdR incorporation.

(2) Illumination of BUdR-cells with near-visible light produces large numbers of chromosomal aberrations, particularly of the multiple interchange type.

(3) Cells in which DNA synthesis has been prevented by the deprivation of proline are impervious to the killing action of BUdR and near-visible light.

(4) A method based on these findings has been devised for detection and isolation of proline-deficient cells in the presence of a large excess of proline-independent cells. The method appears capable of enriching nutritionally defficient forms by several thousandfold.

* From the Eleanor Roosevelt Institute for Cancer Research and the Department of Biophysics (contribution no. 306), University of Colorado Medical Center, Denver, Colorado. This investigation was aided by a grant no. 1 PO1 HD02080-01 from the National Institutes of Health, U.S. Public Health Service.

† American Cancer Society Research Professor.

1 Kao, F. T., and T. T. Puck, *Genetics*, **55**, 513 (1967).

2 (a) Davis, B. D., *J. Am. Chem. Soc.*, **70**, 4267 (1948); (b) Lederberg, J., and N. Zinder, *J. Am. Chem. Soc.*, **70**, 4267 (1948).

3 Drew, R. M., and R. B. Painter, *Radiation Res.*, **11**, 535 (1959).

4 (a) Harris, H., *Biochem. J.*, **72**, 54 (1959); (b) Powell, W. F., *Biochim. Biophys. Acta*, **55**, 969 (1962).

5 Djordjeciv, B., and W. Szybalski, *J. Exptl. Med.*, **112**, 509 (1960).

6 Stahl, F. W., J. M. Crasemann, L. Okun, E. Fox, and C. Laird, *Virology*, **13**, 98 (1961).

7 Ham, R. G., these PROCEEDINGS, **53**, 288 (1965).

8 Ham, R. G., and T. T. Puck, *Proc. Soc. Exptl. Biol. Med.*, **111**, 67 (1962).

9 Ham, R. G., and T. T. Puck, in *Methods in Enzymology*, ed. S. P. Colowick and N. O. Kaplan (New York: Academic Press, 1962), vol. 5, p. 90.

10 Puck, T. T., and J. Steffen, *Biophys. J.*, **3**, 379 (1963).

11 (a) Puck, T. T., and P. I. Marcus, *J. Exptl. Med.*, **103**, 653 (1956); (b) Puck, T. T., D. Morkovin, P. I. Marcus, and S. J. Cieciura, *J. Exptl. Med.*, **106**, 485 (1957); (c) Puck, T. T., *Symp. Intern. Soc. Cell Biol.*, **3**, 63 (1964); (d) Sinclair, W. K., and R. A. Morton, *Biophys. J.*, **5**, 1 (1965).

12 Puck, T. T., these PROCEEDINGS, **44**, 772 (1958).

13 Chu, E. H. Y., *Mutation Res.*, **2**, 75 (1965).

14 (a) Yamada, M., and T. T. Puck, these PROCEEDINGS, **47**, 1181 (1961); (b) Puck, T., and M. Yamada, *Radiation Res.*, **16**, 589 (1962).

[15] De Mars, R., and J. L. Hooper, *J. Exptl. Med.*, **111**, 559 (1960).

[16] (*a*) Lett, J. T., G. Parkins, P. Alexander, and M. G. Ormerod, *Nature*, **203**, 593 (1964); (*b*) Djordjevic, B., *Nature*, **212**, 1469 (1966).

[17] (*a*) Rauth, A. M., *Radiation Res.*, **31**, 121 (1967); (*b*) Aoki, S., R. P. Boyce, and P. Howard-Flanders, *Nature*, **209**, 686 (1966); (*c*) Alexander, P., J. Beer, C. Dean, J. T. Lett, and G. F. Parkins, *Brit. J. Radiol.*, **36**, 860 (1963).

[18] (*a*) Erikson, R. L., and W. Szybalski, *Biochem. Biophys. Res. Commun.*, **4**, 258 (1961); (*b*) Delihas, N., M. A. Rich, and M. L. Eidinoff, *Radiation Res.*, **17**, 479 (1962); (*c*) Mohler, W. C., and M. M. Elkind, *Exptl. Cell Res.*, **30**, 481 (1963).

[19] Puck, T. T., P. Sanders, and D. Peterson, *Biophys. J.*, **4**, 441 (1964).

[20] Puck, T. T., *Progr. Biophys. Biophys. Chem.*, **10**, 237 (1960).

[21] Puck, T. T., *Am. Naturalist*, **94**, 95 (1960).

[22] Phillips, R. A., and L. J. Tolmach, *Radiation Res.*, **29**, 413 (1966). Similar results have also been reported by G. F. Whitmore and S. Gulyas, see *Natl. Cancer Inst. Monograph*, **24**, 141 (1967).

ERRATUM

On page 1232 paragraph 2, insert the following between lines 7 and 8: "reduced the plating efficiency of single cells almost to zero, so that accurate viability"

On page 1232 paragraph 3, the correct reference citation in line 12 should be: "in mammalian cells [17a, c] as in bacteria. [17b]"

41

Reprinted by permission from *Nature* **223**:1380–1381 (1969)

TEMPERATURE SENSITIVE CONDITIONAL MUTANTS
OF MONKEY KIDNEY CELLS

P. M. Naha

THE genetics of phage, bacteria and fungi owe much to the study of conditional lethal mutants, but little is known of the conditional lethal mutants of mammalian cells. I therefore wish to describe the isolation of conditional temperature sensitive mutants of African green monkey kidney cells (BS-C-1) (ref. 1). Because these cells are susceptible to simian virus (SV 40) yet less liable to transformation than other cell lines (hamster kidney cells, mouse embryo cells)[2], I hope that this development will be valuable in the study of, first, control mechanism of virus–host interaction in cases where viral growth is inhibited at a restrictive temperature; second, regulatory mechanism of replication of cellular organelles; and third, cytological and morphological analyses in cultures synchronized by exposure (shock) to limiting conditions of temperature[3].

The work reported here shows that animal cells, in the same way as bacterial cells, can be induced to form temperature sensitive mutants in response to a chemical mutagen. Whether this property is common to all animal cells or to only very few is unknown. It is known that some animal cells can multiply at higher temperatures after infection with viruses; it should now be possible to investigate the conditions of the virus-induced temperature dependence of cells.

The stable line of heteroploid SV 40-sensitive African green monkey (*Cercopithecus aethiops*) kidney cells, BS-C-1 (ref. 1), was obtained from Dr Ian MacPherson. The stock cultures were maintained by planting about 10^6 trypsinized cells in 20 ml. of L-15 medium[4] containing 10 per cent foetal calf serum. A single clone was isolated and cultured through three successive passages in L-15 medium. The cells were then trypsinized and treated with mutagen N-methyl-N′-nitro-N-nitrosoguanidine, and subsequently exposed to 5-bromodeoxyuridine by Puck and Kao's method[5]. Experiments were designed to isolate mutants that are permissive (pm+) at 35° C and restrictive (pm-) at 39° C. Previous control experiments showed that BS-C-1 cell lines grew favourably within this temperature range, the optimum temperature for increased efficiency of plating being about 39° C. All three incubators showed a deviation of $\pm 0.5°$ C from the recorded temperatures of 35° C, 37° C and 39° C. Plastic Petri dishes made for tissue culture were used throughout the experiment.

Temperature sensitive (ts) mutants were isolated as follows. About 200 cells were inoculated into each of a series of Petri dishes in L-15 medium supplemented with 10 per cent foetal calf serum. After 6 h of incubation at 39° C, nitrosoguanidine was added, dissolved in an appropriate concentration of L-15 medium, to make a final concentration of 0.50 µg/ml. of mutagen in contact with the cells. Incubation was continued at that temperature for an additional 16 h, which was a little more than one generation time. The medium was then removed from the plates, the cells were washed twice in L-15 and fresh growth medium was added; incubation was continued for a further 24 h. After this, 5-bromodeoxyuridine was added in solution of L-15 medium so as to make its concentration $4–5 \times 10^{-6}$ M in each plate. The culture plates were again incubated at 39° C for 5–7 days before being exposed to visible light. After exposure to light the medium treated with 5-bromodeoxyuridine was removed, fresh L-15 medium was added and plates were transferred to a 35° C incubator for 12–15 days, until clones of dividing cells were visible under the microscope. These clones were isolated, cultured and tested for efficiency of plating at different temperatures. Control culture plates kept at 39° C failed to produce any visible clone.

The combination of the two mutagens was expected to produce mutation of the desired type according to the procedure I have described. It has been recognized that nitrosoguanidine induces a high frequency of mutation

"

in bacterial systems[6], and the incorporation of 5-bromo-deoxyuridine in DNA[7] renders bacteriophage[8] sensitive to visible or near-visible light. A shift in incubation temperature for selection and isolation of temperature sensitive mutants of bacteriophage T4 was successfully performed by Edgar and Lielausis[9]. I have taken advantage of all these established techniques in the isolation of temperature sensitive mutants reported here.

The principle of the technique described is to allow the normal (wild type) non-mutated and temperature independent mutants to divide and grow at 39° C which, in my control experiments, was the optimum temperature for increased efficiency of plating for BS-C-1 (Fig. 1). The incorporation of 5-bromodeoxyuridine in the DNA of growing cells rendered these cells hypersensitive to visible light. Transfer of the culture plates to a lower temperature (35° C), following pre-incubation at an elevated temperature (39° C), selected out the pm+ temperature sensitive cells which were allowed to divide and form clones at 35° C. These clones were isolated, subcultured and studied for efficiency of plating at temperatures of 35° C, 37° C and 39° C. Our initial experiments have provided four temperature sensitive conditional mutants which grew at 35° C but not at 37° C or 39° C. I report here (Figs. 1 and 2) one such mutant (ts 2).

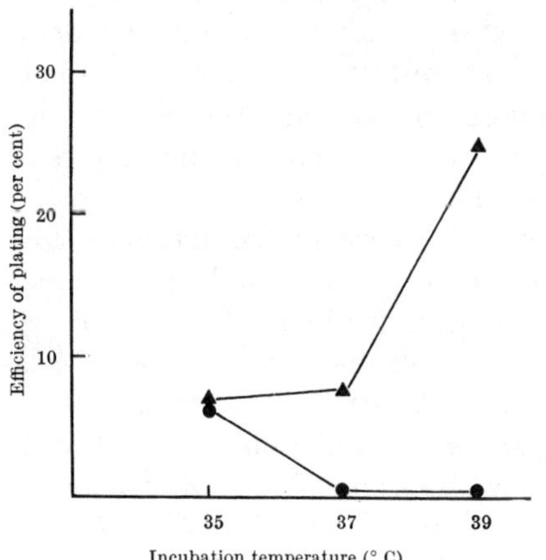

Fig. 1. Graphs showing efficiency of plating expressed in percentage of input cells, of wild type BS-C-1 cells (▲) and the temperature sensitive mutant (●) at temperatures of 35° C, 37° C and 39° C.

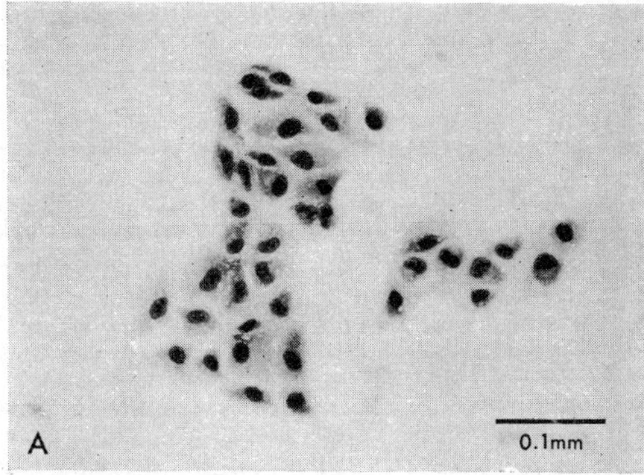

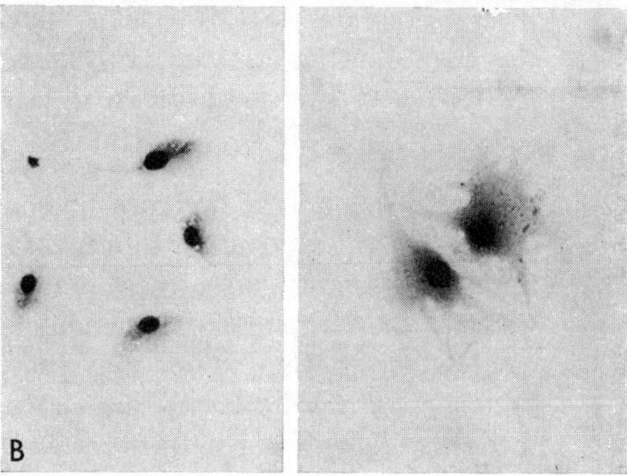

Fig. 2. Pictures showing actual growth of the temperature sensitive mutant (ts 2) after 6 days of incubation at 35° C (*A*) and at 39° C (*B*) and stained with Giemsa.

The temperature sensitive mutant clone (ts 2) was subcultured at 35° C. The cells were trypsinized and plated at different dilutions at temperatures of 35° C, 37° C and 39° C. The efficiency of plating was calculated from the number of input cells. The results are presented in Fig. 1. We observed that the efficiency of plating of normal (wild type) BS-C-1 cells was much higher at 39° C than at 37° C, the increase being by a factor of almost 4 to 5. The mutant (ts 2), however, failed to produce any visible clone at either 37° C or 39° C, but plated at an efficiency almost equal to wild type BS-C-1 at 35° C (about 7–8 per cent). Although the cytology or morphology of the mutant cells has yet to be studied in detail, there was no obvious visible derangement in the structural organization of the clones at 35° C. At 37° C and at 39° C the cells appeared to enlarge (Fig. 2) slightly without undergoing division, but there was no disorganization of nuclear material. The viability of these cells after exposure to higher temperature is now being studied. Attempts are also in progress to isolate similar mutants from SV 40-transformed hamster kidney cells (BHK), in order to investigate the possibility of isolating mutant cell lines which would induce the virus only at elevated temperature, because there is evidence of five to sixty copies of SV 40 DNA per cell[10] which could be activated to produce the virus[11] in certain conditions.

I thank Dr J. S. Porterfield for his help and encouragement.

Received July 4; revised July 30, 1969.

[1] Meyer, jun., H. M., Hopps, H. E., Rogers, N. G., Brooks, B. E., Bernheim, B. C., Jones, W. P., Nisalak, A., and Douglas, R. D., *J. Immunol.*, **88**, 796 (1962).

[2] Sabin, A., and Koch, M. A., *Proc. US Nat. Acad. Sci.*, **49**, 304 (1963).

[3] Newton, A. A., *Synchrony in Cell Division and Growth* (edit. by Zeuthen, E.), 601 (Interscience, London, 1964); Rao, P. N., and Engelberg, J., *Cell Synchrony* (edit. by Cameron, I. L., and Padilla, G. M.), 332 (Academic Press, London, 1966).

[4] Leibovitz, A., *Amer. J. Hyg.*, **78**, 173 (1963).

[5] Puck, T. T., and Kao, F. T., *Proc. US Nat. Acad. Sci.*, **58**, 1227 (1967); Kao, F. T., and Puck, T. T., *ibid.*, **60**, 1275 (1968).

[6] Loveless, A., and Howarth, S., *Nature*, **184**, 1780 (1959); Zamenhof, P. J., *Proc. US Nat. Acad. Sci.*, **55**, 50 (1966); Adelburg, E. A., Mandel, M., and Chien Ching Chen, G., *Biochem. Biophys. Res. Commun.*, **18**, 788 (1965).

[7] Djordjeciv, B., and Szybalski, W., *J. Exp. Med.*, **112**, 509 (1960).

[8] Stahl, F. W., Craseman, J. M., Oknu, L., Fox, E., and Laird, C., *Virology*, **13**, 98 (1961).

[9] Edgar, R. S., and Lielausis, I., *Genetics*, **49**, 649 (1964).

[10] Westphal, H., and Dulbecco, R., *Proc. US Nat. Acad. Sci.*, **59**, 1158 (1968).

[11] Sabin, A., and Koch, M. A., *Proc. US Nat. Acad. Sci.*, **50**, 407 (1963); Watkins, J. F., and Dulbecco, R., *ibid.*, **58** (1967).

42

Reprinted from *Natl. Acad. Sci. (USA) Proc.* **66**:377–384 (1970)

Isolation of Temperature–Sensitive Mutants of L–Cells*

**L. H. Thompson,† R. Mankovitz,‡ R. M. Baker,§ J. E. Till,
L. Siminovitch, and G. F. Whitmore**

DEPARTMENTS OF MEDICAL BIOPHYSICS AND MEDICAL CELL BIOLOGY, UNIVERSITY OF TORONTO;
AND THE ONTARIO CANCER INSTITUTE, TORONTO, ONTARIO

Communicated by François Jacob, March 16, 1970

Abstract. Procedures are described for the isolation of conditional lethal mutants of mouse L-60T cells. The mutant lines were temperature sensitive by the following criteria: (*a*) colony-forming ability, (*b*) growth in suspension culture, and (*c*) rate of uptake of tritiated-thymidine.

The development and use of conditional lethal mutations has considerably facilitated the analysis of function in viruses and bacteria; as a result, the main features of regulation in these materials have been outlined. However, it is not yet clear how applicable these models of regulation are to eucaryotic cells and, in particular, whether or not they provide a basis for understanding the processes involved in the differentiation of mammalian cells. Although techniques have been developed for genetic studies with somatic cell hybrids,[1, 2] progress in this area has been severely inhibited by the limited numbers and kinds of mutants available. Drug resistant lines[3, 4] and a few auxotrophs[5–7] have been reported, and the only other major source of mutants is through the establishment of cell lines derived from the somatic tissues of patients with known genetically determined defects.[8] If techniques were developed for the isolation of conditional lethal mutants of somatic mammalian cells, it should be possible to obtain a much broader spectrum of mutations. In this paper we describe the isolation of temperature-sensitive mutants of mouse L-cells.

Materials and Methods. Mouse L-60T cells[9] were grown in suspension or as monolayer cultures using medium CMRL 1066[10] lacking nucleosides and coenzymes, and supplemented with 10% fetal bovine serum. For most of the experiments detailed here, the permissive temperature was 34°C and the nonpermissive temperature 38.5°C, although temperatures of 33° and 39°C were used in certain early experiments. Temperature regulation in water baths containing suspension cultures was within ±0.1° and in incubators ±0.2°. At 34° and 38.5°C the wild-type population had doubling times of about 18 and 16 hr, respectively.

Incorporation of radioactive precursors into acid-insoluble material was measured for a 30-min pulse in growth medium and the final concentration of these precursors during the pulse was always 1.6 μCi/ml. All radioactive material was obtained from Amersham-Searle, and the specific activities of the stock solutions of radioactive precursors were: tritiated thymidine (^{3}H-TdR), 21.9 Ci/mM; tritiated uridine (^{3}H-Ur), 15.3 Ci/mM; and tritiated lysine, 11.6 Ci/mM. Label incorporated into acid-precipitable material was measured using methods previously described.[11] The drug 1-β-D-arabinofuranosyl-cytosine (ara-C) was obtained from Sigma Chemical Co., and the mutagen *N*-methyl-*N'*-nitro-*N*-nitroso-guanidine (MNNG) was supplied by Mann Research Laboratories.

Cell volume measurements were made on samples removed from suspension cultures at various times, diluted extensively in cold phosphate-buffered saline[12] and passed as single cells through a Coulter-type aperture attached to a pulse height analysis system.[13]

Results.   (1) Isolation of temperature-sensitive mutants: The selection procedure that yielded the initial set of temperature-sensitive clones of L-60T cells is shown in Figure 1a.   An exponentially growing spinner culture at 33°C was treated with the mutagen MNNG (0.5 µg/ml) for 3 hr, which reduced the surviving fraction of cells to approximately 0.2.   After 3 days, the temperature of the culture was raised to 39°C; 24 hr later ^{3}H-TdR (21.9 Ci/mM) was added at a concentration of 2 µCi/ml and left in the culture for 2 days.   Cells cycling during this interval would be expected to incorporate the ^{3}H-TdR and be lethally irradiated[14] whereas temperature-sensitive cells would be arrested and insensitive to the killing agent.   Upon removal of the ^{3}H-TdR, the cells were placed in bottles at 33°C to allow multiplication of the survivors.   After about 2 weeks enough cells had accumulated to start a suspension culture.   The ^{3}H-TdR selec-

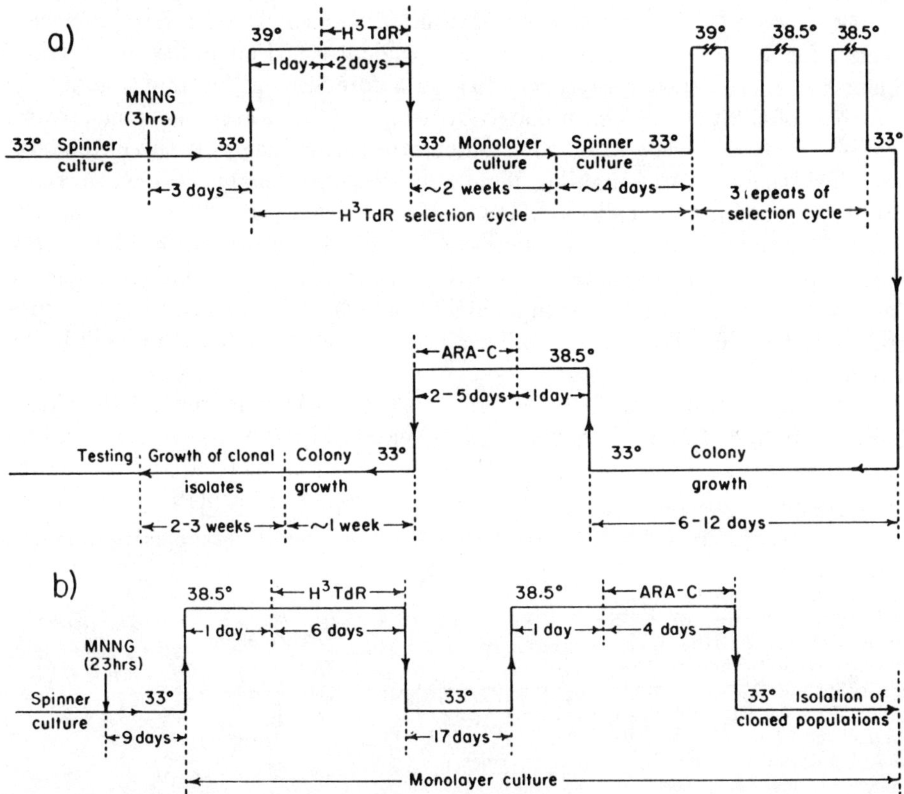

FIG. 1.—Schematic diagrams of the selection procedures used to obtain temperature-sensitive mutants.   (a) A selection which consisted of four selection cycles with ^{3}H-TdR followed by one selection with ara-C.   After the second cycle with ^{3}H-TdR the nonpermissive temperature was lowered from 39° to 38.5°C, which was used as the nonpermissive temperature in all later experiments.   (b) A selection which involved a single treatment with ^{3}H-TdR followed by a treatment with ara-C.

tion cycle was then repeated to enrich further for temperature-sensitive cells. Following a total of four selections with ^{3}H-TdR, single cells were plated and allowed to form small colonies. A final selection was made by treating these small colonies at 38.5° with ara-C at a concentration of 40 μM, to eliminate any clones[15] which were not temperature sensitive but simply resistant to ^{3}H-TdR. After surviving colonies had grown for about 10 days, several were isolated and grown into mass cultures. The testing of each clonal isolation for conditional temperature sensitivity was then performed.

A simplified version of the above selection scheme also produced one temperature-sensitive clone. As shown in Figure 1*b*, a two-step selection was performed with monolayer cultures using a single ^{3}H-TdR treatment followed by an ara-C treatment. Nine days after a spinner culture had been treated with 0.5 μg/ml of MNNG for 23 hr (surviving fraction approximately 0.1), two 32-oz bottles were each inoculated with 2×10^6 cells and incubated at 38.5°C. After 24 hr, ^{3}H-TdR (1 μCi/ml) was added for 6 days. The cultures were then given fresh medium lacking ^{3}H-TdR and incubated at 33°C until several macroscopic colonies had appeared. These colonies were shifted to 38.5°C and a day later were treated for 4 days with ara-C as in the original selection procedure. Those colonies which survived were allowed to grow before being picked and tested.

A total of eight clones were isolated by using the first selection method, seven of which *tsA1-tsA7*, showed obvious temperature-sensitive properties in the tests described below. The remaining clone exhibited questionable temperature sensitivity in preliminary tests and has not been examined further. From the second selection procedure three clones were isolated, at least one of which, *tsB1*, showed pronounced temperature-sensitive properties. Most of the data to be described has been obtained with a single clone, *tsA1*, isolated by the first procedure, but similar data have been obtained for the other clones and will be referred to briefly.

(2) **Temperature sensitivity of clonal isolate *tsA1* with respect to colony-forming ability:** Clones from the above selection experiments were grown into mass cultures and tested for temperature sensitivity by plating and by growth in suspension culture. Figure 2 illustrates the results of a plating experiment

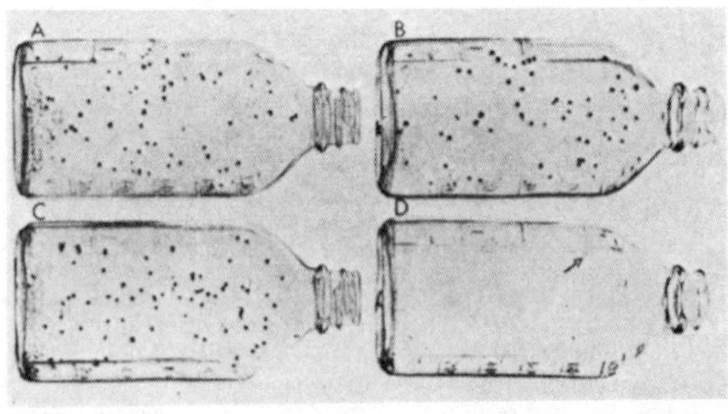

FIG. 2.—Temperature sensitivity of clonal isolate *tsA1* as shown by plating single cells. (*A*) Wild-type cells incubated at 34°C for 16 days; 100 cells inoculated; plating efficiency (P.E.) = 80%. (*B*). Wild type incubated at 38.5°C for 14 days; 100 cells inoculated; P.E. = 67%. (*C*) *tsA1* cells incubated at 34°C for 16 days; 100 cells inoculated; P.E. = 72%. (*D*) *tsA1* incubated at 38.5°C for 14 days; 10^5 cells inoculated; P.E. $\approx 10^{-5}$. The arrow designates the only macroscopic colony.

in which clone *tsA1* was compared with the parental wild-type L-cells.   At the permissive temperature (34°C), both cell types grew well.   The wild-type cells (Fig. 2*a*) had a plating efficiency of 80%, while for *tsA1* cells (Fig. 2*c*) the value was 72%.   There was no appreciable difference in colony size.   In contrast, at the nonpermissive temperature (38.5°C) the wild-type cells grew while the *tsA1* cells did not.   The wild-type cells (Fig. 2*b*) had a plating efficiency of 67%, whereas there were no survivors when as many as 10^4 tsA1 cells were inoculated. Bottles inoculated with 10^5 *tsA1* cells (Fig. 2*d*) showed a background of giant-like cells with an occasional small colony.   These results demonstrate that *tsA1* had the properties of a temperature-sensitive mutant.

(3)   **Temperature sensitivity of clone *tsA1* as measured by growth in suspension culture**.   Tests for temperature-dependent growth of *tsA1* cells in suspension cultures yielded results consistent with the plating experiments, as shown in Figure 3.   The control culture of wild-type cells used in this experiment had

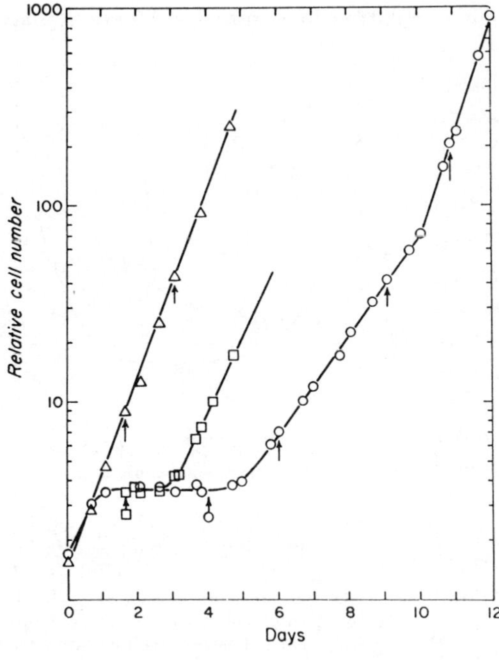

FIG. 3.—Growth curves of wild-type and *tsA1* cells in spinner-flask suspension culture.   At time zero, when both the wild type (△) and *tsA1* (○) were growing exponentially at 34°C, the temperature was raised to 38.5°C. At 40 hr the *tsA1* culture was diluted and split into two cultures.   One of the new cultures was immediately placed at 34°C;   the other was kept at 38.5°C until day 4, when it was returned to 34°C.   The arrows indicate when dilutions were made.

been kept at the permissive temperature since the time the selection experiments were begun, a period of about 4 months.   Raising the temperature from 34° to 38.5°C produced a striking difference in growth between the wild-type and *tsA1* cultures.   The cells in the control culture continued to increase in number with a 16-hr doubling time until the culture was discarded.   In contrast, the number of *tsA1* cells increased approximately twofold during the first 24 hr after the temperature shift to 38.5°C, and then showed no further increase while held at 38.5°C.   After 40 hr the culture of *tsA1* was split into two cultures, one of which was immediately cooled to 34°C where it resumed growth after 1 day (Fig. 3), rapidly achieving a doubling time comparable to that of the wild-type cells at

34°C. The other culture of *tsA1* cells, which remained at 38.5°C for a total of 96 hr, showed a longer-lasting effect of maintenance at the nonpermissive temperature. Although cell division began again within 24 hr, the cells grew more slowly, with a doubling time of about 28 hr, for the next 5 days. Thereafter, a return to a normal doubling time was observed. These results show that *tsA1* cells are temperature sensitive with respect to growth in suspension culture. They also indicate that the kinetics of growth of *tsA1* cells returned to the permissive temperature are dependent on the time interval spent at the nonpermissive temperature.

(4) **Growth of wild-type and mutant *tsA1* cells as a function of temperature:** A further series of growth curves was constructed to test for the effect of different temperatures on the growth of wild-type and *tsA1* cells. All cultures were derived from cells maintained initially at 34°C. After the growth rate at a new temperature had stabilized, a doubling time for the population was obtained from the exponential region of each growth curve if such an exponential portion existed. These values of doubling time are plotted in Figure 4. The doubling time of *tsA1* cells increased abruptly at temperatures above 37°C while a sizable increase in the doubling time of wild-type cells did not become apparent until 40°C.

Stable growth rates were not obtained in the temperature regions indicated by the dashed lines in Figure 4. However, pronounced differences were observed between the shapes of the curves of cell number versus time for the wild type as compared to the mutant populations. As previously shown in Figure 3, *tsA1* cells raised to 38.5°C approximately doubled in number and then showed no division for several days. At 38.0°C *tsA1* cells appeared to terminate division but over a period of about 12 days and after a 14-fold increase in cell number. In contrast to the gradual termination of growth seen for *tsA1* at temperatures from 37.5° to 38.5°C, wild-type cells exhibited exponential growth curves for temperatures up to 40.0°C while at 40.5°C no growth whatever was detectable. From these results it seems most unlikely that the cessation of growth of *tsA1* cells at temperatures just above 37°C is simply the result of an increase in the temperature sensitivity of the same mechanism which prevents the growth of wild-type cells above 40°C.

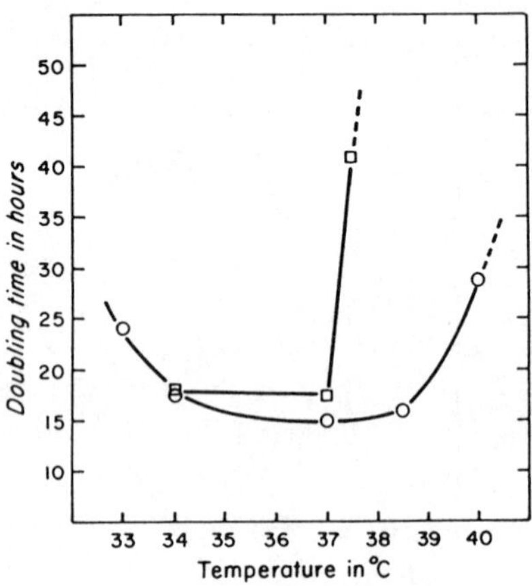

FIG. 4.—Relationship between population doubling time and temperature for wild type (O) and the mutant *tsA1* (□) grown in spinner flasks. Each doubling time value was obtained from a growth curve based on cell counts over periods as long as 14 days.

(5) **Incorporation of radioactive precursors:** The fact that the temperature sensitivity of the *tsA1* cells was apparently not due to a nonspecific loss of metabolic function prompted a preliminary investigation of the rate of uptake of various labeled precursors into acid-insoluble material, during a 30-min pulse,

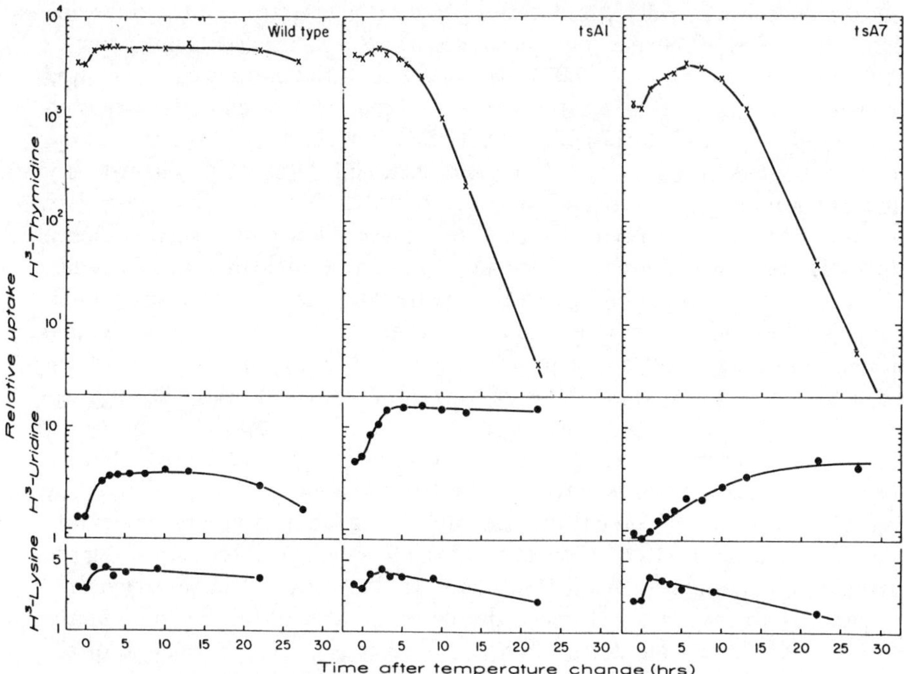

Fig. 5.—The rates of incorporation of ³H-TdR, ³II-Ur, and ³H-lysine into acid insoluble material in wild-type, *tsA1*, and *tsA7* cells. Incorporation per cell was measured over a 30-min period in suspension cultures both before and **after** a temperature shift from 34° to 38.5°C.

as a function of time after raising the temperature to 38.5°C. Figure 5 shows that for the wild-type population the temperature increase produced immediate increases in the rates of thymidine, uridine, and lysine incorporation and that new equilibrium values appeared to be rapidly established. For *tsA1* cells also the temperature increase resulted in an increased uptake of uridine and lysine, but in an immediate and marked decline in the incorporation of thymidine.

In addition to the experiments with *tsA1*, uptake measurements were made also on all of the other temperature-sensitive clones. The third column of Figure 5 shows data obtained for clone *tsA7*. The main differences between the results obtained with clone *tsA7*, as contrasted to *tsA1*, were the initial increase in the uptake of thymidine which preceded the rapid decline, and the lower rate of increase in the uptake of uridine. These data suggest that the temperature sensitivity of *tsA7*, as well as the other isolates, may result from changes in nucleotide metabolism or nucleic acid synthesis.

(6) **Cell volume changes:** One very noticeable feature of the growth of all of the temperature-sensitive isolates at 38.5°C was a rapid increase in cell size. Measurements of mean cell volume of *tsA1* over a period of 5 days at 38.5°C showed a linear increase in average cell size to a value 3.3 times that obtained at 34°C. Under similar conditions wild-type cells showed no change in average cell volume.

(7) **Other temperature-sensitive clonal isolates:** In addition to *tsA1*, the seven other mutant clones also have been tested for their plating efficiency,

suspension growth at 38.5°C, and rate of return to normal growth following temperature arrest. The plating efficiencies at 38.5°C ranged from a low of about 10^{-5} for *tsA1* to a high of 0.014 for *tsB1*. All of the mutant clones showed approximately a 1.5- to 2-fold total increase in cell number as suspension cultures, and this occurred during the first 24 hr after the temperature shift up. Cultures incubated at 38.5°C for 72 hr and then shifted to 34°C showed very marked differences in their resumption of normal growth rates. These data would appear to suggest that at least several independent mutations have been isolated. However, an alternative explanation is that a single mutant has undergone a series of alterations during the time required for growth to mass populations and for testing. This poses the question as to what extent an individual clone maintains its phenotype upon recloning. The results of a preliminary investigation of the phenotypic stability of *tsA1* are presented in the following section.

(8) **Subclones of mutant *tsA1*:** After initial testing of the *tiA1* cells, 10 subclones of this mutant were isolated and tested for temperature sensitivity by plating, growth in suspension culture, and uptake of radioactive precursors into acid-insoluble material. In all the tests, 9 of the 10 subclones were very similar to the parental *tsA1* population while the remaining clone, *tsA1S2*, exhibited altered properties. With inocula of 10^4 cells the parental population and nine of the subclones produced no colonies at 38.5°C. With inocula of 10^5 cells all of the subclones produced a variable background of giant cells containing the occasional cell cluster or "pseudocolony" composed of large cells. At this concentration, however, *tsA1S2* produced noticeably more growth than the other subclones, implying that many of the cells had undergone several divisions although no colonies were actually detected. In suspension culture at 38.5°C, only *tsA1S2* failed to yield arrest and recovery curves similar to those depicted in Figure 3. It showed instead an asymptotic type of growth curve with a 50-fold increase in cell number before multiplication ceased after approximately 14 days. In uptake experiments, only subclone *tsA1S2* was different from the parental *tsA1* line in that the decline of thymidine uptake was much less pronounced.

Discussion. The isolation of eight temperature-sensitive clones in two separate selection experiments shows that it is feasible to isolate such mutants from a heteroploid line of mammalian cells. There was a possibility that the procedures simply selected for mutants pre-existing in the populations. However, very recently we have obtained conditional mutants from populations cloned at 38.5°C to select against such pre-existing mutants.

The most extensive analysis of phenotype has been made with clone *tsA1*, but further experiments will be needed to delineate its defective function. In this mutant, as well as the others, cell division at 38.5°C appears to continue for about one generation before arrest. This suggests that a gene product must decay before the phenotype is expressed. It is also of interest that data not presented here suggest that populations maintained at the nonpermissive temperature for as long as 11 days still contain an appreciable fraction of viable cells.

Analysis of the phenotypes of the various mutants indicated noticeable differences among them. At first sight this would suggest that we have isolated lines containing independent mutations. However, such a conclusion may be premature in view of the result that one of the 10 subclones of *tsA1* appeared to differ significantly from the parental line.

The exact chromosomal nature of the conditional lethal mutations which we have isolated cannot as yet be specified. There is, of course, no method of genetic analysis available which would allow one to state whether or not these are point mutations. The problem in this particular case is compounded by the fact that L-cell lines are highly heteroploid and are known to undergo extensive karyotypic fluctuations. Changes in phenotype could occur due to chromosomal rearrangements or to loss of specific chromosomes. It seems unlikely that the observed temperature sensitivity is due to chromosomal deletion, since preliminary results indicate that partial revertants can be isolated from temperature sensitive populations.

The availability of these conditional lethal mutants of L-cells and those reported recently by Naha[16] for a heteropolid line of African green monkey kidney cells should accelerate developments in somatic cell genetics. However, it will also be important to show whether or not the methods can be applied to cells with diploid karyotypes and to cells capable of differentiation. In this regard, it is of interest that we have recently been able to isolate a temperature-sensitive mutant of the Chinese hamster line V79-S171,[17] which appears to have a chromosome complement which is near diploid.

* Supported by the Medical Research Council of Canada and the National Cancer Institute of Canada.

† Requests for reprints may be sent to Dr. Thompson, The Ontario Cancer Institute, 500 Sherbourne Street, Toronto 284, Ontario, Canada.

‡ Supported by a postdoctoral fellowship from the Medical Research Council of Canada.

§ Supported by a postdoctoral fellowship from the U.S. Public Health Service.

[1] Ephrussi, B., and S. Sorieul, *Compt. Rend. Acad. Sci.*, **254**, 181 (1962).

[2] Matsuya, G., H. Green, and C. Basilico, *Nature*, **220**, 1199 (1968).

[3] Gartler, S. M., and D. A. Pious, *Human Genetik.*, **2**, 83 (1966).

[4] Chu, E. H. Y., and H. V. Malling, these PROCEEDINGS, **61**, 1306 (1968).

[5] Chu, E. H. Y., P. Brimer, K. B. Jacobson, and E. V. Merriam, *Genetics*, **62**, 359 (1969).

[6] Kao, F. T., and T. T. Puck, *J. Cell Physiol.*, **74**, 245 (1969).

[7] Kao, F. T., and T. T. Puck, these PROCEEDINGS, **60**, 1275 (1968).

[8] Krooth, R. S., G. A. Darlington, and A. A. Velazquez, *Annual Revs. Genetics*, **2**, 141 (1968).

[9] Till, J. E., G. F. Whitmore, and S. Gulyas, *Biochim. Biophys. Acta*, **72**, 277 (1963).

[10] Parker, R. C., *Methods of Tissue Culture*, 3rd ed. (New York: Hoeber 1961).

[11] Bacchetti, S., and G. F. Whitmore, *Biophys. J.*, **9**, 1427 (1969).

[12] Dulbecco, R., and M. Vogt, *J. Exptl. Med.*, **99**, 167 (1954).

[13] Taylor, W. B., *Med. Biol. Eng.*, in press.

[14] Whitmore, G. F., and S. Gulyas, *Science*, **151**, 691 (1966).

[15] Karon, M., and S. Shirakawa, *Cancer Research*, **29**, 687 (1969).

[16] Naha, P. M., *Nature*, **223**, 1380 (1969).

[17] Han, A., and W. K. Sinclair, *Biophys. J.*, **9**, 1171 (1969). We thank Dr. Sinclair for his kind cooperation in providing us with the V79-S171 Chinese hamster line.

Editor's Comments
on Papers 43 Through 47

43 KAO, JOHNSON, and PUCK
Complementation Analysis on Virus-Fused Chinese Hamster Cells with Nutritional Markers

44 BAKER et al.
Ouabain-Resistant Mutants of Mouse and Hamster Cells in Culture

45 HARRIS
Mutation Rates in Cells at Different Ploidy Levels

46 BEAUDET, ROUFA, and CASKEY
Mutations Affecting the Structure of Hypoxanthine: Guanine Phosphoribosyltransferase in Cultured Chinese Hamster Cells

47 CHASIN and URLAUB
Chromosome-Wide Event Accompanies the Expression of Recessive Mutations in Tetraploid Cells

THE NATURE OF MAMMALIAN CELL MUTANTS

Since the isolation of the first mutants of cultured mammalian cells in the late 1950s, a great number of mutants have been isolated and characterized. These mutants have provided information on a wide range of cellular activities. The preceding sections of this book dealt with the selection and characterization of some types of mammalian cell mutants. The concluding section will focus on studies in which these mutants have been used to provide information on the genetic nature of mutant cells selected in culture. One additional system, involving a membrane mutant with special relevance for cell hybridization studies, also is included.

The use of auxotrophic mutants for establishing complementation groups of mammalian cell mutants was demonstrated by Puck and co-workers (Paper 43). These studies involved the hybridization of different auxotrophic mutants isolated by the *bromodeoxyuridine-light* technique (see Paper 40). Two independently isolated glycine auxotrophs were fused with each other, and hybrids were selected in glycine-deficient medium. Since the hybrids could grow in the absence of glycine, it was concluded that the alterations in the two glycine auxotrophs were in different genetic units, that is, they belonged to different complementation groups.

In a subsequent study, thirteen independent glycine auxotrophs were fused with each other in various combinations (Kao et al., 1969). The results provided evidence for four distinct complementation groups. Mutants of one complementation group were shown to be deficient in serine hydroxymethylase activity, while mutants of the other three groups exhibited normal levels of activity for the enzyme. Beginning with these early studies, complementation analysis has been a very useful technique for the genetic analysis of auxotrophic mutants of mammalian cells.

Mammalian cell mutants involving membrane functions have been extensively analyzed in order to elucidate the structure and function of cell membranes. One type of membrane mutant, selected for resistance to the cardiac glycoside ouabain, has become especially useful in cell hybridization studies. Cells resistant to ouabain, which inhibits the active transport of sodium and potassium ions, were first isolated by Mayhew (1972). In the resistant cells, active transport of Na^+ and K^+ was no longer inhibited by ouabain.

The biochemical basis for ouabain resistance and its mode of inheritance was studied by Siminovitch and co-workers (Paper 44). In these studies, ouabain resistant Chinese hamster cells were isolated and characterized by both biochemical and genetic techniques. The Na/K ATPase activity in isolated plasma membranes of the mutants was found to be much more resistant to ouabain than was the activity from ouabain sensitive cells. Cell hybridization studies were carried out to investigate the inheritance of ouabain resistance. In hybrids between ouabain sensitive and resistant cells, ouabain resistance was expressed as a codominant trait, that is, the hybrids exhibited a level of resistance intermediate between the parental cells. More importantly, it was shown that ouabain could be used to select hybrids between ouabain sensitive and resistant cells, and this provided a new system for hybrid cell selection. If one of the parental cells in a cross carried both a recessive marker (e.g., inability to grow in HAT medium) and also the codominant ouabain resistance marker, the selection of hybrids did not require the presence of a selectable marker in the other parental cell.

Another important feature of the ouabain system is that human cells are inherently several thousand-fold more sensitive to ouabain than are rodent cells. Thus, for crosses between human and rodent cells, the human cells have an *inbuilt* marker against which selection can be applied, and no additional marker needs to be introduced into the human cells in order to permit the selection of hybrids (Kucherlapati et al., 1975).

As discussed in earlier sections of this volume, azaguanine (AZG) resistance (associated with the absence of hypoxanthine guanine phosphoribosyltransferase, HGPRT) was one of the earliest types of drug resistance to be studied, as well as one of the most important. For almost two decades, AZG resistance provided a focal point for studies on mammalian cell mutants. Studies by Harris (Paper 45) on the frequency of AZG resistant cells stimulated much investigation and debate on the genetic nature of *mutants* selected in culture. In these studies, the mutation rate for AZG resistance was determined in near diploid, tetraploid, and octaploid Chinese hamster cells.

Since AZG resistance behaves as a recessive trait (in cell hybrids), the frequency of AZG resistance would be expected to decrease exponentially as the ploidy of the cells increases. (This should be the case even taking into account the X linkage of HGPRT and the phenomenon of X chromosome inactivation, since an increasing number of X chromosomes are active in cells of higher ploidy.) Instead, the frequency of AZG resistance remained approximately constant in the cells at different ploidy levels. It was suggested that at least some drug resistant *mutants* in cell culture arise by stable changes in phenotypic expression (i.e., as epigenetic variants) rather than by changes in genetic information. These studies were not conclusive, since HGPRT deficiency was not demonstrated in the AZG resistant cells (other alterations are known to result in AZG resistance) and the chromosome number in the resistant cells was not analyzed. Nevertheless, these results raised significant questions about the nature of phenotypic variants selected in culture.

The genetic mechanisms responsible for AZG resistance were investigated by Caskey and co-workers (Paper 46) and Capecchi and co-workers (Sharp et al., 1973). The results of these studies demonstrated structural changes in HGPRT molecules in at least some drug resistant cells. In the studies by Caskey and co-workers, an antiserum against HGPRT was used to test for the presence of immunologic activity in AZG resistant cells. Some clones of AZG resistant cells had no HGPRT enzymatic activity but still reacted strongly with the antiserum. These results indicated the presence of defective HGPRT molecules with immunologic but not enzymatic activity, and suggested that at least some AZG resistant cells result from mutations in the HGPRT structural gene.

A similar conclusion was reached in the studies by Capecchi and co-workers. In those studies, the HGPRT activity was characterized in HAT resistant revertants of AZG resistant cells. Some of the revertants exhibited HGPRT activity with properties unlike the wild type enzyme. These results also suggested that changes in the HGPRT structural gene had occurred in at least some cells selected for altered HGPRT activity.

The techniques of cell hybridization were used by Chasin and Urlaub (Paper 47) to further study the genetic nature of HGPRT deficiency. These studies involved the construction of tetraploid hybrids that contained two active HGPRT alleles and were heterozygous at the glucose 6 phosphate dehydrogenase (G6PD) locus. (Only one of the parental cells hybridized expressed G6PD activity; the other parental cell was a G6PD deficient mutant.) Since the G6PD locus, like HGPRT, is X linked, it was possible to follow two linked loci in the hybrids. HGPRT deficient derivatives of the hybrids were selected and tested for G6PD activity, and only about 50% of the HGPRT deficient hybrids expressed G6PD activity. These results suggested that a chromosome-wide event, affecting the G6PD locus as well as HGPRT, was involved in generating HGPRT deficient mutants from the tetraploid hybrids. Thus, mutation in only one, not both, of the HGPRT alleles would be required to produce HGPRT deficiency in these tetraploid cells. The most likely explanation for the nonmutational loss of the other HGPRT allele (and the linked G6PD allele) would appear to be chromosome loss, which could occur at a high frequency. These results could provide an explanation for the apparently aberrant frequency of AZG resistance in cells of higher ploidy, as observed by Harris (see Paper 45).

The studies on the nature of HGPRT deficient cells suggested that at least some of the drug resistant cells selected in culture are the result of true genetic alterations, that is, mutations. Similar results also have been obtained with other types of drug resistant cells. Thus, the questions raised by the results of Harris (see Paper 45) about the involvement of epigenetic changes in the development of drug resistance appeared to be resolved. However, Harris (1982) recently provided evidence for nonmutational alterations (possibly involving changes in DNA methylation), which could yield drug resistant, enzyme deficient cells. Thus, the extent to which nonmutational mechanisms are involved in the generation of mammalian cell *mutants* still remains an open question.

REFERENCES

Harris, M., 1982, Induction of Thymidine Kinase in Enzyme-Deficient Chinese Hamster Cells, *Cell* **29:**483–492.

Kao, F.-T., L. Chasin, and T. T. Puck, 1969, Genetics of Somatic Mammalian Cells, X. Complementation Analysis of Glycine-Requiring Mutants, *Natl. Acad. Sci. (USA) Proc.* **64:**1284–1291.

Kucherlapati, R. S., R. M. Baker, and F. H. Ruddle, 1975, Ouabain as a Selective Agent in the Isolation of Somatic Cell Hybrids, *Cytogenet. Cell Genet.* **14:**362–363.

Mayhew, E., 1972, Ion Transport by Ouabain Resistant and Sensitive Erlich Ascites Carcinoma Cells, *J. Cell Physiol.* **79:**441–452.

Sharp, J. D., N. E. Capecchi, and M. R. Capecchi, 1973, Altered Enzymes in Drug-Resistant Variants of Mammalian Tissue Culture Cells, *Natl. Acad. Sci. (USA) Proc.* **70:**3145–3149.

43

Reprinted from *Science* **164**:312–314 (1969)

COMPLEMENTATION ANALYSIS ON VIRUS-FUSED CHINESE HAMSTER CELLS WITH NUTRITIONAL MARKERS

F.-T. Kao, R. T. Johnson, and T. T. Puck

Abstract. *Cell fusion experiments have been carried out with Chinese hamster cell mutants with different nutritional growth requirements. Conditions have been devised in which approximately 1 to 2 percent of the cell population remaining after fusion are fused, hybrid cells. The all-or-none nature of the genetic markers employed and the extremely low reversion rates insure that no contamination of the hybrid population with parental forms occurs. Hybrids between glycine- and hypoxanthine-requiring mutants are prototrophic, which indicates that both mutations are recessive. Hybrids between a glycine-deficient mutant and a single-step mutant which requires glycine, hypoxanthine, and thymidine are relieved of the glycine dependency, an indication that the two loci associated with glycine dependence are different. This mutation to the triple-supplement requirement as well as a proline deficiency were also shown to be recessive mutations. The system appears applicable to a variety of genetic problems.*

Experiments have been described (*1*) in which single-step gene mutations could be induced in Chinese hamster cells grown in tissue culture for long periods and possessed of reasonably stable karyotypes. These mutations are produced by standard mutagens such as ethyl methanesulfonate and N-methyl-N'-nitro-N-nitrosoguanidine; they have spontaneous reversion frequencies varying between 10^{-6} and 2×10^{-8}; their corresponding reverse mutations can also be elicited by standard mutagens; and the nutritional requirements introduced appear to be absolute, since no growth whatever is exhibited by deficient mutants in the absence of the specific supplement. In the case of a spontaneous proline-deficient mutation, it was demonstrated that the mutant makes no proline, that the block in the biosynthetic chain lies in the step converting glutamic acid to its gamma-semialdehyde, and that the revertant to proline independence obtained from the mutant cell behaves like a hetero- or a hemizygote with respect to this gene (*2*). Availability of these markers makes possible application of the cell fusion technique (*3*) to obtain rapid and convenient quantitation and isolation of hybrid cells carrying combined genetic markers, without contamination by the parental forms. It also permits conclusions about dominance and recessiveness, gene dosage effects, and linkage in relatively well-characterized genetic markers.

In all cases, the growth medium consisted of F12 (*4*) plus the macromolecular component of fetal calf serum, with specific omissions of metabolites as indicated. Mutants of the Chinese hamster ovary cell (CHO) were prepared by chemical or x-ray treatment, and clonal stocks were isolated by means of the 5-bromodeoxyuridine (BUdR)-"visible" light technique previously described (*5*). The parental form, CHO-K1, has a requirement for proline, a modal chromosome number of 20, and a reasonably constant karyotype which is shared by all the biochemical mutants derived from it through mutagenesis. These include glycine-requiring (gly⁻) and hypoxanthine-requiring (hyp⁻) mutants. In addition, a mutant derived from a Chinese hamster lung cell (CHL), in which a triple requirement for glycine, hypoxanthine,

and thymidine was obtained (gly⁻-hyp⁻-thy⁻) in a single treatment with ethyl methanesulfonate, was utilized. This cell and mutants derived from it have no proline requirement, and their growth rate is maximal even in the absence of added proline, an indication that they contain at least two functional proline genes (*2*). All of the mutant cells employed have plating efficiencies of zero in the absence of their specific nutritional supplements, while the wild types have a plating efficiency of about 80 percent either in the presence or absence of supplements.

The following standard cell fusion procedure was adopted: 5×10^5 cells of each desired type are added to a final volume of 1.0 ml of F12 minus the critical nutrilites. Ultraviolet-inactivated Sendai virus is added to produce a final titer of 200 hemagglutinating units. The mixture is kept at 4.0°C for 15 minutes, and then transferred to a shaker bath (37°C) for 15 minutes.

Experiments were designed to provide good yields of fused cells in which

the binucleate forms predominate, thus avoiding the additional complications of the more highly polyploid forms. In a typical experiment, utilizing gly⁻ and hyp⁻ mutants, aliquots were plated out after the fusion period and the percentage of mono- and multinucleate cells present was counted microscopically. The resulting cell population contained 86.5 percent mono-, 8.5 percent bi-, 3.8 percent tri- and 0.9 percent tetranucleate cells.

When the entire cell population remaining after such fusions was plated in F12 lacking both glycine and hypoxanthine, approximately 1 to 2 percent of the plated cells grew into colonies. This represents a relatively high efficiency of hybrid recovery, since fusion is presumably random, so that homologous or heterologous cell fusion is equally likely. Control experiments carried out with cells of a single genotype yielded the same pattern of multinucleate cells as when two different auxotrophs were employed, but no growth whatever occurred when such

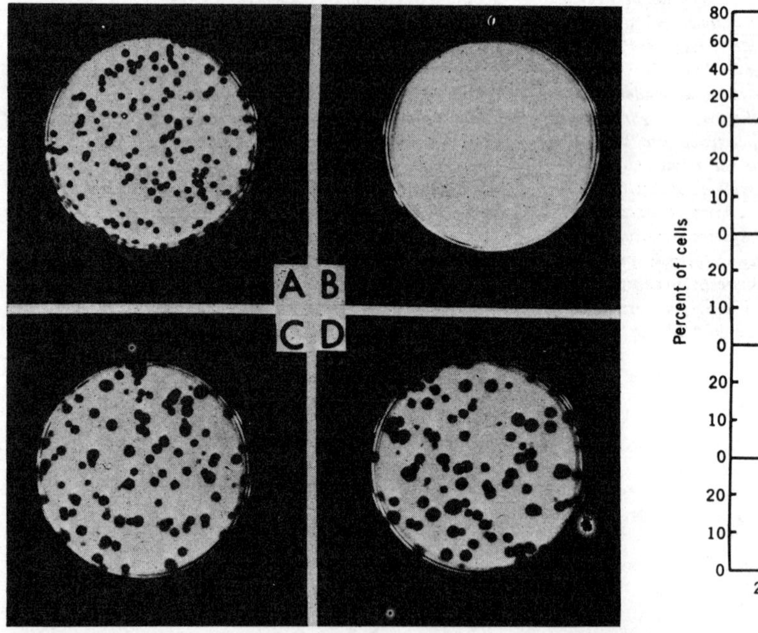

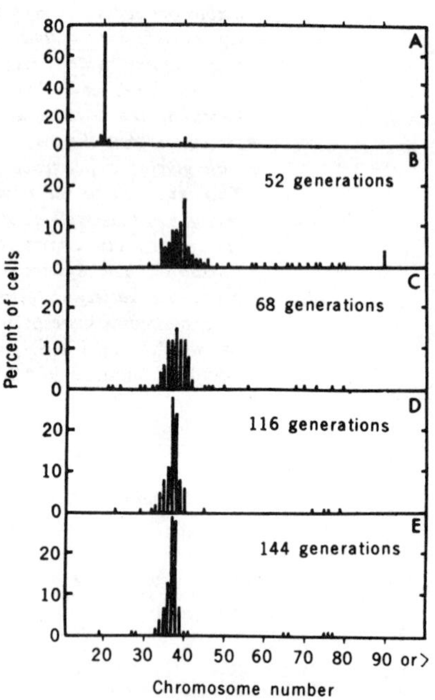

Fig. 1 (left). The appearance of plates seeded with identical numbers of single cells in the various selective media and incubated for 7 days. (A) The gly⁻ mutant in complete medium. (B) The gly⁻ mutant in the glycine-deficient medium. The hyp⁻ mutant exhibits a similar contrast in complete and hypoxanthine-deficient media, respectively. (C) The gly⁻hyp⁻ hybrid in complete medium. (D) The gly⁻hyp⁻ hybrid in medium lacking both glycine and hypoxanthine. Fig. 2 (right). Comparison of the chromosomal number distribution in the parental and hybrid cells. (A) chromosomal distribution of either parental cell. (B–E) Chromosomal distribution of a typical hybrid cell clone, as determined at various periods of growth after the fusion. The initial counts approximate the tetraploid number, but also contain appreciable numbers of more polyploid forms. With the passage of time, the modal number decreases by 5 to 10 percent, and most of the forms with chromosome numbers greater than 40 disappear. By the 116th generation, the distribution of chromosome counts appears to have stabilized, and has remained essentially unchanged after 200 generations. A concomitant increase in plating efficiency from about 50 percent to 70 to 80 percent occurs during this period, presumably because of the population drift to a distribution of forms with more stable chromosomal constitutions.

Table 1. Demonstration that while each of the mutant cells grows only in the supplemented medium, their fused hybrids grow without supplementation.

Clone	Plating efficiency		Generation time (hours)	
	Complete F12	F12 minus (gly + hyp)	Complete F12	F12 minus (gly + hyp)
Wild-type CHO-K1	86.5	84.8	12.2 ± 0.4	12.0 ± 0.3
Gly⁻ mutant	84.3	0	12.1 ± 0.4	∞
Hyp⁻ mutant	85.2	0	12.3 ± 0.5	∞
Hybrid clones:				
807–3	47.0	49.5	12.3 ± 0.4	12.3 ± 0.6
807–4	56.8	59.0	11.8 ± 0.3	12.0 ± 0.5
807–6	55.6	53.5	12.0 ± 0.5	12.1 ± 0.5

fused cells were plated in medium lacking both glycine and hypoxanthine.

Four colonies were picked and clonal cultures were established. All of these exhibited excellent single cell growth in the doubly deficient medium. Since the spontaneous reversion rate of each of the two parental auxotrophs is less than 5×10^{-8}, there is no doubt that all of these clones arose from the fusion process (Fig. 1 and Table 1). These data permit the conclusion that unless strange processes are operating, each of the mutations to glycine and hypoxanthine requirement, respectively, is recessive, since the fused cells show neither deficiency. Table 1 also demonstrates that the growth rate (as revealed by the generation time) of the colonies formed by the hybrids is maximally rapid and identical to the wild type, both in the deficient and the complete media.

In Fig. 2 a comparison of the parental and hybrid chromosomal counts is presented. As expected, the modal chromosome number of the hybrid clearly approximates twice that of the parental forms. With the passage of time, the hybridized cells at first lose chromosomes, eventually reaching a stable modal chromosome number of about 37 to 38. This behavior accords with that noted by previous investigators (6) [except for one case involving fusion of human and mouse cells which has been described, in which the loss of human chromosomes was extensive (6)]. Concomitant with these karyotypic changes, a definite increase in the plating efficiency of the clones occurs, raising their plating efficiency close to that of the parental forms in the supplemented medium.

It has been demonstrated (1) that a class of mutants can be produced which acquire simultaneously the need for supplementation with glycine, hypoxanthine, and thymidine for cell growth. It was postulated that only a single gene change had occurred, perhaps the one involved in the synthesis of tetrahydro-

folic acid. If these three deficiencies result from a single gene change, it would necessarily be a different gene from that responsible for the glycine deficiency alone. In that case fusion between a mutant deficient for glycine only, and a triply deficient mutant, should produce a cell with no glycine requirement for growth.

A standard fusion experiment was performed between a glycine-deficient CHO-K1 cell and the triply deficient gly⁻-hyp⁻-thy⁻ CHL cell. Abundant colony formation from the fused cells was obtained in the absence of any added glycine. Four clones were picked and all exhibited excellent growth (plating efficiencies varying between 21 and 68 percent; generation times of about 12 hours) which was identical in the presence or absence of glycine. Hence the gene responsible for the single glycine deficiency must be different from that involved in the gly-hyp-thy requirement.

The CHO-K1 cell employed in this experiment also requires proline, while the CHL cell does not. The fused cells grew with maximum plating efficiency and growth rate in the presence or absence of any of the following combinations of supplements: proline; glycine; glycine plus thymidine plus hypoxanthine; or all four metabolites combined. Hence the proline gene defect as well as that responsible for the triple deficiency are recessive mutations.

It is of interest that the proline-independent hybrid cell obtained has a maximally high growth rate even in the absence of added proline, which is in contrast to the behavior of the pro⁺ revertant obtained from the pro⁻ cell (2). This result follows expectation, however, because as was shown previously, the pro⁺ revertant behaves like a hetero- or hemizygote, whereas the hybrid cell has presumably received at least two proline genes from the CHL cell so that it should have sufficient proline-synthetic capacity to be inde-

pendent of any exogenous supplementation (2).

The change in chromosome number with time in these clones, together with the gradual increase in plating efficiency during the first several weeks following clone isolation, is consistent with the idea that the initial, chromosomally unstable hybrid cell eventually achieves a stable form.

Since the mutations to deficiency for glycine, hypoxanthine, proline, and the triple glycine-hypoxanthine-thymidine deficiency are all recessive by the complementation test utilized here, it is likely that each of the original mutations involves change such as inactivation of a structural gene, rather than activation of regulatory substances capable of shutting off intact structural genes.

The methodologies applied here can be used for further complementation studies. For example, experiments have shown that fusion of two glycine mutants of independent origin, each requiring glycine alone, has produced a fully glycine-independent hybrid. Therefore, these forms, with exactly the same nutritional requirement, are probably mutated in two different genes of the glycine synthetic chain (or may be exhibiting intragenic complementation, although the latter possibility is unlikely since a high frequency of prototrophs was obtained and these exhibited maximum growth rate in glycine-free medium). The highly selective nature of the genetic markers described here makes this system ideal for other investigations such as gene linkage, and the search for genetic recombination.

References and Notes

1. F. T. Kao and T. T. Puck, *Proc. Nat. Acad. Sci. U.S.* **60**, 1275 (1968).
2. T. T. Puck and F. T. Kao, *ibid.*, p. 561; F. T. Kao and T. T. Puck, *Genetics* **55**, 513 (1967).
3. H. Harris, and J. F. Watkins, *Nature* **205**, 640 (1965); B. Ephrussi and M. C. Weiss, *Proc. Nat. Acad. Sci. U.S.* **53**, 1040 (1965).
4. R. G. Ham, *Proc. Nat. Acad. Sci. U.S.* **53**, 288 (1965).
5. T. T. Puck and F. T. Kao, *ibid.* **58**, 1227 (1967).
6. M. C. Weiss and B. Ephrussi, *Genetics* **54**, 1095 (1966); M. C. Weiss and H. Green, *Proc. Nat. Acad. Sci. U.S.* **58**, 1104 (1967).
7. This is contribution No. 357 from the Eleanor Roosevelt Institute for Cancer Research and the Department of Biophysics, University of Colorado Medical Center, Denver. This investigation, which is No. VIII in a series entitled "Genetics of Somatic Mammalian Cells," was aided by PHS grant No. 5 PO1 HD02080 from the National Institute of Child Health and Human Development. R.T.J. is a postdoctoral fellow, Damon Runyon Memorial Fund for Cancer Research, Inc. T.T.P. is an American Cancer Society research professor.

30 December 1968

44

Copyright © 1974 by MIT
Reprinted from *Cell* 1:9–21 (1974)

OUABAIN-RESISTANT MUTANTS OF MOUSE AND HAMSTER CELLS IN CULTURE

R. M. Baker, D. M. Brunette, R. Mankovitz, L. H. Thompson, G. F. Whitmore, L. Siminovitch, and J. E. Till

Summary

Somatic cell mutants resistant to ouabain, which inhibits the plasma membrane Na/K ATPase, have been isolated from mouse L and Chinese hamster ovary (CHO) cells. Ouabain at concentrations $\geq$ 1 mM with 5–6 mM K$^+$, or $\geq$ 0.1 mM with 0.5 mM K$^+$, inhibits the growth of the wild-type cells and is ultimately cytotoxic. Clones 2- to 100-fold more resistant than wild type in terms of dose can be obtained by single-step selection from a wild-type population in the presence of ouabain. The phenotypes of ouabain-resistant (OUA^R) clones are reproducible with high fidelity and stable over long intervals of growth in the absence of the selecting drug. Wild-type and OUA^R L cell clones were compared with respect to their susceptibility to ouabain inhibition of ^{42}K uptake by whole cells and of Na/K ATPase activity in isolated plasma membranes. In both respects the OUA^R cells are less sensitive to the drug than are wild-type cells. Conditions for optimal ATPase activity in the absence of ouabain were indistinguishable for the wild-type and one OUA^R clone examined in detail and were comparable to requirements reported for ATPase preparations from other source materials. The frequency of OUA^R cells in a wild-type population can be substantially increased, to approximately 10^{-4} per viable cell, by exposure to the chemical mutagen EMS. The spontaneous mutation rate to 10-fold increase in ouabain-resistance is estimated by Luria-Delbruck fluctuation analyses to be 5–6 × 10^{-8} per cell per generation for both L and CHO cells. Cell-cell hybridization experiments utilizing OUA^R and wild-type CHO cells indicate that resistance to ouabain behaves as a codominant trait, and that this marker can be useful for selection of somatic cell hybrids.

Introduction

Because of the significance of the plasma membrane to the integrity and control of cellular processes, efforts to understand its properties and function comprise a major focus of cell biology (Wallach, 1972). With the development of expertise in somatic cell genetics, it is now feasible to isolate mutants in culture and to conduct comparative analyses at the cellular, organelle, and molecular levels of membrane-associated structures and functions in wild-type and specific mutant cells.

This article concerns genetic alterations that affect a property of the Na$^+$ + K$^+$ transport system of mammalian cells. Active Na$^+$ + K$^+$ transport has been associated with the Na$^+$-K$^+$-Mg^{++}-activated ATPase (EC 3.6.1.3, hereafter abbreviated "Na/K ATPase") of the plasma membrane (Skou, 1957; Albers, 1967), an enzyme which is specifically inhibited by the steroid compound ouabain (Glynn, 1964). The growth of cells in culture can be inhibited by ouabain (Mayhew and Levinson, 1968; McDonald et al., 1972); utilizing this observation. Mayhew (1972) was able to select and examine subpopulations of Ehrlich ascites cells that were relatively resistant to the drug. We have found that clones of established cell lines with increased resistance to ouabain cytotoxicity can be obtained readily in single-step selections. The isolation and characteristics of such ouabain-resistant (OUA^R) mouse L cells and Chinese hamster ovary (CHO) cells are described below. Our results indicate that these ouabain-resistant cells are mutant and that they differ from wild type principally in relative resistance to ouabain inhibition of their plasma membrane Na/K ATPase activity.

Preliminary and summary accounts of this work have appeared elsewhere (Baker and Till, 1971; Brunette and Till, 1972; Siminovitch et al., 1972; Till et al., 1973; Thompson and Baker, 1973; Baker and Mankovitz, 1973).

Results

A. Selection of Ouabain-Resistant Cell Lines

The multiplication of wild-type L cells and CHO cells is progressively inhibited with increasing ouabain concentration from 0.1 mM to 1 mM in α medium. Figure 1 illustrates that when exponentially growing wild-type L cells are exposed to 1.0 mM ouabain, cell number increases by approximately 20% and then remains stationary. The inhibition is reversible by resuspension of the cells in medium without ouabain but, as can be seen in Figure 1, the viability of the arrested cells decreases with time so that after 12 days incubation in the presence of drug their plating efficiency (PE) relative to untreated cells has been reduced to nearly 10^{-4}.

334

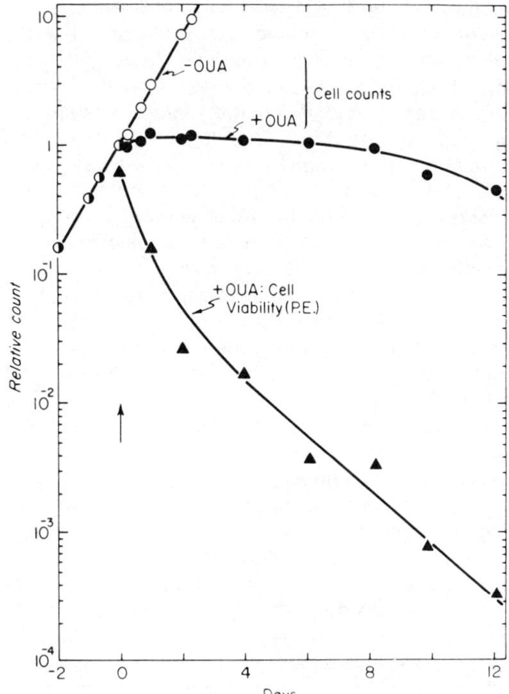

Figure 1. The Inhibition of Growth of Wild-Type L Cells by 1 mM Ouabain (6 mM K+) and Their Loss of Viability with Time

At zero time (arrow) an exponentially growing culture of wild-type cells was divided and maintained with (●) and without (○) 1 mM ouabain present in the medium. Growth as a function of time was followed using a Coulter cell counter. Viability of the cells in ouabain (▲) was assayed by sampling aliquots of cells, centrifuging and resuspending them in fresh drug-free medium, and measuring their plating efficiency in the absence of drug.

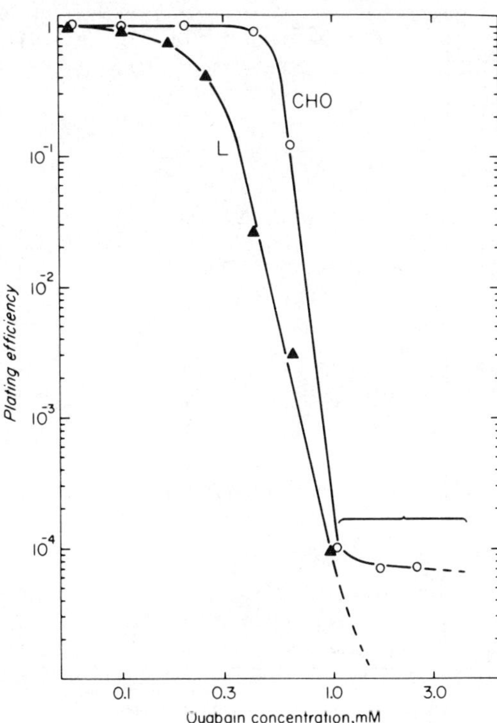

Figure 2. The Plating Efficiencies of Wild-Type L Cells (▲) and CHO Cells (○) as a Function of Ouabain Concentration in Culture Medium with 6 mM K+ at 37°C

The bracket denotes the range of ouabain concentrations selective for ouabain-resistant cells (see text).

The plating efficiencies of wild-type L and CHO cells as a function of ouabain concentration in the culture medium are shown in Figure 2. In both cases there is a diminution in colony size for ouabain concentrations from 0.1–0.3 mM and a striking drop in PE over several orders of magnitude for concentrations greater than 0.3 mM. The colonies scored at ouabain concentrations corresponding to the steep portions of the dose-response curves are generally small and appear to be statistical anomalies associated with incomplete inhibition of wild-type cell multiplication.

For ouabain concentrations $\geq$ 1 mM in the culture medium, the curves for PE of wild-type cells versus drug concentration tail off in the region of 10^{-4} to 10^{-7} colonies per viable cell inoculated (Figure 2; see also Figure 10 below), depending upon the history of the particular culture (cf. Thompson and Baker, 1973). The much reduced slope of the curves in this dose range and the relatively large, though uncommonly heterogeneous, sizes of the colonies that occur indicate a subpop-

ulation of cells with various degrees of resistance to the drug. When isolated and grown to mass population in nonselective medium, these clones retain a ouabain-resistant phenotype. This is illustrated in Figure 3, which shows the plating efficiencies in ouabain of three such clones of ouabain-resistant L cells in comparison to the wild type. In terms of the drug concentration required to reduce PE to 10% of that observed in the absence of drug (the D_{10}), clone OUA^R-LA3 is about six times more resistant than wild type, while clones OUA^R-LA2 and OUA^R-LA7 are fully resistant to ouabain at a concentration 10-fold higher than the D_{10} for wild-type L cells.

In order to examine the behavior of wild-type and ouabain-resistant cells at higher effective doses of ouabain, culture medium with a reduced K+ concentration was utilized. As reported previously, the cytotoxicity of ouabain is accentuated in low potassium medium (Till et al., 1973); this is consistent with the inference that cell killing is due to K+ starvation and observations that K+ is antagonistic to ouabain binding and to inhibition of K+ uptake and Na/K ATPase activity in Hela cells (Baker and Willis, 1970; Lamb and McCall, 1972).

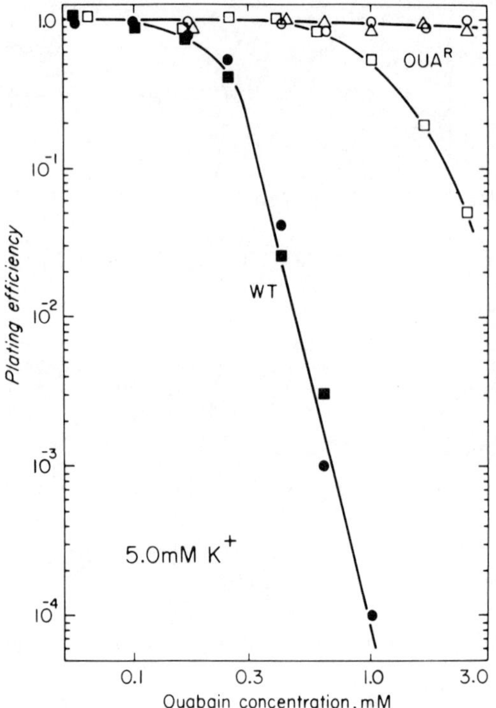

Figure 3. Plating Efficiencies as a Function of Ouabain Concentration in Medium with 5 mM K+ for Three Ouabain-Resistant Clones of L Cells, in Comparison with the Wild-Type (WT)

(●, ■) wild type; (△) OUA^R-LA2; (□) OUA^R-LA3; (○) OUA^R-LA7.

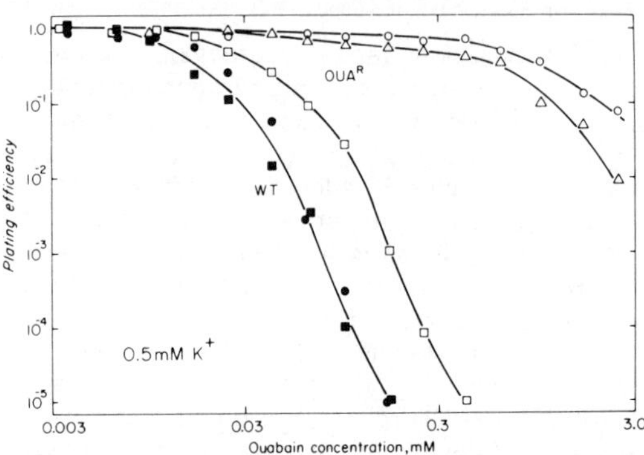

Figure 4. The Dose-Response Curves Describing Plating Efficiency as a Function of Ouabain Concentration in Medium with 0.5 mM K+ for Three Ouabain-Resistant Clones of L Cells and for the Wild-Type

Symbols as described for Figure 3.

Figure 4 shows the dose-response curves for wild-type and the three ouabain-resistant lines of L cells in medium containing only 0.5 mM K+, in contrast to the 5.0 mM K+ used in the experiments described in Figure 3. The 10-fold reduction in K+ concentration results in an approximately 10-fold reduction in the D_{10} for wild-type cells, without reducing their PE or growth rate in the absence of drug. From Figure 4 it is evident that the OUA^R-LA3, OUA^R-LA2, and OUA^R-LA7 clones are about 2-fold, 40-fold, and 70-fold, respectively, more resistant to ouabain than the wild type in terms of their D_{10} in 0.5 mM K+. For ouabain-resistant as well as wild-type cells, with increasing ouabain concentration a decrease in colony size precedes reduction in PE.

We have routinely selected ouabain-resistant colonies of L and CHO cells in monolayer culture by inoculating either mutagenized or unmutagenized wild-type populations of 10^5 to 10^7 cells into plastic culture dishes or glass prescription bottles containing selective medium, and incubating the cultures for the interval normally required for colony formation in the absence of drug. The essential requirement for reliable discrimination of ouabain-resistant cells in monolayer is simply to impose a selection sufficiently stringent that residual growth and/or adherence of wild-type cells to the culture surface is minimized, so that drug-resistant colonies can be unambiguously identified. The effectiveness of the selection depends upon relative drug dose (including K+ concentration) and also upon cell density on the surface. For L and CHO cells, about 10^5 cells is the maximum inoculum per 100 mm culture dish consistent with a clear elimination of wild-type cells by 1 mM ouabain (6 mM K+), while with selective medium containing 3 mM ouabain (6 mM K+) up to 5×10^6 cells can be inoculated per dish. If the effective drug dose is further enhanced by reducing the K+ concentration in the medium, 10^7 or more wild-type cells may be seeded per dish.

Although the effectiveness of the selection for ouabain-resistance is influenced by cell density, the expression of the drug-resistant phenotype is generally independent of cell density. The frequency of ouabain-resistant colonies assayed from any particular wild-type L cell or CHO cell population is independent of the total number of cells inoculated per selection dish, under experimental conditions

Table 1

Number of cells inoculated per dish	Number of OUA^R colonies per number of cells assayed	Frequency of OUA^R clones detected
1×10^5	$7/6 \times 10^5$	1×10^{-5}
1×10^6	$70/6 \times 10^6$	1.2×10^{-5}
3×10^6	$109/9 \times 10^6$	1.2×10^{-5}
1×10^7	$324/3 \times 10^7$	1.1×10^{-5}

The effect of cell density upon the frequency of detection of ouabain-resistant cells in a wild-type CHO cell population plated with 3 mM ouabain, 1 mM K+. Different numbers of cells from the same culture were inoculated into 100 mm culture dishes with the selective medium and incubated for 9 days at 37°C. The frequency of ouabain-resistant colonies observed per cell inoculated into the assay did not depend upon the total cell number inoculated per dish.

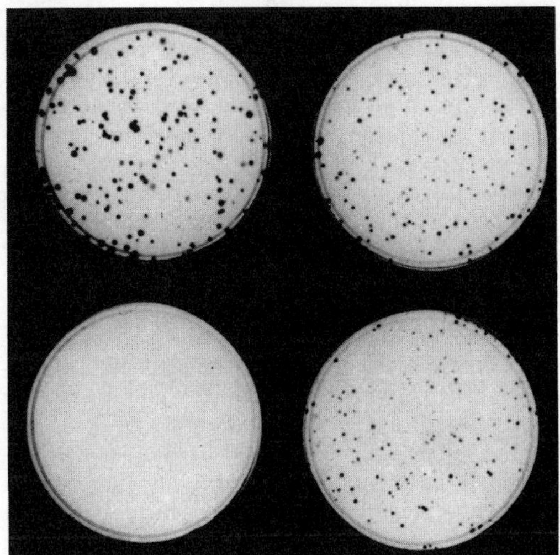

Figure 5. A Photograph of Stained Culture Plates Illustrating the Results from an Experiment to Reconstruct a Selection for Ouabain-Resistant L Cells

The plates were inoculated with:
upper left—200 wild-type cells in drug-free medium (ca. 195 colonies);
lower left—5 × 10⁶ freshly cloned wild-type cells in medium with 3 mM ouabain, 6 mM K+ (no growth);
upper right—200 OUAᴿ-LA2 cells in the ouabain medium (ca. 160 colonies);
lower right—200 OUAᴿ-LA2 plus 5 × 10⁶ wild-type cells in the ouabain medium (ca. 180 colonies).

that minimize background growth of wild-type cells for the largest inoculum. This point is illustrated for CHO cells by the data in Table 1. It is also demonstrated by the results of reconstruction experiments. For example, with L cells essentially identical results were obtained when 200 OUAᴿ-LA2 (Figure 5) or OUAᴿ-LA3 (not shown) cells were plated in 3 mM ouabain (6 mM K+) in the presence and in the absence of 5 × 10⁶ wild-type cells.

We have observed no systematic differences between wild-type and ouabain-resistant clones of L or CHO cells with respect to growth rate in the absence of ouabain or modal chromosome number.

B. Physiological Properties of Wild-Type and Ouabain-Resistant L Cells

Since a primary effect of ouabain upon wild-type cells is to decrease the active transport of potassium, evidently through specific inhibition of the plasma membrane Na/K ATPase, a likely explanation for the difference between wild-type and ouabain-resistant cells is that the latter possess mutant transport sites with an altered response to ouabain. Accordingly, the uptake of K+ by whole wild-type and ouabain-resistant L cells and the Na/K ATPase activities in preparations of plasma membranes iso-

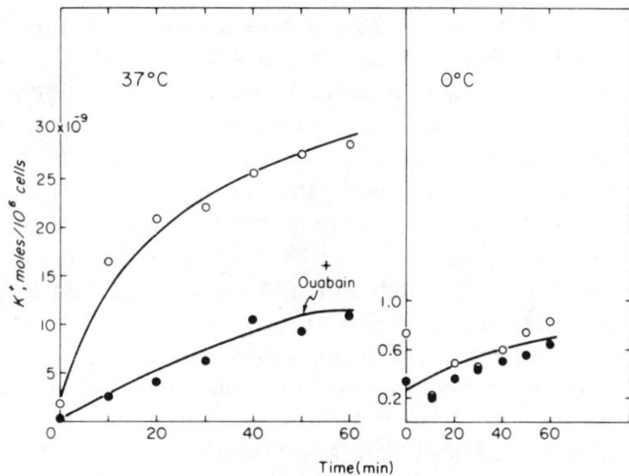

Figure 6. Uptake of K+ by Wild-Type L Cells at 37°C and 0°C in the Presence (●) and Absence (○) of 1 mM Ouabain (5 mM K+), as Followed with ⁴²K as a Tracer

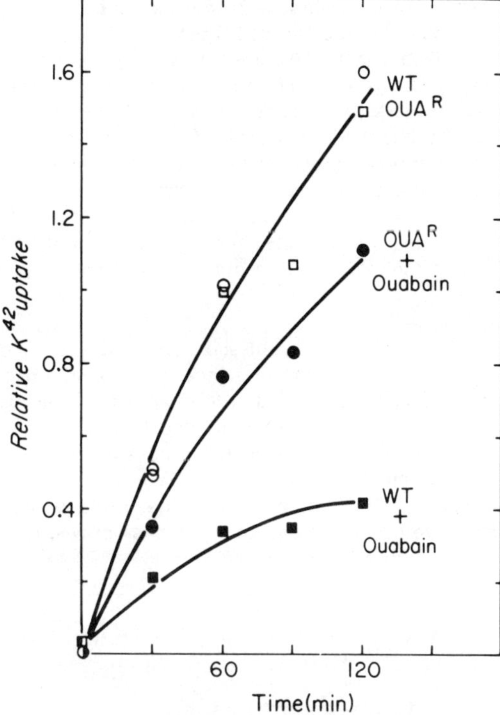

Figure 7. Uptake of ⁴²K as a Function of Time by Wild-Type and OUAᴿ-LA2 Cells in the Presence and Absence of 1 mM Ouabain

(□) wild-type without Ouabain; (■) wild-type + 1 mM ouabain; (○) OUAᴿ-LA2 without ouabain; (●) OUAᴿ-LA + 1 mM ouabain.

lated from these cells were examined (Brunette and Till, 1972; Brunette, 1972).

⁴²K Uptake

The data in Figure 6 describe the effect of 1 mM ouabain on the uptake of exogenous ⁴²K by wild-type L cells at 37°C and 0°C. The uptake at 37°C is principally attributable to active transport, and is inhibited by about 75% in the presence of 1 mM ouabain. The residual uptake at 37°C in the presence of drug may indicate incomplete inhibition of Na/K ATPase at this drug concentration as well as a ouabain-insensitive component of the K+ influx (Dunham and Hoffman, 1971).

The results of an experiment to compare ^{42}K uptake by wild-type and OUA^R-LA2 cells in the presence and absence of 1 mM ouabain (5 mM K+) are shown in Figure 7. The two cell types had similar rates of ^{42}K uptake in the absence of the drug. By contrast, in the presence of 1 mM ouabain uptake occurred at about 30% the normal rate in wild-type cells but at about 70% the normal rate in OUA^R-LA2 cells. Figure 8 demonstrates that throughout the concentration range 0.1 to 1 mM ouabain (5 mM K+) the drug is less effective in inhibiting K+ transport in OUA^R-LA2 cells than in wild-type cells. These results are consistent with the differential toxicity of ouabain on the two cell types in culture (Figure 3). Ouabain at 1 mM is toxic to wild-type cells but not to OUA^R-LA2 cells; ouabain at 0.1 mM is not toxic to wild-type cells but does inhibit their ^{42}K uptake to about the same extent as 1 mM ouabain inhibits uptake by OUA^R-LA2 cells. Analogous results have been obtained for other ouabain-resistant clones of L cells (Brunette, 1972; Till et al., 1973), and Mayhew (1972) has reported a similar finding for wild-type and ouabain-resistant ascites cells.

K+ Content

The potassium concentration in wild-type and OUA^R-LA2 cells in the presence and absence of 1 mM ouabain was estimated from equilibrium levels of ^{42}K taken up during prolonged incubation and cell volume determinations. The cells were grown in medium containing ^{42}K (5 mM total K+) for a period of 70 hr; during this time they underwent at least two doublings except in the case of wild-type

cells in ouabain. In each case, the amount of ^{42}K per cell was relatively constant in five assays over the interval from 22 to 70 hr after exposure to the isotope, and the average value was used to calculate K+ content per cell. Cell volume distributions for the two cell lines under these conditions were measured with an electronic cell-sizing apparatus (Taylor, 1970), and they corresponded to average cell diameters of 14 μ for wild-type and 15 μ for OUA^R-LA2.

From these data, the intracellular K+ concentrations for the two cell types in the absence of ouabain were calculated to be the same within experimental error, 160 mM for wild-type and 150 mM for OUA^R-LA2. The K+ concentration of OUA^R-LA2 cells grown in the presence of 1 mM ouabain was unchanged, but the content of wild-type cells in ouabain was reduced by an order of magnitude to 20 mM. These results offer further support for attributing the different toxic effects of the drug on the two cell types to differential impairment of K+ transport.

Na/K ATPase

The properties of the Na/K ATPase activities in plasma membrane preparations from OUA^R-LA2 and wild-type L cells were examined in some detail. The Na/K ATPase activities of the two cell lines were found to differ in their response to ouabain, but they were indistinguishable with respect to a number of other characteristics.

For both the wild-type and OUA^R-LA2 membranes, the specific activities of Na/K ATPase were in the range 3.9±0.7 (S.E.) μmoles P$_i$/mg protein/ hour. Table 2 summarizes some similar properties of the wild-type and OUA^R-LA2 ATPase activities, which are generally comparable to those which have been reported for Na/K ATPase preparations from other sources. The pH optimum was 7.5 on curves similar to that described by Etheredge, Haaland, and Rosenburg (1971). Variations of enzyme activity with Mg++ concentration exhibited a broad peak with a maximum at about 2 mM. The K$_m$ values for the substrates ATP and Na+ were determined to be 0.4 mM and 20 mM, respectively. The optimum Na+:K+ ratio was found to be 60 mM Na+:5 mM K+. Increasing the ratio of K+ to Na+ eventually resulted in some inhibition of the enzyme, as has been shown in other systems (Skou, 1957). The K$_m$ for K+ activation was approximately 0.35 mM K+. This is comparable to the K$_m$ of approximately 0.5 mM

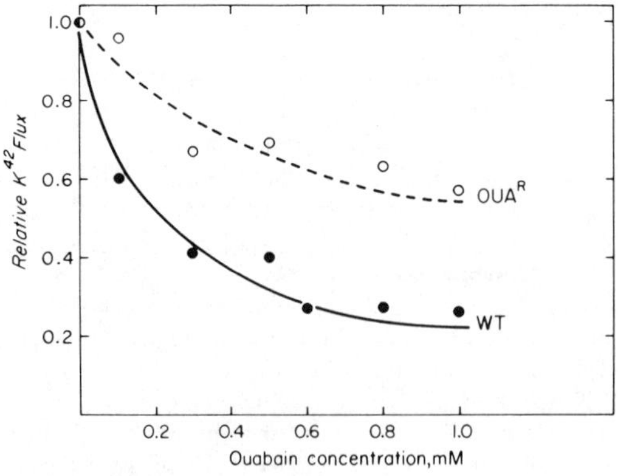

Figure 8. The Effect of Ouabain Concentration on Rate of ^{42}K Uptake (flux) for Wild-Type (●) and OUA^R-LA2 (○) L Cells, as Measured over 10–30 min Uptakes

Table 2 Properties of Wild-Type and OUA^R-LA2 Na/K ATPases

K$_m$ ATP	0.4 mM
K$_m$ Na+ (at 5 mM K+)	20 mM
K$_m$ K+ (at 60 mM Na+)	0.35 mM
Optimum Na+:K+	60 mM Na+:5 mM K+
Optimum Mg++	2 mM
Optimum pH	7.5

K+ for activation of ^{42}K uptake by whole cells, substantiating for this L-cell system the strong correlation between K+ transport and Na/K ATPase activity. In addition to the above properties of the ATPases, membranes isolated from the wild-type and OUA^R-LA2 cells were similar in their abilities to bind the fluorescent probe 1-anilino-8-napthalene sulphonate (Brunette, 1972).

The Na/K ATPase activities of the wild-type and OUA^R-LA2 plasma membrane preparations did differ clearly with respect to their susceptibility to ouabain inhibition. In twelve experiments an average of 14 ± 2% Na/K ATPase activity remained in wild-type membranes in 1 mM ouabain, compared with controls not exposed to ouabain. In concurrent experiments with OUA^R-LA2 membranes an average of 43 ± 7% of the control Na/K ATPase activity was found in the presence of 1 mM ouabain. The results summarized in Figure 9 show that the OUA^R-LA2 ATPase is less sensitive than the wild-type ATPase to ouabain inhibition throughout the dose range 0.1 to 1 mM ouabain, as was observed for K+ transport in whole cells. By contrast, the activities of 5' nucleotidase, a plasma membrane associated enzyme (Bosmann, Hagopian, and Eylar, 1968) presumably unrelated to K+ transport, were the same in the presence and absence of ouabain in both wild-type and OUA^R-LA2 membrane preparations (Brunette, 1972). From the currently available evidence, the alteration in OUA^R-LA2 cells that enables them to survive ouabain concentrations toxic to wild-type cells appears to be specific to the K+ transport system.

C. Genetic Properties of the Ouabain-Resistant Trait

Stability of the Phenotype

Our observations concerning the heritability and stability of the ouabain-resistance phenotype are

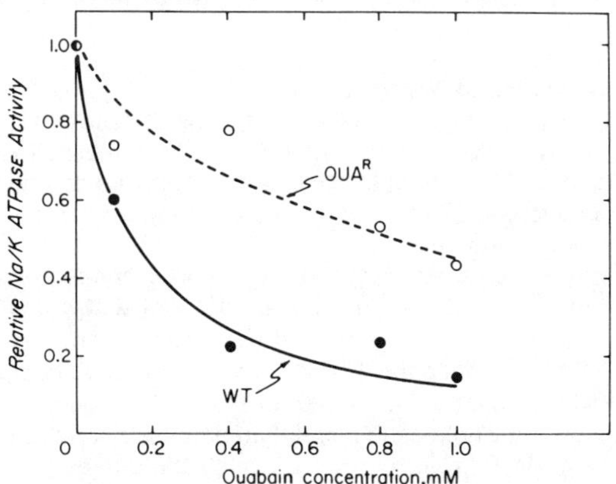

Figure 9. The Variation of Na/K ATPase Activities as a Function of Ouabain Concentration for Plasma Membrane Preparations from Wild-Type (●) and OUA^R-LA2 (○) L Cells

consistent with attributing the drug-resistance to a mutant genotype. Transmission of ouabain-resistance through successive generations of cells has been examined by ascertaining (a) the fidelity with which phenotypes exhibited in the selections are reproducible in subsequent assays, (b) the maintenance of phenotype during long intervals of serial subculture in the absence of the selecting drug, and (c) the range of phenotypic variation shown by subclones of a drug-resistant clone.

(a) The short-term reproducibility of selected ouabain-resistance phenotypes is highly reliable. When individual colonies of L or CHO cells that arise under selective conditions are picked, grown in the absence of ouabain, and retested for drug-resistance, they almost invariably exhibit ouabain-resistance consistent with the selection conditions. Examples of this behavior are furnished by controls that are described in the legends to Tables 4 and 5 below, concerning Luria-Delbruck fluctuation tests. The reliability of a typical selection in distinguishing drug-resistant cells is also illustrated by the following experiment. CHO cells were mutagenized with ethyl methane sulphonate (EMS) and selected in medium with 2 mM ouabain (6 mM K+) at 10^4 cells/ml in Linbro plates. Ignoring the frequent large colonies that developed after the standard incubation time, several hundred relatively small (20–60 cell) to moderate-sized colonies were subcultured in the selective medium. Continued proliferation to a large population was observed from each of 201 colonies of moderate size and from 85 of 104 colonies of smaller size. Thus the fidelity of the selection under these conditions was at least 286/305 = 94%. This figure represents an underestimate because large colonies were not included in the tests and because of the possibility that transfers (by pasteur pipette) of some of the small colonies were unsuccessful. On the basis of accumulated experience, we now consider it virtually certain that any colonies arising under adequate selective conditions (see section A above) will breed true as ouabain-resistant.

(b) In the course of our experiments, various ouabain-resistant clones of L and CHO cells have been routinely maintained in culture for extended periods of time without exposure to ouabain. We have not observed any significant drift in the drug-resistance phenotypes of populations maintained in this way; characteristic responses of the cells to ouabain in terms of plating efficiencies or transport properties have been reproducible over periods of months in culture. For example, Table 3, Part I shows that there was no appreciable difference in the drug-resistance of the OUA^R-LA2 clone assayed after approximately 1 month and after 11 months in nonselective medium.

(c) While general the ouabain-resistant pheno-

Table 3

OUA[R]-LA2	Relative plating efficiency	
Stability of phenotype	2.5 mM ouabain 5 mM K+	2.5 mM ouabain 1 mM K+
I. With time in culture		
1 month	0.98	0.20
11 months	0.79	0.32
II. Among subclones		
18 subclones	0.93 ± 0.15 (0.66 – 1.21)	0.46 ± 0.24 (0.13 – 0.88)
Parental line 19 replicates	—	0.32 ± 0.05 (0.23 – 0.40)

Stability of the ouabain-resistance phenotype of the OUA[R]-LA2 cell line: (I) as a function of time in continuous, nonselective culture; (II) among subclones of this cell line. For experiment I, the cells, freshly cloned at time zero, were cultured at about 7 generations per week at 34°C for the indicated periods. The phenotype was assayed by measuring the PEs observed with 2.5 mM ouabain and 5 mM K+ or 1 mM K+ relative to the PEs in medium with 1 mM K+ without drug. The condition with the lower K+ concentration represents the higher effective dose of ouabain. The errors indicated for experiment II are standard deviations and the quantities in parenthesis are the ranges of observed values. Further details are discussed in the text.

type behaves as a stably heritable trait, its expression is evidently subject to some quantitative variation among subclones. In two experiments with ouabain-resistant L cells we have investigated the range of variation exhibited by subclones in their plating efficiencies (PEs) at marginally toxic ouabain doses. Table 2, Part II summarizes the variability observed among subclones derived from the OUA[R]-LA2 line after 10 months' serial subculture. The average PEs, standard deviations and ranges are shown for 18 subclones in comparison with a set of replicate assays of the parental cell clone. One subclone, not included in these calculations, had a PE < 1% even at the lower dose level tested and may have represented a real "revertant." With this exception, the ouabain-resistance phenotype was qualitatively stable, but variation in the resistances of the subclones to the higher effective dose (2.5 mM ouabain, 1 mM K+) appears to have been significantly greater than could be attributed to normal experimental scatter. In another experiment, 37 subclones were obtained from an L cell clone OUA[R]-LA8 after about 2 months' growth. In 1.5 mM ouabain (6 mM K+), the PE of the parental OUA[R]-LA8 line was approximately 0.05—ranging from 0.02 to 0.08 in four assays. Under the same conditions, each of the subclones was resistant relative to wild-type, but their PEs varied over a 100-fold range from 0.004 to 0.4.

This variation among L cell subclones is small compared to the differences observed between wild-type and most selected drug-resistant cell lines. The differences between subclones seem significant, however, and we believe that they may be

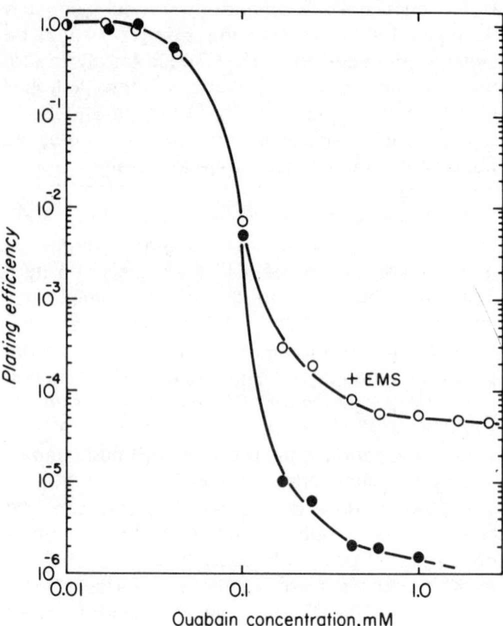

Figure 10. The Plating Efficiencies as a Function of Ouabain Concentration (0.5 mM K+) of Unmutagenized (●) and EMS-Mutagenized (○) Sister Cultures of Wild-Type CHO Cells

3×10^7 cells in suspension culture were exposed to 300 r/ml EMS for 18 hr (40% survival) and assayed for ouabain-resistance after an 8 day recovery and expression period. These cultures were also assayed for the frequencies of cells resistant to 12 γ/ml 8-azaguanine (AG) and of cells prototrophic for proline (PRO+), as described by Thompson and Baker (1973). These frequencies were: unmutagenized, 3.7×10^{-5} AG[R] and 4×10^{-6} PRO+; EMS-mutagenized, 1.6×10^{-3} AG[R] and 1.6×10^{-4} PRO+.

attributable to karyotypic heterogeneity or accumulated alterations in secondary traits relevant to the phenotype. Although analogous subclone experiments have not been carried out with CHO cells, we have noticed small differences in the ouabain-sensitivity displayed by different subclones of the wild type.

Response to Mutagen

The frequency of ouabain-resistant cells in a recently cloned culture can be markedly increased by application of the chemical mutagen ethyl methane sulphonate (EMS). Figure 10 summarizes the results of an experiment to measure the frequencies of CHO cells resistant to different drug concentrations in a culture exposed to 300 γ/ml EMS (40% survival) and in an untreated sister population. The identical results for PEs of the two cultures at ouabain concentrations ≤ 0.1 mM (0.5 mM K+) confirm that as a whole each population had the same drug sensitivity. From the PEs at which the curves plateau for selective drug concentrations, i.e., > 0.1 mM, it can be seen that the mutagen treatment enhanced the frequency of ouabain-resistant colonies by about 40-fold over that in the untreated control. Ouabain-resistance was induced at frequencies around 10^{-4} per viable cell, the frequency of resistant cells declining somewhat with increasing drug dose as expected if there is some heterogeneity with respect to degree of resistance. Analogous assays of a sister culture treated with 100 γ/ml EMS (95% survival) in the same experiment showed a similar high frequency of induced ouabain-resistant clones.

In other experiments in which N-methyl-N'-nitrosoguanidine (MNNG) was utilized as mutagen we have observed no more than 2-fold increases over controls in the frequency of ouabain-resistant cells in treated populations. On the other hand, increases of the order of 10-fold in the frequency of ouabain-resistant clones have been observed with CHO cells following UV irradiation (S. Molnar and M. Rauth, personal communication). These results imply that the ouabain-resistant cells are genotypically mutant, but that the relevant loci may respond preferentially to certain mutagens.

The results of two Luria-Delbruck fluctuation tests to characterize the spontaneous mutation rate of wild-type L cells for resistance to 3 mM ouabain (6 mM K+), in comparison with a replicate-sample control experiment.

Each fluctuation experiment was initiated by inoculating replicate cultures at about 100 cells each in prescription bottles. Following 12 days growth at 37°C, the colonies were dispersed with trypsin. In experiment A, the cultures were then grown in 16-oz prescription bottles without dilution for a further 9 days (to confluence), harvested with trypsin, and assayed. In experiment B, the cultures were maintained in suspension during this second stage of growth until the appropriate population size was attained. Allowing for sampling and cell loss in manipulation, approximately 95% of the cells in each replicate culture were actually inoculated into the assay for ouabain resistance. Cells were seeded in the selective medium at 5×10^6 cells per 100 mm culture dish and incubated for 14 days at 37°C. These selection conditions are identical to those employed in the reconstruction experiment described in Figure 5.

As a further control on the reliability of these selection conditions, 20 (independent) colonies of various sizes were picked from assay plates seeded from different replicate cultures, cultivated in nonselective medium, and subsequently retested at the selecting drug dose. Thirteen of the clones plated with efficiencies of 50–100% at the selecting ouabain concentration and were therefore at least 10-fold more resistant than wild type in terms of drug dose; the remaining 7 clones had plating efficiencies of 4–50% and so were at least 5-fold more resistant than wild type.

The replicate-sample controls were derived by inoculating at the same drug dose two sets of 30 dishes with either 10^5 or 10^6 cells per dish from a single wild-type population, in which spontaneously occurring ouabain-resistant cells had been allowed to accumulate during serial culture.

Luria-Delbruck Fluctuation Analyses; Mutation Rates

We have performed fluctuation tests (Luria and Delbruck, 1943) to determine the pattern and rate of appearance of ouabain-resistant cells in wild-type populations of L and CHO cells

Table 4 shows the results from two fluctuation tests with L cells, in comparison with a control experiment consisting of replicate samplings of a single population. In each fluctuation experiment, about 30 replicate cultures of wild-type cells were grown in nonselective medium from inocula of 100 cells each to populations of approximately 3×10^7 cells each. The entire population of each replicate was then plated in medium with 3 mM ouabain (6 mM K+) at 5×10^6 cells per 100 mm dish. These conditions are reliably selective for clones at least 5-fold, and usually 10-fold, more resistant to ouabain than wild-type in terms of their D_{10} (see Table legend). In experiment A, cells were grown in monolayer culture and trypsinized immediately prior to the selection plating. Since it seemed possible that treatment with trypsin might affect the response of the cells to a surface-active selective drug, in experiment B the cells were cultured in suspension prior to plating. The results in the two experiments were essentially the same.

The data in Table 4 show that the variance in the number of resistant colonies observed among repli-

Table 4. L Cell Fluctuation Tests

	Experiment A (monolayer)	Experiment B (suspension)	Replicate-sample control	
No. replicate cultures	28	32	1	
No. samplings per culture	1	1	30	30
Initial no. cells per replicate	100	100	—	—
Final no. cells per replicate	$3.0 \pm 0.3 \times 10^7$	$3.1 \pm 0.6 \times 10^7$	1×10^5	1×10^6
No. of replicates with N OUA^R colonies:				
N = 0 colonies	6	4	10	0
1	2	10	11	0
2	9	7	4	0
3–4	5	2	5	0
5–8	2	3	0	10
9–16	2	2	0	19
17–32	1	2	0	1
33–64	1	0	0	0
64	0	2	0	0
No. of OUA^R colonies per replicate:				
Range	0–34	0–116	0–4	6–18
Median	2	2	1	11
Mean	4.6	10	1.2	10.5
Ratio $\dfrac{\text{variance}}{\text{mean}}$	20	63	1.2	0.87
Mutation rate, $\times 10^{-8}$				
F_o calculation	5.2 ± 1.3	7.0 ± 2.0	—	—
Median calculation	4.5 ± 1.2	4.5 ± 1.4	—	—

cate cultures in the fluctuation experiments is much greater than that found for replicate platings from a single culture and is clearly skewed toward higher frequencies of resistant clones. This is indicative of random generation of the ouabain-resistant phenotype during proliferation of the wild-type population, as would be expected if it represents a mutant genotype. With these data, a mutation rate can be estimated either from the fraction of replicate cultures in which no resistant cells occurred (F_0) or from the median frequency of resistant cells per replicate culture, using the methods of Lea and Coulson (1949). These calculations indicate a value of approximately 5–6×10^{-8} per cell per generation as the mutation rate for L cells to 10-fold increase in ouabain resistance.

The data from similar fluctuation tests performed with CHO cells are summarized in Table 5. In experiment C the selection was also for cells resistant to 10-fold greater ouabain concentration than are the wild-type; in addition, by reducing the K^+ concentration in the medium, a selection for cells 100-fold more resistant than wild-type was effected in experiment D. The results indicate mutation rates of 5–6×10^{-8} per cell per generation for 10-fold increase in resistance, within experimental error the same as the analogous rate for L cells, and 2–3×10^{-8} per cell per generation for 100-fold increase in ouabain resistance. Again the data are consistent with random appearance of the resistant cells, and with heterogeneity among mutant clones with respect to the degree of resistance.

Expression of Ouabain-Resistance in Hybrid Cells

Hybrids formed from ouabain-resistant and wild-type (i.e., "ouabain-sensitive," OUA^S) CHO cells have been examined in order to ascertain whether the ouabain-resistant phenotype is expressed in circumstances where the wild-type allele is known to be present. By selecting for complementation of recessive auxotrophic markers in the parental cell lines, hybrid clones were derived under conditions nonselective for the ouabain marker (see Experimental Procedures). These hybrids were subsequently assayed in comparison with the parental cell lines for PE in various concentrations of the drug. The results from such an experiment to characterize hybrids between the CHO lines OUA^R-CA1, AUXB1+, PRO− and OUA^S, AUXB1−, PRO+ are shown in Figure 11. The OUA^R-CA1 parental cells, isolated from a selection of unmutagenized wild-type (PRO−) CHO cells in 3 mM ouabain, 1 mM K^+, are more than 100-fold more resistant to the drug than the OUA^S wild-type. (They do show a substantially reduced growth rate at the highest ouabain dose tested.) It is clear from Figure 11 that hybrids of the two cell types exhibit intermediate resistance to the drug, with D_{10}'s for plating efficiency approxi-

Table 5. CHO Cell Fluctuation Tests		
	Experiment *C* (10-fold resistance)	Experiment *D* (100-fold resistance)
Selective medium	3 mM ouabain 6 mM K+	3 mM ouabain 0.5 mM K+
No. replicate cultures	29	30
Initial no. cells per replicate	100	100
Final no. cells per replicate	$2.6 \pm 0.3 \times 10^7$	$5.2 \pm 0.9 \times 10^7$
No. of OUA^R colonies per replicate:		
Range	0–112	0–102
Median	3	2
Mean	10	7.2
Ratio $\frac{\text{variance}}{\text{mean}}$	53	50
Fraction of replicates with no OUA^R colonies (F_0)	0.28	0.33
Mutation rate, $\times 10^{-8}$:		
F_0 calculation	5.0 ± 1.3	2.2 ± 0.6
Median calculation	6.5 ± 1.7	2.6 ± 0.8

The results of two Luria-Delbruck fluctuation tests with CHO cells to measure the spontaneous mutation rates to resistance to effective ouabain doses approximately 10-fold and 100-fold greater than those toxic to wild-type cells (cf. Figures 2 and 11). The procedures resembled that described for experiment A in Table 4 with the following exceptions. For experiment C (10-fold resistance), the replicate cultures were grown to confluence in 8-oz prescription bottles over a total of 11 days. For experiment D (100-fold resistance), the replicates were cultured to confluence in 32-oz bottles, then the cells were dispersed with trypsin, the selective medium was added, and the entire population of each replicate was incubated in the original 32-oz vessel for 12 days at 37°C. The selection conditions were sufficiently stringent to eliminate any background growth of wild-type cells even at this high cell density.

As a check on the reliability of the selection conditions used in experiment C, 11 independent colonies of various sizes were picked from selections of different replicate cultures, grown in nonselective medium, and retested for drug resistance. Ten of the clones displayed 100% plating efficiency at the selecting drug dose; the other, initiated from a small colony that appeared on a selection dish, was quite sensitive to the selecting drug dose but did exhibit significantly greater resistance to drug than wild-type cells when assayed at lower ouabain concentrations.

mately 30- to 60-fold greater than that of a control hybrid formed from corresponding OUA^S parental cell types. Thus the ouabain-resistance marker is codominant, or incompletely dominant.

To date, approximately twenty different hybrid clones prepared with five different ouabain-resistant CHO lines have been examined in similar experiments. In most instances the drug resistance is codominant, in the sense that the hybrid exhibits ouabain resistance intermediate between that of the OUA^S and OUA^R parental cells, although the dose response for PE of the hybrid cells usually approaches that of the OUA^R parent much more closely than that of the OUA^S parent. Some variation is detectable between independent hybrids of the

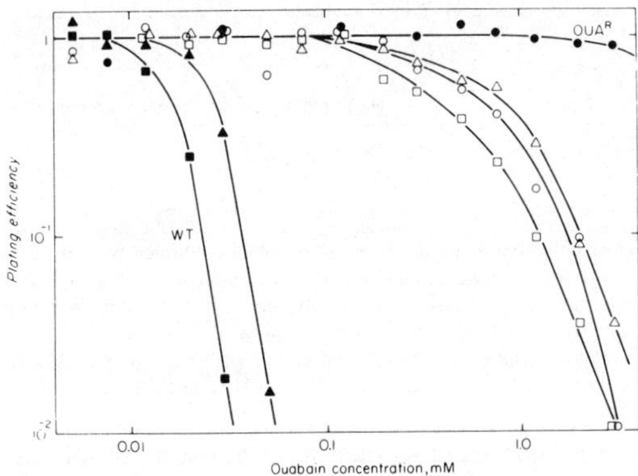

Figure 11. The Dose-Response Curves for Plating Efficiency in Ouabain, 0.5 mM K+ of Parental and Hybrid CHO Cell Lines

(■) pseudo-diploid OUA^S, AUXB1−, PRO+ cells; (●) pseudo-diploid OUA^R-CA1, AUXB1+, PRO− cells; (▲) hybrid clone of OUA^S, AUXB1−, PRO+ with OUA^S, AUXB1+, PRO−, (○, △, □) three independently derived hybrid clones of OUA^S, AUXB1−, PRO+ with OUA^R-CA1, AUXB1+, PRO−, with modal chromosome numbers of 36–39, 34 and 38, respectively. The modal chromosome number of each pseudo-diploid parental cell type was 21.

same cell types, which may reflect inherent fluctuations in expressivity or differences in chromosome constitution. There have been two exceptional cases so far; in one the hybrid clone showed the drug sensitivity characteristic of wild type cells and in the other the hybrid was about as drug resistant as the OUA^R parent. These observations may reflect loss of a chromosome carrying one or the other of the principal alleles.

Dominance of the ouabain-resistance property at certain drug concentrations can be demonstrated utilizing a more direct approach. Table 6, Part I shows the frequency of colony-forming hybrids assayed under several different conditions from fusions of the OUA^R-CA1 and OUA^S parental cell lines described above. The plating conditions indicated in lines a and b were selective for hybrid cells in which the different auxotrophic mutations in the parental lines complemented each other. In addition, the conditions were either nonselective for drug resistance (line a) or selective for a level of drug resistance intermediate between that of the OUA^R and OUA^S parental cell types (line b). Similar frequencies of hybrid colonies were observed in the presence and absence of ouabain, a fact that implies dominance of the ouabain-resistant phenotype at this drug dose. The control experiment summarized in Table 6, Part II (lines a' and b') confirms that hybrids of the corresponding OUA^S parental cell types were viable only under the conditions nonselective for ouabain resistance.

OUA^R as a Marker for Selection of Somatic Cell Hybrids

The ouabain-resistance marker can be used effectively for the selection of hybrids of OUA^R and OUA^S

parental cells. From the immediately preceding discussion, it follows that under conditions in which the ouabain-resistant phenotype is dominant, as in line b of Table 6, further selection imposed for complementation of a recessive marker carried by the OUA^S parental cells is redundant. As shown in line c of Table 6, hybrids can be detected at comparable frequency simply by selecting for the dominant ouabain resistance and against the recessive marker on the OUA^R parental cell line. (Since the different assays reported in Table 6 were inoculated from the same samples of fused cells, the differences in the frequencies of hybrid cells detected are probably meaningful; they suggest that the OUA^R-CA1 marker has greater penetrance in these selections than the PRO+ marker.)

Use of ouabain resistance as a selective marker in conjunction with markers for temperature sensitivity (Thompson and Baker, 1973) and 8-azaguanine resistance with HAT sensitivity (Littlefield, 1964) is demonstrated in Table 7. Hybrid cells were selected with comparable efficiency from fusions of temperature-sensitive (ts) lines and azaguanine-resistant (AG^R) lines whether the ts and AG^R, OUA^R and AG^R, or OUA^R and ts pairs of markers were utilized for selection, and regardless of which parental cell line carried the OUA^R trait.

Discussion

The principal findings presented in this article may be summarized as follows. From wild-type populations of mouse L cells and of Chinese hamster ovary cells, one can select in a single step clones that are viable in ouabain at concentrations ranging up to more than 100-fold greater than the minimal cytotoxic dose for wild-type cells. Clones of ouabain-resistant L cells show decreased sensitivity to ouabain

Table 6

I. Fusion: OUA^R-CA1, AUXB1+, PRO− × OUA^S, AUXB1−, PRO+

Selection	Frequency of hybrid colonies
(a) AUXB1+, PRO+	5.8×10^{-5}
(b) AUXB1+, PRO+, OUA^R	8.9×10^{-5}
(c) OUA^R, PRO+	3.0×10^{-4}

II. Fusion: OUA^S, AUXB1+, PRO− × OUA^S, AUXB1−, PRO+ (Control)

Selection	Frequency of hybrid colonies
(a') AUXB1+, PRO+	7.8×10^{-4}
(b') AUXB1+, PRO+, OUA^R	$<3 \times 10^{-6}$
(c') OUA^R, PRO+	

The frequency of hybrid cells detected under conditions either nonselective or selective for a ouabain-resistance phenotype, for (I) fusions of a OUA^S and a OUA^R CHO auxotroph and (II) two OUA^S CHO auxotrophs. The parental cell types were mixed and treated with Sendai virus as described in the Experimental Procedures and plated in the selective media at $1–2 \times 10^5$ cells per 100 mm culture dish. The selective ouabain dose used for these experiments, and those described in Table 7 below, was 3 mM ouabain, 6 mM K+ (about 10-fold greater than the D_{10} for wild-type cells).

Table 7

Fusion	Frequency of Hybrids		
	HAT^R (AG^S), TEMP^R	HAT^R (AG^S), OUA^R	TEMP^R, OUA^R
OUA^R-CB2, AG^R-A4, TEMP^R x OUA^S, AG^S, *ts*H1	7.0×10^{-4}	$19. \times 10^{-4}$	—
OUA^R-CM1, AG^S, *ts*H1 x OUA^S, AG^R-B11, TEMP^R	2.4×10^{-3}	—	3.1×10^{-4}
OUA^R-CB2, AG^R-A4, *ts*C41 x OUA^S, AG^S, TEMP^R	—	$18. \times 10^{-4}$	8.3×10^{-4}

The use of OUA^R as a marker for selection of somatic cell hybrids, in comparison with 8-azaguanine resistance (accompanied by sensitivity to HAT medium) and with temperature sensitivity. The OUA^R-CB2 and -CM1 markers were selected at 3 mM ouabain, 1 mM K$^+$. The HAT-sensitive AG^R-A4 and -B11 markers were obtained by continuous selection at 5, 10, and 20 γ/ml AG and by single-step selection at 12 γ/ml AG, respectively; the AG^R-B11 line has < 2% normal hypoxanthine guanine phosphoribosyltransferase activity (G. E. Jones, personal communication). The tsH1 line has a temperature-sensitive leucyl tRNA synthetase activity and exhibits a PE $\sim 10^{-7}$ at 38.5°C relative to 34°C (Thompson, Harkins, and Stanners, 1973). The *ts*C41 line has a relative PE $< 10^{-6}$ at 38.5°C (R. Mankovitz, unpublished data).

inhibition of K$^+$ uptake by whole cells and of Na/K ATPase activity in isolated plasma membranes. The property of ouabain resistance behaves as a genetic trait. The phenotype is stable. Treatment of a wild-type CHO cell population with EMS causes a substantial increase in the frequency of ouabain-resistant cells. Luria-Delbruck fluctuation analyses indicate that spontaneous appearance of ouabain-resistant cells in wild-type populations occurs randomly and at a rate of approximately $5-6 \times 10^{-8}$ per cell generation for L cells and for CHO cells. The ouabain-resistance phenotype is codominant in hybrids of resistant and wild-type cells, and it can be utilized effectively for the selection of such hybrids. On the basis of these results, we consider it reasonably certain that ouabain-resistant cells arise by virtue of a genetic mutation that affects the Na/K ATPase activity of the plasma membrane.

Our findings complement and elaborate the observations of Mayhew (1972). He selected ouabain-resistant subpopulations of Ehrlich ascites cells by repeated intermittent exposures of a wild-type population to inhibitory concentrations of the drug. He, also, found that the ouabain-resistant and wild-type populations differed principally in the susceptibility of their Na$^+$—K$^+$ transport functions to inhibition by ouabain.

The selection for ouabain-resistant cells can readily be generalized to different cell lines and to a variety of experimental conditions. In addition to many independent clones of ouabain-resistant L and CHO cells such as those discussed above, it has also been possible to isolate relatively infrequent ouabain-resistant clones of mouse 3T3 cells (R. M. Baker and M. Cescon, unpublished data), of the heteroploid human Hela cell line (R. M. Baker, unpublished data; Mankovitz, Baker, and Buchwald, 1973), and of human diploid fibroblasts (Mankovitz, Baker, and Buchwald, 1973; Buchwald, Mankovitz, and Baker, 1973).

The physiological characteristics of ouabain-resistant cells require further investigation. Phenotypic heterogeneity among different mutant clones is indicated not only by varied dose responses for cell survival, but is also suggested by differences in such parameters as rate of K$^+$ transport in the absence of ouabain (cf. Figure 3 of Till et al., 1973). In different instances the drug resistance may be attributable to alterations in the Na/K ATPase or in adjacent membrane components that result in reduced affinity for the inhibitor or in an altered response to bound ouabain. The properties of a number of independent mutants should therefore be examined. A factor that may complicate such studies is that single-step mutants to ouabain resistance might be heterozygous for the mutant allele and thus harbor both wild-type and drug-resistant transport components. This possibility is suggested by the observations that the aneuploid ouabain-resistant clones examined to date show significantly reduced transport activity at the ouabain doses inhibitory to wild-type cells (e.g., Figure 8, 9) and that the ouabain-resistant phenotype is codominant in hybrid cells. Further physiological characterizations of ouabain-resistant mutants are being pursued with CHO and with human cells. The latter afford the advantage of sensitivity to ouabain concentrations approximately 10^4-fold lower than those necessary to affect rodent cells, and therefore offer the potential for isolation of clones carrying multiple as well as single mutations to ouabain resistance and for refined binding and enzymological studies.

The characteristics of the ouabain-resistant phenotype in cultured cells indicate that it will be valuable as a marker in somatic cell genetic studies —for example, in investigations of somatic cell mutagenesis. Aside from the evident relation of the phenotype to a specific functional alteration, the fidelity of the selection for ouabain resistance and its lack of dependence on cell density at adequate selecting doses are remarkable. Also, the dominant nature of the phenotype suggests that it might be less affected than recessive traits by karyotypic variations in cultured cells.

The ouabain-resistant phenotype also has potential as a selective marker for somatic cell hybrids. Ouabain-resistant cells can be isolated with relative ease and reliability compared with most other mutant phenotypes that have been used for hybridization. Moreover, construction of cell lines carrying both a recessive selective marker (e.g., azaguanine resistance) and the dominant OUA[R] marker enables the isolation of hybrids of that cell line with another cell type that need not carry any selective marker (Tables 6 and 7). Finally, differential ouabain sensitivity can be utilized in the selection of interspecific hybrids between human and rodent cell lines, without resort to selection of mutants. Because human cells are sensitive to 10^4-fold lower concentrations of ouabain than are rodent cells, use of an intermediate dose of ouabain in the selection for hybrid cells ensures effective elimination of the human parental cell type (F. H. Ruddle and R. E. Giles, personal communication).

Experimental Procedures

Cells, Media and Culture Conditions

Wild-type cell cultures were derived from clone WT4 of L cells (Thompson et al., 1970) and from a CHO cell clone (PRO$^-$) kindly provided by W. C. Dewey. The AUXB1$^-$ CHO cell line, which has a triple requirement for glycine, adenosine, and thymidine, was isolated by M. McBurney (McBurney, Ph.D. Thesis, in preparation; Thompson and Baker, 1973). The other marked CHO lines were isolated by the authors according to methodologies described elsewhere (Thompson and Baker, 1973).

CHO cells were maintained in α medium (Stanners, Eliceiri, and Green, 1971) and L cells in CMRL 1066 medium lacking nucleosides or in α medium; these media were routinely supplemented with 10% fetal calf serum (Flow Laboratories). Low-potassium media were prepared by supplementing α or CMRL 1066 media lacking KCl with fetal calf serum containing 10–12 mM K$^+$, as assayed by flame photometry, to final serum concentrations of approximately 5% or 10%, respectively, for 0.5 mM K$^+$ or 1.0 mM K$^+$. In experiments where K$^+$ concentration in the medium was a variable, all media were prepared from a stock lacking KCl and were supplemented with the same concentration of serum and with additional KCl as appropriate. (The K$^+$ concentration in normal media supplemented with 10% serum is approximately 6 mM.) Deficient medium (D α) with supplements for the selection of prototrophs or of hybrid CHO cells was prepared as described by Thompson and Baker (1973). To promote experimental uniformity, medium containing ouabain at 3 mM was prepared in batch quantities, stored at 4°C, and diluted as appropriate for use. As judged by its ability to inhibit K$^+$ transport and cell growth, the drug was stable in solution, even in the presence of serum, over periods of many months. Ouabain, ethyl methane sulphonate, and other drugs, unless otherwise specified, were obtained from Sigma Chemical Co.

Stock cell cultures were incubated at 34°C or 37°C; experiments were performed at 37°C, except as indicated. Sixty mm or 100 mm plastic tissue culture dishes (Falcon Plastics Co.) were utilized for assays of colony-forming ability (plating efficiency) and 96 well Linbro dishes (Linbro Chemical Co., No. ISFB-96TC) were employed in all cloning. Basic cell-culture techniques and procedures employed in our laboratories have been thoroughly discussed elsewhere (Thompson and Baker, 1973).

Measurement of ^{42}K Uptake

The rate of K$^+$ uptake was measured by a slight modification of the method developed by Glynn (1956) for red cells. ^{42}K (New England Nuclear) was added to L cells suspended in CMRL 1066 medium lacking KCl. At various times during subsequent incubation, duplicate samples containing at least 10^6 cells were removed, sedimented at 2200 rpm in a clinical centrifuge in the cold, and washed twice with 10 ml and once with 3 ml of ice-cold 0.15 M KCl. The resulting cell pellet was counted for ^{42}K in a Nuclear Chicago well-type scintillation counter.

Enzyme Assays

Enzyme assays were performed with L cell surface membrane preparations isolated by the method of Brunette and Till (1971).

For ATPase assays, the rate of ATP hydrolysis was measured in a standard medium containing 0.1 M Tris-HCl pH 7.5, 2 mM ATP, 2 mM MgSO$_4$, 60 mM NaCl, 5 mM KCl, and 0.1 mM ethylene diaminetetraacetate (EDTA). Na/K ATPase activity was calculated from the increase in the initial rate of ^{32}P released from (γ -^{32}P) ATP (New England Nuclear) which occurred when K$^+$ was added to a system already containing Mg^{++} and Na$^+$. (There was no detectable difference in ATP hydrolysis in tubes containing Mg^{++} and Na$^+$ + Mg^{++}.) The ATPase reactions were stopped by adding the mixture to an acidified suspension of activated charcoal (British Drug Houses) (Crane and Lipmann, 1971), and the unreacted (γ -^{32}P) ATP (which adsorbs to charcoal) was then removed by centrifugation in the presence of glycerol. Enzyme velocities are expressed either relative to the Na/K ATPase or the total ATPase of a control incubation in the standard medium, defined as 1.0. Experiments in which the effect of ouabain was tested were performed under standard conditions of 150 μg protein/ml, with preincubation of the membranes at the indicated ouabain concentration for 15 min at 37°C. All data presented represents the averages of at least two independent experiments.

The enzyme 5′ nucleotidase was assayed by the method of Heppel and Hilmoe (1951). Inorganic phosphate released in the assay was measured by the method of Lowry and Lopez (1946). Protein determinations were according to the method of Lowry et al. (1951) using bovine serum albumin as a standard.

Cell-Cell Hybridization

For each hybridization, 5×10^5 cells of each parental cell type were inoculated to 2 ml medium in a 16 mm diameter well of a 24-well Linbro dish (Linbro Chemical Co., No. FB-16-24-TC) and incubated overnight. Medium was aspirated from the resulting packed monolayer and it was washed with phosphate buffered saline (PBS) and then covered with 0.2 ml of a suspension of inactivated Sendai virus (diluted with PBS 1:1 to 1:32 from Lot 110, Connaught Medical Research Laboratories) in order to promote cell fusion (Rao and Johnson, 1972). This preparation was incubated in the cold for 30–60 min, then warm medium was added. Following further incubation overnight, the cells were washed with PBS, trypsinized, resuspended, and assayed under the conditions selective for hybrids of the two parental cell types. The characteristics of the selective markers and the necessary selective media used in these experiments have been described by Thompson and Baker (1973).

In parallel for each hybridization, the unmixed parental cell types were fused and assayed under the selective conditions as controls on the frequency of revertants for the selective markers in the hybridization. The frequency of revertants was typically $\leq$ 1% of the frequency of hybrids detected. For each selection condition, the hybrid (i.e., pseudo-tetraploid) character of the scored colonies was confirmed by picking a random selection of 10 or more clones and examining the characteristic cell volumes and, in some cases, modal chromosome numbers.

Acknowledgments

We thank James Kao, Liz Thompson, Ausra Karka, Marie Florian, Scott Hanham, and Gordon Glasgow for technical assistance in various experiments; Aser Rothstein, Gary Jones, and Vic Ling for their comments on the manuscript; and Frank Ruddle for discussion concerning the use of OUA[R] as a selective marker for hybrids. This work was supported by the Medical Research Council of Canada and the National Cancer Institutes of Canada and the United States.

Received August 27 1973

References

Albers, R. W. (1967). Ann. Rev. Biochem. *36*, 727.

Baker, P. F., and Willis, J. S. (1970). Nature *226*, 521.

Baker, R. M., and Mankovitz, R. (1973). 13th Int. Congr. Genetics, Berkeley, p. S15.

Baker, R. M., and Till, J. E. (1971). 15th Ann. Meet. Biophys. Soc., New Orleans, p. 283a.

Bosmann, H. B., Hagopian, A., and Eylar, E. H. (1968). Arch. Biochem. Biophys. *128*, 51.

Brunette, D. M. (1972). Ph.D. Thesis, University of Toronto, Canada.

Brunette, D. M., and Till, J. E. (1971). J. Membr. Biol. *5*, 215.

Brunette, D. M., and Till, J. E. (1972). 16th Ann. Meet. Biophys. Soc., Toronto, p. 58a.

Buchwald, M., Mankovitz, R., and Baker, R. M. (1973). 25th Ann. Meet. Amer. Soc. Human Genetics, Atlanta, p. 17a.

Crane, R. K., and Lipmann, F. (1971). J. Biol. Chem. *201*, 235.

Dunham, P. B., and Hoffman, J. F. (1971). J. Gen. Physiol. *58*, 94.

Etheredge, E., Haaland, J. E., and Rosenburg, M. D. (1971). Biochim. Biophys. Acta *233*, 145.

Glynn, I. M. (1956). J. Physiol. *134*, 278.

Glynn, I. M. (1964). Pharmacol. Revs. *16*, 381.

Heppel, L. A., and Hilmoe, R. J. (1951). J. Biol. Chem. *188*, 665.

Lamb, J. F., and McCall, D. (1972). J. Physiol. *225*, 599.

Lea, D. E., and Coulson, C. A. (1949). J. Genet. *49*, 264.

Littlefield, J. W. (1964). Cold Spr. Harb. Symp. Quant. Biol. *29*, 161.

Lowry, O. H., and Lopez, J. A. (1946). J. Biol. Chem. *162*, 421.

Lowry, O. H., Rosebrough, N. J., Farr, A. L., and Randall, R. J. (1951). J. Biol. Chem. *193*, 265.

Luria, S. E., and Delbruck, M. (1943). Genetics *28*, 491.

Mankovitz, R., Baker, R. M., and Buchwald, M. (1973). 13th Ann. Meet. Amer. Soc. Cell Biol., Miami Beach, p. 214a.

Mayhew, E. (1972). J. Cell. Physiol. *79*, 441.

Mayhew, E., and Levinson, C. (1968). J. Cell. Physiol. *72*, 73.

McDonald, T. F., Sachs, H. G., Orr, C. W. M., and Ebert, J. D. (1972). Exp. Cell. Res. *74*, 201.

Rao, P. N., and Johnson, R. T. (1972). In Methods in Cell Physiology, Vol. 5, David M. Prescott, ed. (New York: Academic Press), p. 75.

Siminovitch, L., Thompson, L. H., Mankovitz, R., Baker, R. M., Wright, J. A., Till, J. E., and Whitmore, G. F. (1972). 9th Canadian Cancer Research Conference, 1971, Honey Harbour (Toronto: University of Toronto Press), p. 59.

Skou, J. C. (1957). Biochim. Biophys. Acta *24*, 394.

Stanners, C. P., Eliceiri, G. L., and Green, H. (1971). Nature New Biol. *230*, 52.

Taylor, W. B. (1970). Med. Bio. Engin. *8*, 281.

Thompson, L. H., and Baker, R. M. (1973). In Methods in Cell Biology, Vol. 6, David M. Prescott, ed. (New York: Academic Press), p. 209.

Thompson, L. H., Harkins, J. L., and Stanners, C. P. (1973). Proc. Nat. Acad. Sci. U.S.A., in press.

Thompson, L. H., Mankovitz, R., Baker, R. M., Till, J. E., Siminovitch, L., and Whitmore, G. F. (1970). Proc. Nat. Acad. Sci. U.S.A. *66*, 377.

Till, J. E., Baker, R. M., Brunette, D. M., Ling, V., Thompson, L. H., and Wright, J. A. (1973). Fed. Proc. *32*, 29.

Wallach, D. F. H. (1972). The Plasma Membrane (New York: Springer-Verlag).

45

Reprinted from *J. Cell. Physiol.* **78**:177–184 (1971)

Mutation Rates in Cells at Different Ploidy Levels [1]

MORGAN HARRIS

*Department of Zoology, University of California,
Berkeley, California 94720*

ABSTRACT   Fluctuation tests of the Luria-Delbrück type have been used to examine the frequency of spontaneous variations in cultures of diploid, tetraploid, and octaploid Chinese hamster cells. Resistance to 8-azaguanine and altered response to thermal shock were chosen as marker systems. Mutation frequencies of approximately 10^{-6} and 10^{-5} have been estimated for heat resistance and azaguanine resistance respectively. Rates of mutation for these markers remain constant or decline slightly in cells with increasing numbers of chromosome sets. This trend is not in accordance with expectations based on the assumption of dominant, co-dominant, or recessive changes at gene or chromosome levels. It is suggested that at least some variations may arise in somatic cells by stable shifts in phenotypic expression rather than by changes in genetic information.

Cytogenetic studies on cell cultures have focussed on variations of two principal types. One class consists of mutant characters first recognized through hereditary defects in the intact organism, and which later were found to be expressed *in vitro* as well. Galactosemia, acatalasia, and orotic aciduria are typical examples of this group, which provides an array of stable markers for study in culture systems (see Krooth and Sell, '70). The second category of variations has a less clearcut genetic background, and is made up of stable changes that arise *de novo* in isolated cell populations. Such variations appear within clonal cultures, and are mitotically transmissable in the absence of selective agents. Drug resistance is one well-known erample; changes in nutritional requirements (Puck, '70) or in thermal sensitivity (Harris, '67) are other illustrations.

Variations that arise *de novo* in somatic cell cultures have usually been interpreted as the result of mutation at gene or chromosome levels. The evidence which underlies this assumption has accumulated primarily through fluctuation tests, patterned after the classic experiments of Luria and Delbrück ('43) in microbial systems. In such assays the variance for any given marker in replicate samples from a single culture is compared with that of samples taken from a series of independently maintained sublines. If in the latter the observed variance is much greater, a process of mutation may be inferred, the frequency of which can be calculated from equations based on statistical considerations. In this manner, mutation rates for mammalian cells in culture have been estimated for resistance to puromycin (Lieberman and Ove, '59), 8-azaguanine (Szybalski, '59; Szybalski, Szybalska and Ragni, '62; Chu, Brimer, Jacobson and Merriam, '69; Morrow, '70), cytosine arabinoside (Bach, '69), and for changes in thymidine transport (Breslow and Glodsby, '69).

It is worth noting, however, that fluctuation tests in actuality merely establish the spontaneous origin of a given change, and do not indicate where the locus of variation may be. Thus, it is not clear whether such variations as drug resistance do in fact arise within differentiated somatic cells by gene or chromosome mutation, or if instead they are the outcome of some other type of spontaneous change, e.g.,— stable shifts in phenotypic expression. Direct evidence for a change in genetic information has not been forthcoming, although experiments have been described which purport to demonstrate DNA-mediated transformation with drug-resistant cells (Szybalska and Szybalski, '62). Other investigators have not been able to sub-

Received Feb. 22, '71. Accepted Mar. 23, '71.

[1] Supported by grant GM 13692, National Institute of General Medical Sciences, and by Cancer Research Funds of the University of California.

stantiate these claims (Ozer, '66). Karotypic shifts have often been invoked to account for changes at the phenotypic level. However, such variations as drug resistance do not in general correlate with chromosome alterations (Harris and Ruddle, '60; Breslow and Goldsby, '69; Morrow, '70). A point of special difficulty is the high frequency of variation within established cell lines, a phenomenon that is difficult to reconcile with generally lower rates for genetic changes in other cell systems.

In view of these uncertainties, we have sought a new perspective on the problem of spontaneous variation in somatic cell populations. One such approach is to predict and test the frequency of change in cell lines at different ploidy levels. Toward this end we have produced a matched set of polyploid sublines from Chinese hamster cells, the members of which were isolated sequentially from a common origin (Harris, '71). With such a model one can make assumptions of variation, based on conventional genetic concepts, and compare these expectations against observed patterns of change. As study systems we have examined changes in heat sensitivity and resistance to 8-azaguanine. Through fluctuation tests the rates of mutation for these markers have been determined for cells at diploid, tetraploid, and octaploid levels. The results obtained are difficult to interpret in terms of gene or chromosome changes, and suggest the need for alternative explanations.

MATERIALS AND METHODS

Cell lines. The populations used in the present study were derived from male Chinese hamster lung cells, strain V79-122D$_1$. Yu and Sinclair ('64) have described the origins of this stock, which was characterized by unusual constancy in karyotypic pattern. Polyploidy was induced in stock cells by colcemid and sublines isolated according to a procedure that has been described elsewhere (Harris, '71). Derivative populations were obtained at diploid, tetraploid, and octaploid levels. These strains are designated V5, V25, and V68 respectively.

Culture techniques. All cultures were maintained in a basic nutrient (10FCSDB) consisting of 10% fetal calf serum plus 90% Dulbecco's modification of Eagle's medium. Stock populations were carried as monolayers in prescription bottles and were gassed with 10% CO_2 in air. A Coulter electronic counter was used for determinations of cell number. For experiments, cells were inoculated into 60 mm petri dishes (Falcon Plastics, Los Angeles) and maintained in a humidified CO_2 incubator. To obtain colony counts, petri dish cultures were rinsed briefly in saline, stained for 30 minutes with a saturated solution of crystal violet in 0.85% NaCl, and air dried. Chromosome numbers were determined on populations used for fluctuation tests, employing air-dry preparations stained with Giemsa.

Derivation of sublines. For estimation of mutation rates, groups of 15 sublines were initiated from small inocula, and grown independently to mass cultures. As a standard procedure, each such experiment was begun with 10–30 well separated cells in a petri dish culture. After four days of incubation a single, isolated colony was trypsinized and the resulting cells (about 200–400 total) distributed into 20 one-ounce prescripton bottles. These cultures were incubated for an additional ten days, when 15 were chosen to inoculate 16 ounce prescription bottles with 0.3–0.6 × 10^6 cells each. This group of 15 independent populations was then used after four to five days of incubation for heat or drug resistance assays.

Fluctuation tests. Cultures to be used for determination of heat or drug sensitivity were trypsinized and counted individually, and the cell number adjusted to 1.0 × 10^6/ml in 10FCSDB. In assays with 8-azaguanine, groups of six petri dish cultures were established for each of the 15 sublines. Individual cultures contained 40,000 cells and azaguanine at a final concentration of 30 μg/ml. Media changes were performed three times per week and the cultures terminated by staining after 12–14 days incubation. For assays of thermal sensitivity, cells from individual sublines were heated in suspension for 60 minutes at 44.5°C. Aliquots in 10FCSDB were placed in one ounce bottles, gassed with 10% CO_2, and immersed in a water bath, the temperature of which was constant ± 0.2°C. Details of apparatus and heat treatment may be found in an earlier paper (Harris, '67). After exposure to heat shock the cells were plated out in petri dishes and incubated at 37°C to determine relative survival. Fluid changes were made two to three times weekly and the cultures terminated by staining after 12–13 days of incubation.

Mutation rates for both heat and drug resistance were estimated by a modification of the original procedures of Luria and Delbrück ('43). In each experiment the data were first evaluated for significance by a chi-square test, after which the mutation rate was determined by the median method outlined by Lea and Coulson ('49). Such estimates were further validated by relating the observed variance to the theoretical ("likely") variance to be expected from the same data (Luria and Delbrück, '43). This comparison was carried out by calculating ratios between standard deviations and the experimental mean.

RESULTS

Changes in heat sensitivity

Quantitative data on survival of pig kidney cells after acute heat shock have been reported in earlier papers (Harris, '66, '67). The viability of these cells declines exponentially when heated in suspension at 43°–47°C, with a slope that is characteristic of the temperature selected for treatment. When the surviving fraction of pig kidney cells is reduced to 10^{-5} or 10^{-6}, recovery colonies are found to exhibit an increased heat resistance. If such survivors are grown to mass populations and again exposed to elevated temperatures, the mortality rate is significantly lower. This altered response to thermal stress is a stable variation which is mitotically perpetuated in populations maintained at 37°C, and does not involve any change in growth rate, cell size, or chromosome pattern.

Pilot studies showed that these observations in general outline are also applicable to the Chinese hamster cells used for the present investigation. In assays, with cell lines V5, V25, and V68, a temperature of 44.5°C and exposure time of 60 minutes served to selected effectively for heat-resistant variants. Viability declined essentially in parallel during heat treatment for cell lines at different ploidy levels. The absolute level of survival varied somewhat in successive experiments, according to the physiological state of test cells. Sensitivity to heat exposure shifts during the population growth cycle, and is minimal at the maximum growth phase. For any single experiment, however, assays of viability with the same cell suspension were highly reproducible.

In order to determine the level of sampling variation, a number of experiments were performed with replicate cell suspensions derived from a single mass culture. For this purpose the culture was first trypsinized, and nine to ten identical aliquots of cells in 10FCSDB were heated for the standard interval of 60 minutes at 44.5°C. Five petri dish cultures were inoculated with the heated cells from each bottle, and viability was determined by colony counts after incubation at 37°C.

Table 1 summarizes the results obtained, which were similar in principle for cells at all three ploidy levels. In each assay, the variance for replicate platings is approxi-

TABLE 1

Variation in heat sensitivity among replicate samples from single cultures

Cell line	V5 (diploid)		V25 (tetraploid)		V68 (octaploid)
Number of samples	10	9	9	10	10
Survivors/10^6 cells	36.4	42.5	12.7	22.2	76.5
Variance	35.7	49.7	12.4	10.8	43.3
Ratio $\dfrac{\text{variance}}{\text{mean}}$	0.98	1.17	0.98	0.49	0.57
χ^2	8.8	10.5	7.8	4.4	5.1
P	0.46	0.24	0.46	0.89	0.82

Samples of 2.0×10^6 cells were suspended in 10 FCSDB, heated at 44.5°C for 60 minutes, and plated at 37°C to determine survival.

mately equal to the mean, as would be expected on the basis of a Poisson distribution for random sampling errors. Alternatively, the observed variation can be evaluated by chi-square tests; these show that differences in heat sensitivity are not significant between replicate samples.

Quite a different picture emerges when cells derived from independent sublines are compared for heat response (table 2). Each such assay is based on comparisons between 15 sublines, carried in parallel through the same procedure. Individual populations differ materially in resistance to heat treatment, resulting in a large variance as compared to the group mean. Chi-square values are similarly elevated, and indicate clearly that variation between sublines exceeds the limits of random error. This pattern of variation is characteristic for groups of sublines at different ploidy levels, and in duplicate experiments performed with any given cell type. The data thus indicate that heat-resistant variants in general originate by spontaneous, random change, rather than by induction during heat treatment.

With these preconditions satisfied, equations similar to those of Luria and Delbrück ('43) may be employed to estimate mutation rates for the observed change. Various modifications can be used to simplify calculation (see Newcombe, '48). The median method of Lea and Coulson ('49) is convenient and was applied throughout the present study. Table 2 gives the values obtained. It can be seen that mutation rates for heat resistance are of the order of 10^{-6}, with all estimates falling within a narrow range. Although the number of determinations is too small to treat statistically, the mutation frequency is not greatly different at the three ploidy levels. From the information available, therefore, it is reasonable to assume that the mutation rate to heat resistance either remains constant or declines slightly in an ascending ploidy series.

Resistance to 8-azaguanine

Fluctuation tests are useful with a drug resistance marker as a basis for comparison with changes in thermal response. Resistance to the purine analogue 8-azaguanine is especially appropriate, since

TABLE 2

Variation in heat sensitivity between independent sublines of common origin [1]

Cell line	V5 (diploid)			V25 (tetraploid)			V68 (octaploid)	
Cells/culture $\times 10^6$	33.5	44.6	25.2	17.6	17.1	28.0	5.7	5.5
Survivors/10^6 cells	63.8	56.6	19.9	12.7	5.8	63.5	2.4	20.5
Variance (corrected for sampling)	252.4	802.5	132.7	185.9	10.0	573.6	4.1	507.7
Ratio $\frac{\text{variance}}{\text{mean}}$	4.0	14.2	6.7	14.6	1.7	9.0	1.7	24.8
χ^2	69.4	212.5	107.4	218.6	38.1	141.0	38.0	283.4
P	<0.001	<0.001	<0.001	<0.001	<0.001	<0.001	<0.001	<0.001
Mutation rate [2] $\times 10^{-6}$	9.6	6.7	3.3	2.2	1.2	9.0	0.64	3.3
Average rate		6.5			4.1			2.0
$\frac{\text{Standard deviation}}{\text{mean}}$ { observed	0.25	0.50	0.58	1.07	0.54	0.39	0.85	1.10
{ calculated	0.58	0.46	0.64	0.67	0.80	0.55	1.03	0.56
Modal chromosome number	23	22	21	44	45	43	85	81

[1] Cells from each subline were suspended in 10 FCSDB, heated at 44.5°C for 60 minutes, and plated at 37°C to determine viability. Fifteen sublines were used for each experiment.
[2] Mutations per cell per generation.

several workers have estimated mutation rates in animal cells with this drug. Thus, Szybalski ('59) reported a figure of 4.9×10^{-4} for the first step change in D98 human cells from azaguanine sensitivity to resistance. In V79 Chinese hamster cells Chu, Brimer, Jacobson, and Merrian ('69) found the frequency to be 1.5×10^{-8}, while in an established mouse cell line, presumably L, Morrow ('70) estimated a rate of 3×10^{-6} for appearance of the aza-guanine resistance marker. Part of this wide differential in mutation rates can be attributed to species differences, but it is likely that variables in assays are significant as well. Albertini and De Mars ('70), for example, observed that certain batches of fetal calf serum contain substances which antagonize the inhibitory effects of azaguanine. Under these conditions increases of five-fold or more in drug concentration were necessary to eliminate sensitive cells.

In the present experiments the selective level of 8-azaguanine was set at 30 μg/ml, which discriminated effectively between variant and wild type cells in all three cell lines used. Resistance to 8-azaguanine may arise by a loss or decrease in activity of the enzyme required for incorporation of analogue to the nucleotide level (Littlefield, '63). However, refractory states can be demonstrated even more frequently in cells with an unimpaired enzyme pattern. These alterations seem to involve a reduction in permeability to or transport of the analogue. No attempt was made in present investigation to discriminate between different types of resistance in the assay procedures.

Table 3 gives the results of background experiments performed with replicate samples from single mass cultures. As in the case of heat resistance the number of survivor colonies after exposure to 8-azaguanine is relatively uniform for platings made from any single cell suspension. This point is documented by chi-square tests and by the similarity of variances with their respective means.

Table 4 summarizes the data obtained on azaguanine resistance in assays with independent sublines. Each experiment is characterized by a relatively large variance, and chi-square tests indicate that variation greatly exceeds sampling error. These findings imply that the change from azaguanine sensitivity to resistance is spontaneous and random, as other investigators have repeatedly maintained (Szybalski, '59; Chu, Brimer, Jacobson and Merriam, '69; Morrow, '70). Table 4 also provides estimates of mutation rates for the three cell types concerned. The frequencies are all of the order of 10^{-5}, and no significant increase or decrease in rate can be found for cells at different ploidy levels. These data are thus consistent with the level or slightly declining trend in mutation rates for heat resistance in the same cell types.

No evidence was found of association between azaguanine resistance and heat resistance within individual sublines. As a check on this possibility, a correlation study was performed with three groups of independent lines that had been used for both types of assays. Of 16 sublines with heat resistance above the group means, nine showed greater than average numbers of azaguanine-resistant colonies, and

TABLE 3

Variation in sensitivity to 8-azaguanine among replicate samples from single cultures

Cell line	V5 (diploid)	V25 (tetraploid)	V68 (octaploid)
Number of samples	10	10	10
Survivors/			
4 × 10⁴ cells	12.4	30.3	1.5
Variance	2.92	15.9	0.98
Ratio $\frac{variance}{mean}$	0.23	0.52	0.66
χ^2	2.12	4.93	5.96
P	0.99	0.84	0.75

Aliquots of 4×10^4 cells were inoculated into 60 mm petri dishes and incubated in 10 FCSDB containing 30 μg 8-azaguanine per ml. Survival data are based on colony counts for three petri dishes per sample.

TABLE 4

Variation in sensitivity to 8-azaguanine between independent sublines of common origin [1]

Cell line	V5 (diploid)	V25 (tetraploid)		V68 (octaploid)	
Cells/culture $\times 10^6$	25.2	28.0	11.0	5.7	5.6
Survivors/10^6 cells	7.4	7.9	27.2	5.9	2.5
Variance (corrected for sampling)	16.2	60.5	182.4	19.5	6.8
Ratio $\dfrac{\text{variance}}{\text{mean}}$	2.2	7.7	6.7	3.3	2.7
χ^2	30.7	121.3	93.9	59.9	48.9
P	0.005	<0.001	<0.001	<0.001	<0.001
Mutation rate [2] $\times 10^{-5}$	2.2	1.9	7.4	2.0	0.65
Average rate	2.2		4.7		1.3
Standard deviation ⎰ observed	0.40	0.99	0.46	0.74	1.06
$\overline{\text{mean}}$ ⎱ calculated	0.46	0.37	0.42	0.52	0.39
Modal chromosome number	21	43	44	85	81

[1] Aliquots of 4×10^4 cells from each of 15 sublines were inoculated into 60 mm petri dishes and incubated in 10 FCSDB containing 30 μg 8-azaguanine per ml. Survival data are based on colony counts for six petri dishes per subline.
[2] Mutations per cell per generation.

seven fewer. Similarly, among 15 sublines with azaguanine resistance above group means, nine were more resistant than average to heat treatment and six were less so. These findings suggest that heat resistance and azaguanine resistance arise by independent variation, and occur together only through random association.

DISCUSSION

The most important finding in the present study is that rates of mutation for two independent markers are essentially constant for cells at different ploidy levels. How does this result fit the preconception of gene or chromosome changes? If a dominant or co-dominant mutation is assumed as a first model, one would expect an increase in mutation rate as the ploidy level rises. This conclusion follows since more alleles at each locus are progressively available for change with each doubling in the number of chromosome sets. However, mutation rates for heat and drug resistance remain level or actually decline in polyploid cells, and the assumption of dominant or co-dominant mutation does not fit the facts here.

A more plausible suggestion might be the occurrence of a recessive mutation at gene or chromosome level, since azaguanine resistance, at least, behaves as a recessive in hybrid combinations (Chu, Brimer, Jacobson and Merriam, '69). If the locus were in a pair of autosomes, the frequency of expression for a recessive

mutation should decline sharply in an ascending polyploid series. This is so because somatic segregation would be needed even in diploid cells to remove the accompanying wild-type allele. Segregation might occur by localized deletions, loss of a whole chromosome, or by such processes as nondisjunction or somatic crossing-over. Events of this type, even if infrequent, could lead to phenotypic expression of the recessive marker. However, in a tetraploid cell there would be three wild-type alleles to be removed by segregation, and in octaploid cells, seven. Thus, the decreasing probability of these multiple events should rapidly reduce the chance for expression of a mutant recessive allele to the vanishing point.

Alternatively, a recessive mutation might occur at a hemizygous locus, in a chromosome represented singly in a permanent cell line, in the single x-chromosome of male cells, or the single functional x of female cells (Lyon, '61). All of these conditions would permit the direct expression of recessive markers in diploid cells. This circumstance, however, does not extend to polyploids, in which multiple alleles are again present for each locus, owing to the duplication of chromosome sets. Segregation would thus be required in tetraploid and especially in octaploid cells, to eliminate the corresponding wild-type alleles.

One might maintain that activity is restricted to an single x-chromosome in polyploids, even though polyploidy is induced well beyond the embryonic period when x-inactivation (in female cells) normally occurs. The work of Siniscalco ('70), however, shows clearly that intergenic complementation takes place between two x-chromosomes in hybrids formed by fusing somatic cells. By implication, the x-chromosomes are individually active as replicates in polyploids produced from differentiated cells. In that case, recessive mutations in the x-chromosomes, as in autosomes, should decline in expression with increasing ploidy level, although less sharply. Here again the essentially level trend in mutation rates for heat and drug resistance is contrary to such expectations in polyploids.

On balance, therefore, the pattern of variation observed in the present experiments is difficult to interpret in terms of conventional models. On the other hand, evidence from other fields shows that azaguanine resistance, at least, can have a clear-cut genetic basis. In the x-linked familial disorder known as the Lesch-Nyhan syndrome, there is a deficiency at the cellular level of the enzyme hypoxanthine-guanine phosphoribosyltransferase (PRT) (E.C.2.4.2.8.). PRT deficiency is found uniformly in cells from affected males, while the tissues from female heterozygotes can be shown to contain normal and PRT-deficient cells in approximately equal numbers (Dancis, Cox, Berman, Jansen and Balis, '69; Fujimoto and Seegmiller, '70). These facts point directly to the presence of a recessive mutant gene located in the x-chromosome.

Cells from Lesch-Nyhan patients fail to incorporate tritiated hypoxanthine and are selectively eliminated in HAT medium (Choi and Bloor, '70). At the phenotypic level, Lesch-Nyhan cells thus resemble *de novo* variants which have become resistant to azaguanine by enzyme deficiency; indeed, the activities referred to in separate papers at PRT or inosinic-guanylic acid pyrophosphorylase are probably identical. However, despite a common mode of phenotypic expression, these two types of variants differ significantly in degree of stability. No reversions to normal enzymatic activity have been observed for Lesch-Nyhan cells when autoradiographs with

tritiated hypoxanthine were employed for assay (Dancis, Cox, Berman, Jansen and Balis, '69; Choi and Bloom, '70) although in mixtures with normal cells, metabolic cooperation may occur (Fujimoto and Seegmiller, '70). In contrast, a spectrum of reversion rates may be observed for *de novo* variants with enzyme deficiency, and in some cell lines the frequency is quite high (Littlefield, '63).

These paradoxes could perhaps be resolved if stable variations in some cases arise *de novo* within somatic cells by stable shifts in phenotypic expression rather than by changes in genetic information. Such changes on a programmed basis are a familiar part of embryonic development, but would not be expected in normal cells, especially as random events, when the phase of differentiation is past. In this connection it is significant that drug resistance and other *de novo* variations are rarely if ever observed in cultures of primary diploid cells (for a possible exception, see Albertini and De Mars, '70). Rather it is the established or permanent cell lines, together with tumor cell populations, in which *de novo* variation is a characteristic and common phenomenon. It seems possible that while many of these changes may resemble mutations they could be in fact the result of impaired control of phenotypic expression, an abnormality that may be distinctive for such cell types. This concept needs to be examined with new techniques, and on a broader basis. It seems fair to suggest that concepts based on normal cells, while useful, may not cover the additional degrees of freedom which appear to exist in tumors and established cell systems *in vitro*.

LITERATURE CITED

Albertini, R. J., and R. de Mars 1970 Diploid azaguanine-resistant mutants of cultured human fibroblasts. Science, *169:* 482–485.

Bach, M. K. 1969 Biochemical and genetic studies of a mutant strain of mouse leukemia L1210 resistant to 1-β-D-arabinofuranosylcytosine (Cytarabine) hydrochloride. Cancer Res., *29:* 1036–1044.

Breslow, R. E., and R. A. Goldsby 1969 Isolation and characterization of thymidine transport mutants of Chinese hamster cells. Exp. Cell Res., *55:* 339–346.

Choi, K. W., and A. D. Bloom 1970 Biochemically marked lymphocytoid lines: establishment of Lesch-Nyhan cells. Science, *170:* 89–90.

Chu, E. H. Y., P. Brimer, K. B. Jacobson and E. V. Merriam 1969 Mammalian cell genetics. I. Selection and characterization of mutations auxothrophic for L-glutamine or resistant to 8-azaguanine in Chinese hamster cells *in vitro*. Genetics, 62: 359–377.

Dancis, J., R. P. Cox, P. H. Berman, V. Jansen and M. E. Balis 1969 Cell population density and phenotypic expression of tissue culture fibroblasts from heterozygotes of Lesch-Nyhan's disease (inosinate pyrophosphorylase deficiency). Biochem. Genetics, 3: 609–615.

Fujimoto, W. Y., and J. E. Seegmiller 1970 Hypoxanthine-guanine phosphoribosyltransferase deficiency: activity in normal, mutant, and heterozygote culture human skin fibroblasts. Proc. Nat. Acad. Sci., 65: 577–584.

Harris, M. 1966 Criteria of viability in heat-treated cells. Exp. Cell Res., 44: 658–661.

——— 1967 Temperature resistant variants in clonal populations of pig kidney cells. Exp. Cell Res., 46: 301–314.

——— 1971 Polyploid series of mammalian cells. Exp. Cell Res. (in press).

Harris, M, and F. H. Ruddle 1960 Growth and chromosome studies on drug resistant lines of cells in tissue culture. In: Cell Physiology of Neoplasia. U. Texas Press, Austin, pp. 524–546.

Krooth, R. S., and E. K. Sell 1970 The action of mendelian genes in human diploid cell strains. J. Cell Physiol., 76: 311–330.

Lea, D. E., and C. A. Coulson 1949 The distribution of the numbers of mutants in bacterial populations. J. Genetics, 49: 264–285.

Lieberman, I., and P. Ove 1959 Estimation of mutation rates with mammalian cells in culture. Proc. Nat. Acad. Sci., 45: 872–877.

Littlefield, J. W. 1963 The inosinic acid pyrophosphorylase activity of mouse fibroblasts partially resistant to 8-azaguanine. Proc. Nat. Acad. Sci., 50: 568–576.

Luria, S. E., and M. Delbrück 1943 Mutations of bacteria from virus sensitivity to virus resistance. Genetics, 28: 491–511.

Lyon, M. F. 1961 Gene action in the x-chromosome of the mouse (*Mus musculus* L.). Nature, 190: 372–373.

Morrow, J. 1970 Genetic analysis of azaguanine resistance in an established mouse cell line. Genetics, 65: 279–287.

Newcombe, H. B. 1948 Delayed phenotypic expression of spontaneous mutations in Escherichia coli. Genetics, 33: 447–476.

Ozer, H. L. 1966 Purine pyrophosphorylase as a selective genetic marker in a mouse lymphoma, P388, in cell culture. J. Cell. Physiol., 68: 61–68.

Puck, T. T. 1970 Biochemical genetics studies on mammalian cells *in vitro*. In: Control Mechanisms in the Expression of Cellular Phenotypes. H. A. Padykula, ed. Academic Press, New York.

Siniscalco, M. 1970 Somatic cell hybrids as tools for genetic studies in man. In: Control Mechanisms in the Expression of Cellular Phenotypes. H. A. Padykula, ed. Academic Press, New York.

Szybalska, E. H., and W. Szybalski 1962 Genetics of human cell lines, IV. DNA-mediated heritable transformation of a biochemical trait. Proc. Nat Acad. Sci., 48: 2026–2034.

Szybalski, W. 1959 Genetics of human cell lines II. Method for determination of mutation rates to drug resistance. Exp. Cell Res., 18: 588–591.

Szybalski, W., E. H. Szybalska and G. Ragni 1962 Genetic studies with human cell lines. Cancer Inst. Monogr., 7: 75–88.

Yu, C. K., and W. K. Sinclair 1964 Homogeneity and stability of chromosomes of Chinese hamster cells *in vitro*. Canad. J. Genetics and Cytol., 6: 109–116.

46

Reprinted from *Natl. Acad. Sci. (USA) Proc.* **70**:320–324 (1973)

Mutations Affecting the Structure of Hypoxanthine: Guanine Phosphoribosyltransferase in Cultured Chinese Hamster Cells

(8-azaguanine/crossreacting material/antiserum/clones/affinity chromatography)

A. L. BEAUDET, D. J. ROUFA, AND C. T. CASKEY

Robert J. Kleberg, Jr. Center for Human Genetics, Departments of Medicine, Biochemistry, and Pediatrics, Baylor College of Medicine, Houston, Texas 77025

Communicated by Alan Garen, November 16, 1972

ABSTRACT   14 Clones resistant to 8-azaguanine, isolated from mutagenically treated cultured cells from Chinese hamsters, were tested for loss of enzymic and immunological activities of hypoxanthine: guanine phosphoribosyltransferase. Three of the clones had no enzymic activity but reacted strongly with antiserum prepared against the enzyme (CRM $^+$), indicating the presence of a defective enzyme protein, probably caused by a mutation in its structural gene. The other 11 mutants had little or no enzymic or immunologic activity. Revertant clones with positive enzymic and crossreacting immunologic activities were derived from both classes of mutants. The antiserum has been used for rapid purification of the phosphoribosyltransferase from Chinese hamster cells by immunoadsorbant affinity chromatography.

Biochemical characterization of induced phenotypic alterations of animal cells is necessary to clarify their genetic basis. We have investigated the hypoxanthine: guanine phosphoribosyltransferase activity (HGPRTase; EC 2.4.2.8.) of 35 clones of Chinese hamster lung cells (CHL) resistant to 8-azaguanine. Resistance was induced by the mutagens N-methyl-N-nitro-N-nitrosoguanidine and ethylmethanesulfonate (1). Sixteen of the 35 resistant clones showed little or no HGPRTase activity (HGPRTase$^-$), as determined directly *in vitro* and indirectly by the incorporation of radioactive hypoxanthine into nucleic acid of whole cells. The remaining 19 mutants had measurable HGPRTase activity. Revertant HGPRTase$^+$ clones were isolated from HGPRTase$^-$ clones by selection in a medium containing hypoxanthine, aminopterin, and thymidine (HAT medium) (2). In this report, we describe the results of studies with anti-HGPRTase serum to test for immunologically crossreactive material (CRM) in 8-azaguanine-resistant clones and in their revertants.

MATERIALS AND METHODS

Cell Growth and Extract Preparation. Growth and selection of 8-azaguanine-resistant clones and extract preparation were as described (1). Resistant clones, designated as N or E, were isolated from the wild-type clone (A-3) after nitrosoguanidine or ethylmethanesulfonate mutagenesis, respectively. Revertants are designated by the parental nomenclature (e.g., N5), followed by the letters N, E, or S (spontaneous) to indicate the treatment used for reversion

---

Abbreviations: HGPRTase, hypoxanthine:guanine phosphoribosyltransferase; CRM, crossreacting material (immunologically).

(e.g., N5N4). Extracts used in precipitation-inhibition studies were dialyzed for 24 hr against 0.01 M potassium phosphate buffer (pH 7.2). Human erythrocyte lysates (3) were dialyzed in an identical manner. The type and source of other cultured cells were: Hela S3 (Dr. Saul Kit); L929 (American Type Culture Collection); NIH Balb 3T3 (Dr. Janet Butel); and human fibroblasts from male foreskin (primary isolate).

Preparation of Antibody. The livers (17.7 g) of 20 Chinese hamsters were homogenized with a Potter–Elvehjem homogenizer in 100 ml of buffer containing: 0.03 M Tris·HCl, pH 7.0–7 mM 2-mercaptoethanol. The extract was centrifuged at 30,000 $\times$ g for 30 min at 4°, and the supernatant was centrifuged at 150,000 $\times$ g for 120 min at 4°. HGPRTase was partially purified from the high-speed supernatant fraction by (a) precipitation with ammonium sulfate (35–70% saturation, 26 g/105 ml = 35%); (b) passage through a carboxymethyl-(CM)-Sephadex column (2.5 $\times$ 25 cm) equilibrated with 0.01 M potassium phosphate (pH 7.0)–7 mM 2-mercaptoethanol; (c) elution from a DEAE-cellulose column (2.5 $\times$ 25 cm) equilibrated with 0.01 M potassium phosphate (pH 7.0) by a 800-ml linear gradient from 0.01 M to 0.2 M potassium phosphate (pH 7.0) (activity was eluted at 0.09 M); and (d) treatment at 80° for 5 min, followed by centrifugation at 30,000 $\times$ g for 20 min and recovery of the supernatant. The HGPRTase specific activity was increased 17.5-fold, with a 17% recovery by these steps.

New Zealand white rabbits were immunized initially and at 5 weeks by subcutaneous injection of the partially purified HGPRTase (0.48 mg in 1 ml) mixed with 1 ml of complete Freund's adjuvant. An immunoglobulin that precipitated HGPRTase activity was detected 4 weeks after the initial immunization. This immunological response was variable, since anti-HGPRTase activity was not detected in the serum of several other rabbits injected with similar preparations of HGPRTase from liver and cultured cells. All sera used in these studies were dialyzed 24 hr against 1,000 volumes of buffer containing 0.01 M potassium phosphate (pH 7.2)–75 mM NaCl.

Double-diffusion analysis (IDF Cell II, Cordis Lab.) of the serum drawn 8 weeks after HGPRTase injection revealed at least three precipitin bands with Chinese hamster liver and cultured cell extracts.

HGPRTase Immunoprecipitation and Precipitation-Inhibition. HGPRTase immunoprecipitation and precipitation-inhibition reactions were performed in 1-ml conical tubes. The 95-μl reaction mixtures contained a source of dialyzed HGPRTase, dialyzed rabbit serum, and a final concentration of 0.01 M potassium phosphate (pH 7.2)–11 or 4 mM NaCl for the precipitation or precipitation-inhibition reactions, respectively. These buffer conditions were rigidly maintained, since interference with precipitation occurred in reactions containing 0.15 M NaCl, more than 5 mM 2-mercaptoethanol, or 20% sucrose. All reactions were incubated for 30 min at 37°, then for 1.5 or 15 hr, as indicated, at 4°. 5 μl Of anti-rabbit gammaglobulin prepared in goats, (Antibodies, Inc.) was then added, followed by 2 hr of incubation at 4°. The HGPRTase–antibody complex was sedimented by centrifugation at 12,000 $\times$ g for 20 min at 4°. All HGPRTase and adenine phosphoribosyltransferase (APRTase, EC 2.4.2.7) activities in the supernatant were determined at 37° for 20 min with 0.02- or 0.01-ml aliquots, respectively (1). The antibody precipitation does not inactivate HGPRTase, since enzyme activity could be recovered quantitatively from the precipitates.

Immunoprecipitation of HGPRTase from cultured cells by rabbit antiserum was enhanced by the presence of anti-rabbit gammaglobulin serum prepared in goats (Table 1). Precipitation of HGPRTase required the anti-HGPRTase serum from rabbit. Anti-rabbit gammaglobulin from goat improved reproducibility, shortened the precipitation time, facilitated precipitation of small amounts of HGPRTase, and conserved rabbit anti-HGPRTase serum. Goat anti-gammaglobulin was used, therefore, in all precipitation reactions. All sera were assayed and found to be free of HGPRTase activity. The apparent stimulation of HGPRTase by small amounts of rabbit antiserum and control serum was observed repeatedly, and we feel that the effect represents HGPRTase stabilization similar to that reported with albumin (4).

Precipitation-inhibition studies used wild-type HGPRTase, from cultured cells, that was partially purified by 35–70% ammonium sulfate fractionation, followed by dialysis against 0.01 M potassium phosphate (pH 7.2)–1 mM 2-mercaptoethanol. Each 95-μl reaction contained 46 μg of HGPRTase and 5 μl of dialyzed rabbit antiserum. Under these conditions, 95% of the HGPRTase activity was precipitated. Different amounts of individual extracts were added to these reactions, and the reaction mixtures were incubated and centrifuged as indicated above. The immunological activity (CRM) of an extract was quantitated by the amount of wild-type HGPRTase activity that was displaced from the antibody complex and appeared in the supernatant fraction. The CRM activity corresponds to nmol of IMP formed per min per mg of extract protein added. Extracts from all HGPRTase$^-$ clones used in precipitation-inhibition assays had no HGPRTase activity, and did not stimulate or inhibit wild-type HGPRTase (data not shown).

Immunoadsorbent Affinity Chromatography. 2 ml Of rabbit anti-HGPRTase serum was precipitated with ammonium sulfate (33%, 0.47 g). The precipitate was dissolved in 1.5 ml of 0.1 M NaHCO$_3$ (pH 8.0) and dialyzed against this solution (1000 volumes) for 16 hr. Immunoglobulin was linked to agarose with cyanogen bromide (5, 6). Cyanogen bromide (500 mg) was added to 5 ml of Sepharose 4B in a total volume of 11 ml of 0.1 M NaHCO$_3$ (pH 8.0). After 8 min of incubation at 20°, during which the pH was maintained at 11 $\pm$ 0.5 with 2 M NaOH, the Sepharose was washed with 450 ml of cold 0.1 M NaHCO$_3$ (pH 8.0). Sepharose (2 ml) was mixed

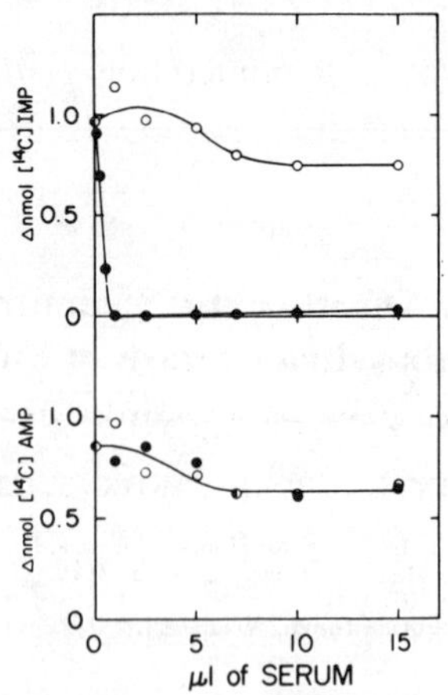

Fig. 1. Anti-HGPRTase enzyme specificity. Precipitation reactions contained 124 μg of unfractionated cultured cell extract and control serum (O) or anti-HGPRTase serum (●) as indicated; they were incubated 15 hr at 4° before addition of anti-rabbit gammaglobulin. The procedures for HGPRTase precipitation and quantitation of HGPRTase and APRTase in the supernatant are described in *Methods*. The values for reaction mixtures incubated without enzyme added (0.016 and 0.031 nmol) were subtracted from HGPRTase (*top*) and adeninephosphoribosyltransferase, (*bottom*) values, respectively. Data are given as nmol per 20 min per 20 μl of supernatant (*top*) or per 10 μl of supernatant (*bottom*).

immediately with the partially purified globulin at 4°, resulting in 80% linkage of the immunoglobulin to the Sepharose.

Immunoglobulin affinity columns containing 1 ml of Sepharose–globulin were equilibrated with 0.01 M potassium phosphate (pH 7.2). Extracts equilibrated in the same buffer were applied to the columns and allowed to adsorb for 10 min. Columns were washed with 0.75 ml of 0.15 M NaCl in the above buffer, and the HGPRTase was eluted with 0.08 M NH$_4$OH in 20% sucrose. 0.75-ml Fractions were collected in tubes containing 0.15 ml of 1 M potassium phosphate (pH 7.2 or 6.5) for the NaCl and NH$_4$OH washes, respectively. These buffers adjust the pH of all fractions to 7.2. The HGPRTase activity of an aliquot of each fraction was further stabilized by addition of bovine-serum albumin and 2-mercaptoethanol to final concentrations of 0.5 mg/ml and 7 mM, respectively. All column operations were conducted at 4°. Protein determinations were performed by the method of Lowry (7).

Materials. Sources of material were: [8-¹⁴C]hypoxanthine (53.7 Ci/mol) and [8-¹⁴C]adenine (41.3 Ci/mol), Schwarz/Mann; Sepharose 4B, Pharmacia; cyanogen bromide, Eastman; p-chloromercuribenzoate, K and K Laboratories; Freund's adjuvant, Difco; all others as described (1).

RESULTS

The enzymic specificity of rabbit anti-HGPRTase serum is demonstrated in Fig. 1. Antiserum from a rabbit immunized with Chinese hamster liver HGPRTase precipitated HGPRTase from cultured cells (shown) and liver (not shown)

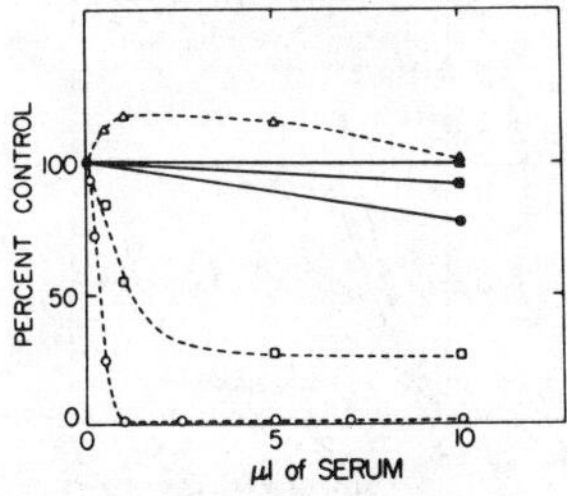

FIG. 2. Specificity of anti-HGPRTase serum for HGPRTase from various organisms. Each precipitation reaction was incubated 15 hr at 4° before addition of anti-rabbit gammaglobulin from goat, and contained—where indicated—124 µg of Chinese hamster lung cell extract (O), 77 µg of L929 extract (□), or 472 µg of human erythrocyte lysate (△). Reactions containing control and anti-HGPRTase sera are indicated by *closed* and *open* symbols, respectively. The 100% values were: Chinese hamster lung cell extract, 0.967 nmol; L929 extract, 1.137 nmol; and human eythrocyte lysate, 1.165 nmol.

equally well. This anti-HGPRTase serum had no effect on adenine phosphoribosyltransferase (APRTase, EC 2.4.2.7), a related purine "salvage" enzyme, from cultured cells (shown) and liver (not shown). Control serum had little effect on immunoprecipitation of either enzyme.

The specificity of antisera for HGPRTase from various organisms is illustrated in Fig. 2. The HGPRTase activities of human erythrocytes, cultured human fibroblasts, and HeLa cells (last two not shown) were not precipitated by the antiserum to Chinese hamster HGPRTase. The discriminatory capacity of this antibody could be useful in studies of human–Chinese hamster cell hybrids. The slopes of the pre-

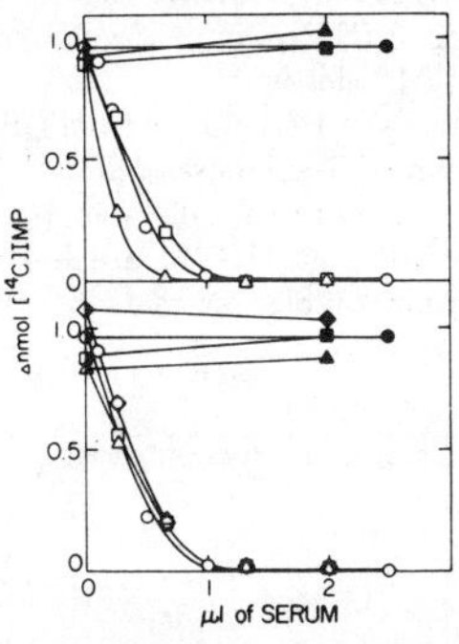

FIG. 3. Immunoprecipitation of HGPRTase from Chinese hamster lung cell mutants. Each precipitation reaction contained extracts as indicated, and was incubated 15 hr at 4° before addition of anti-rabbit gammaglobulin from goat. The quantity and source of protein for each determination was as follows: *upper panel;* 124 µg of wild-type cell extract (O), 63 µg of clone N9 (△), 237 µg of clone E47 (□); *lower panel;* 124 µg of wild-type cell extract (O), 232 µg of clone E31S1 (△), 158 µg of clone N20S1 (□), 119 µg of clone N15N1 (◇). Reactions containing control and anti-HGPRTase sera are indicated by *closed* and *open* symbols, respectively. Extracts in the upper panel were prepared from 8-azaguanine-resistant cell clones, while those in the lower panel were prepared from HGPRTase⁺ revertant clones derived from HGPRTase⁻ mutants.

TABLE 1. *Immunoprecipitation of Chinese hamster lung cell HGPRTase*

µl of anti-HGPRTase	Supernatant HGPRTase	µl of anti-HGPRTase	Supernatant HGPRTase
Minus anti-rabbit gammaglobulin from goat		*Plus anti-rabbit gammaglobulin from goat*	
0	0.777	0	0.787
0.27	1.006	0.27	0.455
0.67	1.094	0.67	0.106
1.33	0.324	1.33	0.041

Each 95-µl precipitation reaction contained 124 µg of unfractionated HGPRTase, 5 µl of anti-rabbit gammaglobulin from goat, rabbit anti-HGPRTase, as indicated, and buffer. These reactions were incubated 1.5 hr at 4° before addition of anti-rabbit gammaglobulin, followed by precipitation and HGPRTase assay of the supernatant fraction, as described in *Methods.* Supernatant HGPRTase is given as ∆nmol of [¹⁴C]IMP per 20 min per 20 µl. The value for reaction mixtures incubated without enzyme added (0.016 nmol) was subtracted from all values.

cipitation curves of mouse L929 and NIH Balb 3T3 (not shown) HGPRTase differ significantly from that of Chinese hamster HGPRTase, indicating their lack of structural homology. The lack of complete precipitation of L929 and 3T3 HGPRTase at high concentrations of antiserum could be a further reflection of the lack of homology, or an indication of the presence of HGPRTase isozymes in the mouse cells. Isozymic forms of HGPRTase have been detected in other species (3, 4, 8).

The immunological precipitation properties of HGPRTase from several 8-azaguanine-resistant CHL clones are given in Fig. 3. The cell lines are of two types: (i) HGPRTase⁺ clones resistant to 8-azaguanine and (ii) HGPRTase⁺ clones (revertants) derived from HGPRTase⁻ clones. The HGPRTase activities in extracts from all HGPRTase⁺ cell lines thus far studied (five 8-azaguanine-resistant clones and six revertants) were precipitated by the antibody; thus, they possess the HGPRTase immunological determinants. Alterations in HGPRTase structure that affect immunological reactivity or enzyme specific activity or stability could theoretically give altered immunoprecipitation. Slight alterations in the HGPRTase precipitation curves for some clones were observed. Because these changes are small as compared with the striking difference between Chinese hamster lung cell and L929 HGPRTase (Fig. 2), they are of questionable significance and require further investigation.

The immunological properties of HGPRTase⁻ clones were compared by a modification of the immunoprecipitation method. Since three immunoprecipitin lines were detected by immunodiffusion analysis, an assay method was used that measures competition for the HGPRTase–antibody complex. This immunologic method, precipitation-inhibition, avoids the study of unrelated antibodies when HGPRTase activity in the supernatant is used as the endpoint (8–10). The inhibition of precipitation of wild-type HGPRTase⁺ by extracts from HGPRTase⁻ clones is shown in Fig. 4. The wild-type HGPRTase was displaced from the antibody by increasing amounts of extract in a linear manner after the slight antibody excess was overcome. The crossreacting activity (CRM) of each extract was calculated from the linear portion

TABLE 2. *Immunologic characteristics of Chinese hamster lung cell mutant extracts*

Phenotypic characteristics	Clone	CRM activity
Wild type	A-3	0.57
8-azaguanine-resistant, *in vitro*	N-1	0.02
HGPRTase⁻, fails to	N-5	0.33
incorporate [¹⁴C]hypoxanthine into	N-6	0.02
whole cells	N-10	0.00
	N-15	0.09
	N-17	0.01
	N-19	0.02
	N-20	0.01
	E-26	0.02
	E-27	0.02
	E-31	0.00
	E-33	0.01
	E-36	0.00
	E-37	0.00
	E-43	0.01
8-azaguanine-resistant, *in vitro*	N-2	0.00
HGPRTase⁻, and incorporates	N-21	0.00
[¹⁴C]hypoxanthine into whole cells	N-22	0.01
	E-28	0.00
	E-39	0.44

CRM specific activity was determined as given in the legend to Fig. 4 and *Methods*. CRM activity is expressed as nmol of [¹⁴C]IMP formed per min per mg of extract added, and was determined from a linear region of each graph. All values were based on total HGPRTase and extract protein added to 0.10-ml precipitation reactions.

of the curve. All mutant extracts used did not have HGPRTase activity, and did not stimulate or inhibit wild-type HGPRTase (see *Methods*). Since attempts to alter wild-type enzyme to an HGPRTase⁻, CRM⁺ state by heating, aging, and treating with *p*-chloromercuribenzoate were not successful, we used enzymically active HGPRTase from the wild type for comparison. Consequently, differences between HGPRTase⁻ extracts are comparable, but absolute differences from the wild-type extract are affected by the enzyme activity present in the latter case. It is clear, however, that there are large differences in the CRM activity of different HGPRTase⁻ clones. The precipitation-inhibition activities of extracts from CRM⁺ HGPRTase⁻ and the wild-type clones were eliminated completely when they were heated for 2 min at 100°, suggesting that inhibition is due to proteins in the extracts.

The CRM activities of several 8-azaguanine-resistant cell extracts are summarized in Table 2. With the exception of the wild-type (A-3), there was no detectable HGPRTase enzymic activity in any of the extracts studied. Furthermore, most clones do not incorporate [¹⁴C]hypoxanthine from culture medium into nucleic acid. A second group of clones listed in the lower portion of Table 2 have no detectable HGPRTase activity *in vitro*, but do incorporate [¹⁴C]hypoxanthine into nucleic acid and, thus, have *in vivo* HGPRTase activity. Three HGPRTase⁻ clones (N-5, N-15, and E-39) are significantly CRM⁺ and are, therefore, felt to represent examples of mutations in the structural gene for HGPRTase. The re-

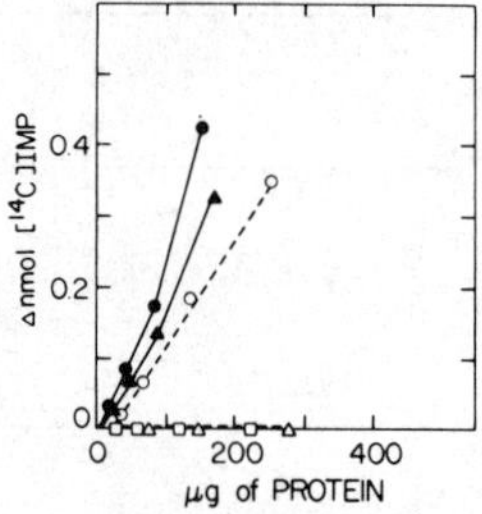

FIG. 4. Precipitation-inhibition by HGPRTase⁻ mutants. All precipitation reactions contained 46 µg of wild-type Chinese hamster lung cell HGPRTase, partially purified by ammonium sulfate precipitation (35–70%); 1.33 µl of anti-HGPRTase; and, as indicated, different amounts of dialyzed extracts from mutants E-39 (▲), N-5 (O), N-21 (□), E-31 (△), and wild-type A-3 (●). All precipitation reactions are incubated 1.5 hr at 4° before addition of anti-rabbit gammaglobulin from goat. The supernatant HGPRTase activity in the absence and presence of anti-HGPRTase serum were 1.094 and 0.062 nmol per 20 min per 20 µl, respectively. The 0.062 nmol background was subtracted from all values.

maining clones have very low or absent CRM activity by this test. The molecular basis for the HGPRTase⁻, CRM⁻ results is discussed below. We have previously concluded that mutants that undergo reversion do not represent large deletions. Revertants (HGPRTase⁺) have been obtained from the following mutants: N-1, N-5, N-6, N-10, N-15, N-17, N-19, N-20, E-26, E-27, E-31, E-37, and E-43 (unpublished results). No revertants of E-33 and E-36 have been isolated. The six revertants selected on HAT medium studied to date are HGPRTase⁺ and CRM⁺. Four of these revertants (N20S1, E31S1, Fig. 3; N14N1, E31N4, not shown) were derived from HGPRTase⁻, CRM⁻ clones, and two revertants (N15N1, Fig. 3; N5N3, not shown) were derived from HGPRTase⁻, CRM⁺ clones.

Structural analysis of HGPRTase from CRM⁺ clones may be facilitated by an additional use of anti-HGPRTase serum. The HGPRTase of cultured cells was purified by affinity chromatography with anti-HGPRTase linked to Sepharose 4B (Fig. 5). Crude cellular extracts of HGPRTase were

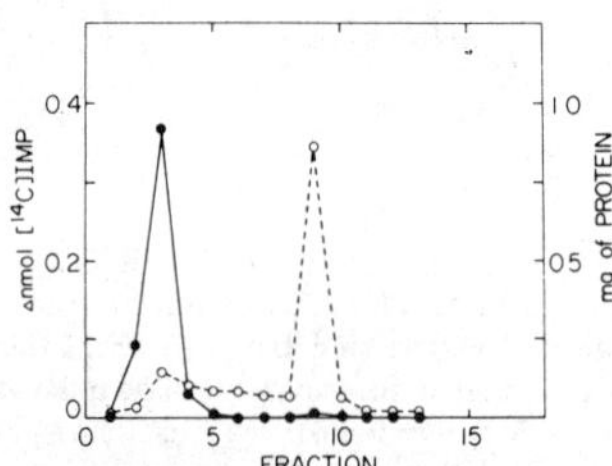

FIG. 5. Immunoadsorbent affinity chromatography of HGPRTase. An unfractionated A-3 cell extract (1.55 mg in 0.75 ml) was applied to and eluted from an anti-HGPRTase–Sepharose 4-B column. 0.03-ml aliquots were assayed (1) for 20 min for HGPRTase activity (O). The total protein in each fraction (●) is indicated. The value for reaction mixtures incubated without enzyme added (0.011 nmol) was subtracted from all HGPRTase values.

applied to and recovered from the immunoadsorbent affinity column with considerable purification. Recovery of HGPRTase activity varied from 19 to 25% in the elution peak, and depended on the method of enzyme elution and stabilization. Because of limited quantities of cellular extract and the extremely small amount of protein present in the elution peak, the overall purification of the HGPRTase protein has not been determined accurately. The HGPRTase purified by this technique was analyzed by Na dodecyl sulfate–acrylamide gel electrophoresis (11); only three protein bands were detected, while innumerable bands were detected in void fractions. With further development, this method should provide a rapid and simple means of purifying HGPRTase of CRM$^+$ azaguanine-resistant clones and revertants for structural analysis.

DISCUSSION

The molecular basis of the stable phenotypic variants of cultured animal cells is currently under investigation in many laboratories. Acquired drug resistance, particularly to 8-azaguanine, has been the most extensively studied phenotype. The data reported here suggest that three of our 8-azaguanine-resistant Chinese hamster clones (N-5, N-15, and E-39) were caused by a mutation in the HGPRTase structural gene. Extracts from these clones showed no detectable HGPRTase enzymic activity, although immunologic procedures detected the presence of HGPRTase protein. Thus, these clones synthesize an immunologically recognizable, but enzymically inactive, form of the HGPRTase molecule.

In humans with the genetic disease Lesch–Nyhan syndrome, cells that are 8-azaguanine resistant and HGPRTase$^-$ have been detected (12). In several Lesch–Nyhan patients, the cells show a HGPRTase$^-$ CRM$^+$ phenotype, similar to the three Chinese hamster clones (8, 13).

For the clones that are CRM$^-$ HGPRTase$^-$, it is difficult to establish the mechanism causing this phenotype, since drug-resistant variants may result from stable epigenetic events–in addition to mutations. Recently, the possible epigenetic origin of tissue culture variants was emphasized by two reports that examined the relationship between the frequency of drug resistance and the chromosomal ploidy of the cell lines. Harris found that the frequency of 8-azaguanine resistance did not vary significantly between diploid, tetraploid, and octaploid clones of Chinese hamster cells (14). Mezger–Freed reported that the frequency of bromodeoxyuridine resistance for diploid frog cells did not approach the square of the frequency for haploid cells (15). These studies suggest that in certain cases the frequency of drug resistance does not correlate readily with a simple mutational model.

Nevertheless, we favor the possibility that at least some of our HGPRTase$^-$ CRM$^-$ clones are mutants, since these were induced with mutagens and isolated with selective techniques identical to those used for the CRM$^+$ HGPRTase$^-$ mutants. It is unlikely that CRM$^+$ mutants should occur exclusively. If any of the HGPRTase$^-$, CRM$^-$ clones are HGPRTase structural gene mutants, we might except that some premature chain-termination (nonsense) mutations have occurred in the group. Studies with *Escherichia coli* alkaline phosphatase indicated that nonsense mutations are frequently CRM$^-$ (16, 17). Similar immunological studies of β-galactosidase mutants from *E. coli* have indicated that this correlation is not absolute, since nonsense mutations closer to the carboxyl terminus may be CRM$^+$ (18). Furthermore, the relative CRM reactivity varies with the type of immunological assay used. Subject to these reservations, some of our CRM$^-$ HGPRTase$^-$ mutants have three characteristics suggestive of nonsense mutants, namely extreme negative phenotype (1), revertibility (unpublished results), and CRM negativity. Our data indicate that the HGPRTase locus of Chinese hamster cells is a genetically useful marker for the study of mutational events in cultured mammalian cells.

This research is supported by Grants GM51598-01 and NCI-NIH-72-2058. A. L. B. and D. J. R. are supported by NIH Fellowships GM51598-01 and F02GM51201-01, respectively. C. T. C. is a Howard Hughes Research Fellow. We thank Miss Susan Seelig for excellent assistance, and Drs. W. Kelley and E. Balis and their coworkers for supplying human anti-HGPRTase serum useful in developmental studies.

1. Gillen, F. D., Roufa, D. J., Beaudet, A. L. & Caskey, C. T. (1972) *Genetics*, **72**, 239–252.
2. Szybalski, W., Szybalska, E. H. & Ragni, G. (1962) *Nat. Cancer Inst. Monogr.* **7**, 75–89.
3. Bakay, B. & Nyhan, W. L. (1971) *Biochem Genet.* **5**, 81–90.
4. Arnold, W. J. & Kelley, W. N. (1971) *J. Biol. Chem.* **246**, 7398–7404.
5. Cuatrecasas, P. & Anfinen, C. B. (1971) *Methods Enzymol.* **XXII**, 343–378.
6. Becker, M., Bocchini, V., Givol, D. & Wilchek, M. (1971) *Biochemistry* **10**, 2828–2834.
7. Lowry, O. H., Rosebrough, N. J., Farr, A. L. & Randall, R. J. (1951) *J. Biol. Chem.* **193**, 265–275.
8. Rubin, C. S., Dancis, J., Lily, C. Y., Nowinski, R. C. & Balis, M. E. (1971) *Proc. Nat. Acad. Sci. USA* **68**, 1461–1464.
9. Kabát, E. A. & Mayer, M. M. (1961) *Exp. Immunochem.* (2nd Ed.) 241–267.
10. Berson, S. A., Yalow, R. S., Glick, S. M. & Roth, J. (1964) *Metabolism* **13**, 1135–1153.
11. Laemmli, U. K. (1970) *Nature* **227**, 680.
12. Fujimoto, W. Y., Subak–Sharpe, J. H. & Seegmiller, J. E. (1971) *Proc. Nat. Acad. Sci. USA* **68**, 1516–1519.
13. Arnold, W. J., Meade, J. C. & Kelley, W. N. (1972) *J. Clin. Invest.* **51**, 1805–1812.
14. Harris, M. (1971) *J. Cell. Physiol.* **78**, 177–184.
15. Mezger–Freed, L. (1972) *Nature New Biol.* **235**, 245–246.
16. Garen, A. & Siddiqi, O. (1962) *Proc. Nat. Acad. Sci. USA* **48**, 1121–1127.
17. Garen, A. (1968) *Science* **160**, 149–159.
18. Lin, S. & Zabin, I. (1972) *J. Biol. Chem.* **247**, 2205–2211.

CHROMOSOME-WIDE EVENT ACCOMPANIES THE EXPRESSION OF RECESSIVE MUTATIONS IN TETRAPLOID CELLS

L. A. Chasin and G. Urlaub

Abstract. *Mutants resistant to 6-thioguanine were induced in pseudotetraploid hybrid Chinese hamster cells homozygous wild-type at the locus for hypoxanthine phophoribosyl transferase but heterozygous for the linked marker glucose-6-phosphate dehydrogenase. About half of these mutants had concomitantly lost the wild-type allele for glucose-6-phosphate dehydrogenase, as expected if mutation plus chromosome segregation had occurred.*

The interpretation that variant clones of cultured animal cells arise from mutations is complicated by the fact that the cells of higher organisms can also give rise to somatically heritable phenotypic changes by epigenetic processes. The latter are presumed to underlie cellular differentiation. If mutation in cultured cells is to be studied as a phenomenon per se, and if it is to be used as a tool to illuminate cell function, it is important to be able to distinguish mutation from epigenetic processes in any particular system.

There are now several systems in which the demonstration of an altered gene product in variant cells provides strong evidence for a mutational event (*1*). In most of these systems, a constitutive and ubiquitous enzyme has been studied, which makes it less likely that variants would arise epigenetically. However, in several of the same or similar systems, predictions of variant frequencies as a function of gene dosage, made on the assumption that purely mutational events were occurring, have not always been borne out. Thus Harris (*2*) found no difference in the spontaneous mutation rate to 8-azaguanine resistance in diploid, tetraploid, and octaploid Chinese hamster cells, and Mezger-Freed (*3*) found no difference between haploid and diploid frog cells in the frequency of 5-bromodeoxyuridine–resistant variants. Resistance to each of

these drugs is usually associated with a recessive loss in enzyme activity (*4–6*), although this point was not tested in the experiments cited. In the case of recessive mutations, it should be necessary to mutate each copy of the gene in question before a mutant phenotype can be expressed and recognized. If each mutation is an independent event, the mutant frequency ·should be an exponential function of the gene dosage; for example, the frequency with two genes should be approximately the square of the frequency in the single ·gene case. The failure of this prediction has led to the idea that mutation is not operating in these systems (*2, 3*).

We have previously examined this question, comparing mutation frequencies induced by ethyl methanesulfonate (EMS) in (pseudo) diploid and tetraploid Chinese hamster ovary (CHO) cells (*7*). Mutants resistant to 6-thioguanine (TG) were induced at a frequency of 2.5×10^{-4} in diploid cells, while in tetraploids the figure was 0.9×10^{-5}. Drug resistance was shown to be recessive and associated with the loss of hypoxanthine phosphoribosyltransferase (HPRT) activity. Since the gene specifying this enzyme is X-linked in humans (*8*) and probably also in Chinese hamster cells (*9, 10*) there is presumably only one active gene in diploid cells and two active genes in tetraploids (*11*). While the 25-fold difference was highly

significant, the mutant frequency in the tetraploids was still about 50 times higher than that expected on the basis of two independently occurring gene mutations. Rather than invoking an epigenetic explanation, we proposed a mechanism combining the spontaneous loss of one chromosome bearing an HPRT gene with mutation of the single remaining gene (*7*). The numerical results were consistent with such an interpretation based on segregation rates of the HPRT$^+$ (TG-sensitive) phenotype from heterozygous tetraploid hybrid cells (*12*).

This explanation leads to at least three predictions: (i) TG-resistant mutants derived from tetraploid cells should all have lost one particular chromosome; (ii) since spontaneous chromosome loss is constantly occurring, segregants should accumulate with time, and the induced mutation rate should increase with clonal age; and (iii) markers linked to the HPRT locus should also have been segregated in mutants derived from tetraploids. The first prediction has been difficult to test, since most of these mutants have lost one to five chromosomes, and even within a clone the distribution of chromosome numbers in tetraploids is not as narrow as that found in diploid CHO cells (*7*).

The second point was examined by cloning a tetraploid subline of CHO cells that had been produced by treatment of the diploid with low doses of Colcemid (*7, 13*). The clonal population was grown in nonselective medium, and the EMS-induced mutation rate was measured at intervals over a period of 3 months. As can be seen in Fig. 1, the general trend (with the exception of one point) indicates increased mutability with clonal age. Although this result is in agreement with the segregation plus

Table 1. Distribution of G6PD phenotypes among TG-resistant mutant colonies derived from a tetraploid hybrid clone. The hybrid clone Y143 was induced to mutate with EMS and TG-resistant cells selected as previously described (7). Colonies that grew in TG were stained directly for G6PD activity (10).

Experiment	Frequency of TG-resistant colonies	Colonies (No.)	
		G6PD$^+$	G6PD$^-$
1	5×10^{-6}	40	59
2	1.1×10^{-6}	19	20
3	4.2×10^{-6}	53	71

mutation hypothesis it does not represent strong evidence in its favor, as almost any proposal involving a spontaneous change in a cell as a prerequisite for mutability would predict such a time-dependent increase.

The third prediction is a much more stringent one, and can be tested by exploiting a recently isolated CHO mutant lacking glucose-6-phosphate dehydrogenase (G6PD) activity (10). The wild-type alleles of G6PD and HPRT are linked in CHO cells, since they cosegregate from heterozygous hybrid cells more than 92 percent of the time (10). If the segregation of one chromosome bearing the HPRT gene is almost always involved in the generation of TG-resistant (HPRT$^-$) mutants from tetraploid cells, then segregation of the linked G6PD locus should also occur. This prediction can be tested by using

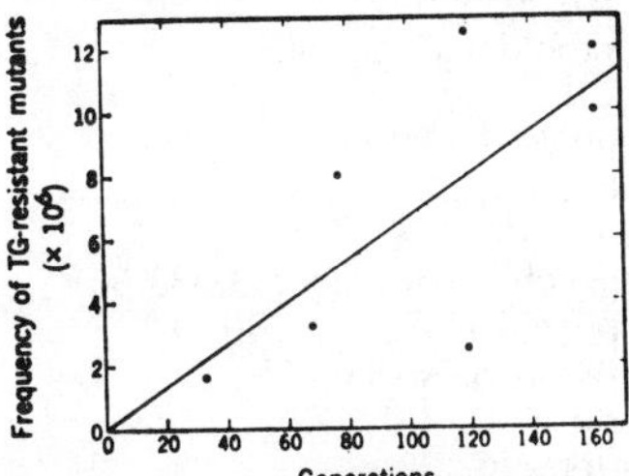

Fig. 1. Frequency of EMS-induced mutation as a function of clonal age. The pseudotetraploid line 5T11 of CHO cells was recloned and grown in monolayer culture in F-12 medium with 10 percent fetal calf serum. Cells were passaged when they reached confluence: at least 2×10^6 cells were transferred at each passage. Subpopulations were periodically induced to mutate and selected for TG-resistant cells as described previously (7). The total number of mutant colonies scored was 114, with a range of 5 to 25 in individual experiments.

as a tetraploid cell a hybrid heterozygous for G6PD and homozygous wild type for HPRT. Such a hybrid was therefore constructed, the parental lines being two mutant subclones of CHO cells, strain 43-64 (glyB$^-$) and strain Y113 (glyA$^-$ G6PD$^-$) (10). Hybrids were selected on the basis of the complementing glycine auxotrophies carried by the two parental lines (14). One hybrid clone (strain Y143) was recloned in selective (glycine-free) medium and used for subsequent experiments. Its genetic constitution can be represented as

$$\frac{\text{glyA}^+\ \text{glyB}^-\ \text{HPRT}^+\ \text{G6PD}^+}{\text{glyA}^-\ \text{glyB}^+\ \text{HPRT}^+\ \text{G6PD}^-}$$

The hybrid has a modal chromosome number of 40, close to the sum of those of the parents (42 chromosomes). Its phenotype is G6PD$^+$, the G6PD$^-$ phenotype being a recessive trait (10). The G6PD phenotype of cells and colonies can be clearly demonstrated by using a histochemical stain for enzyme activity (10).

If random segregation of either of the chromosomes bearing the HPRT and G6PD loci were occurring before mutation, then 50 percent of the HPRT$^-$ mutants derived from this hybrid should be G6PD$^+$ and 50 percent G6PD$^-$. The results of three separate experiments are shown in Table 1. There is good agreement with the predicted 50-50 distribution of G6PD$^+$ and G6PD$^-$ colonies among the mutants. It can also be seen that the average induced mutation frequency in these near-tetraploid hybrid cells is about 1/70 of that usually found in diploid cells (2.5×10^{-4}) (7) and is similar to the frequency found in tetraploid cells of similar clonal age that had been formed by low-Colcemid treatment (Fig. 1).

Although these results do implicate one chromosome-wide event in the generation of TG-resistant mutants from tetraploid cells, they do not clearly discriminate between chromosome loss, homozygosis through mitotic recombination, and X chromosome inactivation as a mechanism. A careful study of the karyotypes of these mutants should be able to demonstrate the first possibility. The results are not consistent with the occurrence of two point mutations as the basis for the TG-resistant phenotype, and they argue against a mechanism that specifically turns off HPRT activity in these cells by an epigenetic process.

There is little doubt that stable, heritable phenotypic changes that represent epigenetic alterations in regulatory programs can be demonstrated in cultured animal cells. The most convincing examples are the extinction (15) or induction (16) of differentiated cell products in hybrids formed between different types of specialized cells. It is equally clear that mutational changes can also be induced in cultured animal cells. If mutation is to be useful as a means to study gene regulation in culture, it will be necessary to distinguish these two types of change within one system.

References and Notes

1. A. Albrecht, J. Biedler, D. Hutchinson, *Cancer Res.* **32**, 1539 (1972); V. Chan, G. Whitmore, L. Siminovitch, *Proc. Natl. Acad. Sci. U.S.A.* **69**, 3119 (1972); A. Beaudet, D. Roufa, C. Caskey, *ibid.* **70**, 320 (1973); L. H. Thompson, J. L. Harkins, C. P. Stanners, *ibid.*, p. 3094; J. Sharp, N. Capecchi, M. Capecchi, *ibid.*, p. 3145; L. Chasin, A. Feldman, M. Konstam, G. Urlaub, *ibid.* **71**, 718 (1974); D. S. Secher, R. G. Cotton, C. Milstein, *FEBS (Fed. Eur. Biochem. Soc.) Lett.* **37**, 311 (1973); B. Birshtein, J.-L. Preud'homme, M. Scharff, in *The Immune System: Genes, Receptors, and Signals*, E. Sircarz and C. F. Fox, Eds. (Academic Press, New York, 1974), pp. 339–351.
2. M. Harris, *J. Cell Physiol.* **78**, 177 (1971).
3. L. Mezger-Freed, *Nat. New Biol.* **235**, 245 (1972).
4. E. Chu, P. Brimer, K. Jacobson, E. Merriam, *Genetics* **62**, 359 (1969).
5. J. W. Littlefield, *Science* **145**, 709 (1964).
6. Y. Matsuya, H. Green, C. Basilico, *Nature (Lond.)* **220**, 1199 (1968).
7. L. Chasin, *J. Cell Physiol.* **82**, 299 (1973).
8. V. McKusick and G. Chase, *Annu. Rev. Genet.* **7**, 435 (1973).
9. A. Westervald, P. Visser, M. Freeke, D. Bootsma, *Biochem. Genet.* **7**, 33 (1972).
10. M. Rosenstraus and L. Chasin, *Proc. Natl. Acad. Sci. U.S.A.*, in press.
11. M. Lyon, *Biol. Rev.* **47**, 1 (1972).
12. L. Chasin, *Nat. New Biol.* **240**, 50 (1972); G. Marin, *Exp. Cell Res.* **57**, 29 (1969).
13. M. Harris, *Exp. Cell Res.* **66**, 329 (1971).
14. F. Kao, L. Chasin, T. Puck, *Proc. Natl. Acad. Sci. U.S.A.* **64**, 1284 (1969).
15. R. Davidson, B. Ephrussi, K. Yamamoto, *J. Cell Physiol.* **72**, 115 (1969); J. Schneider and M. Weiss, *Proc. Natl. Acad. Sci. U.S.A.* **68**, 127 (1971); S. Silagi, *Cancer Res.* **27**, 1953 (1967).
16. G. Darlington, H. Bernhard, F. Ruddle, *Science* **185**, 859 (1974); J. Peterson and M. Weiss, *Proc. Natl. Acad. Sci. U.S.A.* **69**, 571 (1972).
17. We thank Maurice Rosenstraus for valuable assistance. Supported by grants V117A and V117B from the American Cancer Society.

11 November 1974

AUTHOR CITATION INDEX

Abdel-Samie, Y., 83
Abelson, J., 242
Abrahams, P. J., 184, 213
Ackerman, W. W., 255
Adams, M. S., 159
Adelburg, E. A., 317
Adelstein, S. J., 184, 213
Adetugbo, K., 93
Ahkong, Q. F., 62, 63
Air, M., 223
Albers, R. W., 346
Albertini, R. J., 247, 353
Albrecht, A., 252, 260, 361
Aldridge, D. C., 167
Alexander, H., 194
Alexander, P., 315
Alfert, M., 268
Allderdice, P. W., 117, 118
Allfrey, V. G., 233
Allison, W. S., 139
Aloni, Y., 172
Alt, F. W., 260
Amos, D. B., 69, 108, 126
Amos, H., 193
Anderson, E. P., 242
Anderson, R. P., 260
Anderson, W. F., 158, 159, 184
Anfinen, C. B., 359
Aoki, S., 315
Appella, E., 139
Armitage, P., 59
Armstrong, J. J., 167
Arnold, W. J., 359
Aronow, L., 252, 260
Ashkonas, B. A., 93
Asofsky, R., 247
Atanasiu, P., 193
Atkins, L., 159
Attardi, B., 172
Attardi, G., 172
Atwood, K. C., 158
Auer, G., 177
Aula, P., 99, 144
Avery, O. T., 193
Avilès, D., 102
Aviv, H., 158, 159
Avrameas, S., 83
Awdeh, A. L., 93
Axel, R., 212, 223, 234, 237
Axelrad, A. A., 272
Axelrod, D., 233
Aya, T., 99

Bacchetti, S., 233, 325
Bach, M. K., 353
Bachenheimer, S. L., 213
Bakay, B., 144, 242, 247, 359
Baker, P. F., 346
Baker, R. M., 330, 346
Baker, W. W., 247
Baldwin, H. H., 193
Baldwin, R. L., 260
Balis, M. E., 272, 354, 359
Baltimore, D., 159

Bang, F. B., 194
Bank, A., 159
Barile, M. F., 158
Barker, J. E., 158
Barner, H. D., 304
Barnes, D. W. H., 90
Barrell, B. G., 223
Barski, G., 8, 17, 28, 36, 41, 108, 132, 194
Barto, E., 139
Basilico, C., 132, 143, 325, 361
Bateman, A. J., 279
Battips, D. M., 139
Baughan, M. A., 139
Beard, J. W., 290
Bearn, J. G., 193
Beaudet, A. L., 359, 361
Becker, M., 359
Becker, Y., 202
Beer, J., 315
Behrens, K., 99
Bell, E., 172
Benda, P., 69, 78, 83
Bendich, A., 193, 242
Benedeczky, I., 193
Bennett, L. L., Jr., 261, 264, 272
Benoit, J., 194
Bensch, K. G., 193
Benz, E. J., 158, 159
Berg, P., 185, 260
Bergstrom, H., 260
Berman, L., 194, 268
Berman, P. H., 354
Bernhard, H., 361
Bernheim, B. C., 317
Bernini, L. F., 144
Berson, S. A., 359
Bertino, J. R., 255, 260
Bertolotti, R., 83
Biedler, J. C., 17
Biedler, J. L., 252, 260, 361
Biesele, J. J., 17, 261
Biggs, P. M., 202
Biro, P. A., 233
Birshtein, B., 361
Bishop, J. O., 159
Bishop, S. M., 233
Blanco, A., 139
Bloom, A. D., 353
Blumenthal, G. H., 194
Bobek, M., 212
Bobrow, M., 159
Bocchini, V., 359
Bodmer, J. A., 139
Bodmer, W., 118, 132, 139, 143, 172
Boiron, M., 193
Boivin, A., 83
Bond, H. E., 172
Bonnett, H. T., 63
Boone, C., 117, 139
Bootsma, D., 361
Borenfreund, E., 193, 242
Borst, P., 172
Bosch, L., 305

Bosmann, H. B., 346
Botchan, M., 223, 242
Boutwell, R. K., 193
Bower, B. F., 83
Bowman, H. G., 260
Boyce, R. P., 315
Boyer, S. H., 139
Bradley, T. R., 132
Bratton, A. C., 255
Breeze, A. S., 260
Bregula, U., 112
Breitman, T., 202
Breslow, R. E., 202, 353
Brewer, G. J., 139
Brewer, P. A., 268
Briggs, R., 83
Brimer, P., 279, 325, 354, 361
Brinster, R. L., 236
Britten, R. J., 159
Brockman, R. W., 132, 194, 264, 272
Broda, E., 83
Brookes, P., 279
Brooks, B. E., 317
Brown, N. L., 223
Brownlee, G. G., 93
Bruist, M., 260
Brunette, D. M., 346
Bryant, J. C., 28
Bryson, V., 260
Buchanan, J. M., 272
Buchwald, M., 346
Buckland, R. A., 159
Bunn, C. L., 163
Buonassisi, V., 89
Burk, D., 17
Burke, R., 159
Burkholder, G. D., 242

Cahn, M. B., 89
Cahn, R. D., 89
Cantell, K., 144
Capecchi, M. R., 223, 233, 330, 361
Capecchi, N. E., 233, 330, 361
Carlsson, S.-A., 83, 177
Carter, S. B., 167, 172
Carter, T. C., 38
Cartwright, E. M., 93
Caskey, C. T., 359, 361
Cason, J. C., 242
Caspersson, T., 159, 242
Cattanach, B. M., 279
Cedar, H., 223
Chadha, K. C., 212
Chan, V., 361
Chang, A. C. Y., 213
Chapman, V. M., 78, 159, 247
Chase, G., 361
Chasin, L., 223, 233, 234, 260, 330, 361
Chen, S., 159
Chen, T. R., 78, 83, 117, 139, 159
Cheng, Y.-C., 212, 213, 223, 233
Cheung, S. W., 159
Chien Ching Chen, G., 317

Childs, B., 143, 144, 247
Choi, K. W., 353
Chorazy, M. R., 193, 242
Choudary, P. V., 233
Chu, E. H. Y., 17, 28, 279, 314, 325, 354, 361
Churchill, A. E., 202
Ciardi, J. E., 242
Cieciura, S. J., 268, 283, 290, 314
Cinder, C. P., 159
Claflin, A., 139
Clausen, J. J., 59
Clayton, D. A., 172
Clewell, D. B., 223
Clough, D. W., 184
Cocito, C., 193
Cocking, E. C., 63
Coen, D., 172
Coffman, W. D., 290
Cohen, E. P., 69, 93
Cohen, S. N., 213, 260
Cohen, S. S., 304
Cohn, N. K., 264
Colter, J. S., 193
Conconi, F., 159
Condamine, H., 247
Conneally, M., 159
Connor, J. D., 144, 247
Conover, J. H., 159
Constable, F., 63
Cook, P. R., 144, 172, 242
Coombs, R. R. A., 126
Coon, H. G., 83, 89, 118, 172
Cooper, C. M., 184
Cooper, J. A., 172
Corin-Frederic, J., 194
Cornefert, F., 8, 17, 28, 36, 41, 108, 132, 194
Cortner, J. A., 139
Cotton, R. G. H., 69, 93, 361
Coulson, A. R., 223, 346, 354
Courington, D. P., 172
Cowan, N. J., 93
Cox, R. P., 354
Cramp, F. C., 62, 63
Crane, R. K., 346
Crasemann, J. M., 314, 317
Crawford, E. J., 255, 260
Creagan, R., 159
Cress, H. R., 194
Croce, C. M., 63, 112, 162, 172, 237
Crow, J. F., 260
Cruckshank, G., 139
Cuatrecasas, P., 359
Cumming, R. B., 279
Custer, R. P., 247

Dallapiocola, B., 159
Dancis, J., 354, 359
Dannenberg, A., 185
Darlington, A. J., 118, 139
Darlington, C. D., 290
Darlington, G. A., 325, 361
Davey, M. R., 63
Davidson, E. H., 159

Davidson, J. D., 132
Davidson, N., 172
Davidson, R. G., 139, 144
Davidson, R. L., 8, 34, 41, 69, 78, 83, 89, 90, 126, 184, 213, 361
Davidson, S. V., 223, 260
Davies, D., 213
Davis, B. D., 314
Davis, B. J., 132
Dawid, I. B., 172
Dean, C., 315
Defendi, V., 108, 126
Deininger, P. L., 233
Deisseroth, A. B., 158, 159
Deitsch, A., 177
de la Cruz, F., 8, 69
Delbrück, M., 73, 268, 272, 346, 354
Delihas, N., 315
Delius, H., 213
DeLorenzo, R. J., 139
DelSenno, L., 159
De Mars, R., 159, 247, 279, 305, 315, 353
Demerec, M., 260
Der Kaloustian, V. M., 143, 247
Dern, R. J., 139
de Serres, F. J., 279
DeSomer, P., 193
Detter, J. C., 139
Deutsch, J., 172
de Vries, F. A. J., 184, 213
Dewey, M. J., 237
Diacumakos, E. G., 184, 233
Diehl, V., 202
DiMayorca, G. A., 193
Djordjeciv, B., 314, 315, 317
Djordjevic, O., 194
Doctor, B. P., 194
Dohmer, J., 234
Domin, B., 212
Donohue, D. M., 260
Douglas, R. D., 317
Dovey, H., 172
Dow, L., 159
Dressler, D., 260
Drets, M. E., 242
Drew, R. M., 314
Dubbs, D. R., 36, 73, 126, 172, 202, 223, 264
Dulbecco, R., 139, 317, 325
Duncan, C. H., 233
Dunham, P. B., 346
Durrum, E. L., 268
Dym, H., 202

Eagle, H., 63, 167, 202, 242, 261, 268, 304, 305
Earle, W. R., 16, 17, 28, 73, 272
Ebert, J. D., 346
Ebina, T., 242
Eddy, B. E., 193
Edgar, R. S., 317
Edlund, T., 260
Efstradiatis, A., 223

Ege, T., 162, 177
Ehling, U. H., 279
Eicher, E., 158
Eidinoff, M. L., 315
Eisenstadt, J. M., 163, 172
Eliceiri, G. L., 346
Elion, G. B., 272
Elkind, M. M., 315
Ellem, K. A. O., 193
Elliott, W. H., 305
Elsevier, S. M., 159
Emmerson, B. T., 247
Emmons, S. W., 260
Enders, J., 78
Engel, E., 108
Engelberg, J., 317
Enquist, L. W., 233
Ephrati-Elizur, E., 194
Ephrussi, B., 8, 34, 36, 41, 69, 73, 78, 83, 89, 90, 108, 117, 126, 132, 144, 177, 194, 325, 333, 361
Ericsson, J., 177
Erikson, R. L., 315
Eriksson, K. D., 260
Eriksson, T., 63
Errera, M., 99
Espmark, J. A., 108
Etheredge, E., 346
Evans, A. H., 223
Evans, H. J., 159
Evans, M., 247
Evans, V. J., 28
Eylar, E. H., 346

Fagreus, A., 108
Fahmy, M. J., 194
Fahmy, O. G., 194
Fainer, D. C., 139
Fairbanks, G., Jr., 242
Falvey, A., 159
Fan, H., 159
Fan, K., 83
Farabasco, A., 159
Farr, A., 78, 202, 242, 247, 272, 346, 359
Faure-Fremiet, M. E., 17
Feldman, A., 361
Fiddes, J. C., 223
Fiers, W., 184, 213
Fildes, R. A., 139
Fink, G. R., 260
Finlayson, J. S., 247
Fioramonti, M. C., 28
Fiorelli, G., 144
Firkin, F. C., 172
Fischer, G. A., 193, 252, 255, 260
Fisher, D., 62, 63
Fisher, D. C., 272
Fisher, H. W., 304
Flanagan, S. P., 112
Flavell, R. A., 223
Fleischman, R., 305
Flintoff, W. F., 223, 260
Foley, G. E., 261, 304
Folk, W. R., 260

Ford, C. E., 73, 90, 132
Ford, D. K., 17
Forget, B. G., 158, 159
Fougére, C., 69, 83
Fournier, R. E., 117, 162
Fox, E., 314, 317
Fox, M. S., 242
Frangione, B., 93
Fraser, G. R., 144
Frederic, J., 194
Freeke, M., 361
Freeman, A. E., 304
Freeman, M., 305
Freese, E., 279
Freidkin, M., 255, 260
Friedman, D. L., 99
Friedmann, T., 132
Friend, C., 193
Friend, K. K., 159
Fritsch, D., 223
Fritz, P. J., 132
Fujimoto, W. Y., 354, 359
Fukai, K., 52
Fulton, F., 59
Futterman, S., 305

Gabrio, B. W., 255, 260
Gale, G. R., 159
Gall, J., 139
Gandini, E., 159
Ganshow, R., 69
Garen, A., 202, 359
Gartler, S. M., 193, 325
Gaub, J., 177
Gearhart, J. D., 247
Geist, C. E., 158
Georghian, G. P., 260
Gershon, D., 36, 69
Gey, G. O., 17, 290, 304
Giblett, E. R., 139
Gibson, B. H., 172
Gil, L., 260
Gilbert, W., 223
Giles, R. E., 159
Gillen, F. D., 359
Girardi, A. J., 112
Givol, D., 359
Glazer, D., 69, 83
Glick, S. M., 359
Glisin, V. R., 193
Glynn, I. M., 346
Goldberg, B., 83
Goldfarb, M. P., 185
Goldin, A., 255
Goldin, H., 260
Goldsby, R. A., 202, 353
Goldstein, A., 260
Goldstein, D. A., 172
Gordan, G. S., 83
Gordon, J. W., 237
Gordon, R. B., 247
Gorer, P. A., 108
Graessmann, A., 233
Graessmann, M., 233
Graf, L. H., 233, 247

Grafflin, A. L., 255
Graham, F. L., 184, 213, 233
Graybill, F. A., 132
Green, C. D., 247
Green, H., 73, 83, 118, 126, 132,
    143, 159, 233, 260, 325, 333,
    346, 361
Greenberg, D. M., 194, 252
Griffith, F., 193
Grigoio, A., 260
Grobstein, C., 89
Gropp, A., 159, 247
Gross, R. T., 143
Grundstrom, S., 260
Gruter, V., 99
Grzeschik, A., 117
Grzeschik, K. H., 117
Gulyas, S., 325
Gurdon, J. B., 83
Gurner, B. W., 126
Guttes, E., 99
Guttes, S., 99

Haaland, J. E., 346
Haff, R. F., 261, 283
Hagopian, A., 346
Hakala, M. T., 252, 255, 260, 261,
    272, 304
Haley, E. E., 255
Hall, M. R., 172
Halpern, M., 83
Ham, R. G., 83, 213, 242, 314, 333
Hamberg, H., 177
Hamerman, D., 73, 83
Hamerton, J. L., 90
Han, A., 325
Hanggi, U. J., 260
Hanks, J. H., 59
Harbers, E., 305
Hardison, R. C., 223
Harkins, J. L., 346, 361
Harris, A. W., 93
Harris, H., 41, 59, 62, 69, 73, 78,
    93, 108, 112, 126, 132, 139,
    144, 172, 177, 242, 314, 333
Harris, M., 83, 252, 279, 329, 354,
    359, 361
Harrison, T. M., 247
Haslam, J. M., 172
Hauschka, T. S., 28, 59, 69, 108, 126
Hausen, H. zur, 99
Hayer, B. H., 126
Hayflick, L., 126, 163
Haynes, G. R., 202, 213
Healy, G. M., 272
Hearst, J. E., 242
Heidelberger, C., 261, 305
Heijneker, H. L., 184, 213
Held, K. R., 159
Hellman, K. B., 202
Hellström, I., 108
Henderson, A. S., 158
Henderson, I. C., 233
Henderson, N. S., 139
Heneen, W., 99

Henle, W., 202
Heppel, L. A., 346
Herbst, E. J., 194
Herrington, K. A., 272
Herriot, R. M., 193
Hewer, T. F., 194
Hicks, J. B., 260
Hill, M., 193
Hillcoat, B. L., 260
Hilmoe, R. J., 346
Hilwig, I., 159
Himmelhoch, S. R., 242
Hinnen, A., 260
Hirschhorn, K., 159
Hirt, B., 223
Hitchings, G. H., 272
Hnilica, L., 194
Hobbs, G. L., 17
Hoffman, J. F., 346
Holden, M., 202
Holland, J. J., 126
Holland, S., 184
Holoubek, V., 194
Homma, M., 242
Honess, R. W., 213
Hooper, J. L., 315
Hopkinson, D. A., 139
Hopper, J. L., 279
Hopps, H. E., 317
Horak, I., 172
Horibata, K., 93
Horne, M. K., 202
Horton, C. L., 305
Hosaka, Y., 52, 194
Hoshino, A., 194
Hotchkiss, R., 223
Houck, C. M., 233
Hourcade, D., 260
Housman, D., 159
Howard-Flanders, P., 315
Howarth, S., 317
Howell, J. I., 62, 63
Hsu, T. C., 17, 36, 73, 126, 172,
    202, 213, 223, 264
Hudnik-Plevnik, T. A., 193
Huebner, K., 112
Huennekens, F. M., 255, 260, 305
Hughes, H. D., 172
Hughes, R. G., 212
Hughes, S. H., 233
Humphreys, S. R., 255, 260
Hungerford, D. A., 139
Hunter, W. S., 193
Hutchinson, C. A., 223
Hutchinson, D. J., 252, 260, 361
Hutchison, D. J., 193, 242, 260,
    272

Ikeuchi, T., 159
Illmensee, K., 237, 247
Imamura, T., 202
Irie, R. F., 108
Isensee, H., 139
Ishida, N., 242
Ito, Y., 193

Ittensohn, O. L., 242
Iwafuchi, M., 193

Jacob, F., 28, 89
Jacobson, K. B., 132, 279, 325, 354, 361
Jakubickova, J., 193
Jami, J., 102
Jansen, V., 354
Jansson, J. A. T., 260
Jarvis, J. M., 93
Jeffreys, A. J., 223
Jelinek, W. R., 233
Jensen, F. C., 126
Jerada, M., 159
Jerne, N. K., 93
Johansson, C., 159, 242
Johnson, L. A., 247
Johnson, R. T., 99, 172, 346
Johnsson, J., 108
Jones, C., 117
Jones, E., 202
Jones, K. W., 159
Jones, W. P., 317
Joseph, R., 139

Kabat, E. A., 359
Kacian, D., 158, 159
Käfer, E., 36
Kahan, B., 159
Kahn, G., 202
Kahn, R., 202
Kaighn, M. E., 83
Kalman, S. M., 260
Kamo, I., 242
Kanazir, D. T., 194
Kang, K. W., 159
Kantoch, M., 194
Kantor, J. A., 159
Kao, F.-T., 117, 264, 279, 314, 317, 325, 330, 333, 361
Kao, K. N., 34, 63
Kareh, A. E., 237
Karon, M., 325
Kato, H., 99, 242
Kaufman, R. J., 260
Kawana, T., 108
Kay, E. R. M., 193
Kazazian, H. H., 159
Keele, D. K., 144, 247
Kellems, R. E., 260
Kelley, W. N., 132, 143, 247, 264, 359
Kellner, G., 83
Kellogg, D. S., Jr., 17
Kelly, T. J., Jr., 242
Kelus, A., 126
Kemp, T. D., 260
Kieff, E. D., 213
Kihlman, B. A., 144
Killos, L., 184
Kimura, K., 194
King, D. W., 193
King, J. C., 260
King, T. J., 83

Kirby, K. S., 193
Kislev, N., 172
Kit, S., 36, 73, 126, 172, 202, 213, 223, 264
Klatt, O., 17
Klebe, R. J., 78, 83, 139, 159
Klein, E., 108, 290
Klein, G., 108, 112
Klemperer, H. G., 202, 213
Klobutcher, L. A., 233
Klomfass, M., 234
Knezevitch, Z. A., 194
Knowles, B. B., 112
Koch, G., 194
Koch, M. A., 317
Koenig, S., 194
Kohne, D. E., 172
Konigsberg, I. R., 89
Konstam, M., 361
Koprowski, H., 63, 108, 112, 126, 162, 172
Korn, E. D., 272
Kornberg, A., 272
Kostic, L., 194
Koyama, S., 193
Kozak, C. A., 159
Kraiselburd, E., 213
Kraus, A. P., 132, 139
Kraus, L. M., 194
Kretschmer, P. J., 184
Krondahl, U., 162, 177
Krooth, R. S., 325, 354
Krsmanovic, V., 99
Krüger, A., 237
Kubicek, M. T., 290
Kucherlapati, R. S., 158, 159, 330
Kudo, R. R., 17
Kuff, E. L., 172
Kung, M. J., 233
Kunkel, H. G., 268
Kunkel, L. M., 184
Kurita, S., 194, 260
Kusano, T., 118, 159

Lacy, E., 223, 234
Laemmli, U. K., 359
Lai, C. J., 223
Laird, C., 314, 317
Lakocy, A., 260
Lamb, J. F., 346
Lange, R., 237
Lauer, J., 223
Laurent, C., 159
Law, L. W., 132
Lawley, P. D., 279
Lawn, R. M., 223
Lawrence, J. B., 159
Lea, D. E., 346, 354
Leclerc, J. C., 108
Leddy, C. L., 118, 143
Leder, P., 158, 159
Lederberg, J., 28, 290
Lee, L.-S., 213, 223, 233
Leibovitz, A., 317
Leinwand, L., 233

Lennette, E. H., 112
Lerner, R. A., 172
Leroy, P., 194
Lesch, M., 132, 247
Lett, J. T., 315
Lettré, R., 17
Leuchtenberger, C., 194
Leuchtenberger, R., 194
Levan, A., 17, 59, 99, 144
Levinson, C., 346
Levinthal, C., 242
Levis, A. G., 99
Levy, J. P., 108
Levy, M., 304, 305
Lewis, W. H., 17
Lewis, W. H. P., 139
Lieberman, I., 264, 272, 354
Lielausis, I., 317
Likely, G. D., 16, 28, 73, 272
Lily, C. Y., 359
Lin, S., 359
Ling, V., 346
Linnane, A. W., 172
Lipmann, F., 346
Lissak, K., 193
Littlefield, J. W., 36, 38, 41, 63, 73, 78, 83, 93, 108, 132, 139, 144, 159, 172, 202, 213, 242, 260, 346, 354, 361
Litwack, G., 112
Livingston, D. M., 233
London Conference Study Group, 139
Long, C., 159
Lopez, J. A., 346
Lorkiewicz, Z., 194
Loutit, J. F., 90
Loveless, A., 317
Lowry, O. H., 78, 202, 242, 247, 272, 346, 359
Lubs, H. A., 139
Lucy, J. A., 62, 63
Ludwig, E. H., 194
Lukens, L. N., 272
Lukins, H. B., 172
Luria, S. E., 73, 268, 272, 346, 354
Lyon, M. F., 38, 354, 361
Lytle, C. D., 202

McAuslan, B., 202
McBride, O. W., 233, 242, 260
McCaffrey, R. P., 159
McCall, D., 346
McCarty, M., 193
MacCoshan, V., 260
McCulloch, E. C., 272
McDonald, T. F., 346
McDougall, J. K., 159, 185, 223, 233
McEnroe, W. D., 260
McGee, B. J., 108
McGilvery, R. W., 255
McKusick, V. A., 118, 159, 361
McLaren, L. C., 126
MacLeod, C. M., 193
McMorris, F. A., 159

Macnab, J. C. M., 213
McQuilkin, W. T., 28
McRoberts, J. A., 247
Madert, E., 159
Maio, J. J., 242
Maitland, N. J., 185, 223, 233
Mallette, M. F., 194
Malling, H. V., 279
Mandel, B., 304
Mandel, M., 317
Maniatis, T., 223, 234
Mankovitz, R., 346
Mann, J., 213
Mannucci, P. M., 144
Marcus, P. I., 268, 283, 290, 314
Marin, G., 361
Marinetti, G. V., 52
Markert, C. L., 132, 139
Marks, J. F., 144, 247
Marks, P. A., 143, 159
Marmur, J., 159
Marshall, E. K., 255
Martin, D. W., Jr., 247
Martin, G., 247
Martinovitch, P. N., 194
Mather, K., 290
Mathias, A. P., 193
Matsuya, G., 325
Matsuya, Y., 132, 143, 361
Maurer, B., 202
Maxam, A. M., 223
Mayer, M. M., 359
Mayhew, E., 330, 346
Maynard-Smith, S., 159
Mazia, D., 268
Meade, J. C., 359
Medawar, P. B., 194
Meek, E. S., 194
Meera Kahn, P., 78, 144
Meinke, W., 172
Mellman, W. J., 139
Meltzer, H. L., 261
Mendelsohn, J., 242
Merchant, D., 202
Merriam, E. V., 279, 325, 354, 361
Merwin, R. M., 17
Meselson, M., 194
Metafora, S., 159
Meuth, M., 260
Meyer, H. M., Jr., 317
Mezger-Freed, L., 359, 361
Michayluk, M. R., 34, 63
Mierau, R., 234
Migeon, B. R., 118, 132, 143, 247
Miggiano, V., 118, 132, 139, 143
Miles, C. P., 99
Miller, C. G., 260
Miller, C. L., 233
Miller, C. S., 143
Miller, D., 118
Miller, O. J., 112, 117, 118
Miller-Faures, A., 99
Milstein, C., 69, 93, 361
Minna, J. D., 69, 83, 118, 159
Mintz, B., 237, 247

Mirsky, A. E., 83
Misra, D. K., 255
Mitchell, H. K., 261
Mohit, B., 83
Mohler, W. C., 315
Mokrasch, L. C., 255
Moller, F., 139
Möller, G., 108
Monod, J., 89
Moore, D. E., 242
Moore, G. E., 73, 99
Moorhead, P. S., 126, 139
Morbery, M. L., 52
Morgan, C., 202
Morgan, H. F., 52
Morgan, J. F., 17, 28, 59
Morkovin, D., 314
Morrison, W. J., 139
Morrow, J., 354
Morton, H. J., 17, 28, 59
Moser, H., 268
Motulsky, A. G., 144
Mueller, G. C., 99
Mukherjee, B. B., 242
Mulder, C., 184, 213
Mulligan, R. C., 185
Munro, A. J., 93
Munyon, W., 213
Murnane, M., 78
Murphy, W., 202
Myerson, D., 163, 177

Nabholz, M., 118, 132, 139, 143, 172
Nagley, P., 172
Naha, P. M., 325
Nance, W. E., 139, 159
Nathans, D., 223
Neel, J. V., 159
Neely, C. L., 132, 139
Neff, J., 78
Nell, M. B., 34, 62
Nelson, P., 69, 83
Netter, P., 172
Newcombe, H. B., 354
Newton, A. A., 317
Nichol, C. A., 252, 255, 260, 272
Nichols, E. A., 159
Nichols, W. W., 99, 144
Nienhuis, A. W., 158, 159
Nirenberg, M., 69, 83
Nisalak, A., 317
Nishioka, K., 108
Nitowsky, H. M., 144
Nordenskjold, B., 108
Nordin, A. A., 93
Normark, S., 260
Norrby, E., 99
Nowell, P. C., 139
Nowinski, R. C., 359
Nunberg, J. H., 260
Nyhan, W. L., 132, 143, 144, 242, 247, 359

Ochoa, S., 255
O'Connell, C., 223

O'Gorman, P., 108
Ohanian, S. H., 83
Okada, Y., 34, 52, 59, 194
Okenquist, S., 184
Okun, L., 314, 317
Olenov, Y. M., 194
O'Malley, K. A., 34
O'Neill, F., 99
Opara-Kubinska, Z., 194
Opitz, J. M., 117
Oppenheim, S., 108
Oppenworth, F. J., 260
Ormerod, M. G., 315
Ornstein, L., 132
Orr, C. W. M., 346
Orth, G., 193
Osborn, M. J., 305
Ostertag, W., 279
Ostrander, M., 212
Ove, P., 264, 272, 354
Oxman, M. N., 184, 213
Oyama, V. I., 261, 304, 305
Ozer, H. L., 233, 242, 260, 354

Padgett, P. A., 260
Painter, R. B., 314
Paoletti, C., 193
Pardue, M., 139
Parker, R. C., 17, 28, 59, 223, 272, 325
Parkins, G. F., 315
Parks, R. E., Jr., 272
Parrington, J. M., 139
Peacock, J., 69, 83
Pearson, P. L., 159
Pecora, P., 184
Peehl, D. M., 102
Pellicer, A., 223, 233, 234, 237
Penrose, L. S., 159
Perkins, J. P., 260
Perlman, D., 260
Perry, T. L., 193
Peters, J., 255
Peters, T., Jr., 83
Peterson, D., 315
Peterson, E. A., 242
Peterson, J., 361
Petrochilo, E., 172
Pettersson, U., 213
Philips, R. A., 315
Phillips, D. M., 177
Phillips, R. J. R., 38
Phillips, S. G., 177
Picciano, D. J., 159
Piekarski, L. J., 36, 73, 126, 172, 202, 213, 223, 264
Pierce, G. B., 89, 247
Piez, K. A., 305
Pious, D. A., 325
Pires-Soares, J. M., 17
Podgayetskaya, D. Ja., 194
Pollack, R., 118
Pontecorvo, G., 36, 63, 118
Pope, J. H., 202
Porter, K. R., 163, 172

Poste, G., 162
Potter, M., 93, 247
Potter, V. R., 255
Povey, S., 139
Povlsen, C. O., 112
Powell, W. F., 314
Prescott, D. M., 163, 172, 177
Preston, C. M., 233
Preud'homme, J.-L., 361
Price, P. M., 159
Prince, A. M., 83
Prinzie, A., 193
Proudfoot, N. J., 93
Puck, T. T., 83, 117, 264, 268, 279, 283, 290, 304, 314, 315, 317, 325, 330, 333, 354, 361

Quon, D., 223

Ragni, G., 36, 193, 264, 354, 359
Randall, R., 78, 202, 242, 247, 272, 346, 359
Rao, P. N., 99, 172, 317, 346
Rappaport, H., 159
Rattazzi, M. C., 144
Rauth, A. M., 315
Reber, T., 237
Rebiere, J. P., 193
Reed, T., 159
Reeder, R. H., 242
Reeve, P., 162
Reilly, B. E., 223
Remy, C. N., 272
Renwick, J. H., 139
Ressler, E. O., 139
Reuser, A. J., 237
Ricciuti, F. C., 78, 118, 159
Rich, M. A., 315
Rieke, W. O., 193
Riggs, J. L., 112
Rigler, R., 177
Ringertz, N. R., 8, 69, 83, 162, 177
Ringold, G. M., 223
Ris, H., 83
Ritz, E., 102
Roane, D., 202
Robson, E. B., 139
Roderick, T. H., 139
Rogers, N. G., 317
Roizman, B., 202, 213
Roosa, R. H., 132
Rose, H., 202
Rose, S., 159
Rosebrough, N. J., 78, 202, 242, 247, 272, 346, 359
Rosenbloom, F. M., 132, 143, 247, 264
Rosenburg, M. D., 346
Rosenstraus, M., 361
Ross, J., 159
Roth, J., 359
Roth, J. R., 260
Rothfels, K. H., 59, 272
Roufa, D. J., 359, 361
Rous, P., 290

Rousset, J.-P., 102
Rowley, P. T., 159
Rownd, R. H., 260
Rubin, C., 233
Rubin, C. S., 359
Rubin, H., 83
Ruddle, F. H., 78, 83, 117, 118, 139, 144, 158, 159, 162, 177, 233, 234, 236, 260, 330, 354, 361
Ruiz, F., 69, 83
Runner, C. C., 247
Rusch, H. P., 99
Rushforth, N. B., 73, 126, 144
Russell, L. B., 247
Russell, W. C., 202
Rygaard, J., 112

Sabin, A., 317
Sachs, H. G., 346
Sachs, L., 36, 69
Sachsenmaler, W., 99
Safwat, F., 63
Saksela, E., 144
Salser, J. S., 272
Salzman, N. P., 242
Sambrook, J., 223
Sandberg, A. A., 99
Sanders, P., 315
Sanders-Haigh, L., 184
Sanford, K. K., 16, 17, 28, 73, 272
Sanger, F., 223
Santachiara, A. S., 118, 139
Sarker, P. F., 36
Sarov, I., 202
Sasaki, M., 159
Satake, S., 242
Sato, G., 89
Savage, R. E., 8, 69, 83, 177
Scaletta, L. J., 8, 34, 73, 90, 108, 117, 126, 144
Scandlyn, B. J., 172
Scharff, M., 361
Scherer, W. F., 290, 304
Schildkraut, C. L., 242
Schimke, R. T., 260
Schmid, C. W., 233
Schmidt, G. B., 159
Schmizu, T., 193
Schneider, J. A., 83, 361
Schneider, W. C., 172
Schoefl, G. I., 73, 132
Schull, W. J., 159
Schulte-Holthausen, H., 202
Schultz, J., 17
Schwaber, J., 69, 93
Schwartz, A. G., 172, 242
Scolnick, E., 159
Scott, W., 202
Seabright, M., 112
Sebring, E. D., 242
Secher, D. S., 93, 361
Seegmiller, J. E., 132, 143, 247, 264, 354, 359
Sekiguchi, F., 242

Sekiguchi, T., 242
Sekiya, K., 242
Sell, E. K., 354
Senguptu, S., 159
Service, M. W., 260
Shank, P. R., 223, 233
Shapiro, H. S., 159
Sharma, R. A., 212
Sharp, J. D., 233, 330, 361
Sharp, P. A., 213
Shaw, C., 139
Shaw, M. W., 242
Shay, J. W., 163, 172
Shedden, W. I. H., 202, 213
Shelokov, A., 59
Shih, C., 185
Shilo, B.-Z., 185
Shin, S., 144
Shirakawa, S., 325
Shows, T. B., 139, 247
Shroder, T., 159
Siddiqi, O., 359
Siebs, W., 17
Silagi, S., 69, 361
Silber, R., 260
Silverman, L., 184
Silverstein, S., 223, 233, 234, 237
Sim, G. K., 223, 234
Simic, M. M., 193, 194
Siminovitch, L., 59, 223, 233, 260, 272, 346, 361
Simmons, B., 260
Simms, E. S., 272
Simpson, M. S., 264, 272
Sims, P., 260
Sinclair, W. K., 279, 325, 354
Sinensky, M., 260
Singer, S., 272
Siniscalco, M., 117, 144, 354
Sirotnak, F. M., 193, 260
Sjögren, H. O., 108
Skeel, R. E., 260
Skipper, H. E., 261, 264, 272
Skou, J. C., 346
Slocombe, P. M., 223
Slonimsky, P. P., 172
Smiley, J. R., 233
Smith, A. G., 194
Smith, C. A., 172
Smith, C. A. B., 159
Smith, E. W., 159
Smith, G. D., 260
Smith, M., 223
Smith, M. J., 194, 268
Smith, M. S., 272
Smith, S. W., 118, 143
Smithies, O., 139
Smull, C. E., 194
Snell, E. E., 261
Snyder, P. C., 159
Sofuni, T., 99
Sonneborn, T. M., 290
Sorieul, S., 8, 17, 28, 36, 41, 132, 194, 325
Southern, E. M., 223, 233

Spandidos, D. A., 233, 260
Sparks, M. C., 272
Speake, R. N., 167
Spector, D. H., 233
Spencer, N., 139
Spencer, R. A., 69, 108, 126
Spengler, B. A., 260
Spiegelman, S., 159
Spindler, S. M., 260
Spizizen, J., 223
Spolsky, C. M., 172
Sprinson, D. B., 261
Stacey, D. W., 233
Stacey, K. A., 260
Stahl, F. W., 194, 314, 317
Stanbridge, E. J., 102
Stanners, C. P., 346, 361
Stark, G. R., 260
Steffen, J., 314
Steggles, A. W., 158, 159
Stenchever, M. A., 117
Steplewski, Z., 126
Sternbury, J., 260
Stevens, L. C., 90, 247
Steward, F. C., 83
Stewart, S. E., 193
Stubblefield, E., 99, 163
Stulberg, C. S., 194, 268
Sturtevant, A. H., 260
Subak-Sharpe, J. H., 132, 359
Suerinc, E. F., 159
Summers, W. C., 213, 233
Summers, W. P., 213
Sumner, A. T., 159
Suzuki, T., 52, 194
Svasti, J., 93
Sveda, M., 233
Sweet, R., 223, 234
Swift, M. C., 233
Swim, H. E., 261, 283
Swyryd, M., 260
Syverton, J. T., 59, 126, 290, 304
Szybalska, E. H., 132, 143, 193,
    194, 354, 359
Szybalski, W., 36, 132, 143, 193,
    194, 260, 264, 268, 314, 315,
    317, 354, 359

Tachibana, T., 112
Takagi, N., 99
Takahashi, K., 242
Takemori, N., 112
Takemura, C., 194
Takenchi, S., 108
Tartoff, K. D., 260
Tashian, R. E., 139
Taubman, S. B., 83
Taylor, E., 272
Taylor, W. B., 325, 346
Telles, N. C., 202
Temin, H. M., 89
Temple, G. F., 159
Teplitz, R. L., 172
Ternynck, T., 83
Terriere, L. C., 260

Thomas, C. A., Jr., 242
Thomas, R., 223
Thomas, V., 260
Thomasz, A., 223
Thompson, G., 139
Thompson, J. R., 264, 272
Thompson, L. H., 346, 361
Thorbecke, G. J., 83
Thoren, M. M., 242
Thoren, S., 260
Tigyi, A., 193
Till, J. E., 325, 346
Timbury, M. C., 213
Tischfield, J., 159
Tiselius, A., 268
Tishler, P. V., 159
Tjio, J. H., 83, 242
Todaro, G. J., 83, 126, 233
Tolmach, L. J., 315
Tomizawa, K., 268
Toomey, T. P., 233
Topp, W., 223
Torbert, J. V., 159
Trilling, D. M., 233
Trinkhaus, J. P., 89
Tripp, M. B., 139
Tsukamotom, M., 260
Turner, W. B., 167

Uehlinger, V., 83
Uriel, J., 83
Urlaub, G., 223, 233, 234, 260, 361
Uyeki, E., 194

Van Benedez, R. J., 159
Vandepitte, J., 144
van der Eb, A. J., 184, 213, 233
Vande Woude, G. F., 233
Varet, B., 108
Varmus, H. E., 223, 233
Velasquez, A. A., 325
Velez, R., 158, 159
Vendrely, C., 83, 194
Vendrely, R., 83, 194
Veomett, G., 163
Verma, I. M., 159
Verney, E. L., 89
Vinograd, J., 172, 194
Visfeldt, J., 112
Visser, P., 361
Vogel, J., 59
Vogt, M., 139, 272, 325
Vonder Haar, R. A., 233

Wagner, E. F., 237
Wagner, M., 213, 233
Wahl, G., 233, 260
Walker, C. R., 260
Walker, D., 193
Wallace, D. C., 163
Wallace, J., 163, 177
Wallach, D. F. H., 346
Wang, R., 118
Warnaar, S. O., 184, 213
Wasilejko, H. C., 272

Watkins, J. F., 41, 62, 73, 78, 93, 108,
    126, 132, 144, 177, 317, 333
Watson, D. H., 202, 213
Watson-Williams, E. J., 139
Watts, J. W., 59
Way, J. L., 272
Ways, P. O., 139
Weil, R., 193
Weinberg, R. A., 185
Weinstock, R., 223
Weisberger, A. S., 194
Weiss, G. B., 159
Weiss, M., 223, 361
Weiss, M. C., 34, 69, 73, 83, 90,
    126, 132, 143, 333
Weiss, P., 17
Weissman, S. M., 233
Weitkamp, L. R., 159
Welch, A. D., 255
Welshons, W. J., 172, 247
Werkheiser, W. C., 255, 260
Westervald, A., 361
Westfall, B. B., 28
Westphal, H., 317
Whang-Peng, J., 242
Wheeler, T. B., 34
Whitmore, G. F., 325, 346, 361
Whitt, G. S., 139
Wiener, F., 112
Wigler, M., 223, 233, 234
Wilbert, S., 202
Wilchek, M., 359
Wilcox, F. H., 247
Wilczok, T., 193
Wildy, P., 202
Willecke, K., 159, 234, 237, 260
Williamson, A. R., 93
Willis, J. S., 346
Willmer, E. N., 17
Wilson, A. R., 264, 272
Wilson, G. N., 159
Wold, B., 234
Wolff, S., 260
Wolfson, J., 260
Wolf-Watz, H., 260
Wollman, E. L., 28
Wolpert, L., 167
Wood, J. S., 172
Woodhead, A. P., 159
Woods, M. W., 17
World Health Organization, 144
Worst, P., 112
Wright, J. A., 346
Wright, J. E., 139
Wright, W. E., 163
Writter, R. F., 52
Wu, M., 172

Yaffe, D., 177
Yalow, R. S., 359
Yamada, M., 314
Yamamoto, K., 78, 83, 89, 126,
    223, 361
Yasumura, Y., 89
Yerganian, G., 34, 62

Yoshida, M. C., 73, 83, 108, 117, 126, 159
Yosida, T. H., 242
Young, J. M., 17
Young, W. J., 143, 247
Yu, C. K., 354
Yu, M. T., 158

Zabin, I., 359
Zakrzewski, S. F., 252, 255, 260
Zamenhof, P. J., 317
Zamenhof, S., 194
Zech, L., 159, 242

Zetterberg, A., 177
Zinder, N., 202
Zinkham, W. H., 139
Zischka, W., 83
Zur Hausen, H., 202

SUBJECT INDEX

Amethopterin, 250, 251, 253–255, 256–260, 261
Aminopterin, 29, 35–36, 186–194, 263, 269–272,
    281, 293–307
Antibodies, monoclonal, 68, 91–93
Auxotrophic mutants. *See* Nutritional mutants
Azaguanine, 30, 35–36, 37–38, 186–194, 243–247,
    262–264, 265–268, 269–272, 273–279, 328,
    329, 347–354, 355–359

Blastocyst injection, 236, 243–247
Bromodeoxyuridine, 30, 35–36, 114, 119–126,
    195–202, 263, 282, 308–315, 316–317, 326

Cell cycle, 68, 69, 94–99
Chimeric mice, 236, 243–247
Chloramphenicol, 161, 162, 168–172
Chromosomes
    breakage, 116, 140–144
    premature condensation, 69, 94–99
    segregation, in hybrids, 100, 101, 103–108,
        109–112, 113–116, 119–126, 127–132,
        133–139, 140–144, 145–159, 329, 360–361
    translocations, 116
Complementation, in hybrids, 326, 327, 331–333
Cybrids, 161, 168–172
Cytochalasin B, 160–162, 164–167, 168–172, 173–177
Cytoplasmic inheritance, 161, 162, 168–172

Differentiation, in hybrids
    albumin production, 66, 67, 79–83
    immunoglobulin production, 67, 68, 91–93
    melanin production, 65, 66, 70–73, 74–78
    potential, 67, 84–90
Dihydrofolate reductase, 250, 251, 253–255, 256–260
DNA synthesis
    in heterokaryons, 53–59, 68, 69, 94–99
    in mutants, 283, 318–325

Enucleation, 160–162, 164–167, 168–172, 173–177

Folic acid antagonists. *See* Amethopterin;
    Aminopterin
Fusion, of cells
    polyethylene glycol induced, 33, 60–63
    Sendai virus induced, 31–33, 42–52, 53–59

Gene amplification, 251, 256–260
Gene dosage, 66, 67, 74–78, 79–83
Gene mapping, in hybrids, 101, 109–112, 113–117
    119–126, 127–132, 133–139, 140–144, 145–159
Gene transfer
    chromosome-mediated, 235, 238–242
    DNA-mediated, 180–184, 186–194, 203–213,
        214–223, 224–234
    virus-mediated, 181, 195–202
Globin genes, 117, 145–159, 183, 184, 214–223, 236
Glucose-6-phosphate dehydrogenase, 114–116,
    127–132, 140–144, 329, 360–361

HAT medium, 29–31, 35–36, 37–38, 181, 186–194,
    195–202, 251, 252, 261, 263, 269–272

Herpes simplex virus, 181–184, 195–202, 203–213,
    214–223, 224–234
Heterokaryons
    chromosomes, 69, 94–99
    DNA synthesis, 53–59, 68, 69, 94–99
    induction, 31–33, 42–52, 53–59
Hybridomas, 68, 91–93
Hybrids, somatic cell
    chromosome segregation. *See* Chromosomes
    differentiation. *See* Differentiation, in hybrids
    discovery, 6, 7, 9–26, 27–28
    gene mapping. *See* Gene mapping, in hybrids
    induction, 33, 60–63
    interspecific, 31, 32, 39–41
    isozymes, 31, 32, 65, 127–132, 133–139
    malignancy. *See* Malignancy
    selection. *See* Selective systems
    senescence, 7, 37–38
    surface antigens, 64, 65
Hypoxanthine-guanine phosphoribosyltransferase,
    30, 35–36, 37–38, 114–116, 127–132,
    140–144, 186–194, 235, 238–242, 243–247,
    262, 263, 269–272, 328, 329, 355–359, 360–361

Immunoglobulin production, 67, 68, 91–93

Lesch-Nyhan syndrome, 262, 263
Liver cells, 66, 67, 79–83

Malignancy, 7, 9–26, 100, 101, 103–108, 109–112, 182
Melanoma cells, 65, 66, 70–73, 74–78
Membrane mutants, 327, 334–346
Methylation, of DNA, 182, 329
Microcells, 116, 162, 173–177
Microinjection, of DNA, 183, 184, 224–234, 236
Mitochondria, 161, 168–172
Mutagenesis, 264, 273–279
Mutants
    auxotrophic. *See* Nutritional mutants
    drug resistant. *See* Amethopterin; Azaguanine;
        Bromodeoxyuridine; Chloramphenicol;
        Ouabain
    membrane. *See* Membrane mutants
    nutritional. *See* Nutritional mutants
    temperature sensitive. *See* Temperature sensitive
        mutants
Mutation rates, 262, 265–268, 273–279, 328, 329,
    347–354
Myeloma cells, 67, 68, 91–93

Neuroblastoma cells, 67
Nutritional mutants, 264, 280–282, 284–292,
    293–307, 308–315, 326, 327, 331–333

Oncogenes, 182
Ouabain, 327, 328, 334–346

Ploidy effects, 328, 329, 347–354, 360–361
Polio virus, 115
Polyethylene glycol, 33, 60–63
Premature chromosome condensation, 69, 94–99

Recombinant DNA, 117, 182–184, 203–213,
    214–223, 224–234
Regulation, genetic. *See* Differentiation, in hybrids;
    DNA synthesis

Selective systems
    hybridization, 29–31, 35–36, 37–38, 115,
        326–328, 331–333, 334–346
    transformation, 181, 186–194, 195–202
Sendai virus, 32, 33, 42–52, 53–59
Senescence, 7, 37–38
Simian virus 40, 101, 109–112, 161, 184, 224–234
Spleen cells, 68, 91–93
Surface antigens, 64, 65

Temperature sensitive mutants, 282, 283, 316–317,
    318–325
Teratocarcinoma cells, 67, 84–90, 243–247
Thymidine kinase, 30, 35–36, 114, 119–126,
    181–184, 195–202, 203–213, 214–223,
    224–234, 235, 236, 263
Transformation. *See* Gene transfer

Viruses. *See* Herpes simplex virus; Polio virus;
    Sendai virus; Simian virus 40

Xanthine-guanine phosphoribosyltransferase, 183
X chromosomes, 114–116, 127–132, 140–144, 328,
    329, 360–361

About the Editor

RICHARD L. DAVIDSON is the Benjamin Goldberg Professor of Genetics and director of the Center for Genetics at the University of Illinois College of Medicine at Chicago. He joined the University of Illinois in 1981.

Dr. Davidson completed his undergraduate work at Western Reserve University in 1963 and received the Ph.D. from Western Reserve University in 1967. His graduate and postdoctoral studies were carried out in the laboratories of Boris Ephrussi, first in Cleveland and then in Gif-sur-Yvette, France. Dr. Davidson received a National Academy of Sciences postdoctoral fellowship for his studies in France. In 1970 he returned to the United States as a member of the faculty of Harvard Medical School and Children's Hospital Medical Center, Boston. In Boston, Dr. Davidson also was the codirector of the Massachusetts Institute of Technology Cell Culture Center.

Dr. Davidson has written numerous articles on mammalian cell genetics. He is the founder and editor-in-chief of the journal *Somatic Cell Genetics,* and also editor of the book *Somatic Cell Hybridization* (Raven Press, 1974).